Lecture Notes in Artificial Intelligence   3276

Edited by J. G. Carbonell and J. Siekmann

Subseries of Lecture Notes in Computer Science

T0181360

Daniele Nardi   Martin Riedmiller
Claude Sammut   José Santos-Victor (Eds.)

RoboCup 2004:
Robot Soccer
World Cup VIII

 Springer

Volume Editors

Daniele Nardi
Università di Roma "La Sapienza"
E-mail: Daniele.Nardi@dis.uniroma1.it

Martin Riedmiller
University of Osnabrück
E-mail: martin.riedmiller@uos.de

Claude Sammut
University of New South Wales
E-mail: claude@cse.unsw.edu.au

José Santos-Victor
Lisbon Technical University
E-mail: jasv@isr.ist.utl.pt

The picture on the cover was taken by Yukiko Matsuoka, © RoboCup Federation.

Library of Congress Control Number: Applied for

CR Subject Classification (1998): I.2, C.2.4, D.2.7, H.5, I.5.4, J.4

ISSN 0302-9743
ISBN 3-540-25046-8 Springer Berlin Heidelberg New York

Springer is a part of Springer Science+Business Media

springeronline.com

© Springer-Verlag Berlin Heidelberg 2005
Printed in Germany

Typesetting: Camera-ready by author, data conversion by Scientific Publishing Services, Chennai, India
Printed on acid-free paper   SPIN: 11399162   06/3142   5 4 3 2 1 0

Preface

These are the proceedings of the RoboCup 2004 Symposium, held at the Instituto Superior Técnico, in Lisbon, Portugal in conjunction with the RoboCup competition. The papers presented here document the many innovations in robotics that result from RoboCup. A problem in any branch of science or engineering is how to devise tests that can provide objective comparisons between alternative methods. In recent years, competitive engineering challenges have been established to motivate researchers to tackle difficult problems while providing a framework for the comparison of results. RoboCup was one of the first such competitions and has been a model for the organization of challenges following sound scientific principles. In addition to the competition, the associated symposium provides a forum for researchers to present refereed papers. But, for RoboCup, the symposium has the greater goal of encouraging the exchange of ideas between teams so that the competition, as a whole, progresses from year to year and strengthens its contribution to robotics.

One hundred and eighteen papers were submitted to the Symposium. Each paper was reviewed by at least two international referees; 30 papers were accepted for presentation at the Symposium as full papers and a further 38 were accepted for poster presentation. The quality of the Symposium could not be maintained without the support of the authors and the generous assistance of the referees.

The Symposium was co-located with the 5th IFAC/EURON International Symposium on Intelligent Autonomous Vehicles (IAV 2004) and featured four distinguished plenary speakers: Hugh Durrant-Whyte, Luigia Carlucci Aiello, James Albus, and Shigeo Hirose. The program included a discussion panel on Applications of RoboCup Research whose members were Hiroaki Kitano, Christian Philippe, and M. Isabel Ribeiro. The panel was organized and moderated by Hans-Dieter Burkhard.

RoboCup has grown into a major international meeting with many people contributing their time and effort. The Symposium organization was made easy for us by the RoboCup Organizing Committee. We particularly thank the RoboCup General Chairs, Pedro Lima and Luis Custódio and their team. Thanks also to the Institute for Systems and Robotics for their support of the Symposium.

October 2004

Daniele Nardi
Martin Riedmiller
Claude Sammut
José Santos-Victor

Organization

RoboCup Federation

The RoboCup Federation, the governing body of RoboCup, is an international organization that promotes science and technology using soccer games by robots and software agents.

Humanoid League
Nobert Mayer, Osaka University, Japan
Changjiu Zhou, Singapore Polytechnic, Singapore
RoboCup Rescue Robot League
Adam Jacoff, National Institute of Standards and Technology, Intelligent
  Systems Division, USA
RoboCup Rescue Simulation League
Tomoichi Takahashi, Meijo University, Japan
RoboCup Junior
Luis Almeida, University of Aveiro, Portugal
Jeffrey Johnson, Open University, UK
RoboCup Websites
Ansgar Bredenfeld, AIS, Germany

RoboCup 2004 Organization and Support

General Chairs
Luis Custóio, Instituto Superior Técnico, Portugal
Pedro Lima, Instituto Superior Técnico, Portugal

Local Arrangements
Paulo Oliveira, Instituto Superior Técnico, Portugal

E-League Committee
Jacky Baltes (chair), University of Manitoba, Canada
Brett Browning, Carnegie Mellon University, USA
Gerhard Kraetzschmar, University of Ulm, Germany
Raúl Rojas, Free University of Berlin, Germany
Elizabeth Sklar, Columbia University, USA

Simulation League Committee
Oliver Obst (chair), University of Koblenz, Germany
Luis Seabra Lopes (local chair), University of Aveiro, Portugal
Jafar Habibi, Sharif Institute of Technology, Iran
Jelle Kok, University of Amsterdam, The Netherlands
Gregory Kuhlmann, University of Texas at Austin, USA

Small-Size Robot League Committee
Beng Kiat Ng (chair), Ngee Ann Polytechnic, Singapore
Paulo Costa (local chair), University of Porto, Portugal
David Chelberg, University of Ohio, USA
Tadashi Naruse, Aichi Prefectural University, Japan

Middle-Size Robot League Committee
Yasutake Takahashi (chair), Osaka University, Japan
Fernando Ribeiro (local chair), University of Minho, Portugal
Thorsten Buchheim, University of Stuttgart, Germany
Pieter Jonker, Delft University of Technology, The Netherlands

4-Legged Robot League Committee
Thomas Röfer (chair), University of Bremen, Germany
A. Paulo Moreira (local chair), University of Porto, Portugal
Dan Lee, University of Pennsylvania, USA
Alessandro Saffiotti, Örebro University, Sweden

Humanoid League Committee
Changjiu Zhou (chair), Singapore Polytechnic, Singapore
Vitor Santos (local chair), University of Aveiro, Portugal
Thomas Christaller, Fraunhofer Institute for Autonomous Intelligent
    Systems, Germany
Lars Asplund, Uppsala University, Sweden

Rescue Simulation League Committee
Levent Akın (chair), Boğaziçi University, Turkey
Ana Paiva (local chair), Instituto Superior Técnico, Portugal
Jafar Habibi, Sharif Institute of Technology, Iran
Tomoichi Takahashi, Meijo University, Japan

Rescue Robot League Committee
Adam Jacoff (chair), NIST, USA
Jorge Dias (local chair), University of Coimbra, Portugal
Andreas Birk, International University of Bremen, Germany
Satoshi Tadokoro, University of Kobe, Japan

Junior Committee
Gerhard Kraezschmar (chair), University of Ulm, Germany
Luis Almeida (local chair), University of Aveiro, Portugal
Carlos Cardeira (local chair), Instituto Superior Técnico, Portugal
Jeffrey Johnson, Open University, United Kingdom
Elizabeth Sklar, Columbia University, USA

Symposium Committee
Daniele Nardi, Universita di Roma "La Sapienza", Italy
Martin Riedmiller, University of Osnabrück, Germany
Claude Sammut, University of New South Wales, Australia
José Santos-Victor, Instituto Superior Técnico, Portugal

Symposium Program Committee

Tamio Arai, Japan
Hélder Araújo, Portugal
Tucker Balch, USA
Nick Barnes, Australia
Michael Beetz, Germany
Carlos Beltran, Italy
Fabio Berton, Italy
Andreas Birk, Germany
Alan Blair, Australia
Pierre Blazevic, France
Patrick Bonnin, France
Ansgar Bredenfeld, Germany
Thorsten Buchheim, Germany
Magdalena Bugajska, USA
Brad Clement, USA
Peter Corke, Australia
Gamini Dissanayake, Australia
Alessandro Farinelli, Italy
Paul Fitzpatrick, USA
Claudia Goldman, Israel
Giorgio Grisetti, Italy
Joao Hespanha, USA
Andrew Howard, USA
Huosheng Hu, United Kingdom
Vincent Hugel, France
Luca Iocchi, Italy
Mansour Jamzad, Iran
Ray Jarvis, Australia
Jeffrey Johnson, United Kingdom
Pieter Jonker, The Netherlands
Gal A. Kaminka, USA
Frederic Kaplan, France
Gerhard Kraetzschmar, Germany
Kostas Kyriakopoulos, Greece
Gerhard Lakemeyer, Germany
Nuno Lau, Portugal

Gregory Lionis, Greece
Michael Littman, USA
Emanuele Menegatti, Italy
Giorgio Metta, Italy
Rick Middleton, Australia
Bernhard Nebel, Germany
Itsuki Noda, Japan
Urbano Nunes, Portugal
Masayuki Ohta, Japan
Eugénio Oliveira, Portugal
Francesco Orabona, Italy
Simon Parsons, USA
Maurice Pagnucco, Australia
Xanthi Papageorgiou, Greece
Panajotis Pavlou, Greece
Mitchell Potter, USA
David Pynadath, USA
Luis Paulo Reis, Portugal
Paulo Quaresma, Portugal
Thomas Röfer, Germany
Stuart Russell, USA
Thorsten Schmitt, Germany
Alan Schultz, USA
Sandeep Sen, USA
Onn Shehory, Israel
Roland Siegwart, Switzerland
Eduardo Silva, Portugal
Peter Stone, USA
Katya Sycara, USA
Yasutake Takahashi, Japan
Ashley Tews, Australia
Nikos Tsokas, Greece
Kostas Tzafestas, Greece
Nikos Vlassis, The Netherlands
Changjiu Zhou, Singapore

Table of Contents

Posters

RoboCup 2004 Overview

Pedro Lima and Luis Custódio

Institute for Systems and Robotics, Instituto Superior Técnico,
Av. Rovisco Pais, 1049-001 Lisboa, Portugal
{pal, lmmc}@isr.ist.utl.pt

1   Introduction

RoboCup is an international initiative with the main goals of fostering research
and education in Artificial Intelligence and Robotics, as well as of promoting
Science and Technology to world citizens. The idea is to provide a standard
problem where a wide range of technologies can be integrated and examined,
as well as being used for project-oriented education, and to organize annual
events open to the general public, where different solutions to the problem are
compared.

Fig. 1. A view of some of the RoboCup2004 participants, at the entrance of the venue

The RoboCup Federation stated the ultimate goal of the RoboCup initiative as follows: "By 2050, a team of fully autonomous humanoid robot soccer
players shall win a soccer game, complying with the official FIFA rules, against
the winner of the most recent World Cup of Human Soccer." [1]. This main
challenge lead robotic Soccer matches to be the main part of RoboCup events,
from 1997 to 2000. However, since 2000, the competitions include Search and

D. Nardi et al. (Eds.): RoboCup 2004, LNAI 3276, pp. 1–17, 2005.

Rescue robots as well, so as to show the application of Cooperative Robotics and Multi-Agent Systems to problems of social relevance [2]. Also in 2000 was introduced RoboCup Junior, now a large part of any RoboCup event, aiming at introducing Robotics to children attending primary and secondary schools and including undergraduates who do not have the resources yet to take part in RoboCup senior leagues [3].

RoboCup2004 was held in Lisbon, Portugal, from 27 June to 5 July. As in past years, RoboCup2004 consisted of the 8th Symposium and of the Competitions. The competitions took place at the Pavilion 4 of Lisbon International Fair (FIL), an exhibition hall of approximately 10 000 m^2, located at the former site of Lisbon EXPO98 world exhibition. The Symposium was held at the Congress Center of the Instituto Superior Técnico (IST), Lisbon Technical University. Together with the competitions, two regular demonstrations took place on a daily basis: SegWay soccer, by a team from Carnegie-Mellon University, and SONY QRIO robot, by a team from SONY Japan.

Portugal was chosen as the host of the 2004 edition due to its significant representation in RoboCup committees, competitions and conferences, as a result of the effort of the country in recent years to attract young people to Science and Technology and also because EURO2004^{TM}, the 2004 European Soccer Cup, took place in Portugal, therefore improving the chances to bring the media to cover the event.

RoboCup2004 was locally organized by a Portuguese committee composed of 15 researchers and university professors from several Universities, therefore underlining the national nature of the event organization. This committee worked closely with the international organizing and technical committees to set up an event with the record number of 1627 participants from 37 countries, and an estimated number of 500 robots, split by 346 teams. Twenty technicians from FIL were involved in the preparation of the competition site, and 40 student volunteers supported the event realization. The event was hosted by the Institute for Systems and Robotics, a research institute located in the campus of IST.

In the following sections we will briefly overview the main research progresses this year, the technical challenges and the competition results by league, with a brief report on the Symposium, whose accepted papers are the core of this book. More details on competitions, photos, short video clips and other related information can be found in the official web page of the event at www.robocup2004.pt.

2   Soccer Middle-Size Robot League

In this league, two teams of 4-6 mid-sized robots with all sensors on-board play soccer on a field. Relevant objects are distinguished by colours. Communication among robots (if any) is supported on wireless communications. No external intervention by humans is allowed, except to insert or remove robots in/from the field. There were 24 teams in Lisbon selected by the league technical committee from the 34 teams that submitted the qualification material.

2.1   Research Issues

The 2004 rules introduced a set of changes in the games:

1. the field was enlarged to 8m × 12m;
2. the number of players is now flexible and determined by the area occupied by the whole team, even though it can only range from 4 to 6 robots per team;
3. the light used was the artificial light of the exhibition hall, *i.e.*, no special overhead illumination as in past events, therefore with possible non-uniformity and less strong illumination;
4. a referee box was used, but only to start and stop the games for now.

The purpose of 1 and 2 was mainly to improve teamwork, as larger fields tend to encourage passing the ball among robots, as well as cooperative localization of the ball and cooperative navigation among teammates, since relevant objects and landmarks are less often seen during the game. Change 3 was common to almost all the leagues and intended to be a step towards vision under natural illumination in RoboCup. Finally, the introduction of a referee box, already existing in the Small Size and 4-Legged leagues, brings further autonomy and requires further intelligence and team-play to the robots.

After the usual initial adaptation phase, most teams handled the new rules quite well. This year, most teams had their robots running well from the beginning of the tournament, without so many technical problems as in the past, except those concerning wireless communications, which is the main unsolved technical problem in RoboCup events so far.

2.2   Technical Challenges

Every year, the league technical committee (TC) prepares technical challenge competitions where teams show specific skills and technical achievements. There were two technical challenges this year:

– *Ball Control and Planning*,
– *Free Demonstration* of scientific or engineering achievements.

In the *Ball Control and Planning Challenge*, several obstacles are arbitrarily positioned in the field, and the robot must take the ball from one goal to the other with minimum or no contact with the obstacles and within a limited time.

In the *Free Challenge*, teams are free to pick their most relevant technical and/or scientific recent achievement and demonstrate it. The demonstration is evaluated by the members of the TC.

2.3   Results

The 24 teams were organized in 4 groups of 6 teams each, which played a round-robin tournament. Then, the 4 best teams in each group, in a total of 16, were grouped in 4 groups of 4 teams each, for another round-robin tournament. Finally, the first and second place teams from each group were qualified for the

Table 1. Soccer Middle-Size League top three teams

rank	team
1	EIGEN (Keio University, Japan)
2	WinKIT (Kanazawa Institute of Technology, Japan)
3	CoPS Stuttgart (University of Stuttgart, Germany)

Table 2. Overall rank for Soccer Middle-Size League technical challenges

rank	team
1	Persia (Isfahan University of Technology, Iran)
2	AllemaniACs (Technical University of Aachen, Germany)
3	Clockwork Orange (Delft University of Technology, The Netherlands)

playoff phase, which consisted of quarter-finals, semi-finals and final (as well as third-fourth place game). The top three teams of the soccer competition are listed in Table 1.

The winners of the technical challenges were:

- *Ball Control and Planning* challenge: Persia (Isfahan University of Technology, Iran)
- *Free Demonstration* challenge: Persia (Isfahan University of Technology, Iran)

The overall rank for the Middle-Size League technical challenges is shown in Table 2.

3   Soccer Small-Size Robot League

Two teams of 5 small robots without on-board sensors play soccer on a field with an overhead camera which provides feedback to an external computer of the game state (e.g., ball, own and opponent player locations). Relevant objects are distinguished by colour and coloured coded markers on the top of the robots. Commands are sent by the external computer to the robots using wireless communications. No external intervention by humans is allowed, except to insert or remove robots in/from the field. There were 21 teams in Lisbon selected by the league technical committee from the 39 teams that submitted the qualification material.

3.1   Research Issues

The main research challenges for 2004 resulted from three main changes in the rules:

1. the light used was the artificial light of the exhibition hall, i.e., no special overhead illumination as in past events. In this league, this is a particularly troublesome issue, due to the shadow cast by the camera mounting structure on the field;

Table 3. Soccer Small-Size League top three teams

rank	team
1	FU Fighters (Freie Universität Berlin, Germany)
2	Roboroos (University of Queensland, Australia)
3	LuckyStar (Ngee Ann Polytechnic, Singapore)

2. the field was almost doubled to 4 × 5.5 m;
3. the field boundary walls were removed.

The purpose was, similarly to what happened in the Middle-Size League, to encourage more cooperation among robot teammates (especially passes) and to move towards a closer-to-reality perception scenario. The illumination issue was particularly effective in this league, as some teams were relying on their good quality top cameras and had not invested on advanced vision algorithms, required to overcome non-uniform light conditions and low-light illumination. On the other hand, the improvement in teamwork, with many passes and robot formations, was visible as expected, and interesting to follow. The need for better ball control was also noticeable, both for pass reception improvement and to avoid the ball going out of the field most of the time. Some teams in this league show very interesting kicking devices, including some which are capable to raise the ball above the ground.

3.2 Results

In the round-robin phase, the teams were split in 4 groups. The top 2 teams from each group proceeded to the playoff phase. The top three teams are listed in Table 3.

4 Soccer 4-Legged Robot League

Two teams of up to 4 four-legged robots (SONY's specially programmed AIBO robots), with all sensors on-board, play soccer on a field. Relevant objects are distinguished by colours. Communication among robots (if any) is supported on wireless communications. No external intervention by humans is allowed, except to insert or remove robots in/from the field. There were 23 teams in Lisbon selected by the league technical committee from the 32 teams that submitted the qualification material.

4.1 Research Issues

This is the real robot league with the most standardized hardware, as all the platforms are SONY's specially programmed AIBOs. Consequently, teams share their code every year and the advances in software are considerably faster than for other real robot leagues. Nevertheless, this year there were two types of robots: the new ERS-7 AIBOs and the old ERS-210 AIBOs. The former are significantly faster robots, and this made a difference in terms of competition, as only one team with ERS-210 robots made it to the quarter-finals.

The 4-Legged League was the only real robot league which could not use yet the natural light of the hall, still requiring strong local illumination from projectors located around the fields. This is mainly due to the low sensitivity to light of the AIBOs cameras.

The AIBO's 3-degree-of-freedom single camera forces the teams to work on selective directed vision problems, leading to research advances in active vision (e.g., where to look), cooperative world modelling and navigation. Also, some of the rule changes for 2004 fostered the introduction of cooperative localization algorithms, as a consequence of removing the two central beacons of the field, therefore reducing the frequency of landmarks visibility by the robots.

Another rule change concerned obstacle avoidance, less enforced in the past in this league. Gait optimization was also a hot topic among teams, so as to speed up the robots, including the utilization of learning techniques. The fastest gait speed increased from 27 cm/s in 2003 to 41 cm/s this year.

4.2 Technical Challenges

Three technical challenges were held in the 2004 edition of the Four-Legged League:

- The *Open Challenge*, similar to the free challenge in the Soccer Middle-Size League, where free demonstrations were assessed by the other teams. The demonstrations included robot collaboration, ball handling, object recognition, and tracking by vision or sound.
- The *Almost SLAM Challenge*, where a landmark-based self-localization problem involving learning initially unknown landmark colours was the goal.
- The *Variable Lighting Challenge* involved light changing conditions over a 3-minutes time interval, during which a robot had to score as many goals as possible. This was surely hard for 4-Legged teams, and the winner only scored twice.

4.3 Results

In the round-robin phase, the teams were split in 4 groups. The top 2 teams from each group proceeded to the playoff phase. The top three teams are listed in Table 4.

Regarding the technical challenges, the winners were:

- *Open Challenge*: GermanTeam, demonstrating four robots moving a large wagon.
- *Almost SLAM Challenge*: rUNSWift (University of New South Wales, Australia).
- *Variable Lighting Challenge*: ASURA (Kyushu Institute of Technology, Japan).

The top three teams from the overall result for technical challenges are listed in Table 5.

Table 4. Soccer 4-Legged League top three teams

rank	team
1	GermanTeam (HU Berlin, U. Bremen, TU Darmstadt, U. Dortmund, Germany)
2	UTS Unleashed! (University of Technology, Sydney, Australia)
3	NUBots (University of Newcastle, Australia)

Table 5. Soccer 4-Legged League top three teams in the overall technical challenge

rank	team
1	UTS Unleashed! (University of Technology, Sydney, Australia)
2	ARAIBO (University of Tokyo, Chuo University, Japan)
3	ASURA (Kyushu Institute of Technology, Japan)

5 Soccer Humanoid Robot League

Humanoid robots show basic skills of soccer players, such as shooting a ball, or defending a goal. Relevant objects are distinguished by colours. So far, no games took place, penalty kicks being the closest situation to a 1-on-1 soccer game. There were 13 teams in Lisbon selected by the league technical committee from the 20 teams that submitted the qualification material.

5.1 Research Issues

This league made its debut in RoboCup2002, and its main research challenge is to maintain the dynamic stability of robots while walking, running, kicking and performing other tasks. Moreover, perception must be carefully coordinated with biped locomotion to succeed.

This year, significant advances were observed in the humanoids, namely on the technological side. Some teams showed progresses on features such as the more ergonomic mechanical design and the materials used, the ability to walk on uneven terrain, the walking speed, the ability to kick towards directions depending on sensing (e.g., the goal region not covered by the goalie), body coordination, cooperation among robots (a pass was demonstrated by Osaka University) and omnidirectional vision (used by Team Osaka ViSion robot). Also relevant is the fact that most robots came equipped with an internal power supply and wireless communications, thus improving autonomy. Tele-operation of the robots was not allowed this year.

5.2 Technical Challenges

In the humanoid league, since no games are played yet, the main events are the technical challenges: *Humanoid Walk, Penalty Kick* and *Free Style*. This year, humanoid walk included walking around obstacles and balancing walk on a slope. In the free style challenge, a pass between two robots and robot gymnastics could be observed, among other interesting demonstrations. Next year,

Table 6. Soccer Humanoid Robot League technical challenges

Soccer Humanoid Walk technical challenge

rank	team
1	Team Osaka (Systec Akazawa Co., Japan)
2	Robo-Midget (Singapore Polytechnic, Singapore)
3	Senchans (Osaka University, Japan)

Soccer Humanoid Free Style technical challenge

rank	team
1	Team Osaka (Systec Akazawa Co., Japan)
2	Robo-Erectus (Singapore Polytechnic, Singapore)
3	NimbRo (U. of Freiburg, Germany)

Soccer Humanoid Penalty Kick H80 technical challenge

rank	team
1	Senchans (Osaka University, Japan)
2	Robo-Erectus 80 (Singapore Polytechnic, Singapore)

Soccer Humanoid Penalty Kick H40 technical challenge

rank	team
1	Team Osaka (Systec Akazawa Co., Japan)
2	Robo-Erectus 40 (Singapore Polytechnic, Singapore)

challenges will attempt to promote the current weakest points in the humanoid league, by improving battery autonomy, onboard computing, locomotion and real-time perception.

5.3   Results

The winner of the Best Humanoid Award was Team Osaka ViSion humanoid, from Systec Akazawa Co., Japan. The results for the other technical challenges are listed in Table 6.

6   Soccer Simulation League

In this league, two teams of eleven virtual agents each play with each other, based on a computer simulator that provides a realistic simulation of soccer robot sensors and actions. Each agent is a separate process that sends to the simulation server motion commands regarding the player it represents, and receives back information about its state, including the (noisy and partial) sensor observations of the surrounding environment. There were 60 teams in Lisbon selected by the league technical committee from the 196 teams that submitted the qualification material.

6.1   Research Issues

The main novelty in the Soccer Simulation League in 2004 was the introduction of the 3D soccer simulator, where players are spheres in a three-dimensional environment with a full physical model. Besides that, two other competitions already running in past tournaments were present: the 2D and the Coach competitions.

The best teams from the past 2D competitions were able to quickly adapt their code to face the new challenges of the 3D competition quite well. Those challenges included the possibility to move in 3 directions, the motion inertia and delayed effects of motor commands. In the 2D competition, remote participation through Internet was possible for the first time. Participants in the Coach competition must provide a coach agent that can supervise players from a team using a standard coach language. Coaches are evaluated by playing matches with a given team against a fixed opponent.

The main research topics in the league are reinforcement learning, and different approaches to select hard-coded behaviours, such as evolutionary methods or rule based systems.

6.2   Results

The top three teams in the three competitions are listed in Tables 7-9.

Table 7. Soccer Simulation League top three teams in the 3D competition

rank	team
1	Aria (Amirkabir University of Technology, Iran)
2	AT-Humboldt (Humboldt University Berlin, Germany)
3	UTUtd 2004 (University of Tehran, Iran)

Table 8. Soccer Simulation League top three teams in the 2D competition

rank	team
1	STEP (ElectroPult Plant Company, Russia)
2	Brainstormers (University of Osnabrück, Germany)
3	Mersad (Allameh Helli High School, Iran)

Table 9. Soccer Simulation League top three teams in the Coach competition

rank	team
1	MRL (Azad University of Qazvin, Iran)
2	FC Portugal (Universities of Porto and Aveiro, Portugal)
3	Caspian (Iran University of Science and Technology, Iran)

7   Rescue Real Robot League

The RoboCupRescue Real Robot League competition acts as an international evaluation conference for the RoboCupRescue Robotics and Infrastructure Project research. The RoboCupRescue Robotics and Infrastructure Project studies future standards for robotic infrastructure built to support human welfare. The U.S. National Institute of Standards and Technology (NIST) Urban Search and Rescue (USAR) arena has been used in several RoboCupRescue and AIAA competitions and was used in Portugal as well. A team of multiple (autonomous or teleoperated) robots moves inside this arena, divided in 3 regions of increasing difficulty levels, searching for victims and building a map of the surrounding environment, to be transmitted and/or brought back by the robot(s) to the human operators. There were 20 teams in Lisbon selected by the league technical committee from the 37 teams that submitted the qualification material.

7.1   Research Issues

The competition requires robots to demonstrate capabilities in mobility, sensory perception, planning, mapping, and practical operator interfaces, while searching for simulated victims in unstructured environments. The actual challenges posed by the NIST USAR arena include physical obstacles (variable flooring, overturned furniture, and problematic rubble) to disrupt mobility, sensory obstacles to confuse robot sensors and perception algorithms, as well as a maze of walls, doors, and elevated floors to challenge robot navigation and mapping capabilities. All combined, these elements encourage development of innovative platforms, robust sensory fusion algorithms, and intuitive operator interfaces to reliably negotiate the arena and locate victims.

Each simulated victim is a clothed mannequin emitting body heat and other signs of life including motion (shifting or waving), sound (moaning, yelling, or tapping), and carbon dioxide to simulate breathing. They are placed in specific rescue situations (surface, lightly trapped, void, or entombed) and distributed throughout the arenas in roughly the same percentages found in actual earthquake statistics.

This year, two new league initiatives were introduced:

1. a high fidelity arena/robot simulation environment to provide a development tool for robot programming in realistic rescue situations;
2. a common robot platform for teams to use if they choose, based on a standard kit of components, modular control architecture, and support for the simulation mentioned above.

7.2   Results

The competition rules and scoring metric focus on the basic USAR tasks of identifying live victims, assessing their condition based on perceived signs of life, determining accurate victim locations, and producing human readable maps to enable victim extraction by rescue workers — all without damaging the environment or making false positive identifications.

After several rounds of competitive missions, the scoring metric produced three awardees that demonstrated best-in-class approaches in each of three critical capabilities:

1. Toin Pelicans team (University of Toin, Japan) for their multi-tracked mobility platform with independent front and rear flippers, as well as an innovative camera perspective mounted above and behind the robot that significantly improved the situational awareness by the operator.
2. Kurt3D team (Fraunhofer Institute for Artificial Intelligence Systems, Germany) for their application of state-of-the-art 3D mapping techniques using a tilting line scan lidar.
3. ALCOR team (University of Rome "La Sapienza", Italy) for their intelligent perception algorithms for victim identification and mapping.

8   Rescue Simulation League

The main purpose of the RoboCupRescue Simulation League is to provide emergency decision support by integration of disaster information, prediction, planning, and human interface. A generic urban disaster simulation environment was constructed based on a computer network. Heterogeneous intelligent agents such as fire fighters, commanders, victims, volunteers, etc. conduct search and rescue activities in this virtual disaster world. There were 17 teams in Lisbon selected by the league technical committee from the 34 teams that submitted the qualification material.

8.1   Research Issues

The main research objective of this league is the introduction of advanced and interdisciplinary research themes, such as behaviour strategy (e.g. multi-agent planning, real-time/anytime planning, heterogeneity of agents, robust planning, mixed-initiative planning) for AI/Robotics researchers or the development of practical comprehensive simulators for Disaster Simulation researchers.

In 2004, the league was split in two competitions:

- **Agent Competition**, where a team has a certain number of fire fighters, police, and ambulance agents and central stations that coordinate each agent type. The agents are assumed to be situated in a city in which a simulated earthquake has just happened, as a result of which, some buildings have collapsed, some roads have been blocked, some fires have started and some people have been trapped and/or injured under the collapsed buildings. The goal of each team is to coordinate and use its agents to minimize human casualties and the damage to the buildings.
- **Infrastructure Competition**, where the performance of the simulator components developed by the teams is tested. The awarded team is requested

to provide the component for the next year's competition. For this reason teams are expected to accept the open source policy before entering the competition.

8.2   Results

In the Agent competition, the preliminaries consisted of two stages. In the first stage, the teams competed on six maps with different configurations. The first 6 teams went to the semi-final. The remaining 11 teams competed in the second stage which was designed to test the robustness of the teams under varying perception conditions. The latter stage top 2 teams went to the semi-finals too. The top 4 teams of the semi-finals competed in the final.

The final standings were:

1. ResQ (University of Freiburg, Germany), with platoon agents that have reactive and cooperative behaviours which can be overridden by deliberative high-level decisions of the central station agents.
2. DAMAS-Rescue (Laval University, Canada), with a special agent programming language. Using this language, their Fire Brigade agents choose the best fire to extinguish based on the knowledge they have learned with a selective perception learning method.
3. Caspian (Iran University of Science and Technology, Iran).

In the Infrastructure competition, only the ResQ Freiburg team competed, presenting a 3D-viewer and a Fire Simulator. The 3D-viewer is capable of visualizing the rescue simulation both online and offline. The Fire Simulator is based on a realistic physical model of heat development and heat transport in urban fires. Three different ways of heat transport (radiation, convection, direct transport) and the influence of wind can be simulated as well as the protective effects of spraying water on buildings without fire.

9   RoboCup Junior

RoboCupJunior is a project-oriented educational initiative that sponsors local, regional and international robotic events for young students. It is designed to introduce RoboCup to primary and secondary school children.

RoboCupJunior offers several challenges, each emphasizing both cooperative and competitive aspects. In contrast to the one-child-one-computer scenario typically seen today, RoboCupJunior provides a unique opportunity for participants with a variety of interests and strengths to work together as a team to achieve a common goal. Several challenges have been developed: dance, soccer and rescue.

By participating in RoboCupJunior, students especially improve their individual and social skills (building self-confidence, developing a goal-oriented, systematic work style, improving their presentation and communication abilities, exercising teamwork, resolving conflicts among team members). RoboCupJunior has spread in more than 20 countries around the world. We estimate that this

Fig. 2. The RoboCup2004 Junior area

year more than 2000 teams world-wide adopted the RoboCupJunior challenges and prepared for participation in RoboCup in local, regional, or national competitions. The largest RoboCupJunior communities are China (approximately 1000 teams), Australia (approximately 500 teams), Germany, Japan, and Portugal (over 100 teams each).

Lisbon hosted the largest RoboCup Junior event so far, with 163 teams from 17 countries, 677 participants, and about 300 robots.

9.1   Competitions

The Lisbon RoboCupJunior event featured competitions in eight leagues, covering four different challenges: RoboDance, RoboRescue, RoboSoccer 1-on-1, and RoboSoccer 2-on-2 - and in each challenge two age groups - Primary for students aged under 15, and Secondary for students aged 15 and elder. The teams qualifying for the playoffs were interviewed in order to scrutinize their ability to explain their robot designs and programs.

The RoboRescue challenge is performed in an environment mimicking an urban search and rescue site. Robots have to follow a curved path, marked by a black line, through several rooms with obstacles and varying lighting conditions. The task is to find two kinds of victims on the path, marked by green and silver icons. Points are awarded for successful navigation of rooms and for detecting and signalling victims, and the time for executing the task is recorded when it is completed. Perhaps surprisingly, the vast majority of teams demonstrated perfect runs and quickly navigated through the environment while finding and signalling all victims, so that the timing was the decisive factor for making it to the finals and winning.

The RoboSoccer challenge play soccer on a pitch which is covered by a large grayscale floor and surrounded by a black wall. The only difference is that the 1-on-1 field is smaller. Goals can be detected by their walls coloured gray, and the well-known infrared-emitting ball is used for play. In both 1-on-1 and 2-on-2 Primary leagues, teams were split by three groups and played a single round of round-robin games. Teams placed first and second after round-robin directly qualified for the playoffs, and the remaining two playoff spots were determined among the three teams placed third. In 2-on-2 Secondary, we had 6 groups in round-robin and teams placed first and second advanced to the second round. On playoff day, four groups of three teams each played a second round of round-robin games, and the best team from each group advanced directly to the semifinals. Even seasoned RoboCupJunior organizers were stunned by sophisticated robots and the spectacular level of play the teams demonstrated across all of the four Junior soccer leagues.

The RoboDance challenge asks students to design some kind of stage performance which involves robots. Students may engage themselves as part of the per-

Table 10. Junior Leagues top three teams

RoboDance Primary		RoboDance Secondary	
1	Coronation Quebec 1 (Canada)	1	Kao Yip Dancing Team (China)
2	The Rock (Germany)	2	Mokas Team (Portugal)
3	Peace of the World (Japan)	3	Gipsies (Israel)
RoboRescue Primary		**RoboRescue Secondary**	
1	Chongqing Nanan Shanh (China)	1	Dunks Team Revolution (Portugal)
2	Dragon Rescue 100% (Japan)	2	Ren Min (China)
3	Chongqing Nanan Yifen (China)	3	Across (USA)
RoboSoccer 1-on-1 Primary		**RoboSoccer 1-on-1 Secondary**	
1	Shanghai Road of Tianjin (China)	1	Liuzhou Kejiguang (China)
2	Shenzhen Haitao (China)	2	I Vendicatori (Italy)
3	Wuhan Yucai (China)	3	TianJin Xin Hua (China)
RoboSoccer 2-on-2 Primary		**RoboSoccer 2-on-2 Secondary**	
1	NYPSTC1 (Singapore)	1	Kao Yip 1 (China)
2	Ultimate (Japan)	2	Espandana Juniors (Iran)
3	Red and Blue (South Korea)	3	Kitakyushu A.I. (Japan)

Table 11. Junior Dance League award winners

Category	RoboDance Primary	RoboDance Secondary
Programming	ChaCha (Japan)	Godzillas (Portugal)
Construction	The Rock (Germany)	Pyramidical Dragon (Portugal)
Costume	Turtles (Portugal)	Hunan Changsha Yali (China)
Choreography	Crocks Rock (Australia)	Joaninhas (Portugal)
Creativity	Hong Kong Primary Dancing Team (China)	Bejing No. 2 Middle School (China)
Originality	Ridgment Pearl (UK)	Mokas Team (Portugal)
Entertainment Value	RoCCI Girls (Germany)	The Rocking Robot (UK)

formance, or give a narrative to the audience while the robots perform on stage. There is a two minute time limit for the performance, and an international judge committee assesses the performance in seven categories. RoboDance is without doubt the RoboCupJunior activity allowing most flexibility in the design and programming of the robots, and challenges students' inspiration and creativity. All teams of the same age group performed on stage on one the preliminaries, and the best three teams advanced to the finals.

9.2   Results

The top three teams for the different Junior leagues, as well as the winners of the Dance League awards are listed in Tables 10 and 11, respectively.

10   Symposium

The 8th RoboCup International Symposium was held immediately after the RoboCup2004 Competitions as the core meeting for the presentation of scientific contributions in areas of relevance to RoboCup. Its scope encompassed, but was not restricted to, the fields of Artificial Intelligence, Robotics, and Education.

The IFAC/EURON 5th Symposium on Intelligent Autonomous Vehicles (IAV04) took also place at Instituto Superior Técnico, Lisbon from 5 to 7 July 2004. IAV2004 brought together researchers and practitioners from the fields of land, air and marine robotics to discuss common theoretical and practical problems, describe scientific and commercial applications and discuss avenues for future research.

On July 5, the IAV04 Symposium ran in parallel with the RoboCup Symposium and both events shared two plenary sessions:

- James Albus, NIST, USA, "RCS: a Cognitive Architecture for Intelligent Multi-agent Systems".
- Shigeo Hirose, Tokyo Institute of Technology, Japan, "Development of Rescue Robots in Tokyo Institute of Technology".

The other two plenary sessions specific to the RoboCup2004 Symposium were:

- Hugh Durrant-Whyte, U. Sydney, Australia, "Autonomous Navigation in Unstructured Environments".
- Luigia Carlucci Aiello, Universitá di Roma "La Sapienza", Italy, "Seven Years of RoboCup: time to look ahead".

118 papers were submitted to the RoboCup2004 Symposium. Among those, 68 were accepted and are published in this book: 30 as regular papers, 38 as shorter poster papers.

This year, the awarded papers were:

Scientific Challenge Award: "Map-based Multi Model Tracking of a Moving Object", Cody Kwok and Dieter Fox.

Engineering Challenge Award: "UCHILSIM: A Dinamically and Visually Realistic Simulator for the RoboCup Four Legged League", Juan Cristóbal Zagal Montealegre and Javier Ruiz-del-Solar.

11   Conclusion

Overall, RoboCup2004 was a successful event, from a scientific standpoint. The main technical challenge of holding the competitions under a reduced artificial light of the exhibition hall, instead of having special illumination per field as in the past, was overcome by most teams without significant problems, thus showing the evolution on perception robustness within the RoboCup community. Another noticeable improvement is the increase in teamwork across most real robot soccer leagues, from passes to dynamic behaviour switching, including formation control and cooperative localization. Even in the humanoid league a pass between biped robots was demonstrated by one of the teams.

On the educational side, RoboCup Junior was a tremendous success, despite the increased organizational difficulties brought by the fact that the number of participants almost doubled that of 2003.

The next RoboCup will take place in Osaka, Japan, in July 2005.

Acknowledgements

The authors would like to thank the contributions from the chairs of the League Organizing Committees for RoboCup2004: Yasutake Takahashi, Beng Kiat Ng, Thomas Röfer, Changjiu Zhou, Oliver Obst, Adam Jacoff, Levent Akin and Gerhard Kraezschmar.

References

1. Kitano, H., Kuniyoshi, Y., Noda, I., Asada, M., Matsubara, H., Osawa, E.: RoboCup: A challenge problem for AI. AI Magazine **18** (1997) 73–85
2. Kitano, H., Takokoro, S., Noda, I., Matsubara, H., Takahashi, T., Shinjou, A., Shimada, S.: RoboCup rescue: Search and rescue in large-scale disasters as a domain for autonomous agents research. In: Proceedings of the IEEE International Conference on Man, System, and Cybernetics. (1999)
3. Lund, H.H., Pagliarini, L.: Robot soccer with lego mindstorms. In Asada, M., Kitano, H., eds.: RoboCup-98: Robot Soccer World Cup II. Springer Verlag, Berlin (1999) 141–152
4. Kitano, H., ed.: Proceedings of the IROS-96 Workshop on RoboCup, Osaka, Japan (1996)
5. Kitano, H., ed.: RoboCup-97: Robot Soccer World Cup I. Springer Verlag, Berlin (1998)

6. Asada, M., Kitano, H., eds.: RoboCup-98: Robot Soccer World Cup II. Lecture Notes in Artificial Intelligence 1604. Springer Verlag, Berlin (1999)
7. Veloso, M., Pagello, E., Kitano, H., eds.: RoboCup-99: Robot Soccer World Cup III. Springer Verlag, Berlin (2000)
8. Stone, P., Balch, T., Kraetszchmar, G., eds.: RoboCup-2000: Robot Soccer World Cup IV. Lecture Notes in Artificial Intelligence 2019. Springer Verlag, Berlin (2001)
9. Birk, A., Coradeschi, S., Tadokoro, S., eds.: RoboCup-2001: Robot Soccer World Cup V. Springer Verlag, Berlin (2002)
10. Kaminka, G.A., Lima, P.U., Rojas, R., eds.: RoboCup-2002: Robot Soccer World Cup VI. Springer Verlag, Berlin (2003)
11. Polani, D., Browning, B., Bonarini, A., Yoshida, K., eds.: RoboCup-2001: Robot Soccer World Cup VII. Springer Verlag, Berlin (2004)
12. Noda, I., Suzuki, S., Matsubara, H., Asada, M., Kitano, H.: RoboCup-97: The first robot world cup soccer games and conferences. AI Magazine **19** (1998) 49–59
13. Asada, M., Veloso, M.M., Tambe, M., Noda, I., Kitano, H., Kraetszchmar, G.K.: Overview of RoboCup-98. AI Magazine **21** (2000)
14. Coradeschi, S., Karlsson, L., Stone, P., Balch, T., Kraetzschmar, G., Asada, M.: Overview of RoboCup-99. AI Magazine **21** (2000)
15. Veloso, M., Balch, T., Stone, P., Kitano, H., Yamasaki, F., Endo, K., Asada, M., Jamzad, M., Sadjad, B.S., Mirrokni, V.S., Kazemi, M., Chitsaz, H., Heydarnoori, A., Hajiaghai, M.T., Chiniforooshan, E.: Robocup-2001: The fifth robotic soccer world championships. AI Magazine **23** (2002) 55–68
16. Stone, P., (ed.), Asada, M., Balch, T., D'Andrea, R., Fujita, M., Hengst, B., Kraetzschmar, G., Lima, P., Lau, N., Lund, H., Polani, D., Scerri, P., Tadokoro, S., Weigel, T., Wyeth, G.: RoboCup-2000: The fourth robotic soccer world championships. AI Magazine **22** (2001)

Map-Based Multiple Model Tracking of a Moving Object

Cody Kwok and Dieter Fox

Department of Computer Science & Engineering,
University of Washington, Seattle, WA
{ctkwok, fox}@cs.washington.edu

Abstract. In this paper we propose an approach for tracking a moving target using Rao-Blackwellised particle filters. Such filters represent posteriors over the target location by a mixture of Kalman filters, where each filter is conditioned on the discrete states of a particle filter. The discrete states represent the non-linear parts of the state estimation problem. In the context of target tracking, these are the non-linear motion of the observing platform and the different motion models for the target. Using this representation, we show how to reason about physical interactions between the observing platform and the tracked object, as well as between the tracked object and the environment. The approach is implemented on a four-legged AIBO robot and tested in the context of ball tracking in the RoboCup domain.

1 Introduction

As mobile robots become more reliable in navigation tasks, the ability to interact with their environment becomes more and more important. Estimating and predicting the locations of objects in the robot's vicinity is the basis for interacting with them. For example, grasping an object requires accurate knowledge of the object's location relative to the robot; detecting and predicting the locations of people helps a robot to better interact with them. The problem of tracking moving objects has received considerable attention in the mobile robotics and the target tracking community [1, 2, 9, 11, 5, 6]. The difficulty of the tracking problem depends on a number of factors, ranging from how accurately the robot can estimate its own motion, to the predictability of the object's motion, to the accuracy of the sensors being used.

This paper focuses on the problem of tracking and predicting the location of a ball with a four-legged AIBO robot in the RoboCup domain, which aims at playing soccer with teams of mobile robots. This domain poses highly challenging target tracking problems due to the dynamics of the soccer game, coupled with the interaction between the robots and the ball. We are faced with a difficult combination of issues:

- **Highly non-linear motion of the observer:** Due to slippage and the nature of legged motion, information about the robot's own motion is extremely noisy, blurring the distinction between motion of the ball and the robot's ego-motion. The rotation of these legged robots, in particular, introduces non-linearities in the ball tracking problem.

D. Nardi et al. (Eds.): RoboCup 2004, LNAI 3276, pp. 18–33, 2005.

- **Physical interaction between target and environment:** The ball frequently bounces off the borders of the field or gets kicked by other robots. Such interaction results in highly non-linear motion of the ball.
- **Physical interaction between observer and target:** The observing robot grabs and kicks the ball. In such situations, the motion of the ball is tightly connected to the motion or action of the robot. This interaction is best modeled by a unified ball tracking framework rather than handled as a special case, as is done typically.
- **Inaccurate sensing and limited processing power:** The AIBO robot is equipped with a 176×144 CMOS camera placed in the robot's "snout". The low resolution provides inaccurate distance measurements for the ball. Furthermore, the robot's limited processing power (400MHz MIPS) poses computational constraints on the tracking problem and requires an efficient solution.

In this paper, we introduce an approach that addresses all these challenges in a unified Bayesian framework. The technique uses Rao-Blackwellised particle filters (RBPF) [3] to jointly estimate the robot location, the ball location, its velocity, and its interaction with the environment. Our technique combines the efficiency of Kalman filters with the representational power of particle filters. The key idea of this approach is to sample the non-linear parts of the state estimation problem (robot motion and ball-environment interaction). Conditioning on these samples allows us to apply efficient Kalman filtering to track the ball. Experiments both in simulation and on the AIBO platforms show that this approach is efficient and yields highly robust estimates of the ball's location and motion.

This paper is organized as follows. After discussing related work in the next section, we will introduce the basics of Rao-Blackwellised particle filters and their application to ball tracking in the RoboCup domain. Experimental results are presented in Section 4, followed by conclusions.

2   Related Work

Tracking moving targets has received considerable attention in the robotics and target tracking communities. Kalman filters and variants thereof have been shown to be well suited for this task even when the target motion and the observations violate the linearity assumptions underlying these filters [2]. Kalman filters estimate posteriors over the state by their first and second moments only, which makes them extremely efficient and therefore a commonly used ball tracking algorithm in RoboCup [10]. In the context of maneuvering targets, multiple model Kalman filters have been shown to be superior to the vanilla, single model filter. Approaches such as the Interacting Multiple Model (IMM) and the Generalized Pseudo Bayesian (GPB) filter represents the target locations using a bank of Kalman filters, each conditioned on different potential motion models for the target. An exponential explosion of the number of Gaussian hypotheses is avoided by merging the Gaussian estimates after each update of the filters [2]. While these approaches are efficient, the model merging step assumes that the state conditioned on each discrete motion model is unimodal. However, our target tracking problem depends heavily on the uncertainty of the observer position in addition to the motion model.

These two factors can interact to produce multimodal distributions conditioned on each model, and an example will be provided in section 3.1.

Particle filters provide a viable alternative to Kalman filter based approaches [4]. These filters represent posteriors by samples, which allows them to optimally estimate non-linear, non-Gaussian processes. Recently, particle filters have been applied success-fully to people tracking using a mobile robot equipped with a laser range-finder [11, 9]. While the sample-based representation gives particle filters their robustness, it comes at the cost of increased computational complexity, making them inefficient for complex estimation problems.

As we will describe in Section 3.2, Rao-Blackwellised particle filters (RBPF) [3] combine the representational benefits of particle filters with the efficiency and accuracy of Kalman filters. This technique has been shown to outperform approaches such as the IMM filter on various target tracking problems [5, 6]. Compared to our method, existing applications of RBPFs consider less complex dynamic systems where only one part of the state space is non-linear. Our approach, in contrast, estimates a system where several components are highly non-linear (observer motion, target motion). Furthermore, our technique goes beyond existing methods by incorporating information about the environment into the estimation process. The use of map information for improved target tracking has also been proposed by [9]. However, their tracking application is less demanding and relies on a vanilla particle filter to estimate the joint state space of the observer and the target. Our RBPFs are far more efficient since our approach rely on Kalman filters to estimate the target location.

3   Rao-Blackwellised Particle Filters for Multi Model Tracking with Physical Interaction

In this section, we will first describe the different interactions between the ball and the environment. Then we will show how RBPFs can be used to estimate posteriors over the robot and ball locations. Finally, we will present an approximation to this idea that is efficient enough to run onboard the AIBO robots at a rate of 20 frames per second.

3.1   Ball-Environment Interactions

Fig. 1(a) describes the different interactions between the ball and the environment:

- None: The ball is either not moving or in an unobstructed, straight motion. In this state, a linear Kalman filter can be used to track its location and velocity.
- Grabbed: The ball is between the robot's legs or grabbed by them. The ball's position is thus tightly coupled with the location of the robot. This state is entered when the ball is in the correct position relative to the robot. It typically exits into the Kicked state.
- Kicked: The robot just kicked the ball. This interaction is only possible if the ball was grabbed. The direction and velocity of the following ball motion depend on the type of kick. There is also a small chance that the kick failed and the ball remains in the Grabbed state.

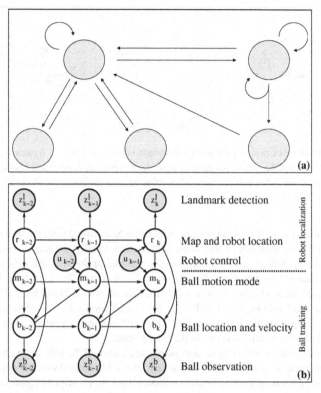

Fig. 1. (a) Finite state machine describing the transitions between different ball motion models. All transitions are probabilistic, the probabilities are enclosed in parentheses. (b) Graphical model for tracking a moving ball. The nodes in this graph represent the different parts of the dynamic system process at consecutive points in time, and the edges represent dependencies between the individual parts of the state space. Filled circles indicate observed nodes, where z_k^l are landmark and z_k^b ball observations

- Bounced: The ball bounced off one of the field borders or one of the robots on the field. In this case, the motion vector of the ball is assumed to be reflected by the object with a considerable amount of orientation noise and velocity reduction.
- Deflected: The ball's trajectory has suddenly changed, most likely kicked by another robot. In this state, the velocity and motion direction of the ball are unknown and have to be initialized by integrating a few observations.

The transition probabilities between states are parenthesized in the figure. From the None state, we assume that there is a 0.1 probability the ball will be deflected at each update. When the ball is somewhere close in front of the robot, the ball will enter the Grabbed state with probability defined by a two-dimensional linear probability function. When the ball collides with the borders or other robots, it will always reflect and move into the Bounced state. This transition is certain because each ball estimate is

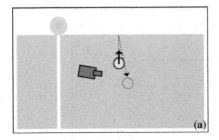 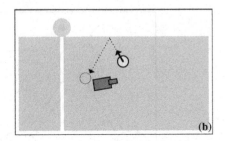

Fig. 2. Effect of robot position on ball-environment interation. The two figures represent two different estimates of the robot's position, differing by less than 20 degrees and 20cm. They result in very different predicted ball positions after collision with the border, with (a) in front of the robot and (b) behind

conditioned on a sampled robot position, which can be assumed to be the true robot position. This will be elaborated further in the next section. Finally, the None state transition back to itself by default, hence it takes up the residual probability after computing the previously mentioned transitions. For the states Kicked and Grabbed, the transitions are associated with whether these actions succeed or not. Kicking the ball has a 0.9 success rate and Grabbing 0.8. Finally, the Bounced and Deflected states are used to initiate changes in the Kalman filters. Once the changes are made, they transition immediately to the normal updates in the None state.

While most of these interactions depend on the location of the ball relative to the robot, the ball's interactions with the environment (*e.g.* the field borders) strongly depend on the ball location on the field, *i.e.* in global coordinates. In order to estimate global coordinates from relative observations, we need to associate relative ball positions with the robot's location and orientation on the field. Hence, the problem of tracking a ball requires the joint estimation of the ball location, the robot location, and the ball-environment interaction. Figure 2 shows an example of their interdependence. The robot is tracking a ball travelling towards the border. It is uncertain about its own location, and Figures 2(a) and (b) are both possible locations. In the figures, the robot has the same estimates of the relative position and velocity of the ball. However, their slight difference in positions leads to very different predicted ball positions after collision with the border, with (a) in front and (b) behind the robot. If we ignore the uncertainty in robot position, this interaction cannot be modeled correctly. In the next section we will see how RBPF can be used to perform this joint estimation.

3.2   Rao-Blackwellised Posterior Estimation

Let $\langle m_k, b_k, r_k \rangle$ denote the state of the system at time k. Here, $m_k = \{$None, Grabbed, Kicked, Bounced, Deflected$\}$ are the different types of interaction between the robot and the environment. $b_k = \langle x_b, y_b, \dot{x}_b, \dot{y}_b \rangle$ denotes the ball location and velocity in global coordinates and $r_k = \langle x_r, y_r, \theta_r \rangle$ is the robot location and orientation on the field. Furthermore, z_k are observations of the ball and landmarks, provided in relative bearing and distance.

A graphical model description of the ball tracking problem is given in Fig. 1(b). The graphical model describes how the joint posterior over $\langle b_k, m_k, r_k \rangle$ can be computed efficiently using independencies between parts of the state space. The nodes describe different random variables and the arrows indicate dependencies between these variables. The model shows that, just like in standard robot localization, the robot location r_k only depends on the previous location r_{k-1} and the robot motion control u_{k-1}. Landmark observations z_k^l only depend on the current robot location r_k. The location and velocity of the ball, b_k, typically depend on the previous ball state b_{k-1} and the current ball motion model m_k. The arc from m_k to b_k describes, for example, the change in ball prediction if the ball was kicked or bounced off a field border. If $m_k = \texttt{Grabbed}$, then the ball location depends on the current robot location r_k, as indicated by the arrow from r_k to b_k. Relative ball observations z_k^b only depend on the current ball and robot position. Transitions of the ball motion model m_k are probabilistic versions of those described in Fig. 1(a).

Now that the dependencies between different parts of the state space are defined, we can address the problem of filtering, which aims at computing the posterior over $\langle b_k, m_k, r_k \rangle$ conditioned on all observations made so far. A full derivation of the RBPF algorithm is beyond the scope of this paper; see [3] for a thorough discussion of the basic RBPF and its properties. RBPFs represent posteriors by sets of weighted samples, or particles:

$$S_k = \{s_k^{(i)}, w_k^{(i)} \mid 1 \leq i \leq N\}.$$

In our case, each particle $s_k^{(i)} = \langle b_k^{(i)}, m_{1:k}^{(i)}, r_{1:k}^{(i)} \rangle$, where $b_k^{(i)}$ are the mean and covariance of the ball location and velocity and $m_{1:k}^{(i)}$ and $r_{1:k}^{(i)}$ are the histories of ball motion models and robot locations, respectively. The key idea of RBPFs is to condition the ball estimate $b_k^{(i)}$ on a particle's history of ball motion models $m_{1:k}^{(i)}$ and robot locations $r_{1:k}^{(i)}$. This conditioning turns the ball location and velocity into a linear system that can be estimated efficiently using a Kalman filter.

To see how RBPFs recursively update posterior estimates, we factorize the posterior as follows:

$$p(b_k, m_{1:k}, r_{1:k} \mid z_{1:k}, u_{1:k-1}) = p(b_k \mid m_{1:k}, r_{1:k}, z_{1:k}, u_{1:k-1})$$
$$\cdot p(m_{1:k} \mid r_{1:k}, z_{1:k}, u_{1:k-1}) \, p(r_{1:k} \mid z_{1:k}, u_{1:k-1}) \quad (1)$$

The task is to generate samples distributed according to (1) based on samples drawn from the posterior at time $k-1$, represented by the previous sample set S_{k-1}. We generate the different components of each particle $s_k^{(i)}$ stepwise by simulating (1) from right to left. In the first step, a sample $s_{k-1}^{(i)} = \langle b_{k-1}^{(i)}, m_{1:k-1}^{(i)}, r_{1:k-1}^{(i)} \rangle$ is drawn from S_{k-1}. Through conditioning on this sample, we first expand the robot trajectory to $r_{1:k}^{(i)}$, then the ball motion models to $m_{1:k}^{(i)}$ conditioned on $r_{1:k}^{(i)}$, followed by an update of $b_k^{(i)}$ conditioned on both the robot trajectory and the motion model history. Let us start with expanding the robot trajectory, which requires to draw a new robot position $r_k^{(i)}$ according to

$$r_k^{(i)} \sim p(r_k \mid s_{k-1}^{(i)}, z_{1:k}, u_{1:k-1}). \quad (2)$$

The trajectory $r_{1:k}^{(i)}$ resulting from appending $r_k^{(i)}$ to $r_{1:k-1}^{(i)}$ is then distributed according to the rightmost term in (1). The distribution for $r_k^{(i)}$ can be transformed as follows:

$$p(r_k \mid s_{1:k-1}^{(i)}, z_{1:k}, u_{1:k-1}) \tag{3}$$

$$= p(r_k \mid s_{k-1}^{(i)}, z_k, u_{k-1}) \tag{4}$$

$$\propto p(z_k \mid r_k, s_{k-1}^{(i)}, u_{k-1}) \, p(r_k \mid s_{k-1}^{(i)}, u_{k-1}) \tag{5}$$

$$= p(z_k \mid r_k, r_{k-1}^{(i)}, m_{k-1}^{(i)}, b_{k-1}^{(i)}, u_{k-1}) \, p(r_k \mid r_{k-1}^{(i)}, m_{k-1}^{(i)}, b_{k-1}^{(i)}, u_{k-1}) \tag{6}$$

$$= p(z_k \mid r_k, m_{k-1}^{(i)}, b_{k-1}^{(i)}, u_{k-1}) \, p(r_k \mid r_{k-1}^{(i)}, u_{k-1}) \tag{7}$$

Here, (4) follows from the (Markov) property that r_k is independent of older information given the previous state. (5) follows by Bayes rule, and (7) from the independencies represented in the graphical model given in Fig. 1(b). To generate particles according to (7) we apply the standard particle filter update routine [4]. More specifically, we first pick a sample $s_{k-1}^{(i)}$ from S_{k-1}, then we predict the next robot location $r_k^{(i)}$ using the particle's $r_{k-1}^{(i)}$ along with the most recent control information u_{k-1} and the robot motion model $p(r_k \mid r_{k-1}^{(i)}, u_{k-1})$ (rightmost term in (7)). This gives the extended trajectory $r_{1:k}^{(i)}$. The importance weight of this trajectory is given by the likelihood of the most recent measurement: $w_k^{(i)} \propto p(z_k \mid r_k^{(i)}, m_{k-1}^{(i)}, b_{k-1}^{(i)}, u_{k-1})$. If z_k is a landmark detection, then this likelihood is given by $p(z_k \mid r_k^{(i)})$, which corresponds exactly to the particle filter update for robot localization. If, however, z_k is a ball detection z_k^b, then total probability gives us

$$p(z_k \mid r_k^{(i)}, m_{k-1}^{(i)}, b_{k-1}^{(i)}, u_{k-1}) \tag{8}$$

$$= \sum_{M_k} p(z_k \mid r_k^{(i)}, m_{k-1}^{(i)}, b_{k-1}^{(i)}, u_{k-1}, M_k) p(M_k \mid r_k^{(i)}, m_{k-1}^{(i)}, b_{k-1}^{(i)}, u_{k-1}) \tag{9}$$

$$= \sum_{M_k} p(z_k \mid r_k^{(i)}, b_{k-1}^{(i)}, u_{k-1}, M_k) p(M_k \mid r_k^{(i)}, m_{k-1}^{(i)}, b_{k-1}^{(i)}, u_{k-1}) \tag{10}$$

where M_k ranges over all possible ball motion models. The second term in (10) can be computed from the transition model, but $p(z_k \mid r_k^{(i)}, b_{k-1}^{(i)}, u_{k-1}, M_k)$ is the likelihood obtained from a Kalman update, which we need to perform for each M_k (see below). In the next section, we will describe how we avoid this complex operation.

At the end of these steps, the robot trajectory $r_{1:k}^{(i)}$ of the particle is distributed according to the rightmost term in (1). We can now use this trajectory to generate the ball motion model part $m_{1:k}^{(i)}$ of the particle using the second to last term in (1). Since we already have $m_{1:k-1}^{(i)}$, we only need to sample

$$m_k^{(i)} \sim p(m_k \mid m_{1:k-1}^{(i)}, r_{1:k}^{(i)}, b_{k-1}^{(i)}, z_{1:k}, u_{1:k-1}) \tag{11}$$

$$\propto p(z_k \mid m_k, r_k^{(i)}, b_{k-1}^{(i)}, u_{k-1}) \, p(m_k \mid m_{k-1}^{(i)}, r_k^{(i)}, b_{k-1}^{(i)}, u_{k-1}). \tag{12}$$

(12) follows from (11) by reasoning very similar to the one used to derive (7). The rightmost term in (12) describes the probability of the ball motion mode at time k given the

previous mode, robot location, ball location and velocity, and the most recent control. As described above, this mode transition is crucial to model the ball's interaction with the environment. To generate motion model samples using (12), we predict the mode transition using reasoning about the different ball-environment interactions (see Fig. 1(a)). The importance weight of the sample $s_k^{(i)}$ has then to be multiplied by the observation likelihood $p(z_k | m_k^{(i)}, r_k^{(i)}, b_{k-1}^{(i)}, u_{k-1})$, which is given by the innovation of a Kalman update conditioned on $m_k^{(i)}, r_k^{(i)}, b_{k-1}^{(i)}$, and u_{k-1}.

To finalize the computation of the posterior (1), we need to determine the leftmost term of the factorization. As mentioned above, since we sampled the non-linear parts $r_{1:k}^{(i)}$ and $m_{1:k}^{(i)}$ of the state space, the posterior

$$b_k^{(i)} \sim p(b_k | m_{1:k}^{(i)}, r_{1:k}^{(i)}, z_{1:k}, u_{1:k-1}) \tag{13}$$

can be computed analytically using a regular Kalman filter. The Kalman filter prediction uses the motion model $m_k^{(i)}$ along with the most recent control u_{k-1}. The correction step is then based on the robot location r_k along with the most recent observation z_k. The Kalman correction step is not performed if z_k is a landmark detection.

To summarize, we generate particles at time k by first drawing a particle $s_{k-1}^{(i)} = \langle b_{k-1}^{(i)}, m_{1:k-1}^{(i)}, r_{1:k-1}^{(i)} \rangle$ from the previous sample set. In the first step, we expand this particle's robot trajectory by generating a new robot location using (7), which gives us $r_{1:k}^{(i)}$. Conditioning on $r_{1:k}^{(i)}$ allows us to expand the history of ball motion models by predicting the next motion model using (12). Finally, $r_{1:k}^{(i)}$ and $m_{1:k}^{(i)}$ render the ball location and velocity a linear system and we can estimate $b_k^{(i)}$ using regular Kalman filter updating. The importance weight of the new particle $s_k^{(i)} = \langle b_k^{(i)}, m_{1:k}^{(i)}, r_{1:k}^{(i)} \rangle$ is set proportional to

$$w_k^{(i)} \propto p(z_k | r_k^{(i)}, m_{k-1}^{(i)}, b_{k-1}^{(i)}, u_{k-1}) p(z_k | m_k, r_k^{(i)}, b_{k-1}^{(i)}, u_{k-1}). \tag{14}$$

3.3   Efficient Implementation

We implemented the RBPF algorithm described above and it worked very well on data collected by an AIBO robot. By computing the joint estimate over the robot location and the ball, the approach can handle highly non-linear robot motion, predict when the ball bounces into field borders, and eliminates inconsistent estimates (*e.g.*, when the ball is outside the field). Unfortunately, the approach requires on the order of 300–500 particles, which is computationally too demanding for the AIBO robots, especially since each particle has a Kalman filter attached to it. The main reason for this high number of samples is that for each robot location, we need to estimate multiple ball motion models. Furthermore, in RBPF the ball and robot estimates are coupled, with the weights of each sample depending on the contributions from the ball and the robot position. One consequence is that ball estimates can influence the robot's localization, since a sample can be removed by resampling if its ball estimate is not very accurate. This influence can be useful in some situations, such as invalidating a sample when the ball estimate is out-of-bounds, since we can infer that its robot position must also be

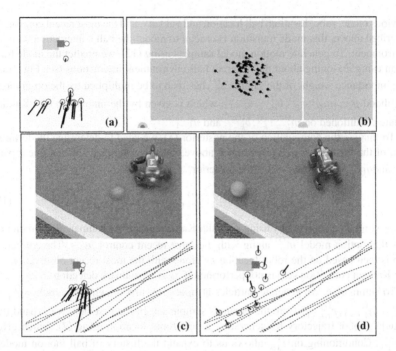

Fig. 3. (a) Robot-centric view of predicted ball samples. The robot kicked the ball to its right using its head, indicated by the small rectangle. If field borders are not considered, the ball samples travel in the kicked direction (ball motion is illustrated by the length and orientation of the small lines). (b) Particles representing robot's estimate of its position at the beginning of the kick command. (c) The robot has kicked the ball towards the border. The ball samples are attached to the robot particles shown in (b) in order to estimate the relative locations of borders. The sampled borders are shown as dashed lines. (d) Most ball samples transition into the Bounced state. Due to the uncertainty in relative border location, ball samples bounce off the border at different times, with different directions and velocities. The ball sample distribution predicts the true ball location much better than without considering the borders (compare to (a)). Note that ball samples can also bounce off the robot

erroneous. However, this artifact is undesirable in general; while the global position of the ball is conditioned on the robot position, the ball does not really provide significant information for localization. This problem is further accentuated by the fact that while ball estimates need to be updated about 20 times per second, the robot location on the field needs to be updated about only once per second. The frequent ball updates result in importance weights that outweigh the contribution of (rather rare) landmark detections. This results in the deletion of many robot location hypotheses even if the robot was not seeing any landmarks.

Our efficient approximation to the full Rao-Blackwellised approach described in the previous section is based on the observation that ball detections do not provide significant information about the robot's location on the field. The key idea is we partitioned the state space into robot and ball positions, which are updated separately. We recombine them into Rao-blackwellised particles when needed. The set of samples S_k is now the pair $\langle R_k, B_k \rangle$, where R_k is the set of robot position samples, and B_k is the set of ball samples, each sample consists of the ball's position and the model it is conditioned on. At the beginning of each update, instead of sampling $s_{k-1}^{(i)} = \langle b_{k-1}^{(i)}, m_{1:k-1}^{(i)}, r_{1:k-1}^{(i)} \rangle$ from S_{k-1}, we sample robot positions $r_{1:k-1}^{(j)}$ from R_{k-1} and ball model-position pairs $b_{k-1}^{(i)}, m_{1:k-1}^{(i)}$ from B_{k-1}. With the decoupled state space, we can use M samples to estimate the robot's position and N samples to estimate the ball's location. Usually, the distribution of ball motion model is less complex than the robot position, so $N < M$. As before, we extend the robot trajectory to $r_k^{(i)}$, but this time we can drop the dependency of the robot position on the ball since $r_k^{(i)}$ is not coupled with a ball position. Thus we have the following approximation of equation (2)

$$r_k^{(j)} \sim p(r_k | s_{k-1}^{(j)}, z_{1:k}, u_{1:k-1}) \tag{15}$$

$$\approx p(r_k | r_{1:k-1}^{(j)}, z_{1:k}, u_{1:k-1}). \tag{16}$$

Thus the estimation of the robot position becomes regular particle filter-based localization [4], with ball observations having no influence on r_k. The distribution of $r_k^{(j)}$ is simply

$$p(r_k | r_{1:k-1}^{(j)}, z_{1:k}, u_{1:k-1}) \propto p(z_{1:k} | r_k, u_{1:k-1}) p(r_k | r_{1:k-1}^{(j)}, u_{1:k-1}) \tag{17}$$

which has the exact form of a robot localization update. We obtain $r_k^{(j)}$ by sampling from the robot motion model $p(r_k | r_{1:k-1}^{(j)}, u_{1:k-1})$ using the most recent control information u_{k-1}, as usual. This gives us an updated robot sample set R_k.

We now turn our attention to the ball estimates. Since they are no longer coupled with robot positions, they are estimated in a pseudo-global coordinate system. Each ball sample $b_k^{(i)}$ has an associated observer position $\rho_{1:k}^{(i)}$ which is initialized to the origin $\rho_1 = (0, 0, 0)$. At each iteration, $\rho_k^{(i)}$ is computed from $\rho_{1:k-1}^{(i)}$ by sampling from the robot motion model and u_{k-1}. When we compute the ball's interaction with the environment, we need the ball estimates in global coordinates. We obtain this by sampling a robot position $r_k^{(j)}$ from R_k for each ball sample, and applying the offset $b_k^{(i)} - \rho_k^{(i)}$ to $r_k^{(j)}$. With this scheme, we approximate (12) as follows:

$$p(z_k | m_k, r_k^{(i)}, b_{k-1}^{(i)}, u_{k-1}) \approx p(z_k | m_k, \rho_k^{(i)}, b_{k-1}^{(i)}, u_{k-1}) \tag{18}$$

$$p(m_k | m_{k-1}^{(i)}, r_k^{(i)}, b_{k-1}^{(i)}, u_{k-1}) \approx p(m_k | m_{k-1}^{(i)}, r_k^{(j)}, b_{k-1}^{(i)}, u_{k-1}) \tag{19}$$

In (18) we compute the likelihood of the ball observation based on $\rho_k^{(i)}$ rather than the joint robot position $r_k^{(i)}$. This approximation is fairly faithful to the original RBPF since the trajectory represented by $\rho_{1:k}^{(i)}$ is generated from $u_{1:k-1}$, similar to $r_{1:k}^{(i)}$. In (19) we predict the motion model of the ball $b_k^{(i)}$ using the global position obtained from the sampled robot position $r_k^{(j)}$ instead of the paired $r_k^{(i)}$ in RBPF. While the variance of this approximation is higher, the expected distribution resulting from interacting with the environment is the same as RBPF's. Note that we can set $j = i$ to avoid extra sampling without loss of generality. In this case, $r_k^{(i)}$ is interpreted as the i-th sample in R_k, not the coupled robot position in RBPF.

Finally, the posterior for the ball positions can be computed by a Kalman filter update using $\rho_k^{(i)}$ instead of $r_k^{(i)}$ in (13)

$$b_k^{(i)} \sim p(b_k | m_{1:k}^{(i)}, \rho_k^{(i)}, z_{1:k}, u_{1:k-1}). \tag{20}$$

The complete approximation algorithm is shown in Table 1. At each iteration, a different relative offset is generated for each ball by sampling from the robot motion model using u_k. Then the ball samples are translated back to global coordinates by attaching their relative ball estimates to the most recent particles of the robot localization. These particles are selected by sampling N robot positions from the M in the set. Thus, the ball and its motion model are estimated exactly as described before, with sampling from the highly non-linear robot motion and the ball motion models. Furthermore, since ball estimates are in global coordinates, the ball-environment interaction can be predicted as before. The only difference is that information about the ball does not contribute to the estimated robot location. However, our approximation drastically reduces the number of robot and ball samples needed for good onboard results (we use 50 robot and 20 ball particles, respectively). The key idea of this algorithm is summarized in Fig. 3. As can be seen in Fig. 3(c) and (d), each ball particle $\langle b_k^{(i)}, m_k^{(i)} \rangle$ uses a different location for the border extracted from the robot location particles $r_k^{(j)}$ shown in (b). These borders determine whether the ball motion model transitions into the Bounced state.

3.4   Tracking and Finding the Ball

Since our approach estimates the ball location using multiple Kalman filters, it is not straightforward to determine the direction the robot should point its camera in order to track the ball. Typically, if the robot sees the ball, the ball estimates are tightly focused on one spot and the robot can track it by simply pointing the camera at the mean of the ball samples with the most likely mode. However, if the robot doesn't see the ball for a period of time, the distribution of ball samples can get highly uncertain and multi-modal. This can happen, for instance, after the ball is kicked out-of-sight by the robot or by other robots.

To efficiently find the ball in such situations, we use a grid-based representation to describe where the robot should be looking. The grid has two dimensions, the pan and the tilt of the camera. Each ball sample is mapped to these camera parameters using inverse kinematics, and is put into the corresponding grid cells. Each cell is weighted by the sum of the importance weights of the ball samples inside the cell. To find the ball, the robot moves its head to the camera position specified by the highest weighted cell. In order to represent all possible ball locations, the pan range of the grid covers $360°$. Cells with pan orientation exceeding the robot's physical pan range indicate that the robot has to rotate its body first.

An important aspect of our approach is that it enables the robot to make use of *negative information* when looking for the ball: Ball samples that are not detected even though they are in the visible range of the camera get low importance weights (visibility

Table 1. The efficient implementation of the Rao-blackwellised particle filter algorithm for ball tracking

---

1. **Inputs:**
   $S_{k-1} = \langle R_{k-1}, B_{k-1} \rangle$ representing belief $Bel(s_{k-1})$, where
   $R_{k-1} = \{\langle r_{k-1}^{(j)}, w_{k-1}^{(j)} \rangle \mid j = 1, \ldots, N\}$ represents robot positions,
   $B_{k-1} = \{\langle b_{k-1}^{(i)}, m_{k-1}^{(i)}, \rho_{k-1}^{(i)}, \omega_{k-1}^{(i)} \rangle \mid i = 1, \ldots, M\}$ represents ball positions
   control measurement u_{k-1},
   observation z_k

2. $R_k := \emptyset, B_k := \emptyset$ // *Initialize*

3. **for** $j := 1, \ldots, N$ **do**   // *Generate N robot samples*

4.   Sample an index l from the discrete distribution given by
   the weights in R_{k-1} // *Resampling*

5.   Sample $r_k^{(j)}$ from $p(r_k \mid r_{k-1}, u_{k-1})$ conditioned on $r_{k-1}^{(l)}$ and u_{k-1}

6. $w_k^{(j)} := p(z_k \mid r_k^{(j)})$ // *Compute likelihood*

7. $R_k := R_k \cup \{\langle r_k^{(j)}, w_k^{(j)} \rangle\}$ // *Insert sample into sample set*

8. **end do**

9. Normalize the weights in R_k

10. **for** $i := 1, \ldots, M$ **do**   // *Update M ball samples*

11.   Sample an index l from the discrete distribution given by
    the weights ω_{k-1} in B_{k-1}

12.   Sample $m_k^{(i)}$ from $p(m_k \mid m_{k-1}^{(i)}, r_k^{(i)}, b_{k-1}^{(l)}, u_{k-1})$

13.   Sample $\rho_k^{(i)}$ from $p(r_k \mid r_{k-1}, u_{k-1})$ conditioned on $\rho_{k-1}^{(l)}$ and u_{k-1}

14. $b_k^{(i)} :=$ Kalman update using $b_{k-1}^{(l)}, m_k^{(i)}, z_k$ and u_{k-1}

15. $\omega_k^{(i)} := p(z_k \mid m_k, \rho_k^{(i)}, b_{k-1}^{(i)}, u_{k-1})$ // *Compute importance weight*

16. $B_k := B_k \cup \{\langle b_k^{(i)}, m_k^{(i)}, \rho_k^{(i)}, \omega_k^{(i)} \rangle\}$ // *Insert sample into sample set*

17. **end do**

18. Normalize the weights in B_k

19. **return** $S_k = \langle R_k, B_k \rangle$

---

considers occlusions by other robots). In the next update step of the Rao-Blackwellised particle filter, these ball samples are very unlikely to be drawn from the weighted sample set, thereby focusing the search to other areas. As a result, the robot scans the whole area of potential ball locations, pointing the camera at the most promising areas first. When none of the ball particles are detected, the ball is declared lost.

4 Experiments

We evaluated the effectiveness of our tracking system in both simulated and real-world environments. We first illustrate the basic properties of our algorithm by comparing it with the traditional Kalman Filter. Then we evaluate how well the approach works on the real robot.

4.1 Simulation Experiments

In the RoboCup domain, robots often cannot directly observe the ball, due to several reasons such as looking at landmarks for localization, or the ball is occluded by another robot. The goalkeeper robot in particular has to accurately predict the trajectory of the ball in order to block it. Hence, accurate *prediction over multiple camera frames* is of utmost importance. To systematically evaluate the prediction quality of our multiple model approach, we simulated a robot placed at a fixed location on the soccer field, while the ball is kicked randomly at different times. The simulator generates noisy observations of the ball. The observation noise is proportional to the distance from the robot and constant in the orientation, similar in magnitude to the information available to the real robot. Prediction quality is measured using the RMS error at the predicted locations.

In this experiment, we measure the prediction quality for a given amount of time in the future, which we call the *prediction time*. Map information is not used, and the ball is estimated with 20 particles (used to sample ball motion models at each iteration). The observation noise of the Kalman filters was set according to the simulated noise. To determine the appropriate prediction noise, we generated straight ball trajectories and used the prediction noise value that minimized the RMS error for these runs. This prediction noise was used by our multiple model approach when the motion model was none. The results for prediction times up to 2 seconds are shown in Fig. 4(a). In addition to our RBPF approach (thick, solid line), we compare it with Kalman filters with different prediction noise models. The thin, solid line shows the RMS error when using a single Kalman filter with prediction noise of the straight line model (denoted KF^*). However, since the ball is not always in a straight line motion, the quality of the filter estimates can be improved by inflating the prediction noise. We tried several noise inflation values and the dotted line in Fig. 4(a) gives the results for the best such value (denoted KF'). Not surprisingly, our multiple model approach greatly improves the prediction quality.

The reason for this improved prediction performance is illustrated in Fig. 5. Our approach, shown in Fig. 5(a), is able to accurately track the ball location even after a kick, which is due to the fact that the particle filter accurately "guesses" the kick at the correct location. The Kalman filter with the straight line motion model quickly diverges, as shown by the dotted line in Fig. 5(b). The inflated prediction noise model (thick, solid line) keeps track of the ball, but the trajectory obviously overfits the observation noise. Further intuition can be gained from Fig. 5(c). It compares the orientation error of the estimated ball velocity using our approach versus the inflated Kalman filter KF^* (KF' shows a much worse performance; for clarity it is omitted from the graph). Clearly, our approach recovers from large errors due to kicks much faster, and it converges to a significantly lower error even during straight line motion.

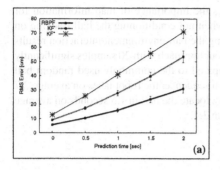
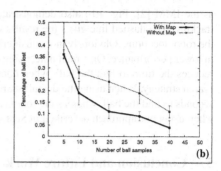

Fig. 4. (a) RMS error of the ball's position for different prediction times. (b) Percentage of time the robot loses track of the ball after a kick for different numbers of particles with and without map information

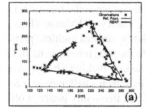
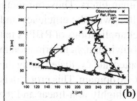
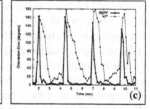

Fig. 5. Tracking of a ball trajectory with multiple kicks, the observer is located on the left. In (a) and (b), the observations are indicated by stars and the true target trajectory is shown as a thin line. (a) shows the estimated trajectory of our RBPF multiple-model approach. (b) shows the estimates using an extended Kalman filter for two different prediction noise settings. The dotted line represents the estimates when using prediction noise that assumes a linear ball trajectory, and the thick, solid line is estimated using inflated prediction noise. (c) Orientation errors over a time period including four kicks. Solid line represents RBPF, dashed line a Kalman filter with inflated prediction noise

4.2   Real-World Experiments

In this section we describe an experiment carried out on the real robot. It demonstrates that the use of map information brings significant improvements to the tracking performance. In the experiment, an Aibo robot and a ball are placed on the soccer field at random locations. The task of the robot is to track the ball and kick it as soon as it reaches it. The kick is a sideway head kick as shown in Fig. 3(c). The robot is not able to see the ball until it recovers from the kick motion. During the experiment, the robot stays localized by scanning the markers on the field periodically.

The solid line in Fig. 4(b) shows the rate of successfully tracking the ball after a kick. As can be seen, increasing the number of samples also increases the performance of the approach. The poor performance for small sample sizes indicates that the distribution of the ball is multi-modal, rendering the tracking task difficult for approaches such as the IMM [2]. Fig. 4(b) also demonstrates the importance of map information for tracking. The dashed line gives the results when not conditioning the ball tracking on the robot locations. Obviously, not considering the ball-environment interaction results in lower performance. On a final note, using our approach with 20 samples significantly reduces the time to find the ball, when compared to the commonly used random ball search strategy. When using the default search sequence, the robot takes on average 2.7 seconds to find the ball, whereas the robot can locate the ball in 1.5 seconds on average when using our approach described in Section 3.4.

5   Conclusion and Future Work

In this paper we introduced a novel approach to tracking moving targets. The approach uses Rao-Blackwellised particle filters to sample the potential interactions between the observer and the target and between the target and the environment. By additionally sampling non-linear motion of the observer, estimating the target and its motion can be performed efficiently using Kalman filters. Thus, our method combines the representational complexity of particle filters with the efficiency and accuracy of Kalman filters. The approach goes beyond other applications of RBPFs in that it samples multiple parts of the state space and integrates environment information into the state transition model.

The technique was implemented and evaluated using the task of tracking a ball with a legged AIBO robot in the RoboCup domain. This problem is extremely challenging since the legged motion of the robot is highly non-linear and the ball frequently bounces off obstacles in the environment. We demonstrate that our efficient implementation results in far better performance than vanilla Kalman filters. Furthermore, we show that taking the environment into account results in additional performance gains. We belief that our approach has applications to tracking problems beyond the RoboCup domain. It can be applied whenever the observer robot performs highly non-linear motion and the environment provides information about the motion of the object being tracked.

In the future we will extend the algorithm to integrate ball information observed by other robots, delivered via wireless communication. Such information can be transmitted efficiently by clustering the ball samples according to the different discrete motion states. The integration of transmitted ball estimates can then be done conditioned on the different discrete ball states. Another important area of future research is the integration of additional information provided by the vision system. Currently, we do not model the location of other robots on the field and the ball transits into the deflected model at random points in time. Furthermore, we estimate the relative location of the field borders using only the robot's location estimates. However, if the robot detects the ball and an object in the same camera image, then this image provides more accurate information about the relative location between the ball and an object.

Finally, we conjecture that further performance gains can be achieved using an unscented Kalman filter [12] to jointly track the position of the robot and the ball. Using the Rao-Blackwellisation described in this paper, the discrete state of the ball would still be sampled. However, each of these samples would be attached with an unscented filter over both robot and ball locations (and velocity). By modeling more dimensions using efficient Kalman filters we expect to be able to track the robot / ball system with far less samples.

See http://www.cs.washington.edu/balltracking for further information about our approach. [8, 7]

References

1. Y. Bar-Shalom and X.-R. Li. *Multitarget-Multisensor Tracking: Principles and Techniques.* Yaakov Bar-Shalom, 1995.
2. Y. Bar-Shalom, X.-R. Li, and T. Kirubarajan. *Estimation with Applications to Tracking and Navigation.* John Wiley, 2001.
3. A. Doucet, J.F.G. de Freitas, K. Murphy, and S. Russell. Rao-Blackwellised particle filtering for dynamic Bayesian networks. In *Proc. of the Conference on Uncertainty in Artificial Intelligence*, 2000.
4. A. Doucet, N. de Freitas, and N. Gordon, editors. *Sequential Monte Carlo in Practice.* Springer-Verlag, New York, 2001.
5. A. Doucet, N.J. Gordon, and V. Krishnamurthy. Particle filters for state estimation of jump Markov linear systems. *IEEE Transactions on Signal Processing*, 49(3), 2001.
6. F. Gustafsson, F. Gunnarsson, N. Bergman, U. Forssell, J. Jansson, R. Karlsson, and P-J. Nordlund. Particle filters for positioning, navigation and tracking. *IEEE Transactions on Signal Processing*, 50(2), 2002.
7. C.T. Kwok, D. Fox, and M. Meilă. Adaptive real-time particle filters for robot localization. In *Proceedings of the 2003 IEEE International Conference on Robotics Automation (ICRA '03), Taipei, Taiwan*, September 2003.
8. C.T. Kwok, D. Fox, and M. Meilă. Real-time particle filters. *IEEE Special Issue on Sequential State Estimation*, March 2004.
9. M. Montemerlo, S. Thrun, and W. Whittaker. Conditional particle filters for simultaneous mobile robot localization and people-tracking. In *Proc. of the IEEE International Conference on Robotics & Automation*, 2002.
10. T. Schmitt, R. Hanek, M. Beetz, S. Buck, and B. Radig. Cooperative probabilistic state estimation for vision-based autonomous mobile robots. *IEEE Transactions on Robotics and Automation*, 18(5), 2002.
11. D. Schulz, W. Burgard, and D. Fox. People tracking with mobile robots using sample-based joint probabilistic data association filters. *International Journal of Robotics Research*, 22(2), 2003.
12. E.A. Wan and R. van der Merwe. The unscented Kalman filter for nonlinear estimation. In *Proc. of Symposium 2000 on Adaptive Systems for Signal Processing, Communications, and Control*, 2000.

UCHILSIM: A Dynamically and Visually Realistic Simulator for the RoboCup Four Legged League

Juan Cristóbal Zagal and Javier Ruiz-del-Solar

Department of Electrical Engineering, Universidad de Chile,
Av. Tupper 2007, 6513027 Santiago, Chile
{jzagal, jruizd}@ing.uchile.cl
http://www.robocup.cl

Abstract. UCHILSIM is a robotic simulator specially developed for the RoboCup four-legged league. It reproduces with high accuracy the dynamics of AIBO motions and its interactions with the objects in the game field. Their graphic representations within the game field also possess a high level of detail. The main design goal of the simulator is to become a platform for learning complex robotic behaviors which can be directly transferred to a real robot environment. UCHILSIM is able to adapt its parameters automatically, by comparing robot controller behaviors in reality and in simulations. So far, the effectiveness of UCHILSIM has been tested in some robot learning experiments which we briefly discuss hereinafter. We believe that the use of a highly realistic simulator might speed up the progress in the four legged league by allowing more people to participate in our challenge.

1 Introduction

A fully autonomous robot should be able to adapt itself to the changes in its operational environment, either by modifying its behaviors or by generating new ones. Learning and evolution are two ways of adaptation of living systems that are being widely explored in evolutionary robotics [5]. The basic idea is to allow robots to develop their behaviors by freely interacting with their environment. A fitness measure determines the degree in which some specific task has been accomplished during behavior execution. This measure is usually determined by the designer. One of the main factors to consider within this approach is the amount of experience that the robot is able to acquire from the environment.

The process of learning through experience is a time consuming task that requires, for real robots, testing a large amount of behaviors by means of real interactions. An alternative consists on simulating the interaction between the robot and the environment. Unfortunately, since simulation is usually not accurate, the acquired behaviors are not directly transferable to reality; this problem is usually referred to as the *reality gap*. There is a large list of experiments in the literature where simulators are used for generating simple robotic behaviors [5][6]. However, there are few examples of the generation of complex robotic behaviors in a simulation with successful transfers to reality.

D. Nardi et al. (Eds.): RoboCup 2004, LNAI 3276, pp. 34–45, 2005.

Simulation can be achieved at different levels, for example: one can simulate the high-level processes of robot behaviors by simplifying the sensor and actuator responses. A more complete representation is obtained when considering the low-level interactions among sensors, actuators and environment. Nevertheless, this usually entails complex models of dynamics and sensor related physical processes. We believe that a fundamental requirement for generating low-level behaviors from simulations is to consider a complete representation which includes low-level physical interactions. Thus, high-level behaviors can be obtained from lower-level behaviors in a subsumption fashion.

Nowadays, generating representative simulators of the dynamics of the interactions of robots might not be an impossible task. Once achieved it allows for the easy generation of a variety of complex behaviors which otherwise would take a very extensive design period. However, we believe that generating a realistic simulator is not just a matter of modeling and design. In order to be fully realistic, the simulator must be adapted through real robot behavior execution as proposed in [12].

The RoboCup four-legged league offers a great challenge and opportunity for exploring low-level behavior acquisition. In this context we have decided to investigate how the use of a very realistic simulator might help the development of new behaviors. Although this league simulation warrants a good degree of attention, we identify a lack of accurate dynamic simulators. Aiming at solving this gap we present UCHILSIM, an accurate simulator in both the dynamic as well as the graphic aspects. The main design goal of the simulator is to become a platform for learning complex robotic behaviors by testing in a virtual environment the same controllers that operate in the real robot environment. UCHILSIM is able to learn its own parameters automatically, by comparing the robot controller behavior fitness values in reality and in simulations. We believe that the extensive use of this kind of tool might accelerate the generation of complex robotic behaviors within the RoboCup domain.

The remainder of this paper is ordered as follows. In section 2 some related work is presented. The UCHILSIM simulator is described in section 3. In section 4 some real learning experiments with AIBO robots using UCHILSIM are shown. Finally, in section 5 the conclusions and projections of this work are presented.

2   Related Work

So far three simulators have been reported for the RoboCup four-legged league, these are the ASURA team simulator [3], the ARAIBO team simulator [1], and the German Team Robot Simulator [7]. All these simulators consider a graphic representation of AIBO robots and soccer environment, but only the German Team Simulator considers as well the dynamics of objects at an elementary level of representation. Table 1 summarizes the main characteristics of these simulators.   We consider that none of them represent with great accuracy both the graphic and the dynamic aspects of the environment. In this context there is a lot of work to be done in order to generate a simulator that accurately mimics the interaction of a robot in a real environment. We believe that it is possible to generate accurate simulators at a level in which it is feasible to learn complex behaviors with the successful transfer of these behaviors to reality.

Table 1. Main characteristics of simulators that have been reported for the RoboCup four legged league

Name of Simulator	Presence of Dynamics	Level of Graphics	Functionalities
ASURA Simulator	No dynamics.	Good graphic representations	Allows capturing AIBO image.
ARAIBO Simulator	No dynamics.	AIBO camera characteristics well treated.	Allows capturing AIBO image.
German Team Simulator	Present at an elementary level.	Elemental graphic representations.	Large set of functionalities, e.g. interfacing with monitor, calibration tools, etc.

3 UCHILSIM

UCHILSIM is a robotic simulator designed for the RoboCup four-legged league. It is built on top of two processing engines: one in charge of reproducing the dynamics of articulated rigid bodies and the other in charge of generating accurate graphic representations of the objects and the soccer setting. The simulator also includes a variety of interfacing functions which allow the core system to be connected with all the UChile1 controlling modules and sub-systems, as well as to some learning mechanisms. The simulator includes a complete graphic user interface. Figure 1 shows a screenshot of the use of UCHILSIM.

The main design goal of UCHILSIM is to allow robots to acquire behaviors learnt in a representative virtual environment. The long-term goal of the simulator is to allow the generation of complex behaviors for RoboCup team players by efficiently combining the acquisition of knowledge in reality as well as simulations [12].

Most of the rigid body dynamics in UCHILSIM are modeled and computed using the Open Dynamics Engine (ODE) library [9], which is an industrial quality library for simulating articulated rigid body dynamics in virtual environments. ODE is fast, flexible and robust; it allows the definition of a variety of advanced joints and contact points under friction. ODE solves equations of motion by means of a Lagrange multiplier velocity model. It uses a Coulomb contact and friction model which is solved by means of a Dantzig LCP solver method. In UCHILSIM special attention has been provided for the modeling of servo motors and ball.

The graphic representation of the objects in UCHILSIM is obtained by using the Open Graphic Library (OpenGL) [10]. The CAD model of the AIBO robots was generated starting from standard data provided by Sony [11]. The changes made to this model were the incorporation of player's red and blue jackets, the renewal of the robot colors, and the modifications in the original model in order to achieve greater accuracy.

The following is a description of the UCHILSIM main components, such as the basic architecture of the system, the dynamic and graphic engines, the user interface, the learning capabilities of the system and the object loader.

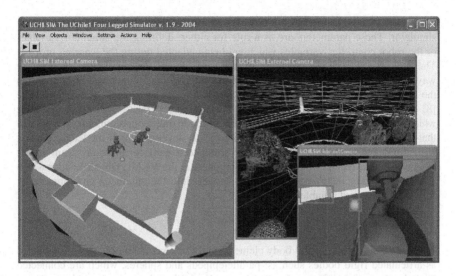

Fig. 1. Illustration of the UCHILSIM software and its user interface. It is possible to observe the display of various viewing options

3.1 Basic Architecture

Figure 2 illustrates the basic architecture of UCHILSIM. It is possible to observe how the core functions of the simulator are interfaced with a variety of applications. In the core of the simulator the dynamic engine closely cooperates with the graphic engine to generate a representation of each object. By means of a learning interface, a set of

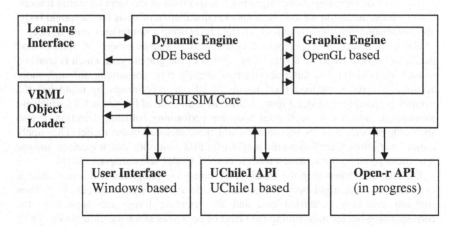

Fig. 2. Illustration of the UCHILSIM architecture. It is possible to observe how the overall system is organized as a set of interfaces around a core subsystem which contains the dynamic and graphic engines

parameters is interchanged with a learning algorithm which runs independently of the simulator. These parameters are as a rule for either: defining the simulator variables or the robot controller variables which are being adapted during the learning process. The user interface allows changing several variables of the simulation, such as the way objects are being rendered as well as the external manipulation of objects within the game field. An application interface allows running the overall UChile1 package [8] on the simulation. We are currently working on an Open-r application interface which will allow the compilation of any open-r code under our simulator. We believe that such kind of tool will be quite relevant for the development of the league since it will allow, for example, to realistically simulate a game against any team in the league. Another recently incorporated component is the VRML Object Loader which allows to quickly incorporate new robot models into the simulated environment by following a fast and reliable procedure.

3.2 Dynamic Engine

In UCHILSIM all the AIBO body elements and the soccer field objects are modeled as articulated rigid bodies such as: parallelepipeds and spheres, which are connected to each other using different types of joints. A joint is a dynamic constraint enforced between two bodies, in order that they can only hold certain positions and orientations in relation to each other. We use universal joint models for defining the relationship among the AIBO torso and its thighs, i.e. rotation is constrained to just two degrees of freedom. Simple hinge joints are used for defining the relation among thighs and taps, i.e. constraining rotation to only one degree of freedom. Fixed joint models are used for defining the relation among the taps and hoofs; in this case one direction of deformation is allowed, and as a result the model accurately represents the rotation of the hoofs. In addition small spheres are attached to the base of each leg by means of fixed joints intended to mimic the effect of the leg tops.

The mass distribution of each rigid body is specified in the form of inertia tensors. For this implementation we assume a uniform distribution of mass for each rigid body. Mass estimation was carried out for each rigid body using a weight measuring device.

Collision detection is performed either with a simple model of spheres and parallelepipeds or by using a simplified version of the graphic grid which is attached to each rigid body. The latter approach is slightly time consuming although more accurate. We have obtained good results in all our experiments by using just the parallelepiped-sphere model. Figure 3 illustrate a diagram of the geometries which are alternatively attached to each rigid body for performing collision detection, it also shows the placement of the servomotors which are included in our model. They apply torque over joints according to the output of a PID controller which receives angular references given by the actuation module of the UChile1 software package [8].

On each simulation step the equation dynamics are integrated and a new state is computed for each rigid body (velocities, accelerations, angular speeds, etc). Then collision detection is carried out, and the resulting forces are applied to the corresponding bodies transmitting the effect of collisions along the entire body. Thus, the friction parameters deserve to be given special attention since they are used for computing the reaction forces between the robot limbs and the carpet. These parameters are under automatic adaptation on each performed experiment.

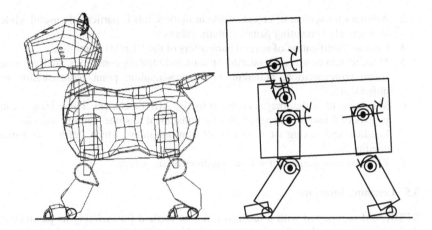

Fig. 3. Collision detection models which are alternatively used in UCHILSIM. The figure on the left corresponds to the graphic grid model used for collision detection. The figure on the right corresponds to the model generated with a set of parallelepipeds and spheres. By using this model we are able to perform accurate dynamic experiments while keeping the simulation at real time speed. The figure also shows the position of the servomotors which are included in the robot model

3.3 Graphic Engine

The dynamic engine computes the corresponding positions and rotation matrixes of each body in the simulation space at the end of each simulation step. Then, for each rigid body, the corresponding graphic object is rendered. This is carried out by efficiently calling the corresponding graphic data. The graphic engine is also in charge of producing the image acquired from the AIBO's cameras. This is quite relevant for producing experiments with vision based systems. Using this system we will specifically intend to produce an extension of the work presented in [14]. We haven't concentrated our efforts on producing extremely realistic images yet, but we estimate that this process will be simple. We will incorporate some of the transformations which were proposed in [1] such as: the camera distortion and CMOS filters. Using a CAD modeler software we gave the AIBO models blue and red jackets, which were originally provided by Sony, we also constructed the corresponding soccer scenario. The process of importing the graphic data into C++ code was quite time consuming before using the object loader. Currently the graphic data is directly obtained from the object loader module.

3.4 User Interface

The user interface of the simulator currently provides the following set of functions:

1. Loading of arbitrary AIBO models in modified VRML format.
2. Placement, at any moment, of different objects within the simulation, such as: robots, balls and other objects.

3. Arbitrary movement of objects while in motion, this is particularly useful while interactively generating games with the robots.
4. On line Modification of several parameters of the UChile1 controller.
5. Modification of several rendering options and viewing conditions, such as: wire frame representation, bounding box representation, point representation of objects, etc.
6. Management of the images captured by the AIBO's camera. These images can be exported into files or automatically transmitted to some learning software.
7. Loading and saving of a variety of configurations which define the game conditions.
8. Efficient management of several windows on the screen.

3.5 Learning Interface

UCHILSIM is powered with a fast and efficient method for updating its parameters during running time. It is designed to communicate with other programs by means of a TCP/IP network. We have considered this, given that the simulator needs to adapt itself in order to perform experiments with the *Back to Reality* approach that will be discussed ahead.

3.6 UChile1 and Open-r API

The entire UChile1 software package, which allows a team of fully autonomous AIBO robots to play soccer, can be compiled for UCHILSIM. We had to carry out several modifications on our code in order to make this possible. However, the simulator is a great tool given that, besides its learning capabilities, it is very useful for debugging any piece of code of our system. We are currently working towards generating an Open-r API for the simulator; the idea is to be capable of compiling any Open-r code for UCHILSIM. We believe that there are several applications for such kind of tool, for example, it might speed up the progress of the league by allowing people around the world to develop and test software without the need of having a real AIBO. Thus, incoming research groups might collaborate with the league by testing their code in a simulated environment. For those who already have a real AIBO it might be interesting to test their systems during long evaluation periods, letting teams play against each other during days. This will allow the accurate analysis of the differences among teams and of course the possibility of learning from these experiences.

3.7 Object Loader

The VRML language allows defining objects into a tree like structure of nodes, each one containing graphic as well as structural information of body elements, such as: mass, articulation points, etc. Although graphic data of AIBO models available to the public does not currently contain dynamic information, we have modified them by incorporating mass and motor data into the VRML files. On earlier versions of our simulator the robot models were hard wired into the simulator and the process of incorporating new models was quite time consuming. On the other hand updating a

VRML file is considerably less time consuming. Using this technique we have added the ERS-220 and the new ERS-7 AIBO robot models into our simulator.

4 Using UCHILSIM for Learning Behaviors with Back to Reality

UCHILSIM is a platform intended for learning complex low level behaviors. The capabilities of UCHILSIM will be illustrated on hands of two different real experiments. We will first briefly describe *Back to Reality*. The Back to Reality paradigm combines into a single framework: learning from reality and learning from simulations. The main idea is that the robot and its simulator co-evolve in an integrated fashion as it can be seen in the block diagram presented on figure 4. The robot learns by alternating real experiences and virtual (in simulator) ones. The evolution of the simulator continuously narrows the differences among simulation and reality (reality–gap). The simulator learns from the implementation of the robots behavior in the real environment, and by comparing the performance of the robot in reality and in the simulator.

The internal parameters of the simulator are adapted using performance variation measures. When the robot learns in reality, the robot controller is structurally coupled to the environment, whereas when it learns in simulation the robot controller is structurally coupled to its simulator. Thus, the simulation parameters are continuously tuned narrowing the reality-gap along the behavior adaptation process.

The *Back to Reality* approach consists of the online execution of three sequential learning processes: *L1* is the learning of the robot controller in the simulated environment. *L2* is the learning of the robot controller in the real environment. *L3* is

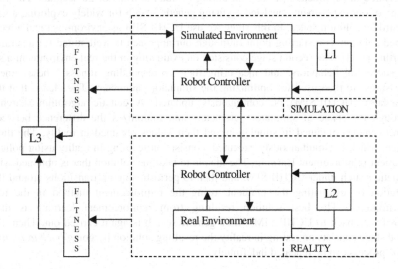

Fig. 4. Back to Reality building blocks

the learning of the simulator parameters. In this process the gap among simulation and reality is narrowed by minimizing, after each run of *L3*, the difference between the obtained fitness in reality and in simulation

During *L1* and *L2* the robot controller adaptation depends on the behavior *B'* observed in the simulated environment, and on the behavior *B* observed in the real environment, respectively. During *L3* the simulated environment is the result of the previous simulated environment and the real environment, as well as the real behavior and the simulated behavior. For implementing *L1* and *L2* any kind of learning algorithm can be used. Although we think that considering the evaluation time limitations of the experiments in reality, a reinforced learning algorithm is more suitable for implementing *L2*. Taking into account the flexibility of simulations, we think a good alternative for implementing *L1* are genetic algorithms. Regarding *L3*, we should consider the fact that the simulator has a large amount of parameters that probably are not explicitly related with aspects of the desired behavior. If this is the case, then *L3* could be implemented using genetic algorithms. Otherwise, reinforced learning could be an alternative. All these issues are addressed in [12].

4.1 Learning to Walk

Since we have a team competing in RoboCup we are particularly motivated on improving the gait speed of our team. One can notice that there is a strong correlation between the speed of a robot-player and the success of a robot soccer team. We considerably improved the gaits of our system by learning with UCHILSIM and the Back to Reality approach. As a behavior fitness measure we used the robot speed measured in centimeters per second during evaluation trials of 20 seconds. The first stage of our experiment consisted on using genetic search for widely exploring a gait controller solution space. In this stage we use UCHILSIM as environment and a hand tuned solution as a starting point (although our approach is well suited for a scratch starting point). The second stage consisted on evaluating in the real environment a set of successful behaviors, and measuring their corresponding fitness. These fitness measures are then used for optimizing the simulator parameters. The idea is that the simulator (UCHILSIM) be continuously updated. A genetic algorithm searches through the space of simulator parameters and minimizes the difference between fitness values obtained in simulation and their values obtained in reality. The third stage, which is simultaneously executed, consists on learning in reality using policy-gradient reinforcement learning. The idea is to take the solution that is obtained with genetic search under UCHILSIM, and then to perform smooth transitions around the solution by estimating the gradient, using the reinforcement method in the real environment. The best solution resulting from reinforcement learning is then transferred back to UCHILSIM where genetic search is again carried out. Then, the final stage consists on testing in reality the resulting solution by going *back to reality*. This process can be repeated indefinitely.

4.1.1 Robot Controller Parameters
The following set of 20 parameters define the AIBO's gait in our experiments (for a detailed explanation see [12]): the locus shape (3 parameters: length, shape factor and

lift factor.); the front locus shape modifier (3 parameters: lift height, air height and descending height); the rear locus shape modifier (3 parameters: lift height, air height and descending height); the front locus operation point (3 parameters: x, y and z); the rear locus operation point (3 parameters: x, y and z); locus skew multiplier in the x-y plane (for turning); the speed of the feet while in the ground; the fraction of time each foot spends on the air; the time spent on the lift stage (its equal to the descending time); the number of points in the air stage of which the inverse kinematics is calculated.

4.1.2 UCHILSIM Parameters

The robot simulator is defined by a set of 12 parameters, which determine the simulator and robot dynamics. These parameters include the ODE values used for solving the dynamic equations and the PID constants used for modeling the leg servos. There are 4 parameters for the mass distribution in the robot: head mass, neck mass, body mass and leg mass; 4 parameters of the dynamic model: friction constant, gravity constant, force dependent slip in friction direction 1 and force dependent slip in friction direction 2; and finally 4 parameters for the joint leg model: proportional, integral and differential constants of the PID controller and maximum joint torque.

4.1.3 Experiments Description

The procedure consists on learning the 20 robot controller parameters in the simulator and in reality, as well as learning the 12 simulator parameters. Genetic algorithms were used for the evolution of the simulator parameters and for the evolution of the robot controller parameters in UCHILSIM. Specifically, a conventional genetic algorithm employing fitness-proportionate selection with linear scaling, no-elitism scheme, two-points crossover with $Pc=0.7$ and mutation with $Pm=0.005$ per bit was employed. Given the set of parameters obtained from the simulator, we continued their adaptation in reality using the Policy Gradient Reinforcement Learning method [4]. The experimental conditions are fully described in [12].

First, walking experiments were carried out in UCHILSIM. The fitness evolution of these individuals is shown on figure 5. From these experiments we extracted a set of the 10 best individuals; they averaged a speed of 18 cm/s. The best individual of this group performed 20 cm/s in reality. The fitness value of each one of these individuals was compared with the resulting fitness that they exhibit in simulation. And the norm of the resulting fitness differences was used as a fitness function to be minimized by genetic search trough the space of simulator parameters. We obtained a minimum fitness of 2 cm/s as discrepancies occurred between the simulation and reality exhibited by these individuals. The best individual was then taken as a starting point for a policy gradient learning process performed in reality. With this method we achieved a speed of 23.5 cm/s. The resulting best individual obtained with reinforcement learning was then taken back to the now evolved adapted simulator where a genetic search took place starting with the population generated with permutations of this best individual. Finally some of the best individuals resulting from the genetic adaptation were tested on reality. Among these trials we found an individual that averaged a speed of 24.7 cm/s in reality. It should be noticed that improvements where done on our own controlling system, and that therefore, the resulting speed is not directly comparable to those obtained by others. Besides the gait locus used by a controller the efficiency on the computations also matter, the low level refinements on motor control, the inverse kinematics models being used, etc.

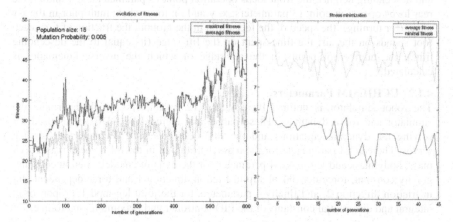

Fig. 5. Left: evolution of fitness for individuals tested with the UCHILSIM simulator in the first stage. Before adapting the simulator the individuals receive larger fitness in simulation than in reality. Right: Adaptation of the simulator, it is possible to observe how the minimization of the average differences of fitness obtained in reality versus simulations takes place

4.2 Learning to Kick the Ball

Since this experiment is presented in [13] we will not offer many details here, however we can say that UCHILSIM was used for learning to kick the ball with AIBO robots using the Back to Reality approach, and that the resulting ball-kick behaviors were quite interesting, performing similarly as the best ball-kicks currently being used in the league. The behaviors which were obtained in the simulator were directly transferable into reality at the end of the adaptation process. Another important aspect to observe is that these behaviors were obtained from scratch.

5   Conclusions and Projections

UCHILSIM is a robotic simulator specially developed for the RoboCup four legged league. It reproduces the dynamics of AIBO motions and its interactions with objects in the game field with a great level of realism. The simulator also has a great amount of detail in the graphic representation of game field objects. The main design goal of the simulator is to become a platform for learning complex robotic behaviors by allowing testing of the same controllers that should operate in the real robot environment within a virtual environment. So far the effectiveness of UCHILSIM has been tested in two robotic behavior learning experiments which we have briefly described.

Currently each simulation step takes about 8*ms* with one AIBO robot being simulated on a Pentium IV 2.5 GHz processor and 512Mb of RAM. Using this computing power we are able to simulate up to two robots at a realistic level. Several computers can be used for running simulations with more than two robots, however we have not implemented this option yet. We are currently working on improving our system; we expect to increase the current frame fate.

The following are some short term projections of UCHILSIM: (1) Consolidate the Open-r Universal UCHILSIM API. (2) Perform experiments where our localization methods will be tested using simulation and reality. (3) Perform experiments on the visual calibration of the simulator virtual cameras with basis on the Back to Reality approach. And (4) perform experiments combining real robots with virtual ones in a single soccer game.

A sample version of UCHILSIM is available at http://www.robocup.cl

References

1. Asanuma, K., Umeda, K., Ueda, R., Arai, T.: Development of a Simulator of Environment and Measurement for Autonomous Mobile Robots Considering Camera Characteristics. Proc. of Robot Soccer World Cup VII, Springer (2003).
2. Google Source Directory Resource of Robotics Simulation Tools http://directory.google.com/Top/Computers/Robotics/Software/Simulation/ (2004).
3. Ishimura, T., Kato, T., Oda, K., Ohashi, T.: An Open Robot Simulator Environment. Proc. of Robot Soccer World Cup VII, Proc. of Robot Soccer World Cup VII, Springer (2003).
4. Kohl, N. and Stone, P. (2004). Policy Gradient Reinforcement Learning for Fast Quadrupedal Locomotion. Submitted to ICRA (2004).
5. Nolfi, S., Floreano, D.: Evolutionary Robotics – The Biology, Intelligence, and Technology of Self-Organizing Machines, In: Intelligent Robotics and Automation Agents. MIT Press (2000).
6. Nolfi, S.: Evolving non-trivial behavior on autonomous robots: Adaptation is more powerful than decomposition and integration. In: Gomi, T. (eds.): Evolutionary Robotics: From Intelligent Robots to Artificial Life, Ontario, Canada, AAI Books (1997).
7. Roefer, T.: German Team RoboCup 2003 Technical Report, Available at http://www.germanteam.de (2003).
8. Ruiz-del-Solar, J., Zagal, J.C., Guerrero, P., Vallejos, P., Middleton, C., Olivares, X.: UChile1 Team Description Paper, In: Proceedings of the 2003 RoboCup International Symposium, Springer, (2003).
9. Smith, Open Dynamics Engine Library, ODE web site available at http://opende.sourceforge.net (2003).
10. The Open Graphics Library. Available at http://www.opengl.org (2004).
11. The Open-r Software Development Kit. Available at http://www.openr.org (2003).
12. Zagal, J.C., Ruiz-del-Solar, J., Vallejos, P.: Back to Reality: Crossing the Reality Gap in Evolutionary Robotics. Proceedings of the IAV 2004, 5th IFAC Symposium on Intelligent Autonomous Vehicles, (in press), (2004).
13. Zagal, J.C., Ruiz-del-Solar, J.: Learning to Kick the Ball Using Back to Reality. Proceedings of RoboCup 2004: Robot Soccer World Cup VIII, Springer (in this volume), (2004).
14. Zagal, J.C., Ruiz-del-Solar, J., Guerrero, P., Palma, R.: Evolving Visual Object Recognition for Legged Robots, In: LNAI Proceedings of RoboCup 2003: Robot Soccer World Cup VII, Springer, (2003).

CommLang: Communication for Coachable Agents

John Davin, Patrick Riley, and Manuela Veloso

Carnegie Mellon University,
Computer Science Department, Pittsburgh, PA 15232
jdavin@cmu.edu, {pfr, mmv}@cs.cmu.edu

Abstract. RoboCup has hosted a coach competition for several years creating a challenging testbed for research in advice-giving agents. A coach agent is expected to advise an unknown coachable team. In RoboCup 2003, the coachable agents could process the coach's advice but did not include a protocol for communication among them. In this paper we present CommLang, a standard for agent communication which will be used by the coachable agents in the simulation league at RoboCup 2004. The communication standard supports representation of multiple message types which can be flexibly combined in a single utterance. We then describe the application of CommLang in our coachable agents and present empirical results showing the communication's effect on world model completeness and accuracy. Communication in our agents improved the fraction of time which our agents are confident of player and ball locations and simultaneously improved the overall accuracy of that information.

1 Introduction

In a multi-agent domain in which agents cooperate with each other to achieve a goal, communication between those agents can greatly aid them. Without communication, agents must rely on observation-based judgments of their collaborator's intentions [1, 2]. Communication enables agents to more explicitly coordinate planning decisions and execute more intelligent group behaviors.

In domains where agents have only partial knowledge of their environment, communication is equally important for sharing state information with other agents. In most robotic applications, agents have only a limited view of the world. However, by communicating with a team, they can achieve a much broader and more accurate view of the world.

Communication is an important aspect of many multi-agent systems, especially those involving teamwork. Recently, several formal models have been proposed to capture the decisions which agents face in communicating [3, 4].

In RoboCup simulation soccer, teams of eleven agents compete against an opposing team. A coach agent with global world information may advise the team members. The soccer players are permitted to communicate through auditory messages, but are limited to ten character messages. In order to maximize the

D. Nardi et al. (Eds.): RoboCup 2004, LNAI 3276, pp. 46–59, 2005.

utility of communication, particularly with messages limited to a short length, we needed an architecture that could flexibly represent a number of different types of information, and at the same time be easy to implement so that it could be adopted as a standard.

Soccer simulation already has a language for communicating between the coach and the players, named CLang, but no standardized language for inter-agent communication has previously existed. While past work has developed general purpose agent communication languages like KQML [5] and FIPA-ACL [6, 7], these languages are not directly applicable here because of the extremely limited bandwidth between the players.

Other soccer simulation teams have used fixed communication schemes in which they always send the same information — such as ball position and player position. While this is beneficial, we believe that communication can be better utilized.

In this paper we will present CommLang, a communication language that we developed that has been adopted as the standard for coachable agents in the 2004 RoboCup competition. We will also describe the algorithms we used to implement CommLang in our coachable team, and show empirical evidence that communication improved the accuracy of the world model in our agents.

2   The CommLang Communication Standard

We developed the CommLang communication standard to provide a means of communicating between coachable agents. However, the protocol is also useful for other simulation agents. It defines representations for different types of information to be transmitted over a limited bandwidth communication channel. The standard specifies how to encode information into character strings for use in the soccer server's character-based communication messages.

CommLang addresses only the composition and encoding of communication messages. We do not specify which specific types of information should be sent, nor how often messages should be sent. We also do not specify how the received information should be used, except that there should be some reasonable utilization of the information. The messages are useful because they contain meaning about what the sending agent's beliefs are.

The soccer simulation server is configured to allow messages of ten characters (out of a set of 73) to be sent each game cycle. Players receive one audio message per cycle as long as a teammate within hearing range (50 meters) uttered a message. We encode communication data into a character representation in order to achieve full use of the 73 character range. After messages are received by a player, they must be decoded to extract the numeric data from the character string.

2.1   Message Types

CommLang defines a set of message types which each represent a specific type of information. A single communication message can be composed of multiple

Table 1. Message types used to encode information

Message Type	Syntax	Cost (characters)
Our Position	[0, X, Y]	3
Ball Position	[1, X, Y, #cycles]	4
Ball Velocity	[2, dX, dY, #cycles]	4
We have ball	[3, player#]	2
Opponent has ball	[4, player#]	2
Passing to player#	[5, player#]	2
Passing to coordinate	[6, X, Y]	3
Want pass	[7]	1
Opponent(player#) Position	[8+player#-1, X, Y, #cycles]	4
Teammate(player#) Position	[19+player#-1, X, Y, #cycles]	4

Table 2. Data types used in composing messages. Each data type uses one character of a message

Data Type	Description	Range	Precision
X	x coordinate of a position	[-53, 53)	1.45
Y	y coordinate of a position	[-34, 34)	0.93
dX	x component of a velocity	[-2.7, 2.7)	0.074
dY	y component of a velocity	[-2.7, 2.7)	0.074
#cycles	number of cycles since data last observed	[0, 72)	1
player#	a player number	[0, 11]	1
msg type ID	the message ID	[0, 29]	1

message types. The available message types and their syntax are listed in Table 1. All message types begin with a message type ID which allows us to identify the type of message and the number of arguments that will follow.

Each message type is made up of discrete units of data referred to as data types. In the interest of simplicity and ease of implementation, each unit of information (data type) uses one character of the message string. The data types are shown in Table 2. Each data type has a precision that is determined by the data range that is represented by the 73 characters. For the data types with floating point ranges, the precision is equal to the maximum loss of accuracy that can occur from encoding the data.

This design is flexible in that auditory messages sent to the server may be variable lengths — they can be composed of any number of message types, as long as they fit within the ten character limit.

Note that the teammate and opponent position message types do not have a single message type ID. Rather, the player number of the player whose position we are communicating is encoded in the message type ID. This allows us to reduce the length of the teammate and opponent message types by including the player number in the message ID number rather than as a separate argument.

Since the message type IDs do not yet use the full 73 character range, it would be possible to modify some of the other message types to encode the

```
char validchars[] = "0123456789abcdefghijklmnopqrstuvwxyz
ABCDEFGHIJKLMNOPQRSUVWXYZ().+-*/?<>_";
NUMCHARS = 73

char getChar( int i ):
    Returns validchars[NUMCHARS-1] if i ≥ NUMCHARS
    Returns validchars[0] if i < 0
    Returns validchars[i] otherwise
int getIndex( char c ):
    Returns the index of character c in the validchars array.
```

Fig. 1. Character conversion functions. The validchars array contains the characters that are permitted in soccer server say messages

player number in the ID. However, we chose not to do this because it would quickly use up the ID range, which may be needed in the future if new message types are added to the standard.

The player number data type ranges from 1 to 11 to indicate a player number. However, in the "We have ball" and "Opponent has ball" message types, a zero may be sent as the player number to indicate that we know which team has the ball, but do not know the player number.

Player and ball positions that are sent in messages should be the position that the player predicts the object is at in the current cycle. The #cycles data type should be equal to the number of cycles ago that the player received data about the object.

2.2   Character Conversions

The task of converting numeric information such as a field position to a character that can be accepted by the soccer server is handled by a character array and lookup functions. A specification for these functions is shown in Fig. 1. These functions are used to assist with encoding and decoding, as described in the next two sections.

2.3   Encoding Messages

We encode communication messages by converting numeric information into character encodings according to the equations in Table 3. The encoded strings of each message type being used are then concatenated into one string and sent to the soccer server as a say message.

The arithmetic in the encoding functions is done using standard floating point operations, and the final integer result is the floor of the float value.

Table 3. Data Type encoding functions

Data type value	Encoded character value
X	getChar((X+53)/106 * NUMCHARS)
Y	getChar((Y+34)/68 * NUMCHARS)
dX	getChar((dX+2.7)/5.4 * NUMCHARS)
dY	getChar((dY+2.7)/5.4 * NUMCHARS)
#cycles	getChar(#cycles)
player#	getChar(player#)
msg type ID	getChar(msg type ID)

Table 4. Data Type decoding functions

Data type of character	Decoded numeric data
X	(getIndex(X)/NUMCHARS * 106) - 53
Y	(getIndex(Y)/NUMCHARS * 68) - 34
dX	(getIndex(dX)/NUMCHARS * 5.4) - 2.7
dY	(getIndex(dY)/NUMCHARS * 5.4) - 2.7
#cycles	getIndex(#cycles)
player#	getIndex(player#)
msg type ID	getIndex(msg type ID)

2.4   Decoding Messages

When a coachable agent receives a message from a teammate, it decodes the message by using the message type IDs to identify which message types are included, and then decodes the data types according to the definitions in Table 4.

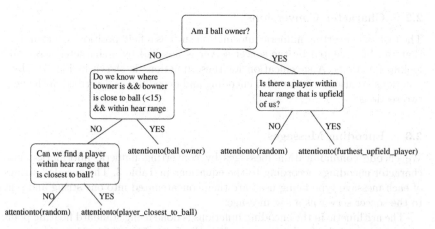

Fig. 2. The decision process used to decide which teammate to listen to. "Attentionto(random)" indicates that we allow the server to randomly choose a message for us

The message must be sequentially processed to extract each message type based on its message type ID.

Since our numeric data was encoded in a character representation, the decoded data may be slightly different from the original data due to the limited precision of each data type (see Fig. 2).

2.5   CommLang in RoboCup 2004

To assist other teams in using CommLang, we have released our encoding and decoding library at: *http://www-2.cs.cmu.edu/~robosoccer/simulator/*. The library is easily extensible — if new types of messages are added to the standard, a new version of the library can be released that is backwards compatible with previous versions so that teams' existing code will still operate.

The RoboCup 2004 simulation league rules require that coachable agents conform to the communication standard as described in this paper. Since simulation games will take place with teams composed of a mix of coachable agents from different research teams, it is necessary for all agents to use the same standard in order to understand each other. It is our belief that use of CommLang will improve cooperation between coachable agents and lead to better performance in the competition.

3   Implementation of Agent Communication

We have implemented coachable agent communication in the CMU Wyverns coachable team which was used in RoboCup 2003. The version of our agents used with our communication implementation also includes a number of improvements that were made after the competition. We had two primary goals in mind while adding communication to our agents:

1. To maximize the utility of agent communication by supporting all of the message types and by increasing the likelihood that communicated information would be useful to teammates.
2. To create a flexible architecture for experimenting with different strategies of communication.

3.1   Sending Message Frequency

In our implementation, every player broadcasts a say message each cycle. We chose this approach because it insures that every player receives a message every cycle. Simulation soccer teams have sometimes used coordinated strategies of having designated players speak each cycle. Our approach has the advantage of allowing us to listen to any player each cycle and attempt to select the most useful player to listen to. The trade-off is that we may not listen to a player for an extended period of time even though it might have something important to say.

Another reason for using this message sending strategy is that the simulation soccer coachable agents are mixed in with agents from other developers to make up a coachable soccer team. Since the CommLang standard does not make any stipulations about message frequency, our agents can not make any assumptions about the sending strategy of teammates.

3.2   Receiving Messages

We use the soccer server "attentionto" command to focus our attention on a teammate of our choosing. This is the player that we will receive say messages from if it is within range. This control gives us the ability to establish preferences towards listening to a certain player. For example, it is often preferable to listen to the ball owner.

The determination of which player to listen to is handled by a decision tree, which is shown in Fig. 2.

3.3   Selecting Message Composition

The most complex part of our implementation is the process for deciding what message types to include in the messages that the player sends. We typically have room to fit three or four message types within the ten character message size limit. The strategy we use to choose those three or four message types is an important factor in determining how useful the message is. For example, if a teammate needs to know where the ball is, but we generally send our player position instead, our messages will have little value.

Our implementation uses a randomized scheme to select the message composition. Each message type is assigned a weight which influences the likelihood of using it in the final message. These weights are intended to reflect the overall utility of the message type. Message types that we think are highly useful in most situations are assigned higher values while less important message types

Table 5. The predicate functions and weights for each message type, as currently configured in our agents

	Predicates	Weight
OurPos	none	0.5
BallPos	Confident of ball position	0.5
BallVel	Confident of ball velocity	0.3
WeHaveBall	Confident of ball info AND We own ball currently	0.2
OppHasBall	Confident of ball info AND Opponent team owns ball	0.2
PassToPlayer	I executed a pass action within the last 5 cycles	1.1
PassToCoord	I executed a pass action within the last 5 cycles	0.8
WantPass	I'm not ball owner AND I'm close to opponent goal AND I have a good goal shot	1.0
OppPos(pnum)	Confident of position of opponent player #pnum	0.4
TeammatePos(pnum)	Confident of position of teammate player #pnum	0.4

```
M := Set of all message types
C := empty message
∀ m ∈ M, m.p = getWeight(m)
remove m_i ∈ M if predicate(m_i) = false Or m_i.p = 0.0
foreach m ∈ M, from largest m.p to smallest
    If( m.p ≥ 1.0 )
        insert m in C
        remove m from M
Normalize such that {m.p | m ∈ M} is a probability distribution
while ( M ≠ ∅ )
    Choose m_i from distribution defined by m_i.p
    If( m_i.size + C.size ≤ MAX_MESSAGE_SIZE )
        insert m_i in C
    remove m_i from M
    Renormalize probability distribution in {m.p | m ∈ M}
return C
```

Fig. 3. Algorithm for choosing message composition

receive lower values. We also define predicates to filter out any message types that are not applicable at the current time. For example, PassToPlayer is only applicable when we have the ball and are passing to a teammate. Table 5 lists the predicates and weights for each message type.

The weights are generally within the range 0.0 to 1.0, but may also go over 1.0. Values over 1.0 serve to guarantee that we use the message type, as long as there is sufficient space. Weights of 0.0 indicate that the message type will never be used.

The main component of the selection algorithm is shown in Fig. 3. We first automatically select any message types that have a weight of 1.0 or greater. Then, the main loop chooses the next message type to include based on the probability distribution over the weights of the remaining message types.

This probabilistic selection method allows us to define preferences towards using particular message types while at the same time insuring that we do not use the same message composition every time.

In addition to the weighting strategy that is in use now, for testing we have also implemented a uniform probability distribution and a distribution for selecting messages from a fixed subset of message types (specifically, it can be used to send only "Our Pos" and "Ball Pos" types).

3.4   Processing Communicated Information

When we receive messages, we process the information using our decoding library routines and then integrate the information into the world model. Data in the world model is replaced with communicated information only if the information is more recent than the knowledge we already have. Therefore, we implicitly trust our teammates' communication information to be accurate and we store it in the same location as our own sensor information. We use the world model from the 2002 UvA Trilearn team [8], which is the team that was used as the original base for our coachable agents.

Three of the message types — PassToPlayer, PassToCoord, and WantPass — are messages that primarily communicate information about a teammate's intentions rather than perceptions of the world's state. Therefore, we respond to these message types by executing new actions as appropriate. If a PassTo-Player message indicates a teammate is passing to us, we immediately attempt to intercept the ball. Similarly, if a PassToCoord message indicates a teammate is passing to a position close to us, we assume the pass is intended for us and attempt to intercept. If a teammate indicates WantPass, we attempt to pass to that player if we possess the ball or obtain possession within the next 5 cycles. These actions can take precedence over coach advice.

4   Empirical Evaluation

After implementing communication in the Wyverns coachable agents, we ran tests of the agents to assess the impact of communication on their world model. Two sets of ten games were run, with each game lasting for 3000 cycles. In the experimental set, communication was used, and in the other set communication was not used.

In both sets, the opponent team consisted of the Wyverns coachable agents as publicly released after RoboCup 2003. The other team, which contained the communication support, was an improved version of the Wyverns players with a number of bug fixes and skill improvements.

Both teams were advised by the CMU Owl coach [9]. Note that the improved Wyverns won most of the games, with 26 total goals, versus 4 goals for the original Wyverns. Therefore the experimental players were often in offensive positions, but they also were sometimes forced into defensive formations.

Since the team's formation can affect the frequency that we see other players, we will note that our offensive players were numbers 2, 7, and 8, our midfielders were 3, 5, and 6, the defensive players were 4, 9, 10, and 11, and player 1 is always the goalie.

4.1   Improvements in Player Confidence

The players' confidence in the ball and player locations was recorded during the games. As currently configured, players are confident in a teammate or opponent location when they have seen the player (or updated the player's position based on communicated data) within the last 12 cycles. Players are confident of the ball's location when they have seen or updated it within the last 6 cycles.

Figure 4 shows the mean percentage of time that the players were confident of the location of the ball and teammates. The communication group had higher confidence rates for all the objects, with most of the increases statisti-

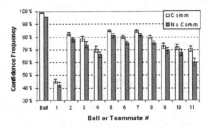
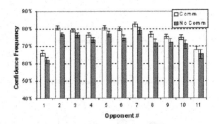

Fig. 4. Confidence frequencies for the ball and teammates. Error bars indicate 95% confidence intervals

Fig. 5. Confidence frequencies for the opponent players

cally significant. Similar results were seen for our perceptions of the opponent players (Fig. 5).

Although this improvement was expected, it is still a useful confirmation of one of communication's benefits. Confidence frequencies play a role in many of an agent's behaviors - for example, our players can not pass to a teammate unless they are confident in its location. Therefore, increases in confidence frequencies can directly influence agent behavior.

4.2   Improvements in Positional Error

The other statistic compiled from the games was the positional error of players in our world model. The positional error values are the difference between where a player thinks a teammate or opponent is, and where that player actually is. The actual positions were obtained using the soccer server's full state mode. Errors were only counted during cycles when the player was confident in the location of the relevant teammate or opponent.

Figure 6 shows the players' mean world model errors for each of their teammates, as well as the ball. The 95% confidence intervals could not be displayed on the figure, but were small enough for the results to be statistically significant.

For most player positions, communicating players had lower mean error than non-communicating players. This was especially noticeable with the estimates of ball position, and the position of player 1 (the goalie).

Since position errors were only collected when the player was confident in the location of the teammate, more error data was sampled in the communication set than the non-communication set because communication increases the frequency at which players are confident of teammate locations. As seen in Fig. 6, the error for communicating players actually increased on a small number of the player-player comparisons. Since communicating players are confident of teammate locations more often, this may cause more error to accumulate in some cases. For example, if an offensive player is out of visible range of the goalie, but receives the goalie's position via communication, the player will still be confident of the goalie's location ten cycles later, but the estimated position could be significantly different from the actual position.

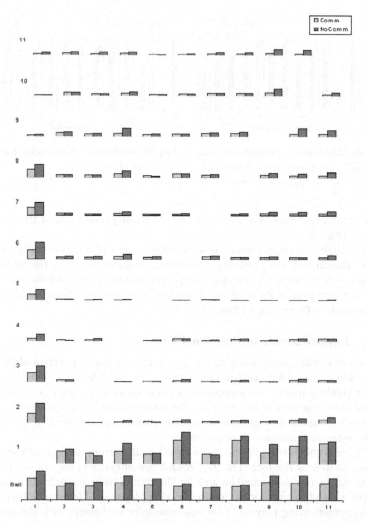

Fig. 6. Mean world model error (in meters) of ball and teammate positions. The numbers along the x axis indicate our players. Each column of bars above represents the positional errors in that player's world model. The players' errors for themselves (the diagonal on the graph) are displayed as zero because self position errors were not recorded (communication does not affect those values). The charts for our perceptions of the teammates listed on the y axis are scaled to a max error value of 20 (meters), and the row for the ball is scaled to a max of 3

In addition, communication can also change the behavior of the agents in subtle ways. For example, if players know the locations of their teammates more often (due to communication), they may decide to scan the field less often.

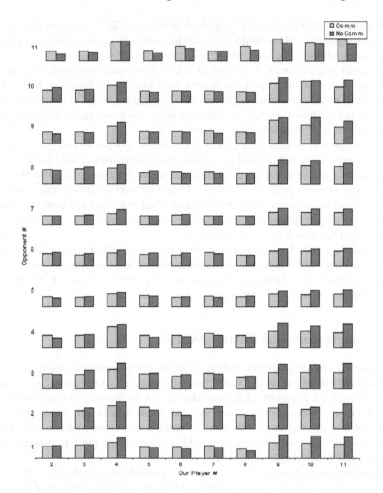

Fig. 7. Mean world model error of opponent positions. The column for our goalie's world model (player 1 on x axis) was omitted because the errors were beyond the scale of the graph — our goalie has high error for the opponents since it usually can not see them. The rows for opponents 2-11 are scaled to a maximum of 4, and the row for opponent 1 is scaled to 18

Therefore, changes in the accuracy of player world models are not necessarily only a direct result of communication — additional factors may include changes to scanning frequency, and other differences.

Communicating players had lower error values for their estimated position of the ball. Their mean ball error was 22.6% lower than non-communicating players. As seen previously in Fig. 4, communicating players were also confident of the ball location 3.3% more often than non-communicating players.

For the agents' estimation of opponent positions (Fig. 7), most of the changes were less significant. This could indicate that our agents were able to get better information about the opponents simply through visual sensors.

However, our players 4, 9, 10, and 11, which are defensive players, all improved their estimation of most of the opponent positions (see columns 4, 9, 10, 11 in the figure). Since our defensive players rarely get good visual information about the opponents since they are far away most of the time, communication helps them a great deal in this regard.

The empirical results with player confidences and positional errors are important because they show that communication improves not only the frequency with which a player is confident of ball and player locations, but also the accuracy of the player's position estimate. This improved accuracy and confidence has implications in many areas. For instance, ball positions are used in calculating interception trajectories, ball handling strategies, and in determining which advice from the coach is applicable.

In summary, coachable agent communication as implemented in the Wyverns agents increases the frequency with which players are confident of the ball and other players. It also decreases the error in the players' estimation of where the ball, teammates, and opponents are.

5   Conclusion and Future Work

The CommLang standard for RoboCup agent communication described in this paper is a flexible and easily extendable language for communicating between agents. It allows a variety of information to be exchanged, expanding the potential for inter-agent cooperation. All coachable agents will need to use this standard and we encourage other soccer simulation teams to do so also. Use of this standard by simulation teams would improve agent interoperability, facilitating interesting mixed-team pickup games.

Our implementation of coachable agent communication in the Wyverns agents is a versatile architecture that provides control over the composition of messages, the source of received messages, and the incorporation of received information into our world model. It is designed to construct communication messages that maximize the usefulness of a limited bandwidth channel of communication.

In past studies of previous versions of the simulated soccer environment [10], communication had an overwhelmingly positive effect. However, at that time, the communication bandwidth was over 50 times the bandwidth allowed currently (512 characters compared to 10 characters). The players in general did not need to reason about what information to include in each communication.

We have presented an algorithm for determining message composition and shown that our initial parameters lead to improvements in world model completeness and accuracy. We have not yet explored what settings (e.g. weights of the message types) yield the largest improvements and this is an interesting avenue for future work. Further, reasoning about the tradeoffs involved in limited bandwidth communication could lead to improvements in some performance measures.

In the future we may wish to consider adding new message types to the communication standard. One general type of message which is not currently in the standard is the information request form (such as exists in KQML [5]). These messages would allow agents to request from their teammates state information such as the ball location.

Another area meriting consideration is the option to maintain information about communication from earlier in the game. We could consider using earlier communication to learn who to listen to — if one of our teammates often sends more useful information than other agents, we could focus on that agent more frequently.

Through our empirical evaluation, we found that communication improved the world model knowledge of our coachable agents. We believe there is significant research potential in this area to determine how communication between agents can be used to the greatest advantage.

References

1. Doyle, R., Atkinson, D., Doshi, R.: Generating perception requests and expectations to verify the executions of plans. In: AAAI-86. (1986)
2. Kaminka, G., Tambe, M.: I'm OK, you're OK, we're OK: Experiments in distributed and centralized socially attentive monitoring. In: Agents-99. (1999)
3. Xuan, P., Lesser, V., Zilberstein, S.: Communication decisions in multi-agent markov decision processes: Model and experiments. In: Agents-2001. (2001) 616–623
4. Pynadath, D., Tambe, M.: The communicative multiagent team decision problem: Analyzing teamwork theories and models. Journal of Artificial Intelligence Research 16 (2002) 389–423
5. Labrou, Y., Finin, T.: Semantics and conversations for an agent communication language. In: IJCAI-97, Morgan Kaufmann publishers Inc.: San Mateo, CA, USA (1997) 584–591
6. FIPA: FIPA ACL message structure specification. http://www.fipa.org/specs/fipa00061/XC00061E.html (2001)
7. FIPA: FIPA communicative act library specification. http://www.fipa.org/specs/fipa00037/XC00037H.html (2001)
8. Kok, J.R., Spaan, M.T.J., Vlassis, N.: Multi-robot decision making using coordination graphs. In: Proceedings of the 11th International Conference on Advanced Robotics. (2003)
9. Riley, P., Veloso, M.: Advice generation from observed execution: Abstract Markov decision process learning. In: AAAI-2004. (2004) (to appear)
10. Riley, P.: Classifying adversarial behaviors in a dynamic, inaccessible, multi-agent environment. Technical Report CMU-CS-99-175, Carnegie Mellon University (1999)

Turning Segways into Robust Human-Scale Dynamically Balanced Soccer Robots

Jeremy Searock, Brett Browning, and Manuela Veloso

School of Computer Science, The Robotics Institute,
Carnegie Mellon University, Pittsburgh, PA, 15213
jsearock@andrew.cmu.edu
{brettb, veloso}@cs.cmu.edu

Abstract. The Segway Human Transport (HT) is a one person dynamically self-balancing transportation vehicle. The Segway Robot Mobility Platform (RMP) is a modification of the HT capable of being commanded by a computer for autonomous operation. With these platforms, we propose a new domain for human-robot coordination through a competitive game: Segway Soccer. The players include robots (RMPs) and humans (riding HTs). The rules of the game are a combination of soccer and Ultimate Frisbee rules. In this paper, we provide three contributions. First, we describe our proposed Segway Soccer domain. Second, we examine the capabilities and limitations of the Segway and the mechanical systems necessary to create a robot Segway Soccer Player. Third, we provide a detailed analysis of several ball manipulation/kicking systems and the implementation results of the CM-RMP pneumatic ball manipulation system.

1 Introduction

Considerable research has been conducted involving human-robot interaction [8], and multi-robot teams [9, 10, 11]. With the inception of RoboCup robot soccer [7], multi-agent team coordination within an adversarial environment has been studied extensively. But, the dual topic of human-robot coordination in an adversarial environment has yet to be investigated. This research involves the intelligent coordination of mixed teams of humans and robots competing in adversarial tasks against one another. The results of this research will further the technology necessary to allow both humans and robots to productively work together in complex environments requiring real time responses.

In order to further investigate human-robot interaction in team tasks, we have developed a new game called Segway soccer. The rules of the game are a combination of soccer and Ultimate Frisbee with an emphasis on fostering human-robot interaction. In order to investigate different perception and cognitive abilities, the humans and the robots need be placed on an equal physical level by utilizing the Segway Human Transporter (HT) and the Segway Robot Mobility Platform (RMP). Several mechanical systems are needed to equip the RMP with the physical abilities necessary to allow a game. We present the Segway and the domain of Segway Soccer

D. Nardi et al. (Eds.): RoboCup 2004, LNAI 3276, pp. 60–71, 2005.
© Springer-Verlag Berlin Heidelberg 2005

in Section 2. We explain the modifications necessary to create a Robot Segway Soccer Player in Section 3 along with the mathematical models necessary to choose among the different ball manipulation systems in Section 4. The implementation details and experimental results of the CM-RMP Robot Soccer Player are in Section 5, and our conclusions and future plans in Section 6.

2 Segway Soccer Platform and Game

The Segway™, developed by Dean Kamen, is a two-wheeled dynamically balanced mobility platform. The Segway has onboard sensors and computer controllers that continually and independently command each wheel in order to maintain balance. The human rider controls the velocity of the Segway by leaning forward, shifting the center of mass, and causing the Segway to drive forward in order to rebalance.

The RMP has provided a robust robotic agent on which to create human-scale robots. The Segway RMP is a 'roboticized' Segway HT consisting of three main modifications. First, a CAN Bus interface is exposed to enable two way, high speed electronic communication with the platform. Second, the Segway's control software is modified to enable a computer to send direct velocity commands to the platform. The third change involves attaching a large mass of about 23kg at a height of 50cm from the robot wheel base. This serves the purpose of raising the robot's center of gravity which slows down the RMP's falling rate in order to allow the control loop to operate effectively.

Using the Segway platform, we have developed a new game called Segway Soccer, which consists of teams of humans and robots competing in a game where the rules are a combination of soccer and Ultimate Frisbee[1]. The objective of the game is to score the most goals by kicking an orange size 5 soccer ball into a goal which is 2.5m wide. One key contribution from ultimate Frisbee is that once a player is declared to have possession of the ball by a referee, the player cannot dribble. Instead, the player has a 1m radius in which to reposition and pass to a teammate. This rule is in place for safety reasons, so that robots and humans will not contest each other for possession. Furthermore, to ensure robots and humans will collectively be involved, a mixed team cannot officially score unless both a robot and a human interact with the ball on the way to the goal.

Placing humans, robots, and robot competitors on an equal physical level using the Segway platform allows their different perception and cognitive abilities to be tested.

Fig. 1. Segway Soccer field   **Fig. 2.** Segway RMP

---

[1] Ultimate Players Association, [*http://www.upa.org*]

The dynamic balancing, speed, and size of the Segway allows this human-robot interaction at a human scale. The Segways can travel at 3.5m/s and have a footprint of 48cm by 64cm. Our Segway RMP also has its camera mounted 1.5m above the ground, providing a human height perspective.

3   Turning the Segway RMP into a Soccer Player

This new domain of human-robot interaction raises the requirements of robot mechanical systems to a more sophisticated level. The challenge becomes designing adequate hardware that will allow a robot to safely and robustly operate in an outdoor environment along with humans in a competitive soccer game.

In meeting this challenge, we have developed 4 key goals. First, the soccer player must be autonomous by perceiving the world, making decisions, and acting without human intervention. Second, the player must be able to interact with human players by recognizing their presence and communicating. Third, the player must be able to manipulate a ball well enough to be competitive with humans. Lastly, safety must be considered in every aspect to prevent injury to humans and damage to equipment.

With these goals in mind, we have to consider the many challenges that accompany designing and implementing a complete robotic system. Cost effectiveness, processing power, perception, weight distribution, and resistance to shock are all important considerations. The unique motion of the Segway also introduces problems not seen with other platforms.

The Segway moves forward by tilting over and driving the wheels in order to rebalance. This motion can lead to the ball becoming stuck underneath the body and wheels of the Segway. This causes the wheels to lose contact or traction with the ground making the Segway unable to sufficiently maintain balance. Any consequential fall could potentially damage equipment. As a solution, guards, consisting of modified rubber mud flaps, were placed in front of the wheels and computer control software was implemented to prevent the RMP from interacting with the ball unless it knows the manipulation system can kick.

Another challenge introduced with Segway Soccer is that there is no unique playing surface; it can be played on grass, Astroturf, or cement. Changes in grass height, ground softness, and surface texture alter the dynamics necessary to manipulate the ball. Unlike other robotic soccer platforms, the Segway also tips up to +/- 20 degrees with respect to the vertical; thus, any attached kicking plate and system will also tip. This requires the manipulation system to be robust enough to manipulate the ball under changing conditions.

With a basic infrastructure in place, we added two laptop computers to interpret the world and control the RMP. One laptop is used to provide the RMP with the capability to process data from a pan/tilt CCD camera and another laptop to quickly decide what action to take and send commands to control the actions of the RMP. Speakers were added to allow the robot to speak to its teammate to help control the flow of the game.

Finally, hardware must be able to protect the components of the Segway from damage during a fall. The laptop computers and ball manipulation system components are mounted as close as possible to the bottom of the Segway reducing their falling distance and the shock they will absorb. The laptops are also securely fastened with straps preventing them from being ejected from the confines of the RMP body. Steel safety stands were added to reduce the total distance the Segway will fall once it is no longer capable of dynamically balancing. The stands mount onto the side of the RMP and only allow it to fall over 30 degrees from the vertical.

4 Ball Manipulation Systems

One of the main challenges to using a Segway RMP to play soccer lies in designing a ball manipulation system that allows a Segway platform to kick a ball to the scale of an outdoor human game. The need for passing in the game and the inability of the robot to safely propel the ball by running into it necessitates the development of a kicking mechanism. Although kickers of all forms have been developed in RoboCup, there have been no formalized comparisons of different mechanisms. We present an analysis of the common kicking mechanisms as well as a detailed implementation and evaluation of a pneumatic kicker as a step towards developing a scientific based method for selecting kicking mechanisms.

4.1 Kicking System Design Considerations

A ball manipulation system can be described as a mechanical manipulator used to accelerate a ball to a desired velocity in a desired trajectory. This can be achieved in many different ways with various actuators. The most common systems come from the realm of robot soccer as seen in RoboCup [7] competitions. These include pneumatic, spring, solenoid, rack and pinion, and rotating plate systems. A careful analysis of the following factors is needed to determine which kicking system best fits a given platform and environment:

Speed-How fast should the ball be kicked?
Accuracy- What distribution is acceptable? Should it only kick straight?
Kick capacity- How many kicks are possible or needed in a game?
Response time-How long will it take for a signal to result in a kick?
Recovery time- How long before a second kick is possible?
Safety- Is the system likely to injure a human player or bystander?
Complexity- What is the build time? What parts are required?
Weight-Can the robot carry the payload?
Size- Will the system fit on the robot and perform adequately?
Price- Is it worth the extra money?
Power-Can the robot carry the power supply? How long will the supply last?
Reliability/Maintenance-How likely will parts break under normal competitive play? How expensive and complex is the maintenance?
Transportability- Can the system be transported in an airplane? Are spare parts readily available?[2]

---

[2] The latter is especially critical given the international nature of RoboCup competition.

With these considerations in mind, the actuating system must be chosen. For each option, we present the basic system components, the mathematical models necessary to properly specify an appropriate actuator, and an example comparing each option to the pneumatic system we implemented and describe in section (5).

4.2 Spring Loaded Mechanisms

Spring kicking mechanisms use an extension or compression spring(s) to store and then release energy to propel the ball. As such, a mechanism is needed to tension the spring(s) and a trigger to release the stored energy. Such mechanisms must be robust, and are non-trivial to design. Apart from the obvious complexity, spring strength is coupled to kicking power, but a more powerful spring is more difficult to retract and hold. This relationship leads to potential problems during a soccer game where the time to reload a powerful spring can take several seconds if a cheaper less powerful motor is used. Springs do provide the best power density out of the given the options [4, 5].

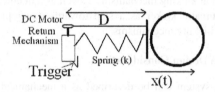

Fig. 3. Spring Kicking Mechanism Schematic

For a spring with a spring constant of k, a kick length of D, and a kicking mass of m, the force equation and resulting equations of motion for the spring mechanism are:

$$\ddot{x} = kx \, m^{-1} \quad x(0) = D \quad \dot{x}(0) = 0. \tag{1}$$

$$x(t) = D\cos\sqrt{k/m}\,t. \tag{2}$$

For the Segway, the pneumatic kicking system model predicts the ball will be kicked at a 4.3 m/s max velocity. Using the above equations, one can determine the spring constant, k, necessary to achieve a similar speed with a similar stroke length to the pneumatic model (0.1524m). Assuming a kicking mass of 0.85kg, a spring constant of 676 N/m would be necessary. This would require a force of at least 103N in order to load and hold the spring at 0.1524m for a kick. Depending on motor size and consequentially cost, the spring would take one to several seconds to reload for another kick. Complexity, reload time, and cost are the liming factors for the spring kicker design.

4.3 Rotating Plate Mechanisms

Rotating plate kickers consist of two or more flat surfaces, bars, or other contacts arranged in a balanced paddle boat configuration. [1] The shaft of the paddle wheel is connected to a DC motor. The angular velocity of the paddle wheel determines the end velocity of the ball, although there is great variability due to the potential variation in the contact point. Pulse Width Modulation can be used to vary the speed of the wheel and thus vary the power of the kick. Rotating plate mechanisms require a significant amount of space to mount the paddle wheel and the drive motor. Furthermore, for larger robots, rotating plates become extremely dangerous to human operators. A rotating plate mechanism scaled to the size of a Segway would have to be approximately 18cm by 38cm. The plates would be rotating fast enough and with enough power to cause injury to humans who happen to fall off of their HT into a kicking device. As a result, we do not consider a rotating plate mechanism in depth.

DC Motor

Rotating Plates

Fig. 4. Rotating Plate Kicker Schematic and Physical Example

4.4 Rack and Pinion Systems

Rack and pinion systems are driven by DC motors and thus the ball velocity is dependent on the output power of the motor. For a rack and pinion motor system with a back emf of k_e, voltage of V, forward torque per amp of K, terminal resistance of R, pinion radius of r, gear ratio of N, and total kicking components mass of m, the following are the equations of motion:

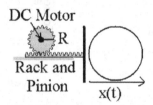

Fig. 5. Rack and Pinion Kicking System Schematic

$$\ddot{x} = \frac{K}{mr^2 R}(Vr - k_e \dot{x}) \quad x(0) = 0 \quad \dot{x}(0) = 0 . \tag{3}$$

$$x(t) = \frac{Vr}{k_e} t - \frac{mr^3 RV}{NKk_e^2}\left(1 - e^{-\frac{k_e KN}{mr^2 R}t}\right) . \tag{4}$$

A rack and pinion system with a powerful motor is comparable to the other options but the price and design requirements are more significant. Using an 80W Maxon motor [12], a 0.015m pinion radius, 1:1 gear ratio, and a total kicking mass of 1.3 kg, the rack and pinion system can accelerate the ball to a theoretical velocity of 6.7 m/s in 0.1524 m (6in). With a motor efficiency around 75% and the friction forces acting against the sliding rack, the actual velocity will be closer to 3.5 m/s.

The Segway does not have enough space to implement a single rack and pinion system. Two rack and pinions would be needed since one rack and pinion could not be placed in the middle of the Segway due to the handle bar mounting. This would require two motors or a much larger single motor to actuate both rack and pinions. This requirement makes this system unfeasible for use on a Segway platform. The two high power motors would also be costly [3].

4.5 Solenoid Systems

A solenoid kicker consists of a solenoid that creates a magnetic field around a shaft that is propelled by the field and accelerated away from the solenoid. The shaft is returned by a built in return spring. Consider a solenoid kicking system with a current of I, ampere turns of N, plunger radius of r, a return spring constant of k, and a total kicking mass of m. The differential equation of motion is given in equation (5). Since this equation can only be solved by numerical means, we approximate the result here for analysis purposes by treating the solenoid force as being constant for the duration. The results are given by equations (6) and (7).

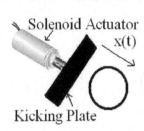

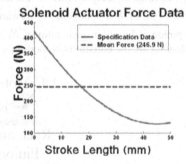

Fig. 6. Solenoid Kicker Schematic   **Fig. 7.** Solenoid Force v. Stroke and Approximate Force

$$\ddot{x} = m^{-1}\left(\frac{1}{2}\mu_0 \left(\frac{rNI}{x}\right)^2 - kx\right), \quad x(0) = 0 \quad \dot{x}(0) = 0. \tag{5}$$

$$\ddot{x} = m^{-1}\left(F_{avg} - kx\right), \quad x(0) = 0 \quad \dot{x}(0) = 0. \tag{6}$$

$$x(t) = \frac{F_{avg}}{k}\left(1 - \cos\sqrt{\frac{k}{m}}t\right). \tag{7}$$

Most commonly available solenoids produce approximately 400N of force and generally have small stroke lengths on the order of 0.0254m (1in), which limit its ability to effectively contact a ball. For the solenoid shown in figure (7) [13], assuming a return spring constant k of 99 N/m, and a kicking mass of 1.5 kg, a ball would be kicked at 4.1 m/s over a 0.0508m stroke length (2in). With the effects of friction and motor efficiency the actual speed would be closer to 3 m/s. This is comparable to the pneumatic system but the smaller stroke length limits is ability to effectively manipulate a ball during a game. The high voltage requirement may also raise safety issues.

4.6 Pneumatic Systems

Pneumatic piston systems usually consist of one or two actuating cylinders, an air reservoir, solenoid valves to control the air flow, a source of compressed gas in the form of compressed air or liquid carbon dioxide (CO_2), and a regulator to maintain a specified pressure. The decision between CO_2 and compressed air depends on the availability of CO_2. Air compressors normally only operate up to 150 psi while CO_2 tanks fill to several thousand psi. This higher pressure allows the cylinders to be fired with a higher output force resulting in a stronger kick. The higher pressure also significantly increases the kick capacity of the system. The major drawback of CO_2 is that it is not easily transportable or available in foreign locations. Additionally, its rapid expansion during each kick results in thermal issues such as the formation of condensation near electronic parts. As a result, a compressed air approach is often used instead.

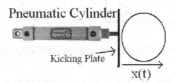

Fig. 8. Pneumatic Kicker Schematic

The pneumatic cylinder pistons connect to a kicking plate that contacts the ball. The plate can vary in material and shape dependent upon application. The one or more pistons can be fired at the same time or in a synchronized order to achieve a directional kick. A pneumatic system offers a wide range of options in its configuration and employment. For a pneumatic system with power factor, f, combined kicking mass, m, return spring constant, k, and operating at a pressure, P, the equations of motion are:

$$\ddot{x} = m^{-1}(fP - kx), \quad x(0) = 0 \quad \dot{x}(0) = 0. \tag{8}$$

$$x(t) = fP k^{-1}\left(1 - \cos\sqrt{k/m}\,t\right). \tag{9}$$

For a Segway, a suitable cylinder would be approximately 0.254m (10in) long, 0.01905m in diameter and produce 274N for force at 140 psi. The pistons are the only moving parts and the air tank consumes the most space. The price of a pneumatic system is also fairly cheap. A functional system can be bought for less than $150. The air used to power the cylinders is naturally accessible and can be refilled quickly during a soccer game with an onboard air compressor. The system has a low chance of malfunctioning and becoming inoperable during a game because the only moving parts are the cylinder shafts. [2, 8]

5 Implementation and Results

With the considerations presented, we chose to use a pneumatic approach due to its relative simplicity, low cost, and transportability. Figure (9) shows the resulting arrangement

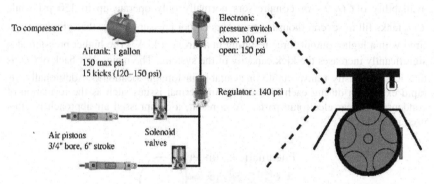

Fig. 9. Schematic diagram of the implemented CM-RMP pneumatic kicking system

Two 0.01905m bore, dual acting pneumatic cylinders were chosen as the main actuating components. This size cylinder provides adequate power with a sturdy shaft that can sustain unexpected stress. Dual acting cylinders do not have a return spring to reset the cylinder shafts back to their original position. This allowed us to implement a return mechanism with just enough force to reset the kicking plate without significantly affecting the output force. We used 4 x No.64 (3.5in x 1/4in) rubber bands to return the kicking plate. Two cylinders also allow for directional kicking.

5.1 Air Reservoir Options

The air reservoir can be designed in two different ways. The reservoir can be large enough to hold enough kicks for the entire game or an onboard air compressor can refill a smaller reservoir. If the robot has enough room to house a larger tank, not having an air compressor allows the overall system to be simpler. We used a 1 gallon tank, which provides a sufficient number of kicks as seen in figure (10). We have an onboard compressor that turns on after 15 kicks and shuts off when the tank pressure reaches 150 psi. The compressor is controlled by a microcontroller that also monitors a mechanical pressure switch that opens at 150 psi. As a result, the operation of the compressor is completely automated.

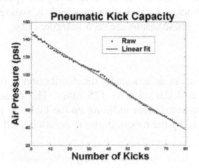

Fig. 10. Number of Kicks v. Reservoir Pressure

5.2 Velocity Test Results

The cylinders accelerate the ball to a max velocity of approximately 3.5 m/s, which is sufficient for a two on two game of Segway soccer. The velocity can increase to 4.5 m/s if the Segway RMP is moving at the ball when it kicks it. The theoretically predicted top speed for a stationary kick is approximately 4.3 m/s. The loss in velocity is due to the efficiency of the pneumatic cylinders, an imperfect impact with

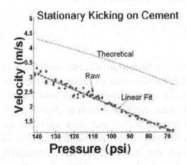

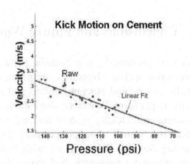

Fig. 11. Experimental and Theoretical Pressure v. Kick Velocity

Fig. 12. Pressure v. Kick Motion Velocity

the ball, and ground friction. An experiment was setup using one of the cylinders, a small kicking plate, and a golf ball. The velocity of the golf ball was measured on a cement floor. This experiment was designed to significantly lower the effects of impact and friction losses. Through these tests, it was determined that the pneumatic cylinder alone had an efficiency of 75%. These losses are due to several factors including cylinder friction, exiting air resistance, and flow rate limitations. Impact losses and ground friction account for an additional 2% loss. The theoretical and experimental kick speed versus cylinder pressure plot is shown in figure (11). Furthermore experiments were conducted measuring the speed of the ball when the Segway RMP played back a kick motion in which it swung its base forward and simultaneously kicked. These results are seen in figure (12).

5.3 Accuracy

The kick is sufficiently accurate as seen by the distribution in figure (13). The mean is 122 mm and the standard deviation is 175 mm. The mean error can be mostly accounted for by experimental error in lining up the kick. In practice, this mean and variance will be modified by the robots ability to position itself next to the ball.

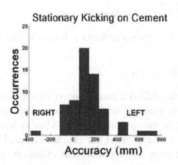

Fig. 13. Histogram of Stationary Kicking on a Cement Surface

6 Conclusions and Future Work

We have presented a new domain, Segway Soccer, in which to investigate human-robot interaction within adversarial environments. Within this domain we have identified the challenges behind placing humans and robots on the same physical level utilizing the Segway platform. We have also established a scientific basis on which to choose a ball manipulation/kicking system for any size robot soccer player and have outlined the other mechanical systems necessary to make a Segway RMP physically capable of playing Segway Soccer along with humans. With our analysis, we have accurately implemented a pneumatic ball manipulation system, which robustly kicks the ball fast enough and with enough accuracy to make passing and goal scoring possible.

Our future work will focus on further developing the concept of Segway Soccer and on additional mechanisms to increase the soccer playing abilities of the Segway, such as recovering the Segway from a fallen state.

Acknowledgements

Thank you to Mike Sokolsky and David Rozner for their help in the construction of the mechanical devices. Finally, we would like to thank the funding sources, who supported and guided this research and development to address this specific problem. This research was sponsored by the United States Army under Grant No. DABT63-99-1-0013. The views and conclusions contained in this document are those of the authors and should not be interpreted as necessarily representing official policies or endorsements, either expressed or implied, of DARPA, the US Army, or the US Government.

References

1. 1.S. Behnke et. al., "Using Hierarchical Dynamical Systems to Control Reactive Behavior." *RoboCup-99:Robot Soccer World Cup III*. Berlin: Springer, 2000, p.189
2. M. Ferraresso et. al., "Collaborative Emergent Actions Between Real Soccer Robots." *RoboCup-2000:Robot Soccer World Cup IV*. Berlin: Springer, 2001, p.297
3. Ng Beng Kiat et. al., "LuckyStar II-Team Description Paper." *RoboCup-2000:Robot Soccer World Cup IV*. Berlin: Springer, 2001, p.543
4. G. Wyeth et. al., "UQ RoboRoos: Achieving Power and Agility in a Small Size Robot." *RoboCup-2001:Robot Soccer World Cup V*. Berlin: Springer, 2002, p.605
5. R. Cassinis et. al., "Design for a Robocup Goalkeeper." *RoboCup-99:Robot Soccer World Cup III*. Berlin: Springer, 2000, p.255
6. A. Bredenfeld et. al., "GMD-Robotst." *RoboCup-2001:Robot Soccer World Cup V*. Berlin: Springer, 2002, pp. 648-9
7. M. Asada *et. al.* "An overview of RoboCup-2002 Fukuoka/Busan". AI Magazine, 24(2): pages 21-40, Spring 2003
8. M. Nicolescu, M. J Mataric, "Learning and Interacting in Human-Robot Domains", Special Issue of IEEE Transactions on Systems, Man, and Cybernetics, Part A: Systems and Humans , Vol. 31, No. 5, pages 419-430, C. C. White and K. Dautenhahn (Eds.), 2001
9. M.B. Dias and A. Stentz. "Opportunistic Optimization for Market-Based Multirobot Control". Proceedings of the IEEE/RSJ International Conference on Intelligent Robots and Systems IROS 2002, 2002
10. M. Ferraresso, et al., "Collaborative Emergent Actions Between Real Soccer Robots." RoboCup-2000:Robot Soccer World Cup IV. Berlin: Springer, 2001, pp.297-300
11. N. Kiat, Q. Ming, T. Hock, Y. Yee, and S. Yoh, "LuckyStar II-Team Description Paper." RoboCup-2000:Robot Soccer World Cup IV. Berlin: Springer, 2001, pp. 543-546
12. Maxon Motors, "F 2260, Graphite Brushes, 80 Watt, No.880," http://www.mpm. maxonmotor.com] P. 95
13. Solenoid City, "Push Type Tubular Solenoid, Series S-70-300-H," [http://www. solenoidcity.com]

A Constructive Feature Detection Approach for Robotic Vision

Felix von Hundelshausen, Michael Schreiber, and Raúl Rojas

Free University of Berlin, Institute of Computer Science,
Takustr. 9, 14195 Berlin, Germany
{hundelsh, schreibe, rojas}@inf.fu-berlin.de

Abstract. We describe a new method for detecting features on a marked RoboCup field. We implemented the framework for robots with omnidirectional vision, but the method can be easily adapted to other systems. The focus is on the recognition of the center circle and four different corners occurring in the penalty area. Our *constructive approach* differs from previous methods, in that we aim to detect a whole palette of different features, hierarchically ordered and possibly containing each other. High-level features, such as the center circle or the corners, are constructed from low-level features such as arcs and lines. The feature detection process starts with low-level features and iteratively constructs higher features. In RoboCup the method is valuable for robot self-localization; in other fields of application the method is useful for object recognition using shape information.

1 Introduction

Robot self-localization is an important problem in the RoboCup domain. Many systems rely on Monte Carlo Localization [19] identifying landmarks such as the colored goals or posts. There have been attempts to detect field line features and to use them for robot self-localization.

In [18] straight lines are detected, however no other features, in particular no curved features are extracted. In [9] straight lines and circles are recognized using the Hough transform [10], but no other features like the corner circle sectors or the rectangle of the penalty area are detected. Although the Hough transform is robust and conceptually elegant, it is inefficient, since for each type of feature a separate parameter space has to be maintained. The search for local maxima in parameter space can be optimized by combining the method with Monte Carlo Localization, however it is still computationally expensive [9].

In this paper, we aim at efficient and robust feature recognition which allows the unique localization of the robot, up to the symmetry of the playing field. We refer to such features as *high-level features* in the following. We concentrate on five different high-level features: The center circle and four different corners which occur at the penalty area, as shown in Fig. 1.

This paper contains two contributions. First, we propose methods which allow the robust and efficient detection of the features mentioned above. Second, and

D. Nardi et al. (Eds.): RoboCup 2004, LNAI 3276, pp. 72–83, 2005.
© Springer-Verlag Berlin Heidelberg 2005

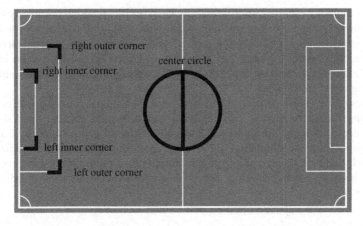

Fig. 1. The center circle and four different corners are recognized by the system. Although the shapes of the different corners are identical, the system is able to identify the position of a detected corner within the penalty area. Within one side of the playing field, each of the corners represents a unique feature

more important, we propose a framework which can be extended to recognize other feature types without doing all the work from scratch.

Feature detection has been addressed in numerous papers: The simplest approach for feature detection is to directly derive parameters of the feature from the data. For instance, for three points (not all collinear) one can easily derive the parameters of the circle containing the points. However, one has to be sure that the points belong to a circle. If one point is wrong, a false circle will be constructed.

The next class of methods is based on least-squares fitting. Instead of deriving the parameters with the minimum number of required points, more points are used and the total error is minimized. There exist fitting methods for lines (i.e see appendix of [13]), circles[17] and ellipses[8], or more generally for conic sections[7]. However it has to be know a priori which points belong to the feature. Outliers affect the result seriously.

Therefore, attempts to develop robust detection methods have been made. The probably most robust are Hough transform[10] based methods. There exist numerous variants for the detection of lines, circles, ellipses and general polygonal shapes. Each input item (i.e a point) votes for all possible parameters of the desired feature and the votes are accumulated on a grid. The parameters with the most votes determine the feature.

The counterpart of Hough transform methods, are techniques that probe parameter space. Instead of beginning with the input data and deriving the parameters in a bottom-up fashion, parameter space is searched top-down way[16, 3].

Other approaches, rely on the initial presence or generation of hypothesis. RANSAC[6], and clustering algorithms such as fuzzy shell clustering (FCS) [5, 4] fall into this category. The advantage of an initial hypothesis is that outliers can be detected easily by rejecting input points which are too distant.

A completely different approach is the UpWrite[1, 14] which iteratively builds small line fragments from points, and higher features like circles and ellipses from line fragments. In [15] the method was compared with the Hough transform and comparable robustness was reported. A similar approach was reported in [11] where ellipses are constructed from arcs, the arcs from line fragments and the line fragments from points.

These ideas have influenced our approach. We follow the principle of constructing higher geometric features, such as circles and corners from smaller components such as lines and arcs. Typically, different types of higher features are composed of the same kind of lower features. This hierarchical organization in which higher features share common components makes the overall recognition process more efficient than approaches which try to detect the individual features separately.

2   Extracting the Field Lines from the Images

We use our region tracking algorithm proposed in [20] to extract the field lines from the images. We determine the boundary of all green regions in the images and we search for green-white-green transitions, perpendicular to the boundary curves. Figure 2 illustrates this process for the omnidirectional images we use. After having extracted the lines they are represented by the pair (P, C) where $P = p_0, ..., p_{n-1}$ is a set of n points with cartesian x,y-coordinates and $C = c_0, ..., c_{l-1}$ supplies connectivity information that partitions P into l point sequences. Here each $c_i = (s_i, e_i)$ is a tuple of indices determining the start and end point in P that belong to the corresponding point sequence. That is, point sequence i consists of the points $p_{s_i}, ..., p_{e_i}$. By manipulating the connectivity information C, point sequences can be split or merged.

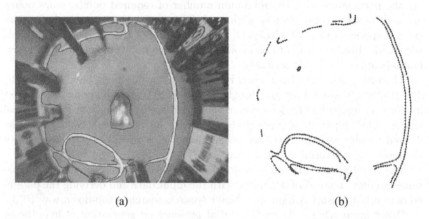

(a)   (b)

Fig. 2. (a) A local line detector is applied along the boundaries of the tracked regions. (b) The resulting line contours consist of a set of lines, where each line is represented by a sequence of points which are marked by small circles

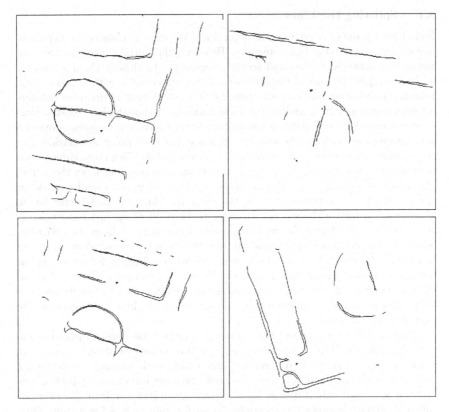

Fig. 3. Four examples of extracted line contours. Some long sequences of points correspond precisely to the shape of the field lines. But there are also outliers, missing lines, and small line fragments due to occlusion and detection errors

The line contours as illustrated in figure 2 b) are distorted due to the mirror in the omnidirectional vision system. In the following we assume that the distortion has been compensated. However, even without removing the distortion correctly, we are still able to detect most of the features. Fig.3 can provide an impression of the initial data. In many cases, there are long point sequences that correspond precisely to the shape of the field lines. However, there are also outliers, missing lines, and small line fragments due to occlusion and detection errors.

3   The Feature Construction Process

In this section, we describe how features are detected in line contours. The overall detection process is performed in several steps which iteratively construct higher features from the components of prior steps. The input for the first step are the line contours extracted from the images.

3.1   Splitting the Lines

Field lines consist of curved and straight lines. We want to classify the perceived line contours into these two categories. However, the initial contours often consist of concatenated curved and straight segments. To classify them separately we have to split the lines at the junctions. Junctions coincide with points of local maximum curvature. We retrieve them by first calculating a curvature measure for each point and then finding the local maxima. Although there exist sophisticated methods to calculate curvature (see for instance [2]), we have adopted a very simple approach for the sake of efficiency. For each point we consider two more points, one before and one after the current point. With these three points we compute two approximative tangent vectors, one reaching from the left to the midpoint and one from the midpoint to the right point. Finally, we define the curvature at the midpoint to be the angle by which the first vector has to be rotated to fall on the second. In order to be resilient against local noise in the curvature, we choose the enclosing points some distance from the midpoint. Since all the points are approximately equally spaced, we can afford to use an index distance instead of a precise geometric distance. That is, for a point $p[i]$ at index i we choose the enclosing points to be $p[i - w]$ and $p[i + w]$ ($w = 4$ in our implementation). At the beginning and end of each line, we continually decrease w in order to be able to calculate the curvature. For the first and last point, the curvature is not defined.

To detect local maxima of the curvature measure, we have adopted the following approach: While traversing the curvature values we detect intervals of values which exceed a given threshold and within each interval, we determine the index with the maximal value. To avoid extrema too close together, a new interval is opened only if it is at least at some given distance from the previous interval. Figure 4 shows the locations found for split points for various lines. Splitting can be performed efficiently by modifying the connectivity information in C (see section 2).

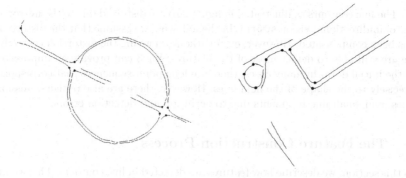

Fig. 4. Some examples that demonstrate the location of the split points

3.2   Classification

After the point sequences have been split, we classify each split sequence either *straight* or *curved*, applying the following test criterion: Similarly as for the curvature measure, we determine the angle between two vectors, but this time those defined by the first, the mid and the last point of the current point sequence. If the absolute value of the angle exceeds a threshold t_ϕ ($t_\phi = 0.4$ radians in our application), then the point sequence is declared to be curved, otherwise straight. Independently of the actual choice of t_ϕ wrong classifications can occur. However, the overall detection process can cope with a limited number of erroneous classifications.

3.3   Constructing Arcs and Straight Lines

For each straight point sequence a line is constructed taking the respective start and end point. Similarly, a circular arc is constructed for each curved point sequence: The start point, the midpoint and the end point define two segments whose perpendicular bisectors intersect at the center of the circular arc. In order to verify, whether the points really form an arc, we calculate the mean deviation of the points from the arc's radius and discard the arc if the distance is above the threshold $t_{rad} = 0.03$. In order to allow larger arcs to have a larger deviation, we divide the mean deviation by the radius before testing against the threshold.

3.4   Grouping Arcs and Detecting the Center Circle

Circular arcs which emerge from the same circle have centers which are close together. In order to group them, we search for clusters of the center points of the arcs. We apply the following cluster algorithm: Initially we have no clusters. The first cluster center is set at the center of the first arc. Then, we traverse the remaining arcs, and calculate the respective distance of their centers to the cluster centers. We choose the closest existing cluster and verify the distance. If it exceeds half of the actual arc's radius, we start a new cluster at the respective position. Otherwise, we adapt the cluster center to represent the weighted mean position of all assigned arc's centers. Here, the weights are the lengths of the arcs, which we approximate by the number of points of the point sequence from which the arc was constructed. We proceed in this way for all arcs and we obtain a set of clusters from which we choose the one with the greatest weight which reflects the number of assigned points. We demand, that at least 20 points have to be assigned and that the radius should be approximately one meter (the radius of the center circle). If these conditions are met, we generate a hypothesis for the center circle which will be refined as described next.

3.5   Refining the Center Circle

The center circle was constructed from the arcs and the arcs where constructed from point sequences which were classified as curved. However, it happens often that point sequences which are part of the center circle are short and almost straight. Typically, they are classified as straight and no arcs are constructed from them. We want to include this data for the precise detection of the center circle. Thus, we traverse all straight point sequences and verify if they could be part of the initial hypothesis for the center circle. We verify the distance of the actual point sequence to the circle and whether the orientation of the point sequence fits the tangent direction of the circle, at the corresponding location. We allow some tolerance, since the initial hypothesis is not perfect. In this way we obtain a set of points, the points of the arcs and the points of misclassified straight lines which belong to the circle. Next, we refine the initial hypothesis of the circle using Landau's method [12], a simple iterative technique which adjusts the center and the radius to the set of points. Typically, only few iterations (1-4) are required. Figure 5 illustrates this step. Later, we will search for the center line which passes through the circle in order to determine the corresponding orientation.

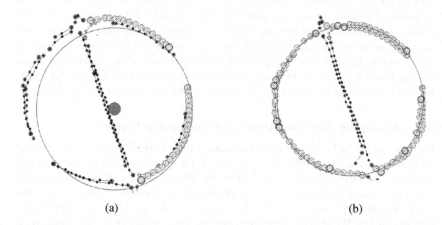

(a)   (b)

Fig. 5. The position and radius of the initial circle is refined. Points which are considered to originate from the circle are shaded. Initially (a), only the points of curved point sequences are considered to be part of the circle. Thus, the initial circle is not optimal. In (b) additional points have been determined by identifying point sequences which are close to the initial circle. The initial circle is iteratively adjusted to the points by Landau's method [12]. Only few iterations are required (two iterations in this example)

3.6   Determining the Principal Directions

Straight field lines of the playing field are either parallel or perpendicular. We want to determine the corresponding two orientations in the extracted contours. We will refer to them as *principal directions*. We will determine them with the straight lines constructed previously. Typically, spurious straight lines are present and we apply a clustering algorithm to cope with the outliers. For each line i, we calculate its orientation ϕ_i, normalizing the orientation to lay

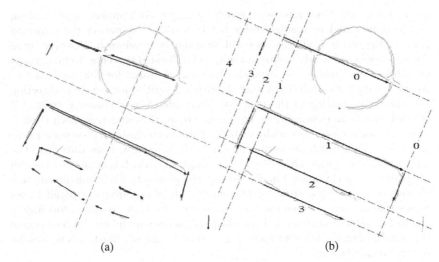

(a)   (b)

Fig. 6. The original line contours are painted gray. In (a) the straight lines are drawn by black arrows and the principal axis' found are shown. Discarded straight lines are drawn in gray. Theresults of grouping the lines into collinear sets are shown in (b). The dashed thin lines reflect the groups. Each group is simplified by a single line. The order of the groups is shown by numbers

within $[0, ..., \pi]$. Each line votes for the angles ϕ_i and $\phi_i + \pi/2$. Here, we apply the same clustering method as for the center circle, but this time working on one-dimensional values. We open a new cluster if the angular difference to the best cluster exceeds 0.3 radians. Otherwise, the cluster center is adjusted, with the weights being the lengths of the contributing lines. Finally, the cluster with the greatest weight determines the first principal direction ψ_0. The second principal direction ψ_1 equals the first, rotated by 90 degree. In the following sections we will consider two lines through the origin having the direction ψ_0 and ψ_1, respectively. We will refer to these lines as the *principal axis* $\mathbf{a_0}$ and $\mathbf{a_1}$. Figure 6a) shows an example.

3.7   Discarding Unreliable and Grouping Collinear Lines

Having determined the main axis' $\mathbf{a_0}$ and $\mathbf{a_1}$ we consider three types of lines. Those which are perpendicular to $\mathbf{a_0}$, those which are perpendicular to $\mathbf{a_1}$, and those whose orientation differs from both ψ_0 and ψ_1 by more than 0.3 radians. We consider the latter lines unreliable and discard them. The following step is performed for both $\mathbf{a_0}$ and $\mathbf{a_1}$ together with the respective perpendicular lines. Therefore, we will write $\mathbf{a}$ instead of $\mathbf{a_0}$ and $\mathbf{a_1}$ in the following.

Let $L = \{l_0, l_1, ..., l_n\}$ denote the set of lines l_i which are perpendicular to $\mathbf{a}$. Furthermore, let $\mathbf{m_i}$ be the mid point of line l_i. We consider the orthogonal projections of all $\mathbf{m_i}$ onto the axis $\mathbf{a}$. Lines which are collinear will yield close projection points. Since **a** passes through the origin, each projection point of $\mathbf{m_i}$ can be represented by a single value t_i which is the distance of the projected point to the origin. Collinear lines will have the same values t_i. However, since the lines are not precisely collinear, the t_i will differ slightly. Thus, to find groups of collinear lines which are perpendicular to **a** we search for clusters of the t_i. Again, we apply the same clustering algorithm described for the circle detection and the determination of the principal directions. A new cluster is opened if the distance to an existing cluster is greater than 20 centimeters. Each cluster, stores the lines which were assigned to it. Thus, each cluster represents a group of collinear lines which are perpendicular to the respective main direction. The lines within each group are replaced by a single line which encompasses the full range of the original lines. Finally, we sort the groups by their one-dimensional cluster centers t_j. Note, that the difference $t_j - t_k$ of two groups of parallel lines is just the distance between the lines. By sorting for t_j, we obtain a topological order of the groups of collinear lines which will be very useful for the detection of the penalty area and the corresponding corners. Figure 6b) illustrates the results of this processing step.

3.8   Detecting the Corners of the Penalty Area

The rectangle marking the goal area and the rectangle marking the penalty area produce three lines parallel to the baseline. The lines are spaced at a distance of 50 and 100 centimeters. Having grouped and sorted the sets of collinear lines, we can easily detect such a structure. Here, we allow a tolerance of 20 centimeters when verifying the distances. Note, that the structure emerges at the start or end of the sequence of sorted collinear line groups, since the lines are the outmost existing lines. Having detected such a structure, the direction towards the goal is now known. Thus, we can distinguish between the left and right side of the lines. In order to find the respective corners, we simply verify whether we find perpendicular lines whose endpoints are close to the given lines. Some additional constraints have been necessary in order to avoid spurious detections. First, if we find three parallel lines that have the given structure in their distances, we calculate the overall length of the structure, in the direction of the lines. This length should not exceed the length of the penalty area, which is 5 meters. We allow a tolerance of 50 centimeters. A second constraint is that no lines which are perpendicular to the three lines are beyond the goal line.

4   Experimental Results

This section describes our experimental results. Feature detection is influenced by many factors: By the preprocessing step which extracts the field lines, by line occlusions, by the region tracking algorithm, by the geometrical distortion calibration and on the lighting conditions.

We tried to focus on the main situations. In order to test the influence of different environments we examined two different playing fields. The first with

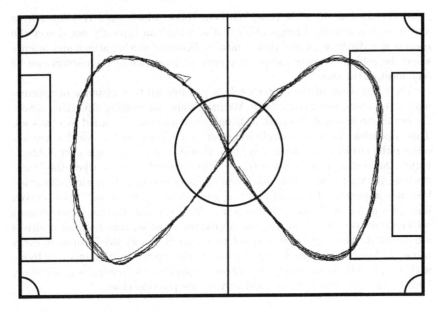

Fig. 7. The path shows a robot moving on the field along a predefined figure. As can be seen, the maximum deviation is lower than 20 cm for a robot driving 0.8 m/s

a green carpet and artificial lighting, the second with a reflecting linoleum floor, with natural light shining from one side through an array of windows. However, since the reflections on the floor where almost white in the images, we reduced the influence of natural light by adding artificial light from above and using venetian blinds. However, we did not shut the blinds completely.

On both playing fields, we let the robot automatically move on a trajectory forming an eight (see figure 7). Localization was achieved using odometry and the recognized features, which yield unique robot poses, up to the symmetry of the playing field. On both fields, the robot did not loose its position. After 10-20 minutes we stopped the experiment. Moreover, when the robot was manually transferred to an unknown position, the robot immediately found its correct position after perceiving a feature. The maximum positional error while driving was about 20 cm.

While the robot moved, we logged all the extracted line contours in a file and later, we manually verified the feature recognition for all frames. On both fields not a single false positive was detected. However, there are situations when features are not recognized.

The corners of the penalty area cannot be detected, if an obstacle is occluding the corner. That is, the system does not infer the intersection point of two perpendicular lines, but rather demands that the endpoints of two lines are close together. Here, some improvements might be possible, however one must take care, not to infer corners which do not exist. However, all four corners will be rarely occluded at the same time. With our robot, which has a very low mirror and thus, has a limited range of sight, the robot can typically not detect the corners in a distance greater than 3 meters. However, during attack and defense when the robots are near the penalty area, at least one of the corners can be seen most of the time.

The recognition of the center circle is possible up to a distance of approximately 3 meters, using our robots. We have some astonishing situations, where the center circle is still detected, although large parts are occluded by obstacles. This is possible, because a single fragment of a curved line can yield a hypothesis for the circle, which is then verified. However, if not a single curved line is found, the center circle is not detected. This typically occurs, when the center circle is distant to the robot, and when the correction of the optical distortion becomes imprecise. Then the splitting procedure tends to split noisy curved lines into many small straight fragments. At this point, further improvements are certainly possible. However, the important point is, that no false positives are detected. Also, note that it is not necessary to detect the features in every frame. Although the current system is able to run the recognition on every frame with 15 fps, this not necessary. It suffices, to perform the recognition, say every 10th frame, since the features yield very strong position clues.

5   Conclusion

We have proposed a method for detecting the center circle and four different types of corners in the neighborhood of the penalty area. Each of these features allows to localize the robot, up to the symmetry of the playing field. The method is robust and efficient and we have tested it thoroughly. Also, we have presented a general framework for feature construction. It should be possible to easily extend the approach to detect other features, such as the corner circle sectors of the playing field, for instance. Detected arcs, groups of straight lines, and the main directions are already available and it should be possible to construct different features from them. We hope that the palette of features will be extended continually by other researchers. Also, the individual algorithm can be improved or extended. In our approach we have used the region tracking algorithm described in [20]. However, maybe more efficient methods can be found that extract chains of points, representing the field lines. Here again, adaptivity to the illumination is the primary difficulty.

References

1. M. D. Alder. Inference of syntax for point sets. In E. S. Gelsema and L. N. Kanal, editors, *Pattern Recognition in Practice IV*, pages 45–58. New York: Elsevier Science, 1994.
2. H. Asada and M. Brady. The curvature primal sketch. In *Proc. 2nd IEEE Workshop Computer Vision: Representation and Control*, pages 8–17, 1984.

3. T.M. Breuel. Fast recognition using adaptive subdivisions of transformation space. In *Proceedings. 1992 IEEE Computer Society Conference on Computer Vision and Pattern Recognition (Cat. No.92CH3168-2)*, pages 445–51, 1992.

4. R. N. Daves and K. Bhaswan. Adaptive fuzzy c-shells clustering and detection of ellipses. *IEEE Transactions on Neuronal Networks*, 3(5):643–662, 1992.

5. R.N. Daves. Fuzzy shell-clustering and applications to circle detection in digital images. *International Journal on General Systems*, 16:343–355, 1990.

6. Martin A. Fischler and Robert C. Bolles. Random sample consensus: a paradigm for model fitting with applications to image analysis and automated cartography. *Communications of the ACM*, 24(6):381–395, 1981.

7. A. Fitzgibbon and R. Fisher, 1995.

8. A. W Fitzgibbon, M Pilu, and R. B Fisher. Direct least square fitting of ellipses. *IEEE Transactions on Pattern Analysis and Machine Intelligence*, 21(5):476–480, 1999.

9. Gerd Mayer Hans Utz, Alexander Neubeck and Gerhard Kraetzschmar. Improving vision-based self-localization. *Gal A. Kaminka, Pedro U. Lima and Raúl Rojas (eds): RoboCup-2002: Robot Soccer World Cup VI, Springer*, 2002.

10. P. V. C. Hough. Method and means for recognizing complex patterns. US Patent 3,069,654, December 1962.

11. Euijin Kim, Miki Haseyama, and Hideo Kitajima. Fast and robust ellipse extraction from complicated images.

12. U M Landau. Estimation of a circular arc center and its radius. *Comput. Vision Graph. Image Process.*, 38(3):317–326, 1987.

13. Feng Lu and Evangelos Milios. Robot pose estimation in unknown environments by matching 2d range scans. *CVPR94*, pages 935–938, 1994.

14. R. A. McLaughlin and M. D. Alder. Recognising cubes in images. In E. S. Gelsema and L. N. Kanal, editors, *Pattern Recognition in Practice IV*, pages 59–73. New York: Elsevier Science, 1994.

15. R. A. McLaughlin and M. D. Alder. The Hough Transform versus the UpWrite. *IEEE Transactions on Pattern Analysis and Matchine Intelligence*, 20(4):396–400, 1998.

16. C. Olson, 2001.

17. Corneliu Rusu, Marius Tico, and Pauli Kuosmanen. Classical geometric approach to circle fitting - review and new developments. *Journal of Electronic Imaging*, 12(1):179–193, 2003.

18. Erik Schulenburg. Selbstlokalisierung im Roboter-Fußball unter Verwendung einer omnidirektionalen Kamera. Master's thesis, Fakultät für Angewandte Wissenschaften, Albert-Ludwigs-Universität Freiburg, 2003.

19. Sebastian Thrun, Dieter Fox, and Wolfram Burgard. Monte carlo localization with mixture proposal distribution. In *AAAI/IAAI*, pages 859–865, 2000.

20. F. v. Hundelshausen and R. Rojas. Tracking regions and edges by shrinking and growing. In *Proceedings of the RoboCup 2003 International Symposium, Padova, Italy*, 2003.

Illumination Insensitive Robot Self-Localization Using Panoramic Eigenspaces

Gerald Steinbauer[1] and Horst Bischof[2]

[1] Institute for Software Technology, Graz University of Technology,
Inffeldgasse 16b/II, A-8010 Graz, Austria
[2] Institute for Computer Graphics and Vision, Graz University of Technology,
Inffeldgasse 16/II, A-8010 Graz, Austria

Abstract. We propose to use a robust method for appearance-based matching that has been shown to be insensitive to illumination and occlusion for robot self-localization. The drawback of this method is that it relies on panoramic images taken in one certain orientation, restricts the heading of the robot throughout navigation or needs additional sensors for orientation, e.g. a compass. To avoid these problems we propose a combination of the appearance-based method with odometry data. We demonstrate the robustness of the proposed self-localization against changes in illumination by experimental results obtained in the RoboCup Middle-Size scenario.

1 Introduction

Mobile robot localization is the problem of determining a robot's pose (i.e. its location and orientation) from sensor data such as odometry, proximity sensors or vision. Self localization is a key problem in autonomous robotics, it has even been referred to as "the most fundamental problem to providing a mobile robot with autonomous capabilities" [1].

Good results have been achieved by using combinations of metric maps of the environment, sensor models, i.e. that model the expected response from the environment, and probabilistic sensor fusion [2, 3]. These approaches show very accurate localization. But generation and storage of the maps and the models are very time and memory consuming. Especially, if the environment gets large.

In contrast, evidence has been provided that topological localization, which recognizes certain spots in the environment, is sufficient to navigate a mobile robot through an environment. This also appears more naturally, if one thinks how humans navigate through a building or a city.

Appearance-based approaches use images to recognize known spots in the environment. This is done by comparing the current image with a set of reference images, previously captured at some reference locations. The approaches differ in the type of the camera used, the representation of the reference images and the calculation of the similarity of different images. Ishiguro et.al. [4] use panoramic cylinder images obtained from an omnidirectional camera and row by

D. Nardi et al. (Eds.): RoboCup 2004, LNAI 3276, pp. 84–96, 2005.

row Fourier transformation as a compact representation. The similarity of images is determined by calculating the sum of the absolute difference of the most significant Fourier coefficients. Menegatti and colleagues [5] extend the method by Monte Carlo Localization to solve the problem of *perceptual aliasing*. That means the response of the current image to several reference images. In [6] topological localization is applied to the RoboCup Middle-Size. A standard camera and an eigenspace based representation are used. The reference images are transformed to eigenimages by Principal Component Analysis and represented by the coefficients in the corresponding eigenspace. The calculation of the similarity of images is done by a k-nearest neighbor algorithm within the eigenspace.

As changes in illumination dramatically effect the appearance of locations in the environment [7], the mentioned methods are sensitive to such changes in illumination. Currently, we have well restricted lighting conditions in the RoboCup Middle-Size, which are a compromise to the vision algorithm applied to robotic soccer today. Due to the decision to introduce a certain amount of natural illumination to the RoboCup Size-Middle and the vision that our robots someday will leave the field and work in more realistic environments, the importance of illumination insensitive algorithms is undoubted.

To cope with such changes in illumination in [8] another representation and calculation of similarity were proposed. An illumination insensitive eigenspace representation and a randomized voting algorithm are used. The illumination insensitive eigenspace approach was originally developed for robust object recognition under varying lighting conditions [9]. The approach exploits the property that the eigenspace representation also holds after linearly filtering the current image and the eigenimages. This filtering is the key to the illumination insensitivity. Additionally, a voting algorithm based on a randomly drawn subset of pixels of the images makes the approach insensitive to highlights, noise and occlusions. The drawback of this method is that it assumes that the reference images and the current image are captured at one certain orientation of the robot. To meet this assumption either the heading of the robot is restricted to that orientation or an additional sensor for orientation, e.g. a compass, is needed. Both are rigorous limitations for the practical use on mobile robots.

In a recent publication [10] a rotation invariant representation of eigenimages was presented. But this representation is not robust, computationally expensive and still sensitive to changes in illumination.

There are also other representations of panoramic images, that are invariant to rotation [11]. But again they are neither robust nor illumination insensitive.

In this work, we propose the combination of the robust illumination insensitive eigenspace approach and sensor fusion with odometry data as a solution to the limitations mentioned above. The basic idea is to use odometry to keep track of the orientation of the robot, to use the predicted orientation to rotate the current image back to the reference orientation and correct the orientation delivered by the odometry by the response of the eigenspace framework.

In the next section we will outline the self-localization based on illumination insensitive eigenspaces. In Section 3 we introduce our extensions and their theoretical foundations. In Section 4 we present preliminary results obtained from experiments in the RoboCup Middle-Size scenario. Finally, in Section 5 some conclusions are drawn and future research perspectives explained.

2   Illumination Insensitive Self-Localization Using Eigenspace

In appearance-based localization the robot is provided with a set of views of the environment taken at several locations in the environment. These locations are called reference locations because the robot will refer to them to locate itself in the environment. The corresponding images are called reference images. When the robot moves, it can compare the current view with reference images captured during a training phase. When the robot finds which of the reference image is more similar to the current view it can infer its position in the environment. The problem of finding the position in the environment is reduced to the problem of finding the best match of the current image among the reference images. A higher localization accuracy can be achieved by interpolation between reference points within the eigenspace, while keeping the number of reference images constant. In the remainder of this section we outline the appearance-based approach presented in [8] and its solution for illumination insensitivity.

2.1   Eigenspace Based Recognition

An eigenspace based representation is used for a compact storage of the reference images and a robust calculation of the similarity of images. To build the eigenspace, we first represent the images from the training set as image vectors, from which the mean image is subtracted, $x_i; i = 0...N-1$, which form an image matrix $X = [x_0 \ x_1 \ ... \ x_{N-1}]$, $X \in \mathbf{R}^{n \times N}$; where n is the number of pixels in the image and N is the number of images. These training images serve as input for the Principal Components Analysis (PCA) algorithm, which results in a set of p eigenimages e_i, $i = 1, ..., p$, that span a low-dimensional eigenspace. Eigenimages are selected on the basis of the variance that they represent in the training set. Every original image x_i can be transformed and represented with a set of coefficients $q_{ij} = x_i e_j$, $j = 1, ..., p$, which represent a point in the eigenspace. That way, every image is approximated as $\tilde{x}_i = \sum_{j=1}^{p} q_{ij} e_j$. Figure 1 depicts the first four eigenimages for a RoboCup Middle-Size field.

The standard approach to localization is to find the coefficient vector q of the momentary input image y by projecting it onto the eigenspace using the dot product $q_i = \langle y, e_i \rangle$, so that $q = [q_1, ..., q_p]^T$ is the point in the eigenspace.

If we want the image y to be recognized as its most similar counterpart in the training set (or in a representation constructed by means of interpolation, see [12]), the corresponding coefficients have to lie close together in the eigenspace. However, in the case when the input image is distorted, either due to occlusion, noise or variation in lighting, the coefficient we get by projecting onto the eigenspace can be arbitrarily erroneous [9].

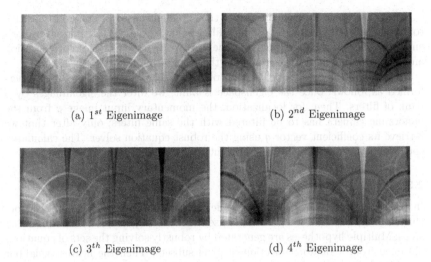

(a) 1st Eigenimage   (b) 2nd Eigenimage

(c) 3th Eigenimage   (d) 4th Eigenimage

Fig. 1. Eigenspace representation of reference images (Experiments on a RoboCup Middle-Size field)

However, it has been shown, that one can also calculate the coefficient vector q by solving a system of k linear equations on $k \geq p$ points $r = (r_1, ...r_k)$

$$y_{r_i} = \sum_{j=1}^{p} q_j e_{jr_i} \ 1 \leq i \leq k \tag{1}$$

using a robust equation solver and multiple hypotheses [13].

In [14] it has been shown how this method can be used to allow robust localization in presence of occlusions. However it does not solve the problem of illumination.

2.2   Illumination Insensitivity

The method presented in [15] takes the computations of parameters one step further. Since Eq. (1) is linear, it also holds that $(f * x)(r) = \sum_{i=1}^{p} q_i (f * e_i)(r)$, where f denotes a filter kernel. This means that if we convolve both sides of the equation with a filter kernel, the equality still holds. Therefore, we can calculate the coefficients q_i also from the filtered eigenimages if we filter the input image.

By using a set of t linear filters $\mathcal{F}$ we can construct a system of equations

$$(f_s * x)(r) = \sum_{i=1}^{p} q_i (f_s * e_i)(r) \ s = 1, ..., t. \tag{2}$$

It is now possible to calculate the coefficients q either by using k points, or using t filter responses at that single point, or a combination of these two.

It is well known from the literature that gradient-based filters are insensitive to illumination variations. By taking a filter bank of gradient filters in several orientations, we can therefore augment the descriptive power of the representation and achieve illumination invariance in the recognition phase.

Illumination invariant localization of a mobile robot can therefore be performed as follows: once the eigenspace is built, we filter the eigenimages by a bank of filters. Then, for localization, the momentary input image y from the panoramic camera has to be filtered with the same filters; only after that we retrieve its coefficient vector q using the robust equation solver. The calculated coefficients are used to infer the momentary location of the robot.

2.3   Robust Voting

In order to robustly recover the coefficients in the presence of noise and occlusions a robust voting algorithm is used. The voting consists of the following steps: Multiple hypotheses are generated by robustly solving the sets of equations obtained from a random selection of pixel subsets. Then the nearest neighbor of each hypothesis in the eigenspace (coefficient vector of the closest point on the parametric manifold) is determined. This selected coefficient vector gets a vote. For voting we use a voting function $v(d) \to [0,1]$ which gives votes that are inversely proportional to the distance d of the hypothesis from its nearest neighbor. Based on the distribution of the votes we decide whether the coefficients should be accepted (e.g. coefficients with accumulated votes above a given threshold). We use the following voting function:

$$v(d_{ij}, \sigma) = e^{\frac{-d_{ij}^2}{2\sigma^2}} , \tag{3}$$

where

$$d_{ij} = \arccos \frac{a_i^T a_j^{(t)}}{\|a_i\|\|a_j^{(t)}\|} \tag{4}$$

is the angle between the estimated coefficient vector a_i and the nearest coefficient vector in the eigenspace $a_j^{(t)}$. We use the angle instead the Euclidean distance between the coefficients as criteria, because coefficient vectors with the same direction but different lengths represent the same image but in different brightness. The parameter σ determines the width of the voting function.

3   Keeping Track of the Orientation

The approach presented in Section 2 has been shown to be robust against changes in illumination, noise and occlusion. The drawback of that approach is that it relies on reference and current images captured in one certain orientation or that the orientation is known trough an additional sensor, e.g. a compass. The reason for that limitation is that the eigenspace-based representation is not invariant to rotation. Therefore, only images taken in the same orientation as the reference images can be recognized in a robust way. Instead of capturing images at one orientation one could take images at several orientations (e.g. all 10°), but this approach has several drawbacks: First, the number of images and therefore the training time is significantly increased. Second, and more important due to the larger set of images which need to be represented, the number of eigenimages needs to be increased, which on the one hand increases the running time and more importantly as experiments have demonstrated decreases the robustness of the whole approach.

In order to overcome this limitation we use a combination of the robust eigenspace-based approach and sensor fusion with odometry data. The basic idea is to use a Kalman filter to keep track of the orientation of the robot. The odometry data are fused into the filter and provide a prediction of the orientation Θ of the robot. This prediction is used to rotate the current image back into the orientation the reference images were captured. This rotated image is used for the localization step. But, instead of using only one rotated image we repeat the localization step with a set of images rotated by angles drawn from the predictive distribution. That image which gets the highest response from the recognition process determines the new position of the robot. Its corresponding rotation is used to correct the Kalman filter.

3.1   Sensor Fusion

Sensor fusion of odometry data and data from other sensors (e.g., vision, proximity sensors) with Kalman filters is a commonly applied method. So we skip a deeper discussion of this topic. An overview on Kalman filters could be found in [16]. But it should be mentioned that we use the fusion method presented in [17]. The method is an extension of the standard Kalman filter. It uses a bank of Kalman filters working on the same state vector to deal with asynchronous measurements. This is a common problem in real robot systems, as also in our system.

3.2   The Extended Localization

In [18] and [19] it has been shown that the distance between the coefficient of a rotated image and the coefficient of its reference image smoothly increase with the absolute rotation angle, while still keeping the shortest distance to its reference coefficient in the eigenspace, assuming moderate rotation. This property is used in the development of an extension to the approach presented in Section 2 which preserves the illumination insensitivity and overcomes the limitations cased by rotated images.

The extended localization algorithm can be outlined as follows: In a first step N omnidirectional reference images are captured around the environment in one certain reference orientation, i.e. $\Theta_R = 0$. The reference images are unwrapped creating panoramic cylinders (see Figure 2). Panoramic cylinders have the advantage that rotating the original omnidirectional image is only a row wise shift on the panoramic cylinder. Using the algorithm of section 2 the set of panoramic cylinders is transfered into the filtered eigenspace representation.

Fig. 2. An omnidirectional image (left) and its panoramic cylinder (right)

The localization cycle consists of the following steps: We assume that the initial orientation is known. Odometry data are fused into a Kalman filter at the time they are available. The Kalman filter provides a prediction for the true orientation of the robot and its uncertainty about this orientation Θ_t at time t, $KF(t) \rightarrow \langle \hat{\Theta}_t, \hat{\sigma}_t \rangle$, where $\langle \hat{\Theta}_t, \hat{\sigma}_t \rangle$ determines a normal distribution for the true orientation Θ_t of the robot. When a new omnidirectional image is captured at time t the image is transfered into its panoramic cylinder and a set of orientation hypotheses $\Sigma_t = [\Theta_{t,1}, ..., \Theta_{t,M}]$ are randomly drawn from the distribution $\langle \hat{\Theta}_t, \hat{\sigma}_t \rangle$. With high probability Σ_t contains the true orientation Θ_t of the robot.

Instead of performing the recognition step only once for the captured image, the recognition step is repeated M times on the captured image shifted by $-\Theta_{t,i}$, $i = 1, ..., M$. This generates M votes for all N reference images: v_{ij}; $i = 1, ..., M$; $j = 1, ..., N$. The shifted images are equivalent to images captured around the reference orientation Θ_R. The property mentioned at the beginning of this section guarantees that the shifted image i with the rotation closest to the true orientation of the robot Θ_t, $i = argmin_j(|\Theta_{t,j} - \Theta_t|)$, will collect the highest votes v_{ij} for its corresponding reference image j. The corresponding reference position of the reference image j is reported as the new determined location of the robot. The corresponding orientation Θ_i of the captured image is used as a measurement update for the Kalman filter.

4   Experimental Results

We evaluated the extended Localization by carrying out several real and simulated experiments in the RoboCup Middle-Size scenario. The experiments were conducted using robots of our RoboCup Middle-Size team [20]. Our Middle-Size test-field served as the test environment. The test-field is a rule-compliant field except of its size (5m x 6m). It is situated under a glass roof in an open hall. Therefore, the field is directly effected by changes of illumination during the day. The change of the appearance of a location on the field is illustrated in Figure 3.

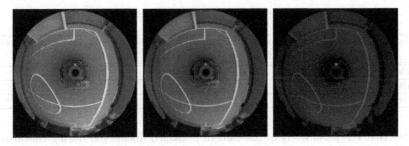

Fig. 3. The appearance of the same location at noon (left), in the afternoon (center) and in the evening (right)

4.1 Real Experiments

The real experiments were carried out in three runs. In a first step the field was divided into a grid with 1 m resolution. The points on the grid served as reference locations on the field. This lead to a total number of 42 locations. At noon for all reference locations a reference image in the reference orientation $\Theta_R = 0$ was captured. All reference images were transformed into a panoramic cylinder with a resolution of 360 x 145 pixels. Using the method described in Section 2 the reference images were transformed into the filtered eigenspace representation. For representation only 15 of the 42 available eigenimages are enough. Due to performance reasons we used a bank of 3 gradient filters with a 3x3 filter kernel instead of the recommended steerable filters.

Within the experiments we used $M = 5$ image orientations randomly drawn from the prediction of the Kalman filter $\langle \hat{\Theta}_t, \hat{\sigma}_t \rangle$ and tested 50 hypotheses with 1000 randomly selected pixel for each rotated image. The σ for the voting algorithm was set to 0.3. The measurement noise of the odometry data was determined during ground truth measurements [21]. The level of measurement noise for the angle measurement was optimized by hand and was kept constant during the experiments.

Three test runs were conducted at noon (immediately after the reference images were captured), in the afternoon and in the evening. Prior to each run the illumination on the field was measured to document the changes in illumination during the runs and its unsteadiness across the field. The illumination was measured on the 42 reference locations and the minimum, the maximum, the mean and the standard deviation was calculated. The results are shown in Table 1.

Table 1. Change of illumination between the test runs

Run	Min/Lux	Max/Lux	Mean/Lux	Std.Dev/Lux
Reference/Noon	335	1288	690.9	248.8
Afternoon	219	857	498.6	157.9
Evening	111	320	203.7	63.1

Table 2. Results of the Localization under different illumination

Test Run	Trials	Reference Positions passed	Reference Positions recognized	Correct Recognitions/%	Tracks Lost
Noon	10	225	225	100	0
Afternoon	10	231	220	95	0
Evening	10	193	161	83	2

Note that the parameters of the camera remained unchanged during capturing the reference images and all runs and no artificial illumination was used.

The test runs were conducted as follows: For each run (noon, afternoon and evening) the robot was randomly placed ten times on a reference location with a random but known orientation. The robot then randomly moved around for a while. Each time the robot passed a reference location (robot entered a cycle of 30 cm diameter around the reference location) it was recorded if the reference location was correctly recognized. The results shown in Table 2 document clearly the performance of the extended localization under different illumination.

In the noon and afternoon runs the robot correctly recognized nearly all reference locations and never lost track of the orientation. The latter is crucial for continuing correct recognition of reference locations. In the evening run the recognition ratio decreased and the robot twice lost track of the orientation due to the very bad illumination (see Figure 3). The lost track was caused by wrong votings on the shifted images and their corresponding rotation, which prevented an adequate correction of the odometry data. Note that we used the Matlab to C++ compiler to convert the Matlab-prototype into a C++ module, executable on the robot. That lead to a frame-rate of only 0.8 Hz. We suppose that a speedup of the implementation will decrease the number of lost tracks, due to the faster correction of erroneous odometry data.

4.2   Offline Experiments

To verify the smooth decreasing of votes for rotated images and the robustness of the method against noise and occlusion we conducted two offline experiments. First we offline calculated the votes for the rotated version of an image captured near a reference location. Figure 4 shows the votes for an image captured in the reference orientation Θ_R near reference location 20. It clearly shows the smooth decreasing of the votes for reference location 20 with increasing absolute rotation angle. For an absolute angle approximately below $10°$ a clear voting for the reference image 20 is provided. This is an encouraging result, as predictions of the orientation with a higher error do not occur very frequently.

In the second offline experiment we evaluated the robustness of the method against noise. Therefore, we used the images and orientations recorded in the real run at noon. We did the same experiment as in the real run except that we introduced a certain amount of noise into the recorded images. This is done by randomly replacing a certain percentage of the pixels in the image by random pixels. Figure 5 depicts an image before and after introducing 50% random pixels.

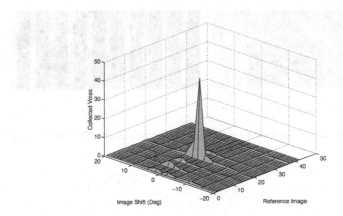

Fig. 4. Collected votes for an image take near reference position 20 in respect to shifts of the image

(a)   (b)

Fig. 5. Original image (a) and the same image after introducing 50% random pixels (b)

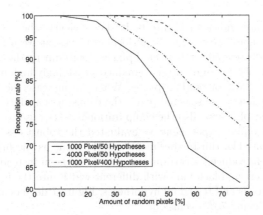

Fig. 6. Recognition rate of reference locations in respect to the amount of introduced random pixels

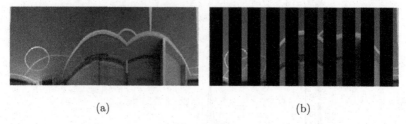

(a)   (b)

Fig. 7. Original image (a) and the same image after introducing 70% occlusion (b)

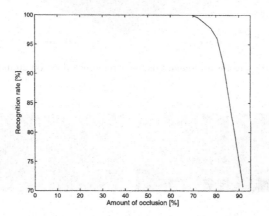

Fig. 8. Recognition rate of reference locations in respect to the amount of introduced occlusion (1000 sampled pixel and 50 hypotheses)

The results are shown in Figure 6. It shows clearly the robustness of the method against noise. Note, due to performance reasons we used a very small number of hypotheses (50) and selected pixels (1000) during the previous experiments. Anyway, the recognition rate remained satisfyingly high until an amount of approximately 25% of random pixels. When we increase the number of hypotheses or the number of selected pixels the robustness against noise increases significantly. But of course also the computational costs raises.

In the third offline experiment we evaluated the robustness of the method against occlusion. The third experiment was the same as the prior experiment except that we introduced a certain amount of occlusion instead of noise. This was done by inserting black bars with different width into the images. We used 1000 sampled pixels and tested 50 hypotheses. Figure 7 depicts an image before and after introducing 70% occlusion.

The results are shown in Figure 8. It shows clearly the robustness of the method against occlusion. Up to 70 % of occlusion the method is robust against occlusion and all positions were recognized correctly.

5   Conclusion and Future Work

In this work we propose an extension of the appearance-based self-localization presented in [8] which overcomes the limitations of that method. The limitations are restrictions of the heading of the robot to a reference orientation or the need of an additional sensor for orientation. The extension is based on a combination of the illumination insensitive localization and sensor fusion with odometry data. A prediction for orientation is used for a robust recognition process and its rotation sensitive feedback is used to correct the odometry data. Preliminary experiments within the RoboCup Middle-Size scenario show that the proposed extended localization is robust against changes of illumination, noise and occlusions while it overcomes the limitations mentioned above. However, a speedup of the implementation is necessary to reliably keep track of the orientation under very bad illumination.

Furthermore, we believe that a combination of the illumination insensitive localization with particle filter methods will further improve the method. Especially, because the robot is unable to recover from dramatic errors in the orientation and it has to know its initial orientation.

Currently, we are working on the application of our extended localization in larger environments, e.g. office buildings. As the recording of reference images and its location by hand is time consuming in large environments, we perform research on map-building and SLAM in combination with the illumination insensitive localization.

Acknowledgements

The research presented in this paper has been supported by Land Steiermark under grand 40Ro03-PE "RoboCup 2004". The authors grateful acknowledge the support of Knapp Automation & Logistik, Sick AG, Kapsch, Saft, Siemens, jumpTec, Farnell In One and BTI for setting up the Mostly Harmless team.

References

1. I.J. Cox. Blanche - An Experiment in Guidance and Navigation of an Autonomous Robot Vehicle. *IEEE Transactions on Robotics and Automation*, 7(2):193–204, 1991.
2. Dieter Fox, Wolfram Burgard, Frank Dellaert, and Sebastian Thrun. Monte carlo localization: Efficient position estimation for mobile robots. In *AAAI/IAAI*, pages 343–349, 1999.
3. F. Dellaert, W. Burgard, D. Fox, and S. Thrun. Using the condensation algorithm for robust, vision-based mobile robot localization. In *Proceedings of the IEEE International Conference on Computer Vision and Pattern Recognition, Fort Collins, CO*. IEEE, 1999.
4. H. Ishiguro and S. Tsuji. Image-based memory of environment. In *Proc. of Int. Conf. on Intelligent Robots and Systems (IROS'96)*, pages 634–639, 1996.

5. E. Menegatti, M. Zoccarato, E. Pagello, and H. Ishiguro. Image-based Monte-Carlo Localisation without a Map. In *Proc. of the 8th Conference of the Italian Association for Artificial Intelligence (AI*IA) Pisa - Italy*, 2003.
6. Goncalo Neto, Hugo Costelha, and Pedro Lima. Topological navigation in configuration space applied to soccer robots. In *RoboCup 2003 International Symposium, Padova, Italy*, 2003.
7. Gerd Mayer, Hans Utz, and Gerhard Kraetzschmar. Playing Robot Soccer under Natural Light: A Case Study. In *RoboCup 2003 International Symposium Padua, Italy*, 2003.
8. M. Jogan, A. Leonardis, H. Wildenauer, and H. Bischof. Mobile robot localization under varying illumination. In *Proc. of 16th International Conference on Pattern Recognition*, 2002.
9. Horst Bischof, Horst Wildenauer, and Ales Leonardis. Illumination Insensitive Recognition using Eigenspaces. *Computer Vision and Image Understanding (to appear)*, 2004.
10. Matjaz Jogan and Ales Leonardis. Robust localization using an omnidirectional appearance-based subspace model of environment. *Robotics and Autonomous Systems*, 45(1):51–72, 2003.
11. Tomas Pajdla and Vaclav Hlavac. Zero phase representation of panoramic images for image based localization. In *8th International Conference on Computer Analysis of Images and Patterns*, pages 550–557, 1999.
12. S. K. Nayar, S. A. Nene, and H. Murase. Subspace methods for robot vision. *IEEE Transaction on Robotics and Automation*, 12(5):750 – 758, 1996.
13. Ale Leonardis and Horst Bischof. Robust recognition using eigenimages. *Computer Vision and Image Understanding: CVIU*, 78(1):99–118, 2000.
14. M. Jogan and A. Leonardis. Robust localization using panoramic view-based recognition. In *15th ICPR*, volume 4, pages 136–139. IEEE Computer Society, 2000.
15. H. Bischof, H. Wildenauer, and A.Leonardis. Illumination insensitive eigenspaces. In *Proc. Int. Conf. on Computer Vision*, volume 1, pages 233–238, 2001.
16. P. S. Maybeck. The Kalman filter, An introduction to concepts. In *Autonomous Robot Vehicles*, pages 194–204, 1990.
17. L. Drolet, F. Michaud, and J. Cote. Adaptable sensor fusion using multiple kalman filters. In *Proceedings IEEE/RSJ International Conference on Intelligent Robots and Systems (IROS)*, 2000.
18. H. Murase and S. K. Nayar. Illumination planning for object recognition using parametric eigenspaces. *IEEE Transactions on Pattern Analysis and Machine Intelligence*, 16(12):1219–1227, 1994.
19. Ales Leonardis Matjaz Jogan, Emil Zagar. Karhunen-loeve transform of a set of rotated templates. *IEEE Trans. on Image Processing*, 12(7):817–825, 2003.
20. Gerald Steinbauer, Michael Faschinger, Gordon Fraser, Arndt Mühlenfeld, Stefan Richter, Gernot Wöber, and Jürgen Wolf. Mostly Harmless Team Description. In *Proceedings of the International RoboCup Symposium*, 2003.
21. Lorenz Mayerhofer. Odometry correction for the keksi omnidrive using an echo state network. Technical report, Institute for Theoretical Computer Science University of Technology Graz, 2003.

A New Omnidirectional Vision Sensor for Monte-Carlo Localization

E. Menegatti[1], A. Pretto[1], and E. Pagello[1,2]

[1] Intelligent Autonomous Systems Laboratory,
Department of Information Engineering, The University of Padua, Italy
emg@dei.unipd.it
[2] also with Institute ISIB of CNR Padua, Italy

Abstract. In this paper, we present a new approach for omnidirectional vision-based self-localization in the RoboCup Middle-Size League. The omnidirectional vision sensor is used as a range finder (like a laser or a sonar) sensitive to colors transitions instead of nearest obstacles. This makes it possible to have a more reach information about the environment, because it is possible to discriminate between different objects painted in different colors. We implemented a Monte-Carlo localization system slightly adapted to this new type of range sensor. The system runs in real time on a low-cost pc. Experiments demonstrated the robustness of the approach. Event if the system was implemented and tested in the RoboCup Middle-Size field, the system could be used in other environments.

1 Introduction

Localization is the fundamental problem of estimating the pose of the robot inside the environment. Several techniques based on the Monte-Carlo localization (MCL) approach was developed. Two kinds of sensors have been used: range finder devices (i.e. lasers and sonars) and vision sensors (i.e. perspective and omnidirectional cameras). The range finders are used to perform scans of the fix obstacles around the robot and the localization is calculates matching those scans with a metric map of the environment [1, 12]. The vision sensors are used to recognize characteristic landmarks subsequently matched within a map [3, 10] or to find the reference image most similar to the image currently grabbed by the robot (without a map) [14, 8, 9]. In our approach we use an omnidirectional vision system as sensor to emulate and enhance the behaviour of range-finder devices. In this work MCL (Monte-Carlo localization) was implemented based on the approach proposed in [1, 12]. We adapted that approach to take into account the new information given by this sensor. Experiments are made in a typical RoboCup environment (a 8x4 m soccer field), characterized by the lack of fix obstacles that could act as reference for a range finder sensor (as it was with the walls surrounding the field until RoboCup 2001). In this situation it is extremely hard to perform robust localization using conventional range finder devices.

D. Nardi et al. (Eds.): RoboCup 2004, LNAI 3276, pp. 97–109, 2005.
© Springer-Verlag Berlin Heidelberg 2005

2   Omnidirectional Vision as an Enhanced Range Finder

RoboCup is a strongly color coded environment: every object has an unique color associated to it. Usually, the image is color segmented before any image processing. In our system only the pixels along the rays depicted in Fig.1 are segmented into the 8 RoboCup colors [1] plus a further class that include all colors not included in the former classes (called *unknown color*). A look-up table is built to obtain a real time color segmentation. The image processing software scan the image for what we called *chromatic transitions of interest*. We are interested in *green-white*, *green-blue* and *green-yellow* transitions. These transitions are related to the structure of the RoboCup fields. In fact, lines are white, goals and corner posts are blue or yellow and the play-ground is a green carpet. To detect a colour transition is more robust with respect to colour calibration than to identify the colour of every single pixel, as reported in [11].

Fig. 1. The scanning algorithm at work: green-white chromatic transitions are highlighted with red crosses, green-yellow transitions with blue crosses, black pixels represent the sample points used for the scan that is performed in a discrete set of distances. No blue-green transitions are detected: robot is far away from the blue goal. Notice the crosses in the outer part of the mirror: this part is used for low distance measures

[1] In RoboCup environment the ball is red, the lines are white, one goal is blue and the other is yellow, the robots are black, the robots' marker are cyan and magenta.

We measure the distance of the *nearest chromatic transitions of interest* along 60 rays as shown in Fig.1. This enable our "range finder" to scan a 360 degree field of view. Our omnidirectional vision sensor is composed by a camera pointed to a multi-part mirror with a custom profile [6]. The inner part of the mirror is used to measure objects farther than 1 m away for the robot, while the outer part is used to measure objects closer than 1 m from the robot. We first scan for color transition close to the robot body in the outer mirror part, and then we scan the inner part of the image up to some maximum distance.

The distances to the nearest color transition are stored in three vectors (in the following called "*scans*"), one for each color transition. During the radial scan, we can distinguish three situations:(1) A chromatic transition of interest is found. The real distance of that point is stored in the corresponding vector; (2) there are no transitions of interest, a characteristic value called *INFINITY* is stored in the vector that mean no transition can be founded along this ray; (3) a not expected transition is found: a *FAKE_RAY* value is stored in the vector. This means something is occluding the vision sensor. All rays with *FAKE_RAY* value are discarded in the matching process (we called this *ray discrimination*). The performances of the system under occlusion are described in [7]. The scanning is not performed in a continuous way along the ray but sampling the image on a discrete subsets of image pixels corresponding to a sampling step of 4 cm in the real world.

3 Monte-Carlo Localization

The Monte-Carlo localization (MCL) is a well-known probabilistic method, in which the current location of the robot is modelled as a posterior distribution (Eq.1) conditioned on the sensor data and represented by a set of weighted particles. Each particle is an hypothesis of the robot pose, and it is weighted according to the posteriors. The posterior probability distribution of the robot pose is called also the robot belief. The belief about the robot position is updated every time the robot makes a new measurement (i.e. it grabs a new image or a new odometry measure is available). It can be described by:

$$Bel(l_t) = \alpha p(o_t|l_t) \int p(l_t|l_{t-1}, a_{t-1}) Bel(l_{t-1}) dl_{t-1} \qquad (1)$$

where $l_t = (x_t, y_t, \theta_t)$ is the robot pose at time t and a_t and o_t are respectively the sensor and the odometry readings at the time t. To calculate Eq. 1, it is necessary the knowledge of two conditional densities, called *motion model* (Sec. 3.1) and *sensor model* (Sec. 3.2). The motion model expresses the probability the robot moved to a certain position given the odometry measures (*kinematics*). The sensor model describes the probability of having a sensor measurement in a certain pose. The motion model and the sensor model depend respectively on the particular robot platform and on the particular sensor. The localization method is performed in 3 steps: (1) All particles are moved according to the motion model of the last kinematics measure; (2) The weights of the particles are determined according to the sensor model for the current sensor reading; (3) A re-sampling step is performed: high probability particles are replicated, low probability ones are discarded. The process repeats from the beginning. For more details please refer to [1, 12].

3.1   Motion Model

The motion model $p(l_t|l_{t-1}, a_{t-1})$ is a probabilistic representation of the robot kinematics, which describes a posterior density over possible successive robot poses. We implemented the MCL system on an holonomic robot, called Barney. The peculiarity of this robot is that it can move in any direction without the need of a previous rotation. Movement between two poses $l_{t-1} = (x_{t-1}, y_{t-1}, \theta_{t-1})$ and $l_t = (x_t, y_t, \theta_t)$ can so be described with (α_u, T, θ_f), where α_u is the difference of heading between the two poses, T is the translation and θ_f is the motion direction. Updating the robot position according only to the kinematics does not take into account errors given by odometry inaccuracy and possible collisions of the robot with other obstacles. Therefore, a random noise term is added to the values given by the last odometry reading. Noise is modelled with Gaussian zero centered random variables $(\Delta_\alpha, \Delta_T, \Delta_{rr}, \Delta_{rT})$. They depend on both the amount of translation and of rotation. So, the motion model can be written as:

$$\alpha'_u = \alpha_u + \Delta_\alpha(\alpha_u);$$
$$T' = T + \Delta_T(T);$$
$$\theta' = \theta + \Delta_{rr}(\theta) + \Delta_{rT}(T).$$

3.2   Sensor Model

The sensor model $p(o_t|l_t)$ describes the likelihood to obtain a certain sensor reading given a robot pose. As introduced in Sec. 3, the sensor model is used to compute the weights of the particles. For each particle j, located in the pose l_t^j, the associated weight is proportional to $p(o_t|l_t^j)$ (i.e. to the likelihood of obtaining the sensor reading o_t when the robot has pose l_t^j). To calculate $p(o_t|l_t^j)$, we need to know the "expected scan" $o(l_t)$. The expected scan is the scan an ideal noise-free sensor would measure in that pose, if in the environment there are no obstacles. Given l the robot pose, the expected scan $o(l)$ for some color transition is composed by a set of expected distances, one for each α_i, that are the angles relative to the robot of an individual sensor ray (Fig. 3): $o(l) = \{g(l, i)|0 \leq i < N\_RAYS\}$. We can compute the expected distances $g(l, i)$ for an ideal noise-free sensor using ray tracing technique considering both metric maps in Fig. 2. The likelihood $p(o_t|l_t)$ can be calculated as $p(o_t|l_t) = p(o_t|o(l_t))$. In other words, the probability $p(o_t|o(l_t))$ models the noise in the scan by the expected scan [1, 12].

When using a sonar or a laser, like in [1, 12], the expected scan is computed from a metric map of the environment. The expected scan is obtained simulating the reflections of the sonar or laser beams against the walls and the fix obstacles. In the RoboCup Middle-Size field, a similar approach was used, very effectively, by the CS Freiburgh Team [13], until RoboCup 2001. However, when in 2002

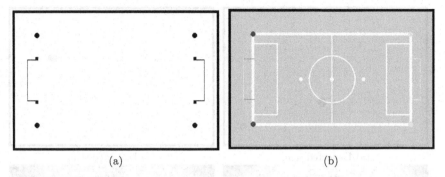

(a)   (b)

Fig. 2. The metric maps used for expected distances computation: in (a) are represented the fix obstacles, in (b) are represented all the chromatic transitions of interest of the environment

the walls surrounding the field were removed, the reliability of this approach was impaired by the lack of fix features detectable by a range-finder sensor. In Fig. 2(a), are presented the fix obstacles that a range-finder sensor could detect. In the Middle-Size field with the 2003 layout, the only detectable objects are the two goals and the four corner-posts. With the new sensor we propose, we can detect not only the fix objects in the field shown in Fig. 2(a), but also all color transitions existing in Fig. 2(b). This enable us to detect much more fix features performing a more reliable *"scan matching"*. The map in Fig. 2(b)shows the chromatic characteristics of the environment. We use this map to compute the expected scan finding with a ray-tracing approach the *nearest chromatic transition of interest* for every pose, as depicted in Fig. 3. Moreover, we use the information about the fix obstacles extracted from the map of Fig. 2(a) to improve the scanning process, e.g. if we find a yellow pixel, this is a goal or a corner-post, so it is not worth looking farther for a white line and so we stop the scanning process along this ray.

Another difference with respect to the usual range-finders is that we do not have just one scan of the environment. We have three scans for every pose of the robot: one for every chromatic transition of interest (green-white, green-blue and green-yellow, see Sec. 2). Moreover, we can filter out some rays when a fake transition is detected (i.e. a chromatic transition that we are not looking for, see Sec. 2). In Fig. 3, two examples in which are compared the expected scans (top) and the real sensor scans (bottom) is presented. In the middle is the image grabbed by the robot. On the left is depicted the scan looking for the *green-white* chromatic transition of interest, on the right the scan looking for the *green-yellow* chromatic transition of interest. Due to the image noise, it might happen that a color transition is not detected or is detected at the wrong distance or is falsely detected (as shown in Fig. 3). So, we need to create a model of the sensor's noise.

Sensor Noise. To compute $p(o|o(l))$, the first step is to model the sensor noise. We implemented a three steps process. First, we modelled the probability a single

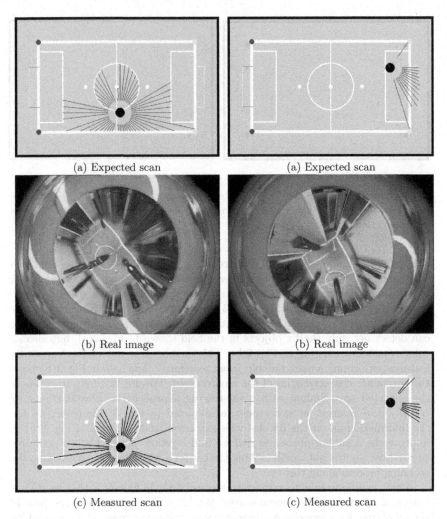

(a) Expected scan   (a) Expected scan

(b) Real image   (b) Real image

(c) Measured scan   (c) Measured scan

Fig. 3. Two examples of expected and measured scans. The one on the left for the green-white transition, the other on the right for the green-yellow transition. Given a pose, in (a) is represented the expected scan for an ideal noise-free sensor in a free environment. In (b) is shown the frame grabbed by the robot in that pose, in (c) is represented the corresponding measured scan

ray of the scan correctly detects the chromatic transition of interest. Second, we take into account all rays and we calculate a single probability value that the measured scan match the expected scan to the corresponding chromatic transitions of interest. Third, the three probability values of the three measured scans are combined to obtain a single value.

Let us describe these steps in more details. The scan performed by the sensor is composed by a set of distances, one for each α_i: $o = \{o_i | 0 \leq i < N\_RAYS\}$. To compute $p(o_i|l)$, i.e. the probability to obtain for a single ray a distance o_i given the pose l, we can consider directly the single expected distance $g(l,i)$, so we can write $p(o_i|l) = p(o_i|g(l,i))$. To calculate $p(o_i|l)$, we collected a large number of omnidirectional images in different known poses in the field (in total about 2.000 images). Then, with the scan algorithm we measure the distance of the chromatic transitions of interest (As an example, the probability density of the measured distance $p(o_i|l)$ for the green-white color transition is plotted in Fig. 4(a)). We described the measured probability density with the mixture of three probability density of Eq. 2. The numerical values of the parameters in Eq. 2 are calculated with a modified EM algorithm iteratively run on the 2000 images [2]. The resulting mixture, for the green-white transition, is plotted in Fig. 4(b). The three terms in Eq. 2 are respectively: an Erlang probability density, a Gaussian probability density and a discrete density. The Erlang variable models wrong readings in the scan caused by image noise and non-perfect color segmentation. The index n depends on the profile of the omnidirectional mirror used in the sensor. The Gaussian density models the density around the maximum likelihood region, i.e. the region around the true value of the expected distance. The discrete density represents the probability of obtaining an *INFINITY* value for the distance, as described in Sec. 2.

$$p(o_i|l) = \zeta_e\left(\frac{\beta^n o_i^{n-1} e^{-\beta o_i} 1(o_i)}{(n-1)!}\right) + \zeta_g \frac{1}{\sqrt{2\pi}\sigma} e^{\frac{-(o_i - g(l,\alpha_i))^2}{2\sigma^2}} + \zeta_d \delta(o_i - \infty) \quad (2)$$

where $\zeta_e, \zeta_g, \zeta_d$ are the mixture coefficients, with $\zeta_e + \zeta_g + \zeta_d = 1$. We computed a different mixture for every different chromatic transition.

Once the $p(o_i|l)$ is computed, it is possible to compute the probability of the whole scan given a pose l multiplying all the $p(o_i|l)$, Eq. 3. To cope with unexpected measures due to occlusion of the scans by the moving objects in the environment (i.e. the other robots and the ball), we filtered out all rays which distance o_i equal the *FAKE_RAY* value (ϕ in the formulas). This is the process called *ray discrimination*, see Sec. 2 (we discussed more in detail this topic in [7]). The detection of occluding obstacles along the rays of a scan is very frequent in the densely crowded environment of the Middle-Size RoboCup field. This rays discrimination allow us to avoid to use other techniques, e.g. *distance filters* [5], that can affect negatively the computational performance of the system.

$$p(o|l) = \prod_{\{i|o_i \neq \phi\}} p(o_i|l) = \prod_{\{i|o_i \neq \phi\}} p(o_i|g(l,i)) \quad (3)$$

3.3   Weights Calculation

Returning to Monte Carlo Localization, we are now able to compute, the weight $w^{(j)}$ associated to each particles j. We first calculate the quantity $\bar{w}^{(j)} = p(o|l_j)$ using (3). Subsequently, all $\bar{w}^{(j)}$ are normalized such that $\sum_j \tilde{w}^{(j)} = 1$

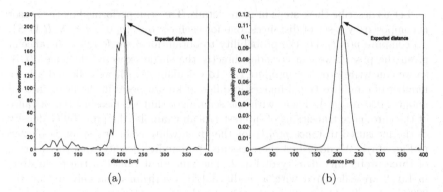

(a)   (b)

Fig. 4. In (a) the distribution of the measured distances for an expected known distance. There is a peak for the expected distance. The measures before the expected one are due to image noise. The high number of maximum distance measures means no chromatic transition was detected. In (b) the density $p(o|l)$ that represent our sensor model computed using EM-algorithm, mathematically described by Eq. 2

$$\tilde{w} = \frac{\bar{w}^{(j)}}{\sum_j \bar{w}^{(j)}} \qquad (4)$$

Our system scans the acquired image for the three chromatic transitions of interest. This ensures three scans for every frame, so three weight values are associated to every particles. To obtain a single weight value, we compute the product of the three weights (Eq. 5), and re-normalize all weights with (4) again.

$$w^{(j)} = \prod_{k=1}^{N} \tilde{w}_k^{(j)} \qquad (5)$$

In Fig. 5, we give a pictorial visualization of the weights calculated by the three different scans of the three chromatic transition of interest. The real pose of the robot is marked with the arrow. Higher weight values are depicted as darker points, lower weight values are depicted as lighter points. In Fig. 5 (a), are represented the weight contributions calculated by the scan looking for the green-white transition. One can notice that, due to the symmetry of the white lines in the field two symmetric positions resulted to have high likelihood. In Fig. 5 (b), are depicted the weight contributions calculated by the scan looking for the green-blue transition. One can notice that all positions far away from the blue goal have a high likelihood, because no green-blue transition was found in the image scan. In Fig. 5 (c), are represented the weight contributions calculated by the scan looking for the green-yellow transition. One can notice there is an approximate symmetry around at the yellow goal. All these contributions are combined with Eq.5 to calculate the overall weights and depicted in Fig. 5 (d). Here, the weights with higher values are clustered only around the actual position of the robot.

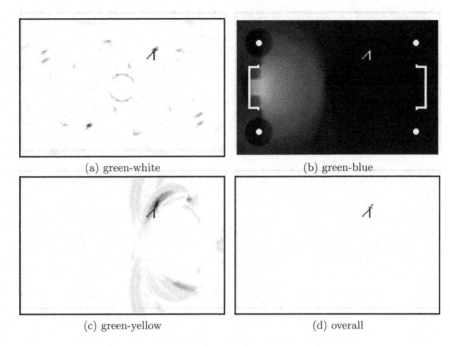

(a) green-white (b) green-blue

(c) green-yellow (d) overall

Fig. 5. Probability distributions $p(o_t|l_t)$ for all possible positions $l = (x, y, \theta)$ of the robot in the field given the scans of a single image. Darker points corresponds to high likelihood. The arrow represents the actual robot pose. In (a) is represented the probability given the scan for transition white, in (b) for transition blue, in (c) for transition yellow, in (d) the three are combined

In order to improve the performance of the system, the distances in the environment are quantized in a grid of 5x5 cm cells, in a way similar to [5]. The expected distances for all poses and the probabilities $p(o_i|g(l, i))$ for all $g(l, i)$ can be pre-computed and stored in six (two *for each* chromatic transition) look-up tables. In this way the probability $p(o_i|l)$ can be quickly computed with two look-up operations, this enables our system to work in real-time at 10 Hz.

4   Experiments

We evaluated our approach on an holonomic custom-built platform, in a 8x4 m soccer field. The robot was equipped with the omnidirectional sensor described in Sec. 2. We tested the system on five different paths (an example path is shown Fig. 6). For each path we collected a sequence of omnidirectional images with the ground truth positions where those images were grabbed and with the odometry readings between two consecutive positions. In order to take into account the odometry errors, robot movements were performed by remote robot control. We

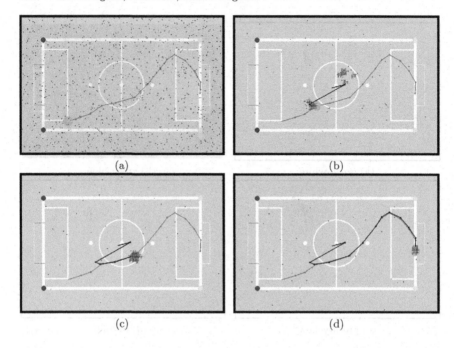

Fig. 6. A sequence of global localization using 1000 particles: the gray circle represents actual robot pose, the red line represents ground-truth path, the black line represents the estimated path of the robot, the black points represent the particles. In (a) particles are uniformly distributed (no knowledge is available on robot position), in (b), after moving 2 meters away and grabbing 4 images and getting 4 odometry readings, the particles are condensed around three possible poses. In (c), after 4 meters, 6 images and 6 odometry readings, uncertainty is solved and particles are condensed around the actual pose of the robot. In (d) after 14 steps: the position of the robot is well tracked. The particles distributed in the environment are the particles scattered to solve the kidnapped robot problem

tested our algorithms using different amount of samples calculating the mean localisation error for the three fundamental localization problems: (1) global localization (the robot must be localized without any a priori knowledge on the actual position of the robot, i.e. Fig. 6 (a)(b)), (2) position tracking (a well localized robot must maintain the localization, i.e. Fig. 6 (c)(d)) and (3) kidnapped robot. The kidnapped robot is the problem in which a well-localized robot is moved to some other pose without any odometry information: this problem can frequently occur in an high populated environment like RoboCup, where often robots push each other attempting to win the ball.

In Fig. 7(a) is shown the error for a global localization sequence using 100, 500, 1000, 10000 samples in the same reference path. The reactivity and the accuracy of the localisation system increase with the number of samples, but

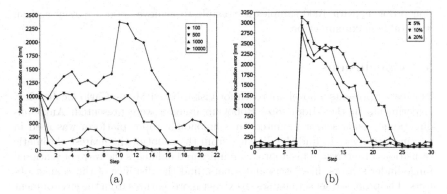

(a)   (b)

Fig. 7. The plots compares the performance of the system varying the number of the samples used in MCL. In (a) global localization errors for a fixed path with different amount of samples, in (b) re-localization after kidnapped robot problem with different rate of uniformly distributed particles. Notice that with 20% the re-localization is faster but the average position tracking error is higher

a large number of samples like 10000 increases dramatically the computational load. A number of 1000 particles is perfectly compatible with real-time requirements and assures a robust and accurate localisation. In Fig. 7(b) is shown the error for a kidnapped robot episode using 1000 samples and different rate of samples uniformly distributed in the environment [4]. With a higher rate of samples scattered in the environment the re-localization is faster (there are more possibility that the samples are distributed close to the robot position), but the average error is higher due to the lower number of sample clustered closed to the robot pose during position tracking.

Finally, we tested our approach in the conventional situation of the positions tracking: in Fig. 8 is shown the average and the maximum error for all our reference paths using different amount of samples. Like in the global posi-

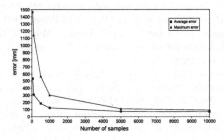

Fig. 8. Statistical evaluation of our system in the position tracking problem for all our reference paths. Accuracy (average error end maximum error) is represented for different amount of samples (50,100,500,1000,5000,10000)

tion problem, with 1000 samples is possible to achieve good accuracy and an acceptable maximum error.

5   Conclusions

This paper presents a novel approach for vision-based Monte Carlo localization. Experiments in the Middle-Size RoboCup domain were presented. An omnidirectional vision sensor mounted on a holonomic robot platform was used in an innovative way to find the distance from the robot of the nearest chromatic transitions of interest. This approach mimics and enhances the way conventional range-finder sensors, like lasers and sonars, find the distance of the nearest objects. The proposed sensor enables to extract more features from the environment thanks to the capability to distinguish different chromatic transitions of interest. The well-known Monte Carlo localization technique was adapted to the characteristics of the sensor. The EM algorithm was used to extract the parameters of the sensor model from experimental data. We presented experiments in a actual Middle-Size RoboCup field to prove the robustness and the accuracy of our technique. We are porting the system to other environment than RoboCup. Depending on the environment, different chromatic transitions of interest can be identified. The system is designed to automatically recalculate the expected scans given the metric and chromatic maps of the new environment.

References

1. F. Dellaert, D. Fox, W. Burgard, and S. Thrun. Monte carlo localization for mobile robots. In *Proc. of the IEEE InternationalConference on Robotics & Automation*, 1999.
2. A. P. Dempster, N. M. Laird, and D. B. Rubin. Maximum likelihood from incomplete data via the em algorithm. In *Journal of the Royal Statistical Society*, volume 39 of *B*, pages 1–38. 1977.
3. S. Enderle, M. Ritter, D. Fox, S. Sablatng, G. Kraetzschmar, and G. Palm. Soccer-robot locatization using sporadic visual features. In E. Pagello, F. Groen, T. Arai, R. Dillman, and A. Stentz, editors, *Proceedings of the 6th International Conference on Intelligent Autonomous Systems (IAS-6)*. IOS Press, 2000.
4. D. Fox, W. Burgard, F. Dellaert, and S. Thrun. Monte carlo localization: Efficient position estimation for mobile robots. In *Proc. of the National Conferenceon Artificial Intelligence*, 1999.
5. D. Fox, W. Burgard, and S. Thrun. Markov localization for mobile robots in dynamic environments. *Journal of Artificial Intelligence Research*, 11, 1999.
6. E. Menegatti, F. Nori, E. Pagello, C. Pellizzari, and D. Spagnoli. Designing an omnidirectional vision system for a goalkeeper robot. In A. Birk, S. Coradeschi, and S. Tadokoro, editors, *RoboCup-2001: Robot Soccer World Cup V.*, pages 78–87. Springer, 2002.
7. E. Menegatti, A. Pretto, E. Pagello, Testing omnidirectional vision-based Monte-Carlo Localization under occlusion In *Proc. of IEEE/RSJ International Conference on Intelligent Robots and Systems (IROS 2004)*, September 2004, Sendai- Japan, pp. (to appear).

8. E. Menegatti, M. Zoccarato, E. Pagello, and H. Ishiguro. Hierarchical image-based localisation for mobile robots with monte-carlo localisation. In *Proc. of European Conference on Mobile Robots (ECMR'03)*, pages 13–20, September 2003.

9. E. Menegatti, M. Zoccarato, E. Pagello, and H. Ishiguro. Image-based monte-carlo localisation with omnidirectional images. In *Robotics and Autonomous Systems, Elsevier*, page (to appear), 2004.

10. T. Röfer and M. Jüngel. Vision-based fast and reactive monte-carlo localization. In *IEEE International Conference on Robotics and Automation*, 2003.

11. T. Rofer and M. Jungel. Vision-based fast and reactive monte-carlo localization. In *Proc. of International Conference on Robotics and Automation (ICRA-2003)*, 2003.

12. S. Thrun, D. Fox, W. Burgard, and F. Dellaert. Robust monte carlo localization for mobile robots. *Artificial Intelligence*, 128(1-2):99–141, 2000.

13. T. Weigel, J.-S. Gutmann, M. Dietl, A. Kleiner, , and B. Nebel. Cs freiburg: Coordinating robots for successful soccer playing. *IEEE Transactions on Robotics and Automation (T. Arai, E. Pagello, L. Parker, (Eds.))*, 18(5):685–699, 2002.

14. J. Wolf, W. Burgard, and H. Burkhardt. Using an image retrieval system for vision-based mobile robot localization. In *Proc. of the International Conference on Image and Video Retrieval (CIVR)*, 2002.

Fuzzy Self-Localization Using Natural Features in the Four-Legged League

D. Herrero-Pérez[1], H. Martínez-Barberá[1], and A. Saffiotti[2]

[1] Dept. Information and Communication Engineering,
University of Murcia, 30100 Murcia, Spain
`dherrero@dif.um.es, humberto@um.es`
[2] Dept. of Technology, Örebro University, 70218 Örebro, Sweden
`asaffio@aass.oru.se`

Abstract. In the RoboCup four-legged league, robots mainly rely on artificial coloured landmarks for localisation. As it was done in other leagues, artificial landmarks will soon be removed as part of the RoboCup push toward playing in more natural environments. Unfortunately, the robots in this league have very unreliable odometry due to poor modeling of legged locomotion and to undetected collisions. This makes the use of robust sensor-based localization a necessity. We present an extension of our previous technique for fuzzy self-localization based on artificial landmarks, by including observations of features that occur naturally in the soccer field. In this paper, we focus on the use of corners between the field lines. We show experimental results obtained using these features together with the two nets. Eventually, our approach should allow us to migrate from landmarks-only to line-only localisation.

Keywords: Autonomous robots, fuzzy logic, image processing, localization, state estimation.

1 Introduction

The current soccer field in the Four-Legged Robot League has a size of approximately $4,5m \cdot 3m$, and the only allowed robot is the Sony AIBO [10]. The exteroceptive sensor of the robot is a camera, which can detect objects on the field. Objects are color coded: there are four uniquely colored beacons, two goal nets of different color, the ball is orange, and the robots wear colored uniforms.

However, in a real soccer field there are not characteristic colored cues. The rules of RoboCup are gradually changed year after year in order to push progress toward the final goal. Removal of the artificial colored beacons will be the next step in this direction. Accordingly, some preliminary development has been done by some teams in this league to allow the robot to self-localize without using the artificial beacons.

For instance, the German Team uses a sub-sampling technique to detect pixels that belong to the field lines. These pixels are used in a Monte-Carlo localisation (MCL) schema [4]. MCL is a probabilistic method, in which the current location

D. Nardi et al. (Eds.): RoboCup 2004, LNAI 3276, pp. 110–121, 2005.

of the robot is modeled as the density of a set of particles. Each particle can be seen as the hypothesis of the robot being located at that position. Using only a small number of samples, it increases the stability of the localization by maintaining separate probabilities for different edge types for each sample. These probabilities are only adapted slowly.

This paper describes the process of self-localization using the field lines as a source of features. In fact, what it is used are the corners produced by the intersection of the field lines, instead of the classical approach of using the line segments. Then, these corners are treated as natural landmarks in a technique based on [5], which uses fuzzy logic to account for errors and imprecision in visual recognition of landmarks and nets, and for the uncertainty in the estimate of robot's displacement. This technique allows for large odometric errors and inaccurate observations with excellent results. However, it should be noted that the idea of corner-based localisation presented here could also be incorporated into other localisation approaches, like MCL.

2 Perception

The AIBO robots [10] use a CCD camera as the main exteroceptive sensor. The perception process is in charge of extracting convenient features of the environment from the images provided by the camera. As the robot will localize relying on the extracted features, both the amount of features detected and their quality will clearly affect the process. Currently the robots rely only on coloured landmarks for localization, and thus the perception process is based on color segmentation for detecting the different landmarks. Soon artificial landmarks will be removed, as it was done in other leagues. For this reason, our short term goal is to augment the current landmark based localization with information obtained from the field lines, and our long term goal is to rely only on the field lines for localisation.

Because of the League rules all the processing must be done on board and for practical reasons it has to be performed in real time, which prevents us from using time consuming algorithms. A typical approach for detecting straight lines in digital images is the Hough Transform and its numerous variants. The various variants have been developed to try to overcome the major drawbacks of the standard method, namely, its high time complexity and large memory requirements. Common to all these methods is that either they may yield erroneous solutions or they have a high computational load.

Instead of using the field lines as references for the self-localization, we decided to use the corners produced by the intersection of the field lines (which are white). The two main reasons for using corners is that they can be labeled (depending on the type of intersection) and they can be tracked more appropriately given the small field of view of the camera. In addition, detecting corners can be more computationally efficient than detecting lines. There are several approaches, as reported in the literature, for detecting corners. They can be broadly divided into two groups:

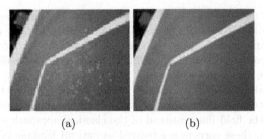

(a)   (b)

Fig. 1. Brightness gradient-based detector. (a) From raw channel. (b) From smoothed channel. The detected corners are indicated by a white small square

- Extracting edges and then searching for the corners. An edge extraction algorithm is pipelined with a curvature calculator. The points with maximum curvature (partial derivative over the image points) are selected as corners.
- Working directly on a gray-level image. A matrix of "cornerness" of the points is computed (product of gradient magnitude and the rate of change of gradient direction), and then points with a value over a given threshold are selected as corners.

We have evaluated two algorithms that work on gray-level images and detect corners depending on the brightness of the pixels, either by minimizing regions of similar brightness (SUSAN) [8] or by measuring the variance of the directions of the gradient of brightness [9]. These two methods produce corners, without taking into account the color of the regions. As we are interested in detecting field lines corners, the detected corners are filtered depending on whether they come from a white line segment or not.

The gradient based method [9] is more parametric, produces more candidate points and requires similar processing capabilities than SUSAN [8], and thus it is the one that we have selected to implement corner detection in our robots.

Because of the type of camera used [10], there are many problems associated to resolution and noise. The gradient based method detects false corners in straight lines due to pixelation. Also, false corners are detected over the field due to the high level of noise. These effects are shown in Fig. 1 (a). To cope with the noise problem, the image is filtered with a smoothing algorithm. Fig. 1(b) shows the reduction in noise level, and how it eliminates false detections produced by this noise (both for straight lines and the field).

The detected corners are then filtered so that we keep only the relevant ones, those produced by the white lines over the field. For applying this color filter, we first segment the raw image. Fig. 2 (b) shows the results obtained using a threshold technique for a sample image. We can observe that this technique is not robust enough for labeling the pixels surrounding the required features. Thus, we have integrated thresholding with a region-based method, namely the Seed Region Growing (SRG), by using the pixels classified by the a conservative thresholding as seeds from which we grow color regions. (See [11] for details.) The resulting regions for our image are shown in Fig. 2 (c). Finally, we apply a

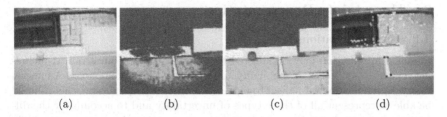

(a)   (b)   (c)   (d)

Fig. 2. Feature detection. (a) RGB image. (b) Segmented image by thresholding. (c) Segmented image by seeded region growing. (d) Gradient-based detected corners (white) and color-based filtered corners (black). The two detected corner-features are classified as a **C** and a **T**, respectively

color filter for all corners obtained with the gradient-based method. We show in Fig. 2 (d) the corners obtained with the gradient-based method (white) and the corners that comply with the color conditions (black).

Once we have obtained the desired corner pixels, these are labeled by looking at the amount of field pixels (carpet-color) and line or band pixels (white) in a small window around the corner pixel. Corners are labeled according to the following categories.

Open Corner. A corner pixel surrounded by many carpet pixels, and by a number of white pixels above a threshold.

Closed Corner. A corner pixel surrounded by many white pixels, and by a number of carpet pixels above a threshold.

Net Closed Corner. A corner surrounded by many white pixels, and by a number of carpet and net (blue or yellow) pixels above a threshold.

Note that in order to classify a corner pixel, we only need to explore the pixels in its neighborhood. From these labeled corners, the following features are extracted.

Type C. An **Open corner** nearby of a **closed corner**. This feature can be detected in the goal keeper area of field.

Type T-field. Two **closed corner**. Produced by the intersection of the walls and the inner field lines.

Type T-net. A **closed corner** nearby of a **net closed corner**. Produced by the intersection of goal field lines and the net.

In Fig. 2 (d), four detected corner pixels have been combined into two corner-features, respectively classified as a **C** and a **T-field**.

The resulting corner-features, together with the landmarks and the nets, are used for localizing a robot in the field. In the rest of this paper, we show how to represent the uncertainty associated to these features, and how to use these features in our fuzzy localization technique in order to obtain an estimate of the robot position.

3   Uncertainty Representation

3.1   Fuzzy Locations

Location information may be affected by different types of uncertainty, including vagueness, imprecision, ambiguity, unreliability, and random noise. An uncertainty representation formalism to represent locational information, then, should be able to represent all of these types of uncertainty and to account for the differences between them. Fuzzy logic techniques are attractive in this respect [6]. We can represent information about the location of an object by a fuzzy subset μ of the set X of all possible positions [12, 13]. For instance, X can be a 6-D space encoding the (x, y, z) position coordinates of an object and its (θ, ϕ, η) orientation angles. For any $x \in X$, we read the value of $\mu(x)$ as the degree of possibility that the object is located at x given the available information.

Fig. 3 shows an example of a fuzzy location, taken in one dimension for graphical clarity. This can be read as "the object is believed to be approximately at θ, but this belief might be wrong". Note that the unreliability in belief is represented by a uniform bias b in the distribution, indicating that the object might be located at any other location. Total ignorance in particular can be represented by the fuzzy location $\mu(x) = 1$ for all $x \in X$.

3.2   Representing the Robot's Pose

Following [5], we represent fuzzy locations in a discretized format in a position grid: a tessellation of the space in which each cell is associated with a number in $[0, 1]$ representing the degree of possibility that the object is in that cell. In our case, we use a 3D grid to represent the robot's belief about its own pose, that is, its (x, y) position plus its orientation θ. A similar approach, based on probabilities instead of fuzzy sets, was proposed in [3].

This 3D representation has the problem of having a high computation complexity, both in time and space. To reduce complexity, we adopt the approach proposed by [5]. Instead of representing all possible orientations in the grid, we use a 2D grid to represent the (x, y) position, and associate each cell with a trapezoidal fuzzy set $\mu_{x,y} = (\theta, \Delta, \alpha, h, b)$ that represents the uncertainty in the robot's orientation. Fig. 3 shows this fuzzy set. The θ parameter is the center, Δ

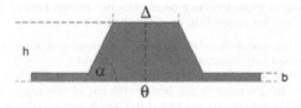

Fig. 3. Fuzzy set representation of an angle measurement θ

is the width of the core, α is the slope, h is the height and b is the bias. The latter parameter is used to encode the unreliability of our belief as mentioned before.

For any given cell (x, y), $\mu_{x,y}$ can be seen as a compact representation of a possibility distribution over the cells $\{(x, y, \theta) \mid \theta \in [-\pi, \pi]\}$ of a full 3D grid. The reduction in complexity is about two orders of magnitude with respect to a full 3D representation (assuming a angular resolution of one degree). The price to pay is the inability to handle multiple orientation hypotheses on the same (x, y) position — but we can still represent multiple hypotheses about different positions. In our domain, this restriction is acceptable.

3.3   Representing the Observations

An important aspect of our approach is the way to represent the uncertainty of observations. Suppose that the robot observes a given feature at time t. The observed range and bearing to the feature is represented by a vector $\vec{r}$. Knowing the position of the feature in the map, this observation induces in the robot a belief about its own position in the environment. This belief will be affected by uncertainty, since there is uncertainty in the observation.

In our domain, we consider three main facets of uncertainty. First, *imprecision* in the measurement, i.e., the dispersion of the estimated values inside an interval that includes the true value. Imprecision cannot be avoided since we start from discretized data (the camera image) with limited resolution. Second, *unreliability*, that is, the possibility of outliers. False measurements can originate from a false identification of the feature, or from a mislabelling. Third, *ambiguity*, that is, the inability to assign a unique identity to the observed feature since features (e.g., corners) are not unique. Ambiguity in observation leads to a multi-modal distribution for the robot's position.

All these facets of uncertainty can be represented using fuzzy locations. For every type of feature, we represent the belief induced a time t by an observation $\vec{r}$ by a possibility distribution $S_t(x, y, \theta | \vec{r})$ that gives, for any pose (x, y, θ), the degree of possibility that the robot is at (x, y, θ) given the observation $\vec{r}$. This distribution constitutes our *sensor model* for that specific feature.

The shape of the $S_t(x, y, \theta | \vec{r})$ distribution depends on the type of feature. In the case of net observations, this distribution is a circle of radius $|\vec{r}|$ in the (x, y) plane, blurred according to the amount of uncertainty in the range estimate. Fig. 4 and 5 show an example of this case. In the figure, darker cells indicate higher levels of possibility. We only show the (x, y) projection of the possibility distributions for graphical clarity.

Note that the circle has a roughly trapezoidal section. The top of trapezoid (core) identifies those values which are fully possible. Any one of these values could equally be the real one given the inherent imprecision of the observation. The base of the trapezoid (support) identifies the area where we could still possibly have meaningful values, i.e., values outside this area are impossible given the observation. In order to account for unreliability, then, we include a small uniform bias, representing the degree of possibility that the robot is "somewhere else" with respect to the measurement.

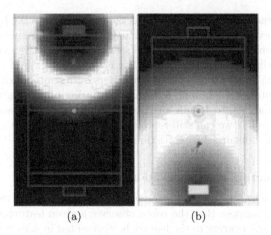

(a)   (b)

Fig. 4. Belief induced by the observation of a blue net (a) and a yellow net (b). The triangle marks the center of gravity of the grid map, indicating the most likely robot localization

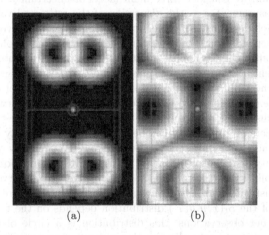

(a)   (b)

Fig. 5. Belief induced by the observation of a feature of type C (a) and T (b). Due to symmetry of the field, the center of gravity is close to the middle of the field

The $S_t(x, y, \theta | \vec{r})$ distribution induced by a corner-feature observation is the union of several circles, each centered around a feature in the map, since our simple feature detector does not give us a unique ID for corners. Fig. 5 shows an example of this. It should be noted that the ability to handle ambiguity in a simple way is a distinct advantage of our representation. This means that we do not need to deal separately with the data association problem, but this is automatically incorporated in the fusion process (see below). Data association is one of the unsolved problems in most current self-localization techniques, and one of the most current reasons for failures.

4   Fuzzy Self-Localization

Our approach to feature-based self-localization extends the one proposed by Buschka *et al* in [5], who relied on unique artificial landmarks. Buschka's approach combines ideas from the Markov localization approach proposed by Burgard in [3] with ideas from the fuzzy landmark-based approach technique proposed by Saffiotti and Wesley in [7].

The robot's belief about its own pose is represented by a distribution G_t on a $2\frac{1}{2}$D possibility grid as described in the previous section. This representation allows us to represent, and track, multiple possible positions where the robot might be. When the robot is first placed on the field, G_0 is set to 1 everywhere to represent total ignorance. This belief is then updated according to the typical predict-observe-update cycle of recursive state estimators as follows.

Predict. When the robot moves, the belief state G_{t-1} is updated to G_t using a model of the robot's motion. This model performs a translation and rotation of the G_{t-1} distribution according to the amount of motion, followed by a uniform blurring to account for uncertainty in the estimate of the actual motion.

Observe. The observation of a feature at time t is converted to a possibility distribution S_t on the $2\frac{1}{2}$ grid using the sensor model discussed above. For each pose (x, y, θ), this distribution measures the possibility of the robot being at that pose given the observation.

Update. The possibility distribution S_t generated by each observation at time t is used to update the belief state G_t by performing a fuzzy intersection with the current distribution in the grid at time t. The resulting distribution is then normalized.

If the robot needs to know the most likely position estimate at time t, it does so by computing the center of gravity of the distribution G_t. A reliability value for this estimate is also computed, based on the area of the region of G_t with highest possibility and on the minimum bias in the grid cells. This reliability value is used, for instance, to decide to engage in an active re-localization behavior.

In practice, the predict phase is performed using tools from fuzzy image processing, like fuzzy mathematical morphology, to translate, rotate and blur the possibility distribution in the grid [1, 2]. The intuition behind this is to see the fuzzy position grid as a gray-scale image.

For the update phase, we update the position grid by performing pointwise intersection of the current state G_t with the observation possibility distribution $S_t(\cdot|r)$ at each cell (x, y) of the position grid. For each cell, this intersection is performed by intersecting the trapezoid in that cell with the corresponding trapezoid generated for that cell by the observation. This process is repeated for all available observations. Intersection between trapezoids, however, is not necessarily a trapezoid. For this reason, in our implementation we actually compute

the outer trapezoidal envelope of the intersection. This is a conservative approximation, in that it may over-estimate the uncertainty but it does not incur the
risk of ruling out true possibilities.

There are several choices for the intersection operator used in the update
phase, depending on the independence assumptions that we can make about the
items being combined. In our case, since the observations are independent, we
use the product operator which reinforces the effect of consonant observations.

Our self-localization technique has nice computational properties. Updating,
translating, blurring, and computing the center of gravity (CoG) of the fuzzy
grid are all linear in the number of cells. In the RoboCup domain we use a
grid of size 36×54, corresponding to a resolution of 10 cm (angular resolution
is unlimited since the angle is not discretized). All computations can be done
in real time using the limited computational resources available on-board the
AIBO robot.

(a)   (b)   (c)

Fig. 6. Feature detection in different states. First the opponent net is detected (upper
row), and then two **C**-type corners (middle and bottom rows). (a) Raw image. (b)
Segmented image. (c) Gradient-based detected corners (white) and color-based filtered
corners (black)

5   Experimental Results

To show how the self-localisation process works, we present an example generated from the goal keeper position. Let's suppose that the robot starts in its own area, more or less facing the opposite net (yellow). At this moment the robot has a belief distributed along the full field – it does not know its own location. Then the robot starts scanning its surroundings by moving its head from left to right. As soon as a feature is perceived, it is incorporated into the localization process.

When scanning, the robot first detects a net and two features of **type C** (Fig.6). The localization information is shown in (Fig.7), where the beliefs associated to the detection are fused. The filled triangle represents the current estimate.

In order to cope with the natural symmetry of the field, we use unique features, like the nets are. When the robot happens to detect the opposite net (yellow), it helps to identify in which part of the field the robot is currently in, and the fusion with the previous feature based location gives a fairly good estimate of the robot position.

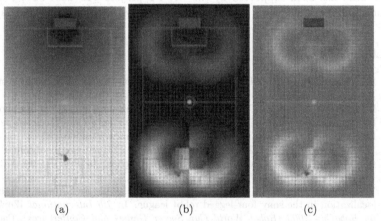

(a)   (b)   (c)

Fig. 7. Belief induced by the observation of: (a) the opponent net, (b) the first **C**-corner feature, and (c) the second **C**-corner feature. The initial position is fully unknown (belief is distributed uniformly over the full field)

6   Conclusions

The fuzzy position grid approach [5] provides an effective solution to the problem of localization of a legged robot in the RoboCup domain. In this domain motion estimates are highly unreliable, observations are uncertain, accurate sensor models are not available, and real time operation is of essence. The fuzzy position grid approach approach has been shown to work in real matches using artificial landmarks and the nets. In this paper, we have extended this approach to use naturally occurring features like corners between the field lines. Corner-feature recognition was based on gray-level image processing for detecting corner pixels, and on colors to reject corners that do not come from field lines and to classify different types of corner-features.

The main advantage of this approach, given the current RoboCup rules, is the ability of having more references for guidance, and thus the amount of time spent for looking for the colored landmarks is reduced because when the robot is playing for the ball or aiming at a net, at the same time it can detect corners. In addition, in the near future colored landmarks will be eliminated from the RoboCup fields, and techniques based on natural features will be of paramount importance. The experimental results presented in this paper show that our technique is suitable in that respect.

Acknowledgments

This work has been supported by the CICYT project TIC 2001-0245-C02-01, Spanish Ministry of Science and Technology, and by the Swedish KK Fundation.

References

1. I. Bloch and H. Maître. Fuzzy mathematical morphologies: a comparative study. *Pattern Recognition*, 28(9):1341–1387, 1995.
2. I. Bloch and A. Saffiotti. Why robots should use fuzzy mathematical morphology. In *Proc. of the 1st Int. ICSC-NAISO Congress on Neuro-Fuzzy Technologies*, La Havana, Cuba, 2002. Online at http://www.aass.oru.se/~asaffio/.
3. W. Burgard, D. Fox, D. Hennig, and T. Schmidt. Estimating the absolute position of a mobile robot using position probability grids. In *Proc. of the National Conference on Artificial Intelligence*, 1996.
4. W. Burgard, D. Fox, D. Hennig, and T. Schmidt. Fast and robust edge-based localization in the sony four-legged robot league. In *7th International Workshop on RoboCup 2003 (Robot World Cup Soccer Games and Conferences)*, Padova, Italy, 2003.
5. P. Buschka, A. Saffiotti, and Z. Wasik. Fuzzy landmark-based localization for a legged robot. In *Proc. of the IEEE/RSJ Intl. Conf. on Intelligent Robots and Systems (IROS)*, pages 1205–1210, Takamatsu, Japan, 2000. Online at http://www.aass.oru.se/~asaffio/.
6. A. Saffiotti. The uses of fuzzy logic in autonomous robot navigation. *Soft Computing*, 1(4):180–197, 1997. Online at http://www.aass.oru.se/~asaffio/.
7. A. Saffiotti and L. P. Wesley. Perception-based self-localization using fuzzy locations. In *Proc. of the 1st Workshop on Reasoning with Uncertainty in Robotics*, pages 368–385, Amsterdam, NL, 1996.

8. S.M. Smith and J.M. Brady. Susan - a new approach to low level image processing. *International Journal of Computer Vision*, 1(23):45–78, 1997.

9. E. Sojka. A new and efficient algorithm for detecting the corners in digital images. In *Proc. 24th DAGM Symposium*, pages 125–132, Springer, LNCS 2449, Berlin, NY, 2002.

10. Sony. Sony AIBO robots. http://www.aibo.com.

11. Z. Wasik and A. Saffiotti. Robust color segmentation for the robocup domain. In *Proc. of the Int. Conf. on Pattern Recognition (ICPR)*, Quebec City, Quebec, CA, 2002. Online at http://www.aass.oru.se/~asaffio/.

12. L. A. Zadeh. Fuzzy sets. *Information and Control*, 8:338–353, 1965.

13. L. A. Zadeh. Fuzzy sets as a basis for a theory of possibility. *Fuzzy Sets and Systems*, 1:3–28, 1978.

A Behavior Architecture for Autonomous Mobile Robots Based on Potential Fields*

Tim Laue and Thomas Röfer

Bremer Institut für Sichere Systeme, Technologie-Zentrum Informatik,
FB 3, Universität Bremen, Postfach 330 440, 28334 Bremen, Germany
{timlaue, roefer}@tzi.de

Abstract. This paper describes a behavior-based architecture which integrates existing potential field approaches concerning motion planning as well as the evaluation and selection of actions into a single architecture. This combination allows, together with the concept of competing behaviors, the specification of more complex behaviors than the usual approach which is focusing on behavior superposition and is mostly dependent on additional external mechanisms. The architecture and all methods presented in this paper have been implemented and applied to different robots.

1 Introduction

Artificial potential fields, developed by [1] and also in detail described by [2, 3], are a quite popular approach in robot motion planning because of their capability to act in continuous domains in real-time. By assigning repulsive force fields to obstacles and an attractive force field to the desired destination, a robot can follow a collision-free path via the computation of a motion vector from the superposed force fields. Especially in the RoboCup domain, there also exist several applications of potential functions for the purposes of situation evaluation and action selection [4, 5, 6].

Most approaches consider potential fields only as a tool being embedded in another architecture, e. g. a planner [4, 7]. In these cases, the potential field implementations are limited to special tasks such as obstacle avoidance or the computation of an appropriate passing position. Due to some limitations, which are discussed in Sect. 2.1, the standard approach is not able to produce a behavior of an adequate complexity for several tasks, e. g. robot soccer.

The approach presented here combines existing approaches in a behavior-based architecture [3] by realizing single competing behaviors as potential fields. The architecture has generic interfaces allowing its application on different platforms for a variety of tasks. The process of behavior specification is realized via a description language based on XML.

---

* The Deutsche Forschungsgemeinschaft supports this work through the priority program "Cooperating teams of mobile robots in dynamic environments".

D. Nardi et al. (Eds.): RoboCup 2004, LNAI 3276, pp. 122–133, 2005.

This paper, which is based on the works of [8], is organized as follows: Section 2 gives an overview of the structure of the architecture, the mechanisms for behavior selection and the way of behavior specification, Section 3 shows the possibilities of modeling the environment, the Sections 4 and 5 describe the used approaches and some extensions for motion planning and action evaluation. Previous applications of the architecture are presented in Section 6 and the paper ends with the conclusion in Section 7.

2   Architecture

The architecture described in this section represents a framework for modeling the environment and a set of behaviors. It is also responsible for the task of behavior selection.

2.1   Competition Instead of Exclusive Superposition

As above-mentioned, potential fields are based on the superposition of force fields. Being a quite smart technique for obstacle avoidance, this approach fails when accomplishing more complex tasks including more than one possible goal position. An obvious example is the positioning of a goalkeeper: The usage of attractive force fields for its standard defense position as well as for a near ball to be cleared would lead to a partial erasement of the fields causing an unwanted behavior. This problem could be solved by using an external entity which selects the most appropriate goal, but this proceeding would affect the claim of a stand-alone architecture. Therefore, different tasks have to be splitted into different competing behaviors. This applies also to tasks based on action evaluation, especially since they use a different computation scheme, as explained in Sect. 5.

The approach of action selection by [9,3], as shown in Fig. 1, has been considered as being most suitable for this architecture. A number of independent behaviors without any fixed hierarchy as in [10] compete for execution

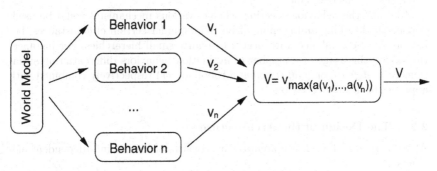

Fig. 1. The basic scheme of behavior selection. Only the behavior the output v of which has the highest activation value $a(v)$ will be executed

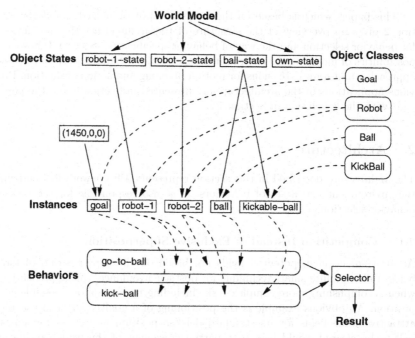

Fig. 2. A simplified example of a behavior for playing soccer, showing the elements of the architecture and their interdependencies

by the respective computation of activation values which represent the current appropriateness.

As potential fields are rather based on functions and vectors than on symbolic representations [11], there are several facile ways to gain activation values from single behaviors, e. g. the value of a potential function at a certain position or the length of a computed motion vector. Constant activation values for standard behaviors are also possible.

To refine the behavior selection scheme, various mechanisms could be used. Among the blocking and keeping of behaviors under certain circumstances, behaviors can be combined with others to realize small hierarchies. For instance, this allows the usage of a number of evaluation behaviors differentiating situations (e. g. *defense* or *midfield play* in robot soccer) respectively combined with appropriate motion behaviors.

2.2   The Design of the Architecture

As to be seen in Fig. 2, the architecture consists of several interdependent elements:

Object States. An object state contains information about an object in the environment. The appropriate data has to be converted from the robot's sensor layer, as described in Sect. 2.3.

Object Classes. An object class is used as a template for object instances. Due to its complex configuration, it is described in a separate section (3).

Object Instances. Obtaining all necessary parameters and data from classes and object states, instances may be assigned to one or more behaviors. In the case of a world model containing absolute positions, their state can also be static.

Behaviors. In this architecture a behavior equals a potential field for motion planning respectively an action evaluation via potential functions. These methods are described in more detail in the Sections 4 and 5.

2.3   Embedding the Architecture in Robot Systems

To use the architecture within a special robot system, two components have to be programmed to connect the generic interfaces with both the robot's sensor layer and the execution layer. Concerning the sensor component, it does not matter whether the robot system uses localization techniques to compute a world model using absolute positions or just reacts on sensor measurements, as long as an appropriate mapping of the data structures may be found. The second component has to be an interpreter which translates the abstract motion parameters or activates chosen actions, e. g. a kick or a ball catching motion, both by setting the corresponding motor parameters.

a)
```
<object name="Opponent-Robot" type="repulsive">
    <asymptotic-function range="200"
                         at-zero="1000"
                         const-interval="1"/>
    <point-field/>
    <circle radius="90"/>
</object>
```

b)
```
<motionfield name="go-to-ball">
    <return-const value="-1"/>
    <include name="ball"/>
    <include name="own-penalty-area"/>
    <include-group name="all-robots"/>
</motionfield>
```

Fig. 3. Extracts from an XML behavior specification: a) An object class describing the attributes of an opponent robot. b) A motion behavior for moving towards a ball

2.4   Using XML for Behavior Specification

As afore mentioned, the architecture consists of several elements. Specifying a behavior is therefore equivalent to specifying all entities and their interdependencies. Such modeling tasks may be managed in a better way by using external configuration files than by programming all needed elements in a programming language such as e. g. C++. Thus, a format for behavior specification has been described in XML. Figure 3 shows extracts from an example behavior. The reasons to use XML instead of defining a new grammar from scratch are its structured format as well as the big variety and quality of existing editing, validation, and processing tools. Many XML editors, for example, are able to check the validity of a behavior specification at run time or even assist the user during the selection of elements and their attributes. Actually, a behavior specification has to be transformed into an intermediate code which may easily be interpreted, since on many embedded computing platforms, XML parsers are not available due to resource and portability constraints.

3   Modeling of the Environment

Since the methods for motion planning and action selection, which will be explained in the following sections, are not directly configurable, the whole behavior depends on the model of the environment, in particular on the parameters of the object classes. The architecture offers various options allowing a detailed description. An object class O may be considered as the following tuple:

$$O = (f_O, G_O, F_O) \tag{1}$$

With f_O being a potential function, which may be chosen among several standard functions [2] and a *social function* introduced by [12]. The range of the field as well as its gradient depend directly on the parameters which have to be assigned to f_O. A geometric primitive G_O is used to approximate an object's shape. The kind of field F_O determines the shape of the region influenced by O, which may, for example, be a circumfluent area around G_O or a tangential field around the position of the instance.

As shown in Fig. 4, an object instance excites in its environment both a potential field and a *charge* [4] based on its potential function. They may be computed separately in the following way: Let P be an arbitrary position in the environment of an object instance and $v_O\,(P, S, G_O, F_O)$ a function computing a vector from P to an instance I of the class O given its geometry and kind of field as well as the object state S assigned to the instance, the value $\varphi_I\,(P)$ of the potential function is to be computed as follows:

$$\varphi_I\,(P) = f(|v_I\,(P, S, G_O, F_O)|) \tag{2}$$

Analogous, the field vector $v_I\,(P)$ of the potential field is:

$$v_I\,(P) = f'(|v_I\,(P, S, G_O, F_O)|)\frac{v_I\,(P, S, G_O, F_O)}{|v_I\,(P, S, G_O, F_O)|} \tag{3}$$

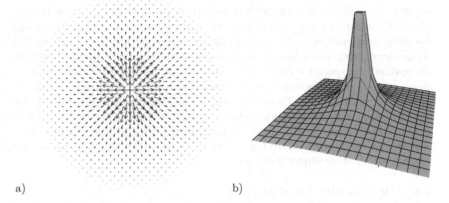

a)   b)

Fig. 4. In the environment of an object instance both a) a potential field and b) the value of the potential function may be computed

Due to this duality, each object instance may be assigned to and used by motion behaviors as well as for the evaluation of actions.

4   Motion Behaviors

All motion behaviors are mainly based on the standard motion planning approach by [1]. Some of the main extensions have been the integration of relative motions which allow the robot to behave in spatial relations to other objects, e.g. to organize in multi-robot formations, and the implementation of a path planner to avoid local minima.

4.1   The General Procedure of Motion Planning

Following the standard approach as described in [1, 2, 3], a vector v can be computed by the superposition of the force vectors v_I of all n object instances assigned to a behavior:

$$v = \sum_{I=1}^{n} v_i\,(R) \tag{4}$$

With R being the current position of the robot, v can be used to determine the robot's direction of motion, rotation and speed. Due to the equal interface of all objects, this method does not have to distinguish between attractive and repulsive objects.

4.2   Relative Motions

Assigning force fields to single objects of the environment allows the avoidance of obstacles and the approach to desired goal positions. Nevertheless, moving to more complex spatial configurations, e. g. positioning between the ball and the penalty area or lining up with several robots is not possible directly. An extension of the *Motor-Schema* approach [13] to dynamically form multi-robot formations has been developed by [14]. A quite similar technique has been used in this architecture, but it is not limited to a set of special formations. Relative motions are realized via special objects which may be assigned to behaviors. Such an object consists of a set of references to object instances and a spatial relation, e. g. *between* or *relative-angle*. These are used to dynamically compute a geometric representation of the desired destination region, which will subsequently be used as the object's G_O. Via an additional assignment of a potential function, such objects may be integrated into the process of motion planning in a transparent way among all other object instances.

4.3   Dealing with Local Minima

Local minima are an inherent problem of potential fields [15] which has, of course, to be discussed when using this approach for motion planning since an optimal standard solution does not exist. In the past, several heuristics have been proposed [3], but they cannot guarantee to be effective. Due to the increasing computing resources even in embedded systems, the usage of path planners becomes practicable in real-time applications [16].

This approach uses the A* algorithm [17] for path planning. To discretize the continuous environment, a search tree with a dynamic number and size of branches, similar to [16] is built up (see Fig. 5). The potential functions of the objects in the environment are used to determine the path costs.

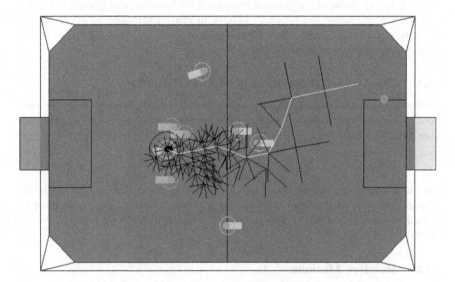

Fig. 5. Planning a path to the ball on a Sony Four-legged Robot League field. The dark lines show the generated search tree. The chosen path is drawn bright

The path planner is integrated in such a way that it may be used by every motion behavior. There also exist mechanisms to detect whether the robot enters or leaves a local minimum and thus to use the path planner only on demand.

5   Behaviors for Action Evaluation

In this architecture, actions are considered to be indivisible entities which have to be executed by the robot after their selection, e. g. the activation of a kick motion. It is also possible that an action evaluation behavior is combined with a motion behavior to determine the appropriateness of its execution.

There exist several approaches using potential functions to evaluate certain situations or the results of planned actions. Most of them rasterize the environment into cells of a fixed size [5, 7] and compute the value of each cell to determine the most appropriate position. A quite different approach is the *Electric Field Approach* [4] which is computationally much less expensive. By computing the anticipated world state after an action, only relevant positions have to be evaluated, as shown in Fig. 6. Due to its efficient computation and the direct mapping of possible actions, a similar approach has been integrated into the architecture.

5.1   The Procedure of Action Evaluation

Analogous to the computation of a field vector, the value $\varphi(P)$ (for which [4] use the term *Charge*) may be determined at an arbitrary position P, being the sum of the potential functions of all object instances assigned to the behavior:

$$\varphi(P) = \sum_{I=1}^{n} \varphi_I(P) \tag{5}$$

To use this method to evaluate a certain action which changes the environment, e. g. kicking a ball, this action has to be mapped to a geometric transformation in order to describe the motion of the manipulated object. A set off different

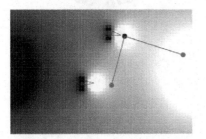

Fig. 6. Two examples of action evaluation from [4]. In both cases a kick to the side as well as a forward kick of a ball are evaluated. Bright regions represent a better evaluation

transformations, inter alia including rotation, translation, and tracing the potential field gradient, has been implemented, together with mechanisms to check for collisions and practicability of the action. External mechanisms as planners are not needed. To describe more complex actions, e. g. turning with a ball and subsequently kicking to the goal, sequences of actions may also be specified.

Having computed an anticipated future state, the value φ may also be determined along with a value φ_d representing the change of the environment which will be caused by an execution of the action. Both values may be used as activation values of the behavior.

6   Applications

Up to now, the architecture has been applied to two different platforms, both being RoboCup teams of the Universität Bremen, to be seen in Fig.7.

The main test and development platform have been the robots of the *Bremen Byters*, which are a part of the *GermanTeam* [18], a team which competes in the Sony Four-legged Robot League. To specify the environment of that domain, about 40 object instances based on 15 different classes have been used. Each player has a different role and therefore a different set of behaviors. For playing soccer, about 10–15 behaviors have been needed, e. g. *Go to Ball, Go to Defense Position* or *Kick Ball Forward*. Due to the large number of degrees of freedom, these robots are able to perform many different motions, resulting in a variety of kicks. After measuring the effect of each kick on the position of the ball, they could directly be specified as behaviors and therefore be evaluated and selected by the architecture. Also sequences of actions have been used, allowing a quite

a)   b)

Fig. 7. The behavior architecture has been used in RoboCup competitions as well on a) Sony ERS-210A robots as to control b) Small-size robots

forward-looking play. Except for the handling of special game states, for which all members of the German Team use XABSL [19], the complete behavior could be realized via the architecture presented in this paper.

The second platform has been the control program of B-Smart [20], running on an external PC and computing the behaviors for a team of omni-directional driven small-size robots, which are able to move with a speed of up to two meters per second. Due to the generic nature of the architecture and the similarity of the domains, large parts of the Bremen Byters' behaviors and their specification of the environment were copied leaving only several changes of parameters to be made, e. g. the dimensions of the field and the ranges of the potential functions. Through that, the portability of this approach of behavior modeling has been shown. Nevertheless, the direct interchange of unchanged behaviors is not a primary goal. Especially considering robot soccer competitions, optimizations for a single system are often more important.

7   Conclusion and Future Works

This paper presents a behavior-based architecture for autonomous mobile robots, integrating several different approaches for motion planning and action evaluation into a single general framework by dividing different tasks into competing behaviors. This approach turned out to be able to specify an overall behavior sufficiently for handling complex tasks in the robot soccer domain. The architecture's abstract design using external behavior specifications in XML files has also appeared to be well manageable.

In the future, the authors intend to port the architecture to other platforms to test and extend the capabilities of this approach. There exist several features already implemented but not adequately tested, e. g. the integration of object instances based on a probabilistic world model. In addition, the behavior selection process is currently extended to deal with a hierarchy of sets of competing behaviors similar to [21, 22], allowing the specification of even more complex overall behaviors.

At the RoboCup 2004, the architecture will be used again by B-Smart and will also be integrated in the behavior control of the German Team.

Acknowledgements

The authors would like to thank the members of the GermanTeam as well as the members of B-Smart for providing the basis of their research.

References

1. Khatib, O.: Real-time Obstacle Avoidance for Manipulators and Mobile Robots. The International Journal of Robotics Research **5** (1986) 90–98
2. Latombe, J.C.: Robot Motion Planning. Kluwer Academic Publishers, Boston, USA (1991)

3. Arkin, R.C.: Behavior-Based Robotics. MIT Press, Cambridge, Massachusetts, USA (1998)
4. Johannson, S.J., Saffiotti, A.: Using the Electric Field Approach in the RoboCup Domain. In Birk, A., Coradeschi, S., Tadokoro, S., eds.: RoboCup 2001: Robot Soccer World Cup V. Volume 2377 of Lecture Notes in Artificial Intelligence., Springer (2002)
5. Meyer, J., Adolph, R.: Decision-making and Tactical Behavior with Potential Fields. In Kaminka, G.A., Lima, P., Rojas, R., eds.: RoboCup 2002: Robot Soccer World Cup VI. Volume 2752 of Lecture Notes in Artificial Intelligence., Springer (2003)
6. Weigel, T., Gutmann, J.S., Dietl, M., Kleiner, A., Nebel, B.: CS-Freiburg: Coordinating Robots for Successful Soccer Playing. IEEE Transactions on Robotics and Automation 18 (2002) 685–699
7. Ball, D., Wyeth, G.: Multi-Robot Control in Highly Dynamic, Competitive Environments. In Browning, B., Polani, D., Bonarini, A., Yoshida, K., eds.: RoboCup 2003: Robot Soccer World Cup VII. Lecture Notes in Artificial Intelligence, Springer (2004) to appear.
8. Laue, T.: Eine Verhaltenssteuerung für autonome mobile Roboter auf der Basis von Potentialfeldern. Diploma thesis, Universität Bremen (2004)
9. Maes, P.: How To Do the Right Thing. Connection Science Journal 1 (1989) 291–323
10. Brooks, R.: A Robust Layered Control System for a Mobile Robot. IEEE Journal of Robotics and Automation 2 (1986) 14–23
11. Brooks, R.: Intelligence without representation. Artificial Intelligence Journal 47 (1991) 139–159
12. Reif, J.H., Wang, H.: Social Potential Fields: A Distributed Behavioral Control for Autonomous Robots. In Goldberg, K., Halperin, D., Latombe, J.C., Wilson, R., eds.: The Algorithmic Foundations of Robotics. A. K. Peters, Boston, MA (1995) 331 – 345
13. Arkin, R.C.: Motor Schema-Based Mobile Robot Navigation. The International Journal of Robotics Research 8 (1989) 92–112
14. Balch, T., Arkin, R.C.: Behavior-based Formation Control for Multi-robot Teams. IEEE Transactions on Robotics and Automation 14 (1999) 926–939
15. Koren, Y., Borenstein, J.: Potential Field Methods and Their Inherent Limitations for Mobile Robot Navigation. In: Proceedings of the IEEE International Conference on Robotics and Automation, Sacramento, California, USA (1991) 1398–1404
16. Behnke, S.: Local Multiresolution Path Planning. In Browning, B., Polani, D., Bonarini, A., Yoshida, K., eds.: RoboCup 2003: Robot Soccer World Cup VII. Lecture Notes in Artificial Intelligence, Springer (2004) to appear.
17. Hart, P., Nilsson, N.J., Raphael, B.: A Formal Basis for the Heuristic Determination of Minimum Cost Paths in Graphs. IEEE Transactions on Systems Science and Cybernetics SSC-4 (1968) 100–107
18. Röfer, T., Brunn, R., Dahm, I., Hebbel, M., Hoffmann, J., Jüngel, M., Laue, T., Lötzsch, M., Nistico, W., Spranger, M.: GermanTeam 2004. The German RoboCup National Team. In: RoboCup 2004: Robot Soccer World Cup VIII. Lecture Notes in Artificial Intelligence, Springer (2004) to appear.
19. Lötzsch, M., Bach, J., Burkhard, H.D., Jüngel, M.: Designing Agent Behavior with the Extensible Agent Behavior Specification Language XABSL. In Browning, B., Polani, D., Bonarini, A., Yoshida, K., eds.: RoboCup 2003: Robot Soccer World Cup VII. Lecture Notes in Artificial Intelligence, Springer (2004) to appear.

20. Kurlbaum, J., Laue, T., Lück, B., Mohrmann, B., Poloczek, M., Reinecke, D., Riemenschneider, T., Röfer, T., Simon, H., Visser, U.: Bremen Small Multi-Agent Robot Team (B-Smart) Team Description for RoboCup 2004. In: RoboCup 2004: Robot Soccer World Cup VIII. Lecture Notes in Artificial Intelligence, Springer (2004) to appear.
21. Behnke, S., Rojas, R.: A hierarchy of reactive behaviors handles complexity. In: Proceedings of Balancing Reactivity and Social Deliberation in Multi-Agent Systems, a Workshop at ECAI 2000, the 14th European Conference on Artificial Intelligence, Berlin (2000)
22. Jäger, H., Christaller, T.: Dual Dynamics: Designing Behavior Systems for Autonomous Robots. In Fujimura, S., Sugisaka, M., eds.: Proceedings of the International Symposium on Artificial Life and Robotics (AROB '97), Beppu, Japan (1997) 76–79

An Egocentric Qualitative Spatial Knowledge Representation Based on Ordering Information for Physical Robot Navigation

Thomas Wagner[1] and Kai Hübner[2]

[1] Center for Computing Technologies (TZI),
Universität Bremen, D-28359 Bremen
twagner@tzi.de
[2] Bremen Institute of Safe Systems (BISS),
Universität Bremen, D-28359 Bremen
khuebner@tzi.de

Abstract. Navigation is one of the most fundamental tasks to be accomplished by many types of mobile and cognitive systems. Most approaches in this area are based on building or using existing allocentric, static maps in order to guide the navigation process. In this paper we propose a simple egocentric, qualitative approach to navigation based on ordering information. An advantage of our approach is that it produces qualitative spatial information which is required to describe and recognize complex and abstract, i.e., translation-invariant behavior. In contrast to other techniques for mobile robot tasks, that also rely on landmarks it is also proposed to reason about their validity despite insufficient and insecure sensory data. Here we present a formal approach that avoids this problem by use of a simple internal spatial representation based on landmarks aligned in an *extended panoramic representation* structure.

1 Introduction

Navigation is one of the most fundamental tasks to be accomplished by robots, autonomous vehicles, and cognitive systems. Most successful approaches in the area of robot navigation like potential fields (see [10] and [7]) are based on allocentric, static maps in order to guide the navigation process (e.g. [9]). This approach has an intuitive appeal and gains much intuition from cognitive science: the *cognitive map* (a good recent overview [16]). The main purpose is to build up a precise, usually allocentric, quantitative representation of the surrounding environment and to determine the robot's position according to this allocentric, quantitative map.

One difficulty results from the fact that the same spatial representation serves as a basis for different tasks often with heterogeneous requirements. For example, more abstract reasoning tasks like planning coordinated behavior, e.g., *counterattack* and *double pass*, and plan recognition usually rely on more abstract,

D. Nardi et al. (Eds.): RoboCup 2004, LNAI 3276, pp. 134–149, 2005.

qualitative spatial representations. Generation of qualitative spatial descriptions from quantitative data is usually a difficult task due to uncertain and incomplete sensory data. In order to fit heterogeneous requirements, we should be able to represent spatial qualitative description at different levels of granularity, i.e., invariant according to translation and/or rotation and based on different scalings.

Based on recent results from cognitive science (see, e.g., [30]), we present a formal, egocentric, and qualitative approach to navigation which overcomes some problems of quantitative, allocentric approaches. By the use of ordering information, i.e., based on a description of how landmarks can shift and switch, we generate an *extended panoramic representation* (EPR). We claim that our representation in combination with path integration provides sufficient information to guide navigation with reduced effort to the vision process. Furthermore the EPR provides the foundation for qualitative spatial descriptions that may be invariant to translation and/or rotation.

Since our approach abstracts from quantitative or metrical detail in order to introduce a stable qualitative representation between the raw sensor data and the final application, it can for example be used in addition to the well-elaborated quantitative methods.

2   Motivation

Modeling complex behavior imposes strong requirements on the underlying representations. The representation should provide several levels of abstraction for activities as well as for objects. For both types of knowledge, different representations were proposed and it was demonstrated that they can be used successfully. Activities can, e.g., be described adequately with hierarchical task networks (HTN) which provide clear formal semantics as well as powerful, efficient (planning-) inferences (see e.g. [4]). Objects can be described either in ontology-based languages (e.g., OWL [21]) or constraint-based languages (e.g., [8]). Both types of representations allow for the representation of knowledge at different levels of abstraction according to the domain and task specific requirements. In physically grounded environments, the use of these techniques requires an appropriate qualitative spatial description in order to relate the modeled behavior to the real world.

2.1   Allocentric and Egocentric Representations

In an egocentric representation, spatial relations are usually directly related to an agent by the use of an egocentric *frame of reference* in terms like, e.g., *left, right, in front, behind.* As a consequence, when an agent moves through an environment, all spatial relations need to be updated. In contrast, representations based on an allocentric frame of reference remain stable but are much harder to acquire. Additionally, the number of spatial relations which have to be taken into account may be much larger because we have to consider the relations between each object and all other objects in the environment, whereas the number of

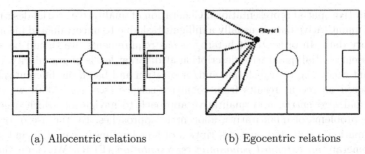

(a) Allocentric relations   (b) Egocentric relations

Fig. 1. Allocentric vs. egocentric spatial relations

relations in egocentric representations can be significantly smaller (see Fig. 1)[1].
An interesting phenomenon, when looking into the didactic literature about, e.g.,
sports [12] we often find that (tactical and strategic) knowledge is described in
both, egocentric and allocentric terms, whereas, e.g., the literature about driving
lessons strongly relies on purely egocentric views. At least one of the reasons
are that the latter representation seems to provide better support for acting
directly in physically grounded environments, since perception as well as the
use of actuators are directly based on egocentric representations. In addition,
egocentric representations provide better support for rotation and translation
invariant representations when used with a qualitative abstraction (see sections
3.3 and 4 for more details).

3   Related Work

3.1   Cognition: Dynamic, Egocentric Spatial Representations

The fact that even many animals (e.g., rodents) are able to find new paths leading to familiar objects seems to suggest that spatial relations are encoded in an
allocentric static *"cognitive map"*. This almost traditional thesis is supported by
many spatial abilities like map navigation and mental movement that humans
are able to perform (beginning with [26] and [14]). Nevertheless, recent results
in cognitive science provide strong evidence for a different view ([30] among
many others). Instead of using an allocentric view-independent map, humans
and many animals build up a dynamic, view-dependent egocentric representation. Although the allocentric interpretation of the *cognitive map* seems to differ
radically from the egocentric representation theory, both theories can account
for many observations and differ mainly in two points: The allocentric, *cognitive
map*-interpretation assumes that the spatial representation is view-independent
and that therefore viewpoint changes do not have any influence on the performance of, e.g., spatial retrieval processes. Many recent experiments provide
evidence for the opposite, they show that viewpoint changes can significantly

---

[1] For reasons of clarity not all allocentric relations are drawn in diagram 1(a).

reduce performance in terms of time and quality (e.g. pointing errors) (among others, [28] and [29]). The second main difference is concerned with the dynamic of the underlying representation. The egocentric interpretation assumes that all egocentric relations have to be updated with each egocentric movement of a cognitive system. The underlying assumption of a sophisticated series of experiments done by Wang ([28] and [29]) was that spatial relations have to remain stable in an allocentric, *cognitive map* independent from egocentric movements. When errors arise, e.g., because of path integration, the error rate (*"configuration error"*) should be the same for all allocentric relations; otherwise they rely on an egocentric representation. The results indicate clear evidence for egocentric representations and have been confirmed in a series of differently designed experiments[2], e.g., [3] and [5].

3.2   Robot Navigation

Navigation and localization is the most fundamental task for autonomous robots and has gained much attention in the robotic research over the last decades. While several earlier approaches addressed this problem qualitatively [9], e.g., topological maps ([11], [15], [1]), more recent approaches focus very successfully on probabilistic methods. Famous examples are RHINO [23], MINERVA [22] and more recently [25]. Currently, the most promising techniques for robust mobile robot localization and navigation are either based on Monte-Carlo-Localization (MCL) (see [18] for RoboCup-application and the seminal paper [24]) or on various extentions of *Kalman-filters* (e.g., [13]) using probabilistic representations based on quantitative sensory data. MCL is based on a sample set of postures; the robot's position can be estimated by probabilities which allow to handle not only the *position tracking*- and the *global localization* problem but also the challenging *kidnapped robot* problem of moving a robot without telling it.

Furthermore, probabilistic methods based on quantitative data play a crucial role in handling the mapping problem, i.e., the SLAM-problem[3]. Very much the same is true for many robotic approaches to navigation, e.g., potential fields for avoiding obstacles by following the flow of superposed partial fields in order to guide the robot to a goal position (see [10] and [7] for a RoboCup-application) based on quantitative data.

According to the *spatial semantic hierarchy* (SSH) [9], these approaches try to address the problems related to robot navigation on the *control level*. Besides the strong computational resource requirements, they usually do not address the problem of generating a discrete, qualitative spatial representation which for

---

[2] Nevertheless, these results do not allow the strict conclusion that humans do not build up an allocentric cognitive map. On the contrary, e.g., Easton and Sholl [3] have shown that under very specific conditions it is possible to build up allocentric maps. Nevertheless, these results indicate, that under more natural conditions human navigation relies on egocentric snapshots and a dynamic mapping between these.

[3] This term is also directly connected to a set of algorithms addressing exactly this problem (e.g., [2]).

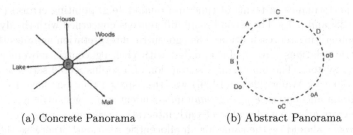

(a) Concrete Panorama   (b) Abstract Panorama

Fig. 2. Panorama-views

instance is required at more abstract levels, e.g., for describing complex coordinated tactical and strategic behavior both on individual- and on team level.

3.3   The Panorama Approach

The concept of panorama representation has been studied extensively in the course of specialized sensors (e.g., omnivision, see, e.g., [31]). We present an extended approach based on the panorama approach by Schlieder ([20] and [19]).

A complete, circular panorama can be described as a 360° view from a specific, observer-dependent point of view. Let P in Fig. 2(a) denote a person, then the panorama can be defined as the strict ordering of all objects: *house, woods, mall, lake*. This ordering, however, does not contain all ordering information as described by the scenario. The *mall* is not only directly between the *woods* and the *lake*, but more specifically between the opposite side of the *house* and the *lake* (the tails of the arrows). In order to represent the spatial knowledge described in a panorama scenario, [20] introduced a formal model of a panorama.

Definition 1 (Panorama). *Let $\Theta = \{\theta_1, \ldots, \theta_n\}$ be a set of points $\theta_i \in \Theta$ and $\Phi = \{\phi_1, \ldots, \phi_n\}$ the arrangement of n-1 directed lines connecting θ_i with another point of Θ, then the clockwise oriented cyclical order of Φ is called the panorama of θ_i.*

As a compact shorthand notation we can describe the panorama in Fig. 2(b) as the string $< A, C, D, Bo, Ao, Co, Do, B >$. Standard letters (e.g., A) describe reference points, and letters with a following o (e.g., Ao) the opposite side (the tail side). As the panorama is a cyclic structure the complete panorama has to be described by n strings with n letters, with n being the number of reference points on the panorama. In our example, the panorama has to be described by eight strings. Furthermore, the panorama can be described as a set of simple constraints $dl(vp, lm_1, lm_2)$[4]. Based on this representation, [19] also developed an efficient qualitative navigation algorithm.

The panorama representation has an additional, more important property: it is invariant with respect to rotation and translation. But evidently, not ev-

---

[4] Short for $direct - left(viewpoint, landmark_1, landmark_2)$.

ery behavior can be described in such an abstract manner. In order to model complex, coordinated behaviors, often more detailed ordinal information is involved. Additionally, different metric information (e.g., distance) is required in some situations. In the following section, we show how the panorama can be extended in a way that more detailed ordinal and metric information can be introduced.

4 An Extended Panorama Representation

Instead of building an allocentric map we provide an egocentric snapshot-based approach to navigation. The most fundamental difference between both approaches is that an egocentric approach strongly relies on an efficient, continuous update mechanism that updates all egocentric relations in accordance with the players' movement. In this section we show that this task can be accomplished by strict use of a simple 1D-ordering information, namely an extended qualitative panorama representation (EPR).

This update mechanism has to be defined with respect to some basic conditions:

- Updating has to be efficient since egocentric spatial relations change with every movement, i.e., the updating process itself and the underlying sensor process.
- The resulting representation should provide the basis for qualitative spatial descriptions at different levels of granularity.
- The resulting representation should provide different levels of abstraction, i.e., rotation and/or translation invariance.
- The process of mapping egocentric views should rely on a minimum of allocentric, external information.

Due to the nature of ordering information, this task has to be divided into two subtasks: (1) updating within a given frame of reference (short notation: FoR), i.e., the soccer field and (2) updating of landmark representations from an external point of view, e.g., the penalty area. In section 4.1 we briefly discuss the key properties of the first task in relation to ordering information from a more theoretical point of view, whereas in section 5 these aspects are investigated in more detail. In section 4.2 we describe the theoretical framework underlying the mapping- and update-mechanism for egocentric views on external landmarks.

4.1 Within a Frame of Reference

A crucial property of panoramic ordering information is that it does not change as long as an agent stays within a given FoR, i.e., the corners of a soccer field, do not change unless the player explicitly leaves the field (see Fig. 3(a)). So in order to use ordering information for qualitative self-localization we have to introduce an egocentric FoR. But even with an egocentric FoR the location within this

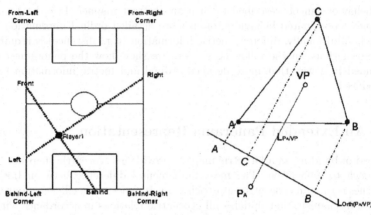

(a) Use of Egocentric Frame of Reference   (b) Triangle panorama construction
                                             by projection (result here: (ACB))

Fig. 3. FoR and Triangle panorama

FoR can only be distinguished into a few different qualitative states (e.g., ego-front between front-left and front-right corner of the field, see Fig. 3(a)). This way of qualitative self-localization is too coarse for many domains as well as for the different RoboCup-domains. In section 5 we demonstrate in more detail how angular distances can be used to overcome this problem[5].

A perhaps even more important property of spatial locations within a given FoR is that they can be used as a common FoR for the position of different landmarks in relation to each other (e.g., the position of the penalty area can be described in within-relation to the soccer field). This property is especially important for an egocentric snapshot-based approach to navigation since it provides the common frame that is required to relate different snapshots to each other (for a more detailed discussion see [27]).

4.2   Updating Outside-Landmark Representations

In a re-orientation task we can resort the knowledge about the previous position of a player. Therefore we concentrate on an incremental updating process, based on the following two assumptions: (1) It is known that the configuration of perceived landmarks $A, B, \ldots \in L$ either form a triangle- or a parallelogram configuration (e.g. either by vision or by use of background knowledge). (2) The position P_{t-1} of an agent A in relation to L at time step $t-1$ is known. The EPR (LP_T) of a triangle configuration can then be defined as follows (see also Fig. 3(b)):

---

[5] An additional approach is to introduce more landmarks that are easy to perceive or to introduce additional allocentric FoR when available (e.g., north, south, etc.).

Definition 2 (Triangle Landmark Panorama). *Let P_A denote the position of an agent A and $C_{T(ABC)}$ the triangle configuration formed by the set of points A, B, C in the plane. The line $L_{P_A/VP}$ is the line of view from P_A to VP, with VP being a fixed point within $C_{T(ABC)}$. Furthermore, $L_{Orth(P_A/VP)}$ be the orthogonal intersection of $L_{P_A/VP}$. The panoramic ordering information can be described by the orthogonal projection $P(P_A, VP, C_{T(ABC)})$ of the points ABC onto $L_{Orth(P_A/VP)}$.*

Therefore, moving around a triangle configuration $C_{T(ABC)}$ results in a sequence of panoramas which qualitatively describe the location of the observer position. A $360°$ movement can be distinguished in six different qualitative states:

Observation 1. *(Triangle Landmark Panorama Cycle)*
The EPR resulting from the subsequent projection $P(P_A, VP, C_{T(ABC)})$ by counter-clockwise circular movement around VP can be described by the following ordered, circular sequence of panoramas:
$(CAB), (ACB), (ABC), (BAC), (BCA), (CBA)$

For each landmark panorama the landmark panorama directly left as well as at the right differ in exact two positions that are lying next to each other (e.g., $(ABC), (BAC)$ differ in the position exchange between A and B). These position changes occur exactly when the view line $L_{P_A/VP}$ intersects the extension of one of the three triangle lines: L_{AB}, L_{AC}, L_{BC}. Starting with a given line (e.g., L_{AB}) and moving either clock- or counter-clockwise, the ordering of line extensions to be crossed is fixed for any triangle configuration (see Fig. 3(b)). This property holds in general for triangle configurations but not, e.g., for quadrangle configurations (except for some special cases as we will see below). Since (almost) each triplet of landmarks can be interpreted as a triangle configuration, this form of qualitative self-localization can be applied quite flexibly with respect to domain-specific landmarks. The triangle landmark panorama, however, has (at least) two weaknesses: The qualitative classification of an agent's position into six areas is quite coarse and, triangle configurations are somewhat artificial constructs that are rarely found in natural environments when we consider solid objects[6]. A natural extension seems to be applying the same idea to quadrangles (see Fig. 4). The most direct approach is to interpret a quadrangle as a set of two connected triangles sharing two points by a common line so that each quadrangle would be described by a set of two triangle panoramas. With this approach, the space around a quadrangle would be separated into ten areas and therefore it would be more expressive than the more simple triangle panorama. It can be shown that eight of the resulting triangle landmark panoramas (one for each triangle of the quadrangle) can be transformed into quadruple tuples that result when we transform, e.g., a rectangle directly into a landmark panorama representation (e.g., the given tuple ((BCA)(CDA)) can be transformed into (BCDA) without

---

[6] The triangle configuration can be applied generally to any triplet of points that form a triangle - also to solid objects. The connecting lines pictured in Fig. 3(b) and 4(a) are used to explain the underlying concept of position exchange (transition).

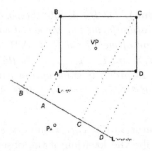
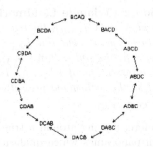

(a) Parallelogram panorama construction   (b) Circular representation of panoramic
by projection (result here: (BACD))   ordering information for parallelograms

Fig. 4. Parallelogram panorama

loss of information)[7]. The expressiveness of the other two landmark panoramas is weaker: they have to be described as a disjunction of two quadruple tuples. Since the expressiveness is weaker and the landmark panorama representation of a quadruple tuple panorama representation is much more intuitive we focus on the latter one (see Fig. 4(a)).

Definition 3 (Parallelogram Landmark Panorama). *Let P_A denote the position of an agent A and $C_{P(ABC)}$ the parallelogram configuration formed by the set of points A, B, C, D in the plane. The line $L_{P_A/VP}$ is the line of vision from P_A to VP, with VP being a fixed point within $C_{P(ABCD)}$. Furthermore, $L_{Orth(P_A/VP)}$ be the orthogonal intersection of $L_{P_A/VP}$. The landmark panoramic ordering information can then be described by the orthogonal projection $P(P_A, VP, C_{P(ABCD)})$ of the points ABCD onto $L_{Orth(P_A/VP)}$.*

Moving around a parallelogram configuration $C_{P(ABCD)}$ also results in a sequence of landmark panoramas which describe the location of the observer position qualitatively. A $360°$ movement can be split into twelve different states:

Observation 2. *(Parallelogram Landmark Panorama Cycle)*
The panoramic landmark representations resulting from the subsequent projection $P(P_A, VP, C_{P(ABCD)})$ by counter-clockwise circular movement around VP can be described by the following ordered, circular sequence of panoramas:

---

[7] The detailed proof will take too much space. However, the basic proof idea is quite straightforward: each panorama transition happens because of the intersection of the landmarks' line extensions with the line of vision of the moving agent, so the number of disjoint lines (multiplied by 2, since each line is intersected twice) specifies the number of transitions and therefore the number of distinguishable areas. The loss of expressiveness of two of the triangle tuples can be explained in the same way: assume that the quadrangle $ABCD$ is defined by the two triangles ABC and ADC sharing the diagonal AC. Position changes of the points B/D cannot be distinguished since they happen in two different triangles, which are not in relation to each other. Alternatively, we can show that the number of resulting ordering constraints is smaller (for more details on the constraint representation see section 3.3).

$((BCAD), (BACD), (ABCD), (ABDC), (ADBC), (DABC),$
$(DACB), (DCAB), (CDAB), (CDBA), (CBDA), (BCDA))$

The two presented landmark panoramas can be mapped flexibly onto landmarks that can be found in natural environments like a penalty area. While solid objects often form rectangle configurations, irregular landmarks can be used in combination as a triangle configuration, since this approach is not strictly restricted to point-like objects. An interesting extension is to build up more complex representations by using landmark configurations as single points in larger landmark configurations. This allows us to build up nesting representations which support different levels of granularity according to the requirements of the domain.

5   Implementation

According to the described scenarios, the EPR is meant to be a qualitative fundament for tasks that are important for mobile robot exploration. Due to the oversimplification of the four-legged league RoboCup scenario (i.e., no penalty area and no goal area to move around), the latter outside-case described in section 4.2 does not find capital application here, but we claim that it will show its features in more complex scenarios which offer a larger number of landmarks to move around. Here, we will show some experimental extraction of EPR sequences to practically point up the idea presented in section 4.1 and the basic idea of building panoramic ordering information from the image data.

For our first experiments, we use the *RobotControl/SimRobot* [17] simulation environment for the simulation of one four-legged robot. This tool is shared with the GermanTeam, which is the German national robotic soccer team participating in the Sony four-legged league in the international RoboCup competitions. The EPR concept presented is not proposed to be restricted to this special domain, as discussed. The tool supports simulated image retrieval and motion control routines that are easy to use and portable to physical robots, while it is possible to encapsulate the EPR and adapted image feature extraction in distinct solutions, letting other modules untouched.

5.1   Visual Feature Extraction

In order to expediently fill the EPR with information, the recognition of landmarks is necessary. Usually, the robot's viewing angle of 57.6^o degrees is not sufficient to get a reasonably meaningful EPR with the feature extraction of goals and flags supported by the *RobotControl* tool (see [18] for a description of these features).

Even if the scene is regarded from one goal straight to the other, there are just three landmarks that can be found. On the other hand, the standard configuration of all landmarks as can be seen in Fig. 5 is of an unfavorable kind for the EPR. The landmarks build a convex structure that the robot can never

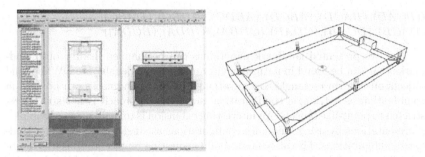

Fig. 5. Simulation environment of the GermanTeam (left); the standard four-legged league field configuration (right)

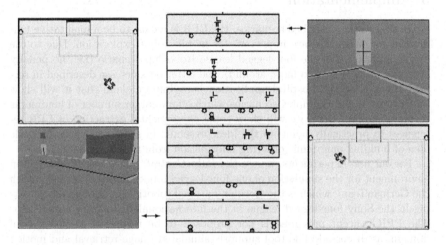

Fig. 6. Landmarks for the EPR. Center column: Landmarks extracted (for six representation between given start position (left) and goal position (right): "L" for L-junctions, "T" for T-junctions, "X" for X-junctions; horizontal lines (circles), vertical lines (squares), goals (light triangles) and flags (dark triangles))

leave, thus the ideal EPR will never allow to reason about the environment by permuted landmarks (see section 4.1). Thus, we further introduced the symmetry line operator proposed by Huebner [6] to extract 2D field lines as additional features from the image data. The method is simple, robust, and works without plenty of parametrization. Additionally, it offers the opportunity to test the approach with natural landmarks (lines) instead of artifacts (colored beacons). After processing the images, lines are distinguished from curves and represented by their start and end point in the image (see Fig. 6).

These lines can be put into the EPR by adopting these points or the center point, for example. Anyway, a classification of edge types is more efficient with respect to the subsequent need of recovering landmarks. To support the panorama with a broader range of landmark types which ideally are points on the field, we classify each pair of lines extracted from an image into different line pair types. In our experiment, we extracted L-junctions, T-junctions and X-junctions (see Fig. 6) representing the additional landmarks that are used for the EPR.

5.2   Qualitative Representation

The simulated environment for the experiment corresponds to the standard four-legged league field configuration with lines instead of the sideboards. One robot is instructed to move a certain path presented by a given sequence of EPRs. Using the EPR representation and a qualititative conversion of the feature angles, we can establish a qualitative EPR sequence of detected landmark configurations for a path. Some samples of such sequences might look like the following, corresponding to the EPR of Fig. 6:

```
[(T_JUNC,VERY_FAR);(L_JUNC,SAME);(T_JUNC,SAME);(X_JUNC,SAME);]
[(T_JUNC,VERY_FAR);(X_JUNC,SAME);(FLAG,SAME);(T_JUNC,SAME);]
[(T_JUNC,FAR);(FLAG,SAME);(L_JUNC,CLOSE);(T_JUNC,MEDIUM);
 (T_JUNC,MEDIUM);(L_JUNC,SAME);(L_JUNC,CLOSE);]
[(FLAG,FAR);(L_JUNC,SAME);(L_JUNC,CLOSE);(T_JUNC,MEDIUM);
 (L_JUNC,SAME);(T_JUNC,CLOSE);(L_JUNC,MEDIUM);(T_JUNC,SAME);]
[(FLAG,FAR);(GOAL,FAR);]
[(FLAG,FAR);(GOAL,FAR);(L_JUNC,CLOSE);]
```

Each of these ordering sequences corresponds to a snapshot-like qualitative description of the robot's location during the path. E.g., in the first sequence, there are four ordered and classified landmarks that are additionally described by their qualitative angular distance to the previous landmark. Caused by the panoramic representation, the first "T"-junction is very far (VERY_FAR) displaced from the previous landmark (the "X"-junction in this case). Including qualitative angular distances like VERY_FAR also allows to convert this angular representation to a number of qualitative location descriptions (e.g., according to the first sequence, "The X-junction is very far LEFT of the T-junction." or "The X-junction is same RIGHT of the T-junction.").

As also can be seen in this example, the line landmarks appear and disappear frequently in the robot's view. This is caused by the landmark feature extraction working on insufficient simulated image data. We are optimistic that real images are more comfortable for the extraction of lines, because they are not supposed to be fragmented like those in simulated images. Although this is error-prone in this regard, we claim to deal with this problem using the EPR. The representation can generally be useful for this re-orientation task, where the agent knows at least to some extent where it has been. Based on this information, the circular panorama landmark representation can tell us which hypotheses are plausible according to previous information.

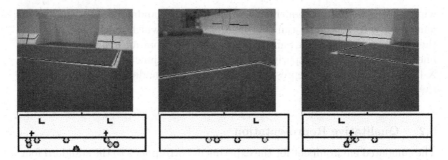

Fig. 7. Landmarks for the EPR on real images. Top row: image data and extracted field / border lines. Bottom row: Landmarks extracted

The same panoramic representation is additionally used in our simulation soccer team *Virtual Werder*. Although sensor problems are neglectable since the world model is more comprehensive and detailed, it provides a simple and intuitive interface for the generation of qualitative descriptions.

5.3   Experiments on Real Images

Finally, some experiments have been made to test the proposed feature extraction and EPR construction on real images (see Fig. 7)[8] using one Sony AIBO ERS-7 model inside a common four-legged league scenario. Without plenty of adaptation, the results are as good as those in the simulation examples. Problems appearing by the line extraction technique (e.g. side walls as lines, lines found over horizon, optional grouping of lines to handle occlusions) will be addressed in future work to increase robustness and performance.

6   Conclusion and Future Work

Navigation, localization, planning, and reasoning for physically grounded robots imposes strong but heterogeneous requirements on the underlying spatial representation in terms of abstraction and precision. In contrast to many other approaches to this topic which try to generate *allocentric* maps, we proposed a new *egocentric* approach based on recent results from cognition. The qualitative EPR is dynamic in a predictable way for outside landmarks as stated in the two observations described above. This representation, however, provides also interesting properties for navigation inside fixed landmarks (e.g., navigating within a room).

Besides the re-orientation task mentioned in the last section, the landmark panorama can help to focus perception in a qualitative self-allocation task. Dur-

---

[8] The difference of size in the corresponding images is caused by the different image sizes between the old AIBO model ERS-210 and the new ERS-7.

ing the transition of one panorama landmark into another exactly one position change is performed. Therefore, in this case the perception of further landmarks is without any use for updating the qualitative position of the agent. Additionally, the panorama landmark representation is not only useful for position updating but also for re-orientation without knowledge about the previous position. The perception of a partial landmark panorama of a triangle configuration is sufficient to provide us with two hypotheses about the current position. In order to validate which hypothesis holds we just have to find out where another landmark appears in the panoramic structure. Additionally, a landmark panorama provides a stable basis for qualitative, spatial descriptions (e.g. left of, right of), since it is, obviously, sensitive to rotation but invariant to transition, it is also interesting for several outstanding applications based on qualitative information.

Although a detailed analysis of the relation to the recent cognitive results is out of the scope in this paper, we want to mention that the EPR shows several properties which are observed in recent experiments: e.g., translation tasks seem to be performed more easily and accurately than rotation tasks.

Several tasks remain to be done. We are currently extending our landmark-based (re-)orientation vision module so that it is not only able to track EPRs but also allows active snapshot-based navigation (first results are available). Thereby we implement the concept of outside-landmarks that formally describes how landmarks can shift and switch during movement (see section 4.2). This should also allow to detect the geometric structure of previously unseen objects. After validating our extended panorama representation in the RoboCup-domain, we consider to transfer this method of the EPR into an omnidirectional vision module for mobile robot tasks.

Acknowledgements

The presented work is being funded by the German Research Foundation (DFG) within the project *Automatic Plan Recognition and Intention Recognition of Foreign Mobile Robots in Cooperative and Competitive Environments* as part of the Priority Research Program SPP-1125 *Cooperative Teams of Mobile Robots in Dynamic Environments*.

References

1. D. Busquets, C. Sierra, and R. L. De Mantaras. A multiagent approach to qualitative landmark-based navigation. *Autonomous Robots*, 15(1):129–154, 2003.
2. H. Durrant-Whyte, S. Majumder, S. Thrun, M. de Battista, and S. Schelling. A bayesian algorithm for simulaneous localization and map building. In *Proceedings of the 10'th International Symposium on Robotics Research (ISRR'01))*, Lorne, Australia, 2001. AAAI Press/MIT Press.
3. R. D. Easton and M. J. Sholl. Object-array structure, frames of reference, and retrieval of spatial knowledge. *Journal of Experimental Psychology: Learning, Memory and Cognition*, 21(2):483–500, 1995.

4. K. Erol, J. Hendler, and D. S. Nau. HTN planning: Complexity and expressivity. In *Proceedings of the Twelfth National Conference on Artificial Intelligence (AAAI-94)*, volume 2, pages 1123–1128, Seattle, Washington, USA, 1994. AAAI Press/MIT Press.

5. B. Garsoffky, S. Schwan, and F. W. Hesse. Viewpoint dependency in the recognition of dynamic scenes. *Journal of Experimental Psychology: Learning, Memory and Cognition*, 28(6):1035–1050, 2002.

6. K. Huebner. *A Symmetry Operator and its Application to the RoboCup.* 7th International Workshop on RoboCup 2003 (Robot World Cup Soccer Games and Conferences). Lecture Notes in Artificial Intelligence. Springer Verlag, 2004. To appear.

7. S. Johansson and A. Saffiotti. Using the Electric Field Approach in the RoboCup Domain. In A. Birk, S. Coradeschi, and S. Tadokoro, editors, *RoboCup 2001: Robot Soccer World Cup V*, number 2377 in LNAI, pages 399–404. Springer-Verlag, Berlin, DE, 2002.

8. U. John. Solving large configuration problems efficiently by clustering the conbacon model. In *IEA/AIE 2000*, pages 396–405, 2000.

9. B. Kuipers. The spatial semantic hierarchy. *Artificial Intelligence*, 119(1-2):191–233, 2000.

10. J. C. Latombe. *Robot Motion Planning*, volume 18. Kluwer Academic Press, 1991.

11. T. S. Levitt and D. T. Lawton. Qualitative navigation for mobile robots. *Artificial Intelligence*, 44:305–360, 1990.

12. M. Lucchesi. *Coaching the 3-4-1-2 and 4-2-3-1.* Reedswain Publishing, edizioni nuova prhomos edition, 2001.

13. C. Martin and S. Thrun. Online acquisition of compact volumetric maps with mobile robots. In *Proceedings of the IEEE International Conference on Robotics and Automation (ICRA)*, 2002.

14. J. O'Keefe and A. Speakman. *The Hippocampus as A Cognitive Map.* Oxford, Clarendon Press, 1978.

15. T. J. Prescott. Spatial representation for navigation in animats. *Adaptive Behavior*, 4(2):85–125, 1996.

16. A. D. Redish. *Beyond the Cognitive Map - From Place Cells to Episodic Memory.* The MIT Press, Cambridge, Mass., 1999.

17. T. Roefer. *An Architecture for a National RoboCup Team*, pages 417–425. RoboCup 2002: Robot Soccer World Cup VI. Lecture Notes in Artificial Intelligence. Springer Verlag, 2003.

18. T. Roefer and M. Juengel. Vision-Based Fast and Reactive Monte-Carlo Localization. In *Proceedings of the IEEE International Conference on Robotics and Automation (ICRA-2003)*, pages 856–861, Taipei, Taiwan, 2003.

19. C. Schlieder. Representing visible locations for qualitative navigation. pages 523–532. 1993.

20. C. Schlieder. Ordering information and symbolic projection, 1996.

21. M. K. Smith, C. Welty, and D. L. McGuinness. Owl web ontology language guide. W3c candidate recommendation 18 august 2003, 2003. http://www.w3.org/TR/owl-guide/.

22. S. Thrun, M. Beetz, M. Bennewitz, W. Burgard, A. B. Cremers, F. Dellaert, D. Fox, D. Hähnel, C. Rosenberg, N. Roy, J. Schulte, , and D. Schulz. Probabilistic algorithms and the interactive museum tour-guide robot minerva. *International Journal of Robotics Research*, 19(11):972–999, 2000.

23. S. Thrun, A. Bücken, W. Burgard, D. Fox, T. Fröhlinghaus, D. Henning, T. Hofmann, M. Krell, and T. Schmidt. *AI-based Mobile Robots: Case Studies of Successful Robot Systems*, chapter Map learning and high-speed navigation in RHINO. MIT Press, 1998.

24. S. Thrun, D. Fox, W. Burgard, and F. Dellaert. Robust monte carlo localization for mobile robots. *Artificial Intelligence*, 128(1-2):99–141, 2000.

25. S. Thrun, D. Hähnel, D. Ferguson, M. Montemerlo, R. Triebel, W. Burgard, C. Baker, Z. Omohundro, S. Thayer, and W. Whittaker. A system for volumetric robotic mapping of abandoned mines. In *Proceedings of the IEEE International Conference on Robotics and Automation (ICRA)*, 2003.

26. E. C. Tolman. Cognitive maps in rats and men. *Psycholoical Review*, 55:189–208, 1948.

27. T. Wagner, C. Schlieder, and U. Visser. An extended panorama: Efficient qualitative spatial knowledge representation for highly dynamic enironments. *IJCAI-03 Workshop on Issues in Desgning Physical Agents for Dynamic Real-Time Environments: World Modeling, Planning, Learning, and Communicating*, pages 109–116, 2003.

28. R. F. Wang. Representing a stable environment by egocentric updating and invariant representations. *Spatial Cognition and Computation*, 1:431–445, 2000.

29. R. F. Wang and E. S. Spelke. Updating egocentric represenations in human navigation. *Cognition*, 77:215–250, 2000.

30. R. F. Wang and E. S. Spelke. Human spatial representation: Insights from animals. *Trends in Cognitive Science*, 6(9):176–182, 2002.

31. J. Y. Zheng and S. Tsuji. Panoramic representation for route recognition by a mobile robot. *International Journal of Computer Vision*, 9(1):55–76, 1992.

Sensor-Actuator-Comparison as a Basis for Collision Detection for a Quadruped Robot

Jan Hoffmann and Daniel Göhring

Institut für Informatik, LFG Künstliche Intelligenz,
Humboldt-Universität zu Berlin, Unter den Linden 6, 10099 Berlin, Germany
http://www.aiboteamhumboldt.com

Abstract. Collision detection in a quadruped robot based on the comparison of sensor readings (actual motion) to actuator commands (intended motion) is described. Ways of detecting such incidences using just the sensor readings from the servo motors of the robot's legs are shown. Dedicated range sensors or collision detectors are not used. It was found that comparison of motor commands and actual movement (as sensed by the servo's position sensor) allowed the robot to reliably detect collisions and obstructions. Minor modifications to make the system more robust enabled us to use it in the RoboCup domain, enabling the system to cope with arbitrary movements and accelerations apparent in this highly dynamic environment. A sample behavior is outlined that utilizes the collision information. Further emphasis was put on keeping the process of calibration for different robot gaits simple and manageable.

1 Introduction

Many research efforts in mobile robotics aim at enabling the robot to safely and robustly navigate and to move about both known and unknown environments (e.g. the rescue scenarios in the RoboCup Rescue League [1], planetary surfaces [13]). While wheeled robots are widely used in environments where the robot can move on flat, even surfaces (such as office environments or environments that are accessible to wheelchairs [5]), legged robots are generally believed to be able to deal with a wider range of environments and surfaces. There are many designs of legged robots varying in the number of legs used, ranging from insectoid or arachnoid with 6, 8 or more legs (e.g. [2]), 4-legged such as the Sony Aibo [3], humanoid: 2-legged (e.g. [8]).

Obstacle avoidance is often realized using a dedicated (360°) range sensor [12]. Utilizing vision rather than a dedicated sensor is generally a much harder task since a degree of image understanding is necessary. For the special case of color coded environments, straight forward solutions exist that make use of the knowledge about the robot's environment (such as the color of the surface or the color of obstacles [6]). If, however, obstacle avoidance fails, robots often are unable to detect collisions since many designs lack touch sensors or bumpers. The robot is unaware of the failure of its intended action and ends up in a

D. Nardi et al. (Eds.): RoboCup 2004, LNAI 3276, pp. 150–159, 2005.
© Springer-Verlag Berlin Heidelberg 2005

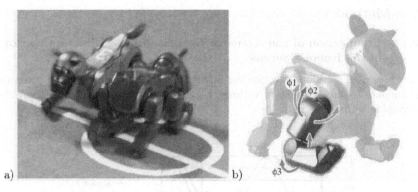

Fig. 1. a) A collision of two robots. Neither robot cannot move into the desired direction. Even worse, robots often interlock their legs which further prevent them from resolving the situation. b) Illustration of the DOFs of the Aibo. Each robot leg has three joints, two degrees of freedom (DOF) in the *shoulder joint* and one DOF in the *knee joint*, denoted Φ_1, Φ_2 and Φ_3. Joints are labeled in the following way: F(ront) or H(ind) + L(eft) or R(ight) + Number of joint $(1, 2, 3)$. Using this nomenclature, the knee joint of the highlighted leg in the above image is *FR3*

situation it is unable to resolve; it is - quite literally - "running into a wall" without noticing it.

Apart from the current action failing, collisions (and subsequently being stuck) have severe impact on the robot's localization if odometry is used to any degree in the localization process (as is the case in [11, 4]). For these approaches to be robust against collisions, they tend to not put much trust in odometry data .

This work investigates the possibilities of detecting collisions of a Sony Aibo 4-legged robot using the walking engine and software framework described in [10]. The robot does not have touch sensors that can be used to detect collisions of it with the environment. As we will show, the servo motor's direction sensors can be used for this task. Work by [9] shows that it is possible to learn servo direction measurements for different kinds of (unhindered) motions and use this to detect slippage of the robot's legs and also to detect collisions of the robot with its environment.

The approach to collision detection using the Aibo presented by [9] stores a large number of reference sensor readings and uses these to detect unusual sensor readings caused by collision and slip. Our approach differs in that we make assumptions about the robot motion that allows the robot to detect collisions by comparing the actuator command (desired motion) to the sensor readings (actual motion). The used set of reference values can be much smaller using this approach. We will show that the method is robust and also quickly adjustable to different walking gaits, robots, and surfaces. Section 4 compares the two approaches in detail.

2   Method

2.1   Comparison of the Actuator Signals and the Direction Sensor of the Robot's Servos

The presented collision detection method is based on the comparison of actuator commands to direction sensor readings. Fig. 2 shows typical sensor measurements alongside actuator signals.

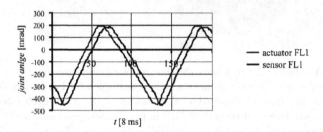

Fig. 2. Sensor and actuator data of freely moving legs (in the air) at a desired ground speed of 75 mm/s. Sensor and actuator curves are almost congruent except for a slight phase shift

It can be seen that for an unhindered period of movement T the sensor and actuator curve of a joint are congruent, i.e. they are of the same shape but shifted by a phase $\Delta\varphi$. If $\Delta\varphi$ was zero, the area in between the two curves becomes minimal:

$$0 \leq \int_{t_0}^{t_0+T} (a(t) - s(t + \Delta\varphi))^2 \, dt \tag{1}$$

Tests using discrete time showed that collisions cause a discrepancy between actuator and sensor data which can be recognized by calculating the area between the sensor and actuator data. It was found that it was not necessary to sum over one complete period of the motion to detect collisions. Shorter intervals yield faster response times. Trading off response time and sensitivity to sensor noise, we found that 12 frames[1] were sufficient. The last 12 frames are used to calculate the the Total Squared Difference (TSD):

$$TSD_{a,s}(\Delta\varphi) = \sum_{i=t_1}^{t_2} (a_i - s_{(i+\Delta\varphi)})^2 \tag{2}$$

Diagram 3 shows the TSD of the *FL1* joint (left shoulder) for a robot colliding with the field boundary. The peaks in the TSD when the robot's leg hits the boundary are clearly distinguishable.

---

[1] A frame is an atomic step of the motion module; there are 125 frames per second, one frame is 8 ms long.

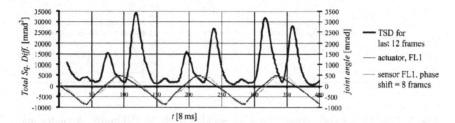

Fig. 3. Sensor and actuator data of a collision with the field boundary walking forward at 150 mm/s. In the TSD the collisions can be seen as peaks in the curve. They occur briefly after the actual collision and can easily be distinguished from unhindered movements

For classification of collisions the TSD is compared to a threshold. If the TSD is larger than this threshold, it is assumed that a collision has occurred. The thresholds for every motion component (i.e. walking forward/backward, walking sideways, rotation) are saved in a lookup table. For combined motions (e.g. walking forward and walking sideways at the same time) the different thresholds for each motion component are summed as described in section 3.3.

2.2 Aligning the Actuator and Sensor Curve

Fig. 4 shows the impulse response of one of the robot's servo motors. It can be seen that the joint doesn't move for about 5 frames (40 ms). Reasons for this are momentum and calculation time; the step height and the load that the joints have to work against also have an influence on the observed phase difference. After 5 frames the joint slowly starts moving and accelerates until it reaches its maximum speed after 8 frames. Just before reaching its destination, the joint angle changes are decreasing. This is due to the joint's P.I.D. controller smoothing the robot's motions.

In figure 5, a) the TSD is shown for a sample motion. The smallest values of the TSD are found at the intersection of the two curves. Collision effects

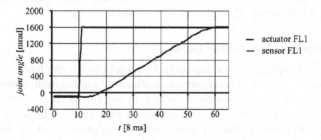

Fig. 4. Sensor and actuator data for a rectangular actuator impulse. The actuator function jumps to its new value. The corresponding servo's direction sensor readings are shown

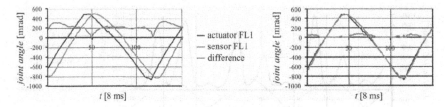

Fig. 5. *Left.* Sensor and actuator data for walking freely at 150 mm/s. Actuator and sensor curve out of phase and the corresponding TSD *Right.* As above but phase shifted. Sensor function is shifted by 8 frames. The corresponding TSD now clearly shows collisions (peaks in the curve)

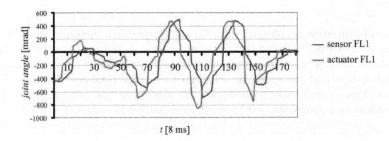

Fig. 6. Actuator commands and sensor measurements during an actual RoboCup game. The robot is changing directions frequently. It can be seen that the servo is unable to perform the requested motions

have little influence on the difference level. In b) actuator and sensor curves are aligned by shifting the sensor data curve left by 8 frames. The calculated TSD shows a strong response to collisions.

Since phase shifts of varying length were observed, the 12 frames wide window of the TSD is calculated for several phase shifts $\Delta\varphi$ ranging from 6 to 15 frames. The smallest TSD is used to detect collisions. This approach eliminates phase shifts which are not caused by collisions and reduces the risk of wrongly recognized collisions (false positives). Due to the small number of possible values of $\Delta\varphi$, real collisions still produce a strong signal.

2.3 Filtering of Actuator Input

The presented approach to collision detection works well under laboratory conditions, i.e. when applied to homogeneous motions with small, well defined motion changes (see sample application described in section 4). In real world applications, motion commands may change rapidly over time as the robot interacts with the environment. In the dynamic, highly competitive RoboCup domain, the robot changes its walking speed and direction quite frequently as determined by the behavior layer of the agent. Figure 6 shows the actuator commands for a robot playing soccer. Most of these changes are relatively small and unproblematic but some are too extreme to be executed by the servos, e.g. when the robot suddenly sees the ball and moves towards it at the highest possible speed. This is compensated by increasing the TSD threshold if the joint acceleration exceeds a certain value. This increased threshold is used only for some tenths of a second and then falls back to its initial level.

2.4   Threshold Calibration

The values of the thresholds are calibrated manually. They are measured for each of the elementary motions (forward/backward, sideways, rotation) in steps of 30 mm/s and 0.5 rad respectively. This adds up to a total of 40 measurements needed for operation.

A threshold is determined by letting the robot walk freely and without collision or slip on the field *freely* for about three seconds while monitoring both motor commands and sensor readings. The TSD is calculated and the maximum TSD is used to derive a threshold value. The maximum TSD value that occurred is tripled; this means that for the robot to detect a collision, the TSD must be 3 times greater than the maximum TSD measured during calibration.

In our experiments the calibration was done by hand since robot gaits do not undergo frequent change and the calibration process is performed quickly. We therefore did not see the need for automating the calibration process (given that an external supervisor has to make sure that no collisions occur during calibration anyway).

3   Detectability of Collisions During Directed Robot Locomotion

For different walking directions, collisions have different effects on the robot's joints depending on how the joints are hindered in their motion. Therefore, the following cases were investigated. In our experiments, only the legs' servos were used for collision detection. However, the robot's head motors can also be used to directly detect whether a robot hits an obstacle with its head (or its head's freedom of motion is otherwise impaired by an obstacle).

In the detection of collisions, a trade off has to be performed between being sensitive to collisions and being robust against false positives (i.e. the detection of a collision where in reality the robot was moving freely). Since we wanted to avoid false positives, the threshold value for detecting collisions was raised at the cost of being less sensitive to detecting collisions. Furthermore, by integrating the information gathered over a short period of time, false positives can be suppressed. This, however, makes the robot less reactive.

3.1   Elementary Motions

Walking Forward or Backward. Collisions are easily detected in the front left or right shoulder joints *FL1* and *FR1* of the robot, depending on which of the legs hits the obstacle (see 1). This way collisions with the field boundary can be detected. Collisions with other robots can also be detected, but not as reliably because this sort of collision is of a much more complex type (the other robot may be moving, etc.). Collisions when walking backwards are slightly harder to recognize because of the particular position of the joints of the hind legs. This is due to the robot's body being tilted forward and the backward motion not being symmetric to the forward motion.

The rate of detection of collisions during forward movement was about 90%; for backward movement it was about 70%. Sometimes collisions would not be detected because the robot would push itself away from obstacles rather than being hindered in its joints' motions. No false positives were observed.

Walking Sideways. Collisions which are occurring while the robot is walking sideways can be recognized best in the sideways shoulder joint θ_2 (e.g. *FL2*) on the side where the robot hits the obstacle. This is not quite as reliable as in the case of forward motions because the Aibo loses traction more quickly when walking sideways for the gait that was used. About 70% percent of the actual collisions were detected. Some phantom collisions were detected at a rate of about 1-2 per minute.

Turning. The same joints that are used to recognize collisions while moving sideways can be used to recognize collisions when the robot is turning. This way, a common type of collision can also be detected: The legs of two robots attempting to turn interlock and prevent the rotation from being performed successfully. How well this can be recognized depends on how much grip the robots have and on the individual turning (or moving) speeds.

The detection rate of collisions and the rate of false positives is of the same order as when the robot is moving sideways. When raising the detection threshold to completely eliminate false positives for a robot rotating at 1.5 rad/sec, the rate of detection drops to about 50%.

3.2   Leg Lock

The before mentioned "leg lock" also occurs in situations where two robots are close to each other (e.g. when chasing the ball). Leg lock is detected in the same way collisions are. Therefore, "leg lock" is detected but cannot be distinguished from other collisions.

3.3   Superposition of Elementary Motions

While it is easy for the robot to recognize the above motions separately, it is harder to recognize collisions when motions are combined, e.g. when the robot walks forward *and* sideways at the same time. For lower speeds, the resulting motions can be viewed as a superposition of the three elementary motions and the resulting threshold is approximated by the sum of the three individual thresholds:

$$T(v, s, r) = T(v, 0, 0) + T(0, s, 0) + T(0, 0, r) \qquad (3)$$

where v is the forward, s the sideways, and r the rotation component of the motion. For high speeds, the requested motions exceed the servos performance. To compensate for this, the collision thresholds are increased by multiplication by a scale factor f which is a function of v and s:

$$f = f(v, s) = \begin{cases} 1 & \text{if } v < 50\text{mm/s and } s < 50\text{mm/s} \\ \frac{v+s}{100} & \text{otherwise} \end{cases} \tag{4}$$

With this extension, the method can be applied to practically all kinds of robot motions and speeds that we observed in a RoboCup game.

4 Application and Performance

Sample Application. A simple behavior was implemented in the XABSL behavior mark up language [7]: The robot walks straight ahead; if it touches an obstacle with one of its front legs, it stops and turns left or right depending on the leg the collision was detected with. The robot turns away from where the collision occurred then continues to walk straight.

This simple behavior was tested on the RoboCup field in our laboratory and it was found to work reliably regardless of the type of collision (e.g. static obstacle or other robots). Collisions were detected with high accuracy. In some rare cases, collisions would not be detected immediately because of slippage of the robot's legs. In these cases, the robot would recognize the collision after a brief period of time (order of tenths of seconds).

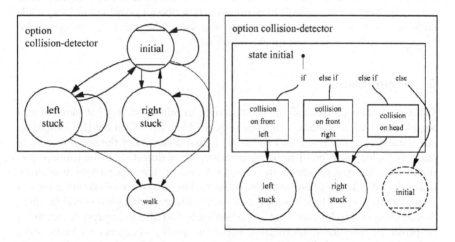

Fig. 7. Simple behavior option graph denoted in XABSL [7]. The robot walks forward until it hits an obstacle. It then turns away from it and continues walking in the new direction

RoboCup. As pointed out in [9], collision detection can be used to have the robot "realize" that an intended action was not successful and to have it act accordingly. It did, however, prove to be a difficult task to find the right action in a situation where two robots run into each other. This usually happens when they pursue the same goal, in our case when both are chasing the ball. Backing off gives the opponent robot an advantage, pushing it makes the situation worse. Current work investigates possible actions.

Other Approaches. A similar approach aimed at traction monitoring and collision detection was presented by another RoboCup team, the "Nubots", in 2003 [9]. The method compares the current sensor data to reference sensor data. It does not use actuator commands for collision detection. The reference data consists of sensor data value and variance for a given motion type and is measured prior to the run. This training is done by measuring the sensor data of possible combinations of elementary motions. A four-dimensional lookup table is used to store the reference data. The four dimensions of the table are: forward/backward motion (backStrideLength), sideward motion (strafe), rotation (turn), and time parameter which stores information about the relative position of the paw in its periodic trajectory. Using this approach, the "Nubots" were able to detect collisions and slip. However, the four-dimensional lookup table requires a considerable amount of memory and training time (according to [9], 20x12x20x20 entries are used to fully describe a gait). During the training it is important that no collisions or slip occur. Using the lookup-table, no assumptions are made about similarities between actuator command and sensor readings.

In contrast, our approach makes the assumption that there is a similarity between intended and actual motion and the variance of the sensor signal is constant for the entire period of the motion. Making these (fair) assumptions, very little memory is needed (40 parameters describe all possible motions) while still achieving good results in detecting obstacles. The parameter table needed for a given gait is generated quickly and easily.

5   Conclusion

With the presented method the robot is able to reliably detect collisions of a 4-legged robot with obstacles on even surfaces (e.g. RoboCup field). Comparing the requested motor command to the measured direction of the servo motors of the robot's legs was found to be an efficient way of detecting if the robot's freedom of motion was impaired. In a sample behavior, the robot turns away from obstacles after having detected the collision. The method was extended for use in RoboCup games. Here it is used to detect collisions (with players and the field boundaries) and to let the robot act accordingly and also to improve localization by providing additional information about the quality of current odometry data (validity). Further work will focus on finding appropriate reactions in competitive situations.

Acknowledgments

The project is funded by the Deutsche Forschungsgemeinschaft, Schwerpunkt-programm "Kooperierende Teams mobiler Roboter in dynamischen Umgebungen" ("Cooperative Teams of Mobile Robots in Dynamic Environments").

Program code used was developed by the GermanTeam, a joint effort of the Humboldt University of Berlin, University of Bremen, University of Dortmund, and the Technical University of Darmstadt. Source code is available for download at http://www.robocup.de/germanteam.

References

1. Robocup rescue web site. http://www.isd.mel.nist.gov/robocup2003. 2003.
2. J. E. Clark, J. G. Cham, S. A. Bailey, E. M. Froehlich, P. K. Nahata, R. J. Full, and M. R. Cutkosky. Biomimetic Design and Fabrication of a Hexapedal Running Robot. In *Intl. Conf. Robotics and Automation (ICRA2001)*, 2001.
3. M. Fujita and H. Kitano. Development of an Autonomous Quadruped Robot for Robot Entertainment. *Autonomous Robots*, 5(1):7–18, 1998.
4. J.-S. Gutmann, W. Burgard, D. Fox, and K. Konolige. An experimental comparison of localization methods. *Proceedings of the 1998 IEEE/RSJ Intl. Conference on Intelligent Robots and Systems (IROS'98)*, 1998.
5. A. Lankenau, T. Röfer, and B. Krieg-Brückner. Self-Localization in Large-Scale Environments for the Bremen Autonomous Wheelchair. In *Spatial Cognition III*, Lecture Notes in Artificial Intelligence. Springer, 2002.
6. S. Lenser and M. Veloso. Visual Sonar: Fast Obstacle Avoidance Using Monocular Vision. In *Proceedings of IROS'03*, 2003.
7. M. Lötzsch, J. Bach, H.-D. Burkhard, and M. Jüngel. Designing agent behavior with the extensible agent behavior specification language XABSL. In *7th International Workshop on RoboCup 2003 (Robot World Cup Soccer Games and Conferences)*, Lecture Notes in Artificial Intelligence. Springer, 2004. to appear.
8. C. L. P. Dario, E. Guglielmelli. Humanoids and personal robots: design and experiments. *Journal of Robotic Systems*, 18(2), 2001.
9. M. J. Quinlan, C. L. Murch, R. H. Middleton, and S. K. Chalup. Traction Monitoring for Collision Detection with Legged Robots. In *RoboCup 2003 Symposium*, Lecture Notes in Artificial Intelligence. Springer, 2004. to appear.
10. T. Röfer, I. Dahm, U. Düffert, J. Hoffmann, M. Jüngel, M. Kallnik, M. Lötzsch, M. Risler, M. Stelzer, and J. Ziegler. GermanTeam 2003. In *7th International Workshop on RoboCup 2003 (Robot World Cup Soccer Games and Conferences)*, Lecture Notes in Artificial Intelligence. Springer, 2004. to appear. more detailed in http://www.robocup.de/germanteam/GT2003.pdf.
11. T. Röfer and M. Jüngel. Vision-Based Fast and Reactive Monte-Carlo Localization. *IEEE International Conference on Robotics and Automation*, 2003.
12. T. Weigel, A. Kleiner, F. Diesch, M. Dietl, J.-S. Gutmann, B. Nebel, P. Stiegeler, and B. Szerbakowski. CS Freiburg 2001. In *RoboCup 2001 International Symposium*, Lecture Notes in Artificial Intelligence. Springer, 2003.
13. K. Yoshida, H. Hamano, and T. Watanabe. Slip-Based Traction Control of a Planetary Rover. In *Experimental Robotics VIII*, Advanced Robotics Series. Springer, 2002.

Learning to Drive and Simulate Autonomous Mobile Robots

Alexander Gloye, Cüneyt Göktekin, Anna Egorova,
Oliver Tenchio, and Raúl Rojas

Freie Universität Berlin, Takustraße 9, 14195 Berlin, Germany
http://www.fu-fighters.de

Abstract. We show how to apply learning methods to two robotics problems, namely the optimization of the on-board controller of an omnidirectional robot, and the derivation of a model of the physical driving behavior for use in a simulator.

We show that optimal control parameters for several PID controllers can be learned adaptively by driving an omni directional robot on a field while evaluating its behavior, using an reinforcement learning algorithm. After training, the robots can follow the desired path faster and more elegantly than with manually adjusted parameters.

Secondly, we show how to learn the physical behavior of a robot. Our system learns to predict the position of the robots in the future according to their reactions to sent commands. We use the learned behavior in the simulation of the robots instead of adjusting the physical simulation model whenever the mechanics of the robot changes. The updated simulation reflects then the modified physics of the robot.

1 Learning in Robotics

When a new robot is being developed, it is necessary to tune the on-board control software to its mechanical behavior. It is also necessary to adapt the high-level strategy to the characteristics of the robot. Usually, an analytical model of the robot's mechanics is not available, so that analytical optimization or a perfect physical simulation are not feasible. The alternative to manual tuning of parameters and behaviors (expensive and error-prone *trial and error*) is applying learning methods and simulation (cheap but effective *trial and error*). We would like the robot to optimize its driving behavior after every mechanical change. We would like the high-level control software to optimize the way the robot moves on the field also, and this can be best done by performing simulations which are then tested with the real robot. But first the simulator must learn how the real robot behaves, that is, it must synthesize a physical model out of observations. In this paper we tackle both problems: the first part deals with the "learning to drive" problem, whereas the second part deals with the "learning to simulate" issue.

D. Nardi et al. (Eds.): RoboCup 2004, LNAI 3276, pp. 160–171, 2005.
© Springer-Verlag Berlin Heidelberg 2005

1.1   Learning to Drive

When autonomous robots move, they compute a desired displacement on the floor and transmit this information to their motors. Pulse width modulation (PWM) is frequently used to control their rotational velocity. The motor controller tries to bring the motor to speed — if the desired rotational velocity has not yet been reached, the controller provides a higher PWM signal. PID (proportional, integral, diferential) controllers are popular for this kind of applications because they are simple, yet effective. A PID controller can register the absolute difference between the desired and the real angular velocity of the motor (the error) and tries to make them equal (i.e. bring down the error to zero). However, PID control functions contain several parameters which can only be computed analytically when an adequate analytical model of the hardware is available. In practice, the parameters are set experimentally and are tuned by hand. This procedure frequently produces suboptimal parameter combinations.

In this paper we show how to eliminate manual adjustments. The robot is tracked using a global camera covering the field. The method does not require an analytical model of the hardware. It is specially useful when the hardware is modified on short notice (adding, for example, weight or by changing the size of the wheels, or its traction). We use learning to find the best PID parameters. An initial parameter combination is modified stochastically — better results reinforce good combinations, bad performance imposes a penalty on the combination. Once started, the process requires no human intervention. Our technique finds parameters so that the robot meets the desired driving behavior faster and with less error. More precise driving translates in better general movement, robust positioning, and better predictability of the robot's future position.

1.2   Learning to Simulate

Developing high-level behavior software for autonomous mobile robots (the "playbook") is a time consuming activity. Whenever the software is modified, a test run is needed in order to verify whether the robot behaves in the expected way or not. The ideal situation of zero hardware failures during tests is the exception rather than the rule. For this reason, many RoboCup teams have written their own robot simulators, which are used to test new control modules in the computer before attempting a field test. A simulator saves hours of work, especially when trivial errors are detected early, or when subtle errors require many stop-and-go trials, as well as experimental reversibility.

The simulator of the robotic platform should simulate the behavior of the hardware as accurately as possible. It is necessary to simulate the delay in the communication and the robot's inertia; heavy robots do not move immediately when commanded to do so. The traction of the wheels, for example, can be different at various speeds of the robot, and all such details have to be taken into account. An additional problem is that when the robots are themselves being developed and optimized, changes in the hardware imply a necessary change in the physical model of the robot for the simulation. Even if the robots does not change, the environment can change. A new carpet can provide better or worse traction and if the model is not modified, the simulation will fail to reflect accurately the new situation. In practice, most simulation systems settle for a simplistic "Newtonian" constant-friction mass model, which does not correspond to the real robots being used.

Our approach to solve this modelling problem is to learn the reaction of the robots to commands. We transmit driving commands to a mobile robot: The desired direction, the desired velocity and desired rotation. We observe and record the behavior of the robot when the commands are executed using a global video camera, that is, we record the instantaneous robot's orientation and position. With this data we train predictors which give us the future position and orientation of the robots in the next frames, from our knowledge of the last several ones [4]. The data includes also commands sent to the robots. The predictor is an approximation to the physical model of the robot, which covers many different situations, such as different speeds, different orientations during movement, and start and stop conditions. This learned physical model can then be used in our simulator providing the best possible approximation to the real thing, short of an exact physical model which can hardly be derived for a moving target.

2   Related Work

We have been investigating learning the physical behavior of a robot for some time [4]. Recently we started applying our methods to PID controllers, using reinforcement learning.

The PID controller has been in use for many decades, due to its simplicity and effectiveness [6]. The issue of finding a good method for adjusting the PID parameters has been investigated by many authors. A usual heuristic for obtaining initial values of the parameters is the Ziegler-Nichols method [18]. First an initial value for the P term is found, from which new heuristic values for the P, I, and D terms are derived. Most of the published methods have been tested with computer simulations, in which an analytical model of the control system is provided. When an analytical model is not available, stochastic optimization through genetic programming [13] or using genetic algorithms is an option. Our approach here is to use reinforcement learning, observing a real robot subjected to real-world constraints. This approach is of interest for industry, where often a PID controller has to tune itself adaptively and repetitively [17].

The 4-legged team of the University of Texas at Austin presented recently a technique for learning motion parameters for Sony Aibo robots [12]. The Sony robots are legged, not wheeled, and therefore some simplification is necessary due to the many degrees of freedom. The Austin team limited the walking control problem to achieving maximum forward speed. Using "policy gradient reinforcement learning" they achieved the best known speed for a Sony Aibo robot. We adapted the policy reinforcement learning method to omnidirectional robots by defining a quality function which takes into account speed and accuracy of driving into account. Another problem is that we learn to drive in all directions but not just forward. This makes the learning problem harder, because there can always be a compromise between accuracy and speed, but we succeeded in deriving adequate driving parameters for our robots.

With respect to simulations, the usual approach is to build as perfect a model of the robot and feed it to a simulation engine such as ODE (Open Dynamics Engine). This is difficult to do, and the simulated robot probably will not behave as the real robot, due to the many variables involved. In an influential paper, for example, Brooks and Mataric identify four robotic domains in which learning can be applied: learning parameters, learning about the world, learning behaviors, and learning to coordinate [5]. We are not aware, at the moment, of any other RoboCup team using *learned* physical behaviors of robots for simulations. We think that our approach saves time and produces better overall results than an ODE simulation.

3   The Control Problem

The small size league is the fastest physical robot league in the RoboCup competition, all velocities considered relative to the field size. Our robots for this league are controlled with a five stages loop: a) The video image from cameras overlooking the field is grabbed by the main computer; b) The vision module finds the robots and determines their orientation [15]; c) Behavior control computes the new commands for the robots; d) The commands are sent by the main computer using a wireless link; e) A Motorola HC-12 microcontroller on each robot receives the commands and directs the movement of the robot using PID controllers (see [7]). Feedback about the speed of the wheels is provided by the motors' impulse generators.

For driving the robots, we use three PID controllers: one for the forward direction, one for the sideward direction and one for the angle of rotation (all of them in the coordinate system of the robot). The required Euclidean and angular velocity is transformed in the desired rotational speed of three or four motors (we have three and four-wheeled robots). If the desired Euclidian and angular velocity has not yet been achieved, the controllers provide corrections which are then transformed into corrections for the motors.

3.1   Microcontroller

The control loop on the robot's microcontroller consists of the following sequence of tasks: The robot receives from the off-the-field computer the target values for the robot's velocity vector v_x, v_y and the rotational velocity ω, in its local coordinate system; the HC-12 microcontroller, which is constantly collecting the current motor speed values by reading the motors' pulse generators, converts them into Euclidian magnitudes (see Section 3.2); the PID-Controller compares the current movement with the target movement and generates new control values (Section 3.3); these are converted back to motor values, which are encoded in PWM signals sent to the motors (Section 3.2).

3.2   From Euclidian Space to Wheel Parameters Space and Vice Versa

The conversion of the robot velocity vector (v_x, v_y, ω) to motor velocity values w_i of n motors is computed by:

$$
\begin{pmatrix} w_1 \\ w_2 \\ \vdots \\ w_n \end{pmatrix} = \frac{1}{r} \begin{pmatrix} x_1 & y_1 & b \\ x_2 & y_2 & b \\ \vdots & \vdots & \vdots \\ x_n & y_n & b \end{pmatrix} \begin{pmatrix} v_x \\ v_y \\ \omega_n \end{pmatrix}.
\tag{1}
$$

The variable r is the diameter of the omnidirectional wheels, b is the distance from the rotational center of the robot to the wheels, and $F_i = (x_i, y_i)$ is the force vector for wheel i. The special case of three wheels at an angle of $120°$ can be calculated easily.[1]

For the opposite direction, from motor velocities to Euclidian velocities, the calculation follows from Eq. (1) by building the pseudo-inverse of the transformation matrix. We map the values of n motors to the three dimensional motion vector. If the number of wheels is greater than three, the transformation is overdetermined, giving us the nice property of compensating the pulse counter error of the wheels (by a kind of averaging).

3.3   PID Controller

As explained above, we have programmed three PID controllers, one for the forward (v_x), one for the sideward velocity (v_y), and one for the desired angular velocity (ω). Let us call $e_x(t)$ the difference between the required and the actual velocity v_x at time t. Our PID controller computes a correction term given by

$$
\Delta v_x(t) = Pe_x(t) + I(\sum_{k=0}^{\ell} e_x(t-k)) + D(e_x(t) - e_x(t-1))
\tag{2}
$$

There are several constants here: P, I, and D are the proportionality, integration, and difference constants, respectively. The value of $\Delta v_x(t)$ is incremented (with respect to the leading sign) by an offset and then cut into the needed range. The correction is proportional to the error (modulated by P). If the accumulated error is high, as given by the sum of past errors, the correction grows, modulated by the integral constant I. If the error is changing too fast, as given by the difference of the last two errors, the correction is also affected, modulated by the constant D. A controller without I and D terms, tends to oscillate, around the desired value. A controller with too high I value does not oscillate, but is slow in reaching the desired value. A controller without D term can overshoot, making convergence to the desired value last longer.

---

[1] http://www-2.cs.cmu.edu/~reshko/PILOT/

The error value used in the above formula is multiplied by a scaling constant before plugging its value in the formula. This extra parameter must also be learned. It depends on the geometry of the robot.

3.4   Learning the PID Parameters

We solve the parameter optimization problem using a policy gradient reinforcement learning method as described in [12]. The main idea is based on the assumption that the PID parameters can be varied independently, although they are correlated. Thus, we can modify the parameter set randomly, calculate the partial error derivative for each parameter, and correct the values. Note that, in order to save time, we vary the whole parameter set in each step and not each parameter separately.

The parameter set $\mathcal{P}$ consists of $6 \times 3 = 18$ elements $(p_1, \ldots, p_{18})$. The number of parameters is independent from the number of wheels, because we use one PID controller for each degree of freedom and not for each wheel. The standard hand-optimized parameters are used as the starting set. In each step, we generate a whole new suite of n parameter sets

$$
\begin{aligned}
\mathcal{P}^1 &= (p_1 + \pi_1^1, \ldots, p_{18} + \pi_{18}^1) \\
\mathcal{P}^2 &= (p_1 + \pi_1^2, \ldots, p_{18} + \pi_{18}^2) \\
&\vdots \\
\mathcal{P}^n &= (p_1 + \pi_1^n, \ldots, p_{18} + \pi_{18}^n).
\end{aligned}
\tag{3}
$$

Whereby the value π_i^j is picked with uniform probability from the set $\{-\epsilon_i, 0, +\epsilon_i\}$ and ϵ_i is a small constant, one for each p_i.

The evaluation of one parameter set consists of a simple test. The robot has to speed up from rest into one particular direction, at an angle of 45° relative to its orientation. It has to drive for some constant time, without rotations and as far as possible from the starting point. Then the robot has to stop abruptly, also without rotating and as fast as possible (see Fig. 1(b)). During this test phase, the robot does not receive any feedback information from the off-the-field computer.

Each test run is evaluated according to the evaluation function $\mathcal{Q}(\mathcal{P}^j)$, which is a weighted function of the following criteria: the deviation of the robot to the predetermined direction, the accumulated rotation of the robot, the distance of the run, and the distance needed for stopping. The only positive criterion is the length of the run; all other are negative.

We evaluate the function $\mathcal{Q}(\mathcal{P}^j)$ for all test parameter sets $\mathcal{P}^j$, where $j = 1, \ldots, n$. The sets are collected according to the π constants for every parameter into three classes:

$$
\begin{aligned}
\mathcal{C}_i^+ &= \{\mathcal{P}^j | \pi_i^j = +\epsilon_i, j = 1, \ldots, n\}, \\
\mathcal{C}_i^- &= \{\mathcal{P}^j | \pi_i^j = -\epsilon_i, j = 1, \ldots, n\}, \\
\mathcal{C}_i^0 &= \{\mathcal{P}^j | \pi_i^j = 0, j = 1, \ldots, n\}
\end{aligned}
\tag{4}
$$

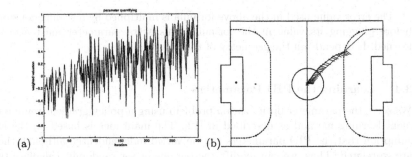

(a)   (b)

Fig. 1. (a) The quality of the tested parameter sets. The graph shows only the evaluated parameter sets which are varying up to ϵ_i from the learned parameter i. Therefore, the values are noisy. (b) Example of a test run. The straight line shows the desired direction and the dotted line shows the real trajectory of the robot on the field

For every class of sets, the average quality is computed:

$$\mathcal{A}_i^+ = \frac{\sum_{\mathcal{P} \in \mathcal{C}_i^+} \mathcal{Q}(\mathcal{P})^x}{\|\mathcal{C}_i^+\|}, \ \mathcal{A}_i^- = \frac{\sum_{\mathcal{P} \in \mathcal{C}_i^-} \mathcal{Q}(\mathcal{P})^x}{\|\mathcal{C}_i^-\|}, \ \mathcal{A}_i^0 = \frac{\sum_{\mathcal{P} \in \mathcal{C}_i^0} \mathcal{Q}(\mathcal{P})^x}{\|\mathcal{C}_i^0\|} \quad (5)$$

This calculation provides us a gradient for each parameter, which shows us, whether some specific variance π makes the results better or not. If this gradient is unambiguous, we compute the new parameter value according to Eq. 6:

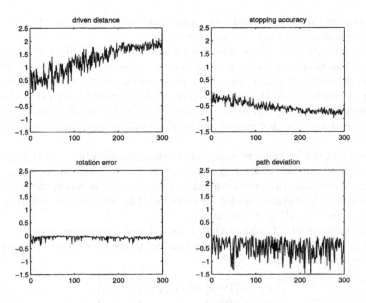

Fig. 2. The four componets of the quality function. The total quality is a weighted average of these four magnitudes

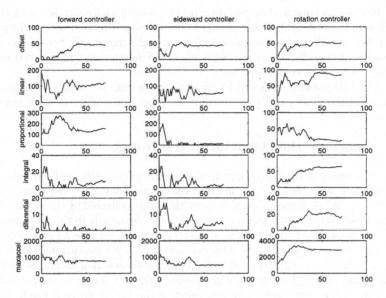

Fig. 3. The learned parameters, started from the hand-optimized set. The parameters are not independent

$$p'_i = \begin{cases} p_i + \eta_i & \text{if } \mathcal{A}_i^+ > \mathcal{A}_i^0 > \mathcal{A}_i^- \text{ or } \mathcal{A}_i^+ > \mathcal{A}_i^- > \mathcal{A}_i^0 \\ p_i - \eta_i & \text{if } \mathcal{A}_i^- > \mathcal{A}_i^0 > \mathcal{A}_i^+ \text{ or } \mathcal{A}_i^- > \mathcal{A}_i^+ > \mathcal{A}_i^0 \\ p_i & \text{otherwise} \end{cases} \qquad (6)$$

Where the learning constant for each parameter is η_i. The old parameter set is replaced by the new one, and the process is iterated until no further progress is detected.

3.5   Results

We made two experiments: In the first, we learned only the P, I, and D values. These values were initialized to 0. The evolution of the learning process is shown in Fig. 1(a) and in Fig. 2. Fig. 1(a) shows the quality of the test runs during the whole learning process and Fig. 1(b) a sample run of the robot. The total quality of a test run can be broken down into the four quality magnitudes. These are shown in Fig. 2. At the beginning, the robot does not move. After 10 iterations of parameter evaluation and adaptation, the robot moves at an average of one meter per second, with acceleration and braking. The average deviation from the desired direction is 10 percent at the end point.

In a second experiment, the PID controller was initialized with the hand-optimized values. The result is shown in Fig. 3. The parameters are not independent, which leads to some oscillations at the beginning.

In the experiment shown in Fig. 3 the rotation controller is also learned, although the task of the robot is only to drive in one direction without rotating. However, rotation control is important for straight movement, because an error caused by wheel slippage at the beginning of the movement — in the acceleration phase — can be compensated later.

The PID parameters learned with our experiments are used now for control of our robots. They can be relearned, if the robot is modified, in a few minutes.

4   Learning the Behavior of the Robot

We reported in a previous paper how we predict the position of our small-size robots in order to cope with the unavoidable system delay of the vision and control system [4]. When tracking mobile robots, the image delivered by the video camera is an image of the past. Before sending the new commands to the robot we have to take into account when the robot will receive them, because it takes some time to send and receive commands. This means that not even the current real position of the robots is enough: we need to know the future position and future orientation of the robots. The temporal gap between the last frame we receive and the time our robots will receive new commands is the *system delay*. It can be longer or shorter, but is always present and must be handled when driving robots at high speed (up to 2 m/s in the small size league). Our system delay is around 100 ms, which corresponds to about 3 to 4 frames of a video camera running at 30 fps.

The task for our control system is therefore, from the knowledge of the last six frames we have received, and from the knowledge of the control commands we sent in each of those six frames, to predict the future orientation and position of the robots, four frames ahead from the past.

The information available for this prediction is preprocessed: since the reaction of the robot does not depend on its coordinates (for a homogeneous floor) we encode the data in the robot's local coordinate system. Obstacles and walls must be handled seperately. We use six vectors for position, the difference vectors between the last frame which has arrived and the other frames in the past, given as (x, y) coordinates. The orientation data consist of the difference between the last registered and the six previous orientations. Each angle θ is encoded as a pair $(\sin \theta, \cos \theta)$ to avoid discontinuity when the angle crosses from 2π to 0. The desired driving direction, velocity and rotation angle transmitted as commands to the robot of the last six frames are given as one vector with (v_x, v_y, ω)-coordinates. They are given in the robots coordinate system. We use seven float values per frame, for six frames, so that we have 42 numbers to make the prediction. We don't use the current frame directly, but indirectly, because we actually use the differences between each last frame and the current frame. The current motor values do not influence the robot motion in the next four frames (because of the delay), so they are irrelevant.

We use neural networks and linear regression models to pretict the future positions and orientations of the robot, one or four frames in advance. For details see [8].

4.1   The Simulator

Once we have trained a neural network to simulate the physical response of the robot to past states and commands, we can plug-in this neural network in our behavior simulation. We play with virtual robots: they have an initial position and their movement after receiving commands is dictated by the prediction of the behavior of the real robots in the next frame. We have here an interesting interplay between the learned physical behavior and the commands. In each frame we use the trained predictors to "move" the robots one more frame. This information, however, is not provided to the behavior software. The behavior software sends commands to the virtual robots assuming that they will receive them with a delay (and the simulator enforces this delay). The behavior software can only ask the neural network for a prediction of the position of the robot in the fourth frame (as we do in real games). But the difference between what the high-level behavior control "knows" (the past, with four frames delay) and how the simulator moves the robot (a prediction, only one frame in advance) helps us to reproduce the effect of delays. Our simulator reproduces playing conditions as nearly as possible. Fig. 4 shows a comparison of the driving paths obtained with different models and reality.

As can be seen in Fig. 4, the "Newtonian" model is too smooth. The real driving behavior is more irregular, because when the robot drives it overshoots or slips on the floor. Our simple "Newtonian" model could be extended by a more realistic one, which includes for examle nonlinear friction terms. The learned behavior (c) reflects more accurately this more problematic and realistic driving behavior, so we don't need any model. This is important for high-level control, because the higher control structures must also learn to cope with unexpected driving noise.

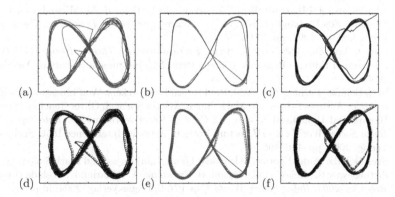

Fig. 4. A comparison of driving paths: (a) real path driven by a real robot, (b) simulation of the same path with a physical model and (c) simulation with the prediction model. (d),(e),(f) show the same paths including the orientation of the robot, drawn as small segments

5   Conclusions and Future Work

Our results show that it is possible to apply learning methods in order to optimize the driving behavior of a wheeled robot. They also show that learning can even be used to model the physical reaction of the robot to external commands.

Optimizing the driving behavior means that we need to weight the options available. It is possible to let robots move faster, but they will collide more frequently due to lack of precision. If they drive more precisely, they will tend to slow down. Ideally, in future work, we would like to derive intelligent controllers, specialized in different problems. The high-level behavior could decide which one to apply, the more aggressive or the more precise. Another piece of future work would be trying to optimize high-level behaviors using reinforcement learning, so that they compensate disadvantages of the robot on-board control.

We have seen in Fig. 4, that the predicted robot motion (c) is more accurate than the real one (a), because the prediction (any prediction) smooths the observed real movements. The driving noise could be also analyzed and its inclusion in the simulator would make the model even more realistic.

This paper is part of our ongoing work on making robots easier to adapt to an unknown and variable environment. We report elsewhere our results about automatic color and distortion calibration of our computer vision system [9]. The interplay of vision and control software will make posible in the future to build a robot, put it immediately on the field, and observe how it gradually learns to drive.

References

1. Karl J. Aeström, Tore Hägglund, C. Hang, and W. Ho, "Automatic tuning and adaptation for PID controllers—A survey", in L. Dugard, M. M'Saad, and I. D. Landau (Eds.), *Adaptive Systems in Control and Signal Processing*, Pergamon Press, Oxford, 1992, pp. 371–376.
2. Karl J. Aeström, and Tore Hägglund, *PID Controllers: Theory, Design, and Tuning*, Second Edition, Research Triangle Park, NC, Instrument Society of America, 1995.
3. Sven Behnke, Bernhard Frötschl, Raúl Rojas, Peter Ackers, Wolf Lindstrot, Manuel de Melo, Andreas Schebesch, Mark Simon, Martin Spengel, Oliver Tenchio, "Using Hierarchical Dynamical Systems to Control Reactive Behavior", *RoboCup-1999: Robot Soccer World Cup III* Lecture Notes in Artificial Intelligence 1856, Springer-Verlag, 2000, pp. 186–195.
4. Sven Behnke, Anna Egorova, Alexander Gloye, Raúl Rojas, and Mark Simon, "Predicting away the Delay", in D. Polani, B. Browning, A. Bonarini, K. Yoshida (Eds.), *RoboCup-2003: Robot Soccer World Cup VII*, Springer-Verlag, 2004, in print.
5. Rodney A. Brooks, and Maja J. Mataric, "Real robots, real learning problems, in Jonathan H. Connell and Sridhar Mahadevan (Eds.), *Robot Learning*, Kluwer Academic Publishers, 1993.
6. Albert Callender, and Allan Brown Stevenson, "Automatic Control of Variable Physical Characteristics U.S. patent 2,175,985. Issued October 10, 1939 in the United States.

7. Anna Egorova, Alexander Gloye, Achim Liers, Raúl Rojas, Michael Schreiber, Mark Simon, Oliver Tenchio, and Fabian Wiesel, "FU-Fighters 2003 (Global Vision)", in D. Polani, B. Browning, A. Bonarini, K. Yoshida (Eds.), *RoboCup-2003: Robot Soccer World Cup VII*, Springer-Verlag, 2004, in print.

8. Alexander Gloye, Mark Simon, Anna Egorova, Fabian Wiesel, Oliver Tenchio, Michael Schreiber, Sven Behnke, and Raùl Rojas, "Predicting away robot control latency", Technical Report B08-03, Free University Berlin, 2003.

9. Alexander Gloye, Mark Simon, Anna Egorova, Fabian Wiesel, and Raùl Rojas, "Plug & Play: Fast Automatic Geometry and Color Calibration for a Camera Tracking Mobile Robots", Institut für Informatik, Freie Universität Berlin, February 2004, paper under review.

10. Barbara Janusz, and Martin Riedmiller, "Self-Learning neural control of a mobile robot", *Proceedings of the IEEE ICNN'95*, Perth, Australia, 1995.

11. Alexander Kleiner, Markus Dietl, Bernhard Nebel, "Towards a Life-Long Learning Soccer Agent", in D. Polani, G. Kaminka, P. Lima, R. Rojas (Eds.), *RoboCup-2002: Robot Soccer World Cup VI*, Springer-Verlag, Lecture Notes in Computer Science 2752, pp. 119-127.

12. Nate Kohl, and Peter Stone, "Policy Gradient Reinforcement Learning for Fast Quadrupedal Locomotion," Department of Computer Science, The University of Texas at Austin, November 2003, paper under review.

13. John R. Koza, Martin A. Keane, Matthew J. Streeter, William Mydlowec, Jessen Yu, and Guido Lanza, *Genetic Programming IV: Routine Human-Competitive Machine Intelligence*, Kluwer Academic Publishers, 2003.

14. Martin Riedmiller and Ralf Schoknecht, "Einsatzmöglichkeiten selbständig lernender neuronaler Regler im Automobilbereich", *Proceedings of the VDI-GMA Aussprachetag*, Berlin, March 1998.

15. Mark Simon, Sven Behnke, Raúl Rojas, "Robust Real Time Color Tracking", in P. Stone, T. Balch, G. Kraetzschmar (Eds.), *RoboCup 2000: Robot Soccer World Cup IV*, Lecture Notes in Computer Science 2019, Springer-Verlag, 2001, pp. 239–248.

16. Peter Stone, Richard Sutton, and Satinder Singh, "Reinforcement Learning for 3 vs. 2 Keepaway", in P. Stone, T. Balch, G. Kraetzschmar (Eds.), *RoboCup 2000: Robot Soccer World Cup IV*, Lecture Notes in Computer Science 2019, Springer-Verlag, 2001, pp. 249-258.

17. K.K. Tan, Q.G. Wang, C.C. Hang and T. Hägglund, *Advances in PID Control*, Advances in Industrial Control Series, Springer Verlag, London, 1999.

18. J. G. Ziegler, and N. B. Nichols, "Optimum settings for automatic controllers, *Transactions of ASME*, Vol. 64, 1942, pp. 759-768.

RoboCupJunior — Four Years Later

Elizabeth Sklar[1] and Amy Eguchi[2]

[1] Department of Computer Science, Columbia University,
1214 Amsterdam Avenue, New York, NY 10027, USA
sklar@cs.columbia.edu
[2] School of Education, University of Cambridge,
17 Trumpington Street, Cambridge, CB2 1QA UK
eae24@hermes.cam.ac.uk

Abstract. In this paper, we report on the status of the RoboCupJunior league, four years after it was founded. Since its inception in 2000, we have been surveying and/or interviewing students and mentors who participate in the international event. Here we present a high-level overview of this data. We discuss demographics of participants, characteristics of preparation and educational value. We highlight trends and identify needs for the future, in terms of event organization, educational assessment and community-building.

1   Introduction

RoboCupJunior (RCJ), the division of RoboCup geared toward primary and secondary school students, was founded in 2000. The focus in the Junior league is on *education*. RCJ offers three challenges (see figure 1) — **soccer**, **rescue** and **dance** — each emphasizing both cooperative and competitive aspects. The stated mission of RoboCupJunior is: "to create a learning environment for today, and to foster understanding among humans and technology for tomorrow". RCJ provides an exciting introduction to the field of robotics, a new way to develop technical abilities through hands-on experience with computing machinery and programming, and a highly motivating opportunity to learn about teamwork while sharing technology with friends. In contrast to the one-child-one-computer scenario typically seen today, RCJ provides a unique opportunity for participants with a variety of interests and strengths to work together as a team to achieve a common goal.

The idea for RoboCupJunior was demonstrated in 1998, with a demonstration at RoboCup-98 in Paris [1]. The first international competition was held in 2000 in Melbourne, Australia at RoboCup-2000, with 25 teams from 3 countries participating [2]. In 2001, in Seattle, USA, there were 25 teams from 4 countries (83 students plus 17 mentors) [3]. In the following year, the initiative exploded and 59 teams from 12 countries came to RoboCupJunior-2002 in Fukuoka, Japan (183 students plus 51 mentors) [4]. Most recently, 67 teams from 15 countries participated at RoboCupJunior-2003 in Padova Italy (233 students plus 56 mentors). The fifth annual international RoboCupJunior event will be held in Lisbon, Portugal in early July 2004 and a similar rate of expansion is expected.

D. Nardi et al. (Eds.): RoboCup 2004, LNAI 3276, pp. 172–183, 2005.

Fig. 1. RoboCupJunior challenges

This paper focuses on reporting on the status of the RoboCupJunior league from several standpoints. Since its inception, we have been involved in evaluating RCJ for the dual purpose of tracking its growth in terms of members and internationalization and examining what it is about RCJ (and robotics in general) that is so exciting and motivating for students and what intrigues mentors and keeps them involved from one year to the next. We have conducted interviews and surveys of students and mentors since the first RCJ event in Melbourne. Our initial report was a pilot study, based on interviews of mentors [2]. A follow-up study was reported in 2002 that included input from students and compared the trends identified in 2000 to data collected in 2001 [3]. The data indicated the possibility of exciting results, if a more comprehensive study were conducted with more participants over a longer time period. This report attempts to take a step in that direction.

Since the initiative exploded in popularity in 2002, we have collected more than three times as much data as in the first two years. Here, we analyze that data and compare it with the initial years. Additionally, we report the demographic statistics and increase in participation world-wide. We describe developments within the RCJ league and close with a brief summary and future plans.

It is important to keep in mind when reading this report that the data was collected at an annual international event hosted by a different country each year. The host regions typically account for approximately 40% of RCJ teams, so as a result, the trends are highly subjective to the norms and characteristics of each region. The interesting piece is to find trends that breach the cultural divide and some are identified below.

2   Participation

We examine rates of participation in terms of the number of teams and the number of students and mentors who have attended each event. Three years are compared: 2001-2003. All the data presented for subsequent years both in terms of registration and evaluation statistics was not collected in 2000, so we consider 2000 to be a pilot year and here restrict our comparisons to 2001-2003. We view the data in three ways, looking at the population distribution across countries, challenges and gender.

We examine the international distribution of teams over the three years. In 2001, teams from 4 countries participated. This rose to 12 countries in 2002 and 15 in 2003.

Table 1. Distribution of teams from different countries

	2001	2002	2003		2001	2002	2003
Australasia				*Europe*			
Australia	10 (40%)	8 (14%)	5 (7%)	Denmark		1 (2%)	
China		2 (3%)	4 (6%)	Finland		1 (2%)	1 (1%)
Japan		29 (49%)	12 (18%)	Germany	5 (20%)	5 (8%)	15 (22%)
Korea		5 (8%)		Italy			7 (10%)
Singapore			4 (6%)	Norway		1 (2%)	1 (1%)
Taiwan			2 (3%)	Portugal			1 (1%)
Thailand		4 (7%)		Slovakia		1 (2%)	3 (4%)
North America				UK	2 (8%)		3 (4%)
Canada		1 (2%)	2 (3%)	*Middle East*			
USA	8 (32%)	1 (2%)	4 (6%)	Iran			3 (4%)

Countries are grouped by region and listed in alphabetical order. Entries contain the number of teams that participated that year. The number in parenthesis indicates what percentage of total participation was represented by that country. Blank entries indicate that a country did not participate in the corresponding year. Bold entries highlight the host country each year.

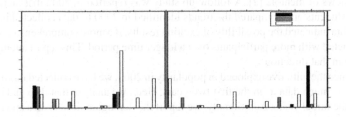

Fig. 2. Distribution each year of students from different countries

key: = Australia;   = Canada;   = China;   = Denmark;   = Finland;   = Germany;   = Iran;   = Italy;   = Japan;   = Korea;   = Norway;   = Portugal;   = Singapore;   = Slovakia;   = Taiwan;   = Thailand;   = UK;   = USA.

Statistics on teams are shown in table 1. The corresponding statistics counting students (instead of teams) is illustrated graphically in figure 2.

Table 2 shows the distribution of participants entering each of the four challenges: 1-on-1 soccer, 2-on-2 soccer, rescue and dance. Note that 1-on-1 soccer was not held in 2001 and rescue was not held in 2002. These were the decisions of local organizers[1]. It is interesting to note that 2-on-2 soccer remains the most popular challenge, involving from about two-thirds to three-quarters of participants. Dance attracts about one-third of participants. Rescue, revamped in 2003, draws just under 20% of participants; however it is anticipated that this percentage will rise (see section 4.2).

---

[1] Subsequently, through efforts to provide a more formal structure for RCJ, these crucial types of decisions were placed in the hands of an international technical committee which includes two members of the RoboCup Federation Executive Committee.

Table 2. Challenges

	2001		2002		2003	
	teams	students	teams	students	teams	students
dance	7 (28%)	29 (35%)	12 (20%)	63 (34%)	18 (27%)	67 (29%)
rescue	4 (16%)	16 (19%)			12 (18%)	39 (17%)
1-on-1 soccer			4 (7%)	10 (5%)	14 (21%)	49 (21%)
2-on-2 soccer	22 (88%)	68 (82%)	45 (76%)	125 (68%)	45 (67%)	166 (71%)

Table 3. Percentage of female student participants, per year and by challenge

	2001	2002	2003
total	10 (12%)	30 (16%)	37 (16%)
dance	5 (17%)	16 (25%)	22 (**33%**)
rescue	0 (0%)		7 (18%)
1x1soccer		0 (0%)	3 (6%)
2x2soccer	7 (10%)	16 (13%)	13 (8%)

We are also interested in studying the gender balance, across years and challenges, for both students and mentors. Here, we present data on rates of female participation within the student population. Note that we are not including mentors because the data we have is sparse and inaccurate. One planned improvement for the immediate future is standardization and centralization of data collection for all RCJ participants (students and mentors).

In table 3, we show the percentage of female student participants over all three years. The total number is presented and is also broken down by challenge. Note that some students participate in more than one challenge, which explains why the sum of the values broken down by challenge exceeds the total number in the top row. Also note that the challenge percentages are the rates of female participation calculated over all students who participated in that challenge (not over all participants).

RoboCupJunior has seen strong growth in the number of female participants. We highlight the fact that the dance challenge, which provides a unique outlet for creativity, attracted 33% female participants in 2003, double the rate of just two years earlier, and also well above typical rates for computer science and most engineering fields, which generally range from 10-20% female. This trend has been duplicated in all of the national open events held thus far, most notably in Australia where over half the RCJ dance participants in 2003 were female. This is impressive, as the scale of the Australian RoboCupJunior effort is such that each state has its own regional championship and on the order of five hundred students participate in the country's national RoboCupJunior event each year.

3   Evaluation

In 2000, we conducted video-taped interviews of mentors with the intent of beginning a longterm study of the effects of RCJ across a wide spectrum of technical, academic and social areas. We transcribed and analyzed this data, which is presented in [2]. The experience informed the creation of a set of surveys for both students and mentors, which we administered in 2001. These were analyzed and presented in [3]. We used these results to modify our survey methodology, shifting from mostly open-ended, qualitative questions to a closed, quantitative questionnaire for 2002 and 2003. In this section, we detail our findings from these last two years, comparing them to the results of 2001. Note that since our data collection methodology from 2000 differed so significantly from the subsequent years, we do not attempt to make any direct comparisons to this data.

It is also worth mentioning that we conducted video-taped interviews of students in 2001 and 2002. Analysis of this data is problematic for several reasons. The data collection methodology was inconsistent, primarily because interviews were conducted by volunteers who were not trained interviewers and so many interviews became conversations and deviated from the prescribed set of questions. This was compounded by language issues (both at the time of the interviews and later, during transcription). In addition, accurate transcription is extremely time-consuming and error-prone. Thus we cannot draw statistical conclusions from these interviews, although they have served a useful purpose in developing the surveys. We did not conduct interviews in 2003 and do not plan any for 2004. Future evaluations may include more structured interviews conducted by researchers in areas such as education and human development.

Table 4 shows the rate of return on the surveys for all three years (2001-2003). Totals are shown; as well, the data is subdivided by gender and country. Except for the totals, the percentages are calculated as the rate of return of all people who responded (not over all participants). The percentages reported for "country" are the percentage of the total responses that were from each nation. This gives an indication of the extent to which overall trends might be attributed to a particular country. This is not the same as the response rate from each country, i.e., the percentage of participants from a particular country who complete the surveys. That is shown in table 5 and gives an indication of to what extent a collective country's response is representative of that country.

3.1   Students' Responses

A total of 192 students participated in the survey (2001, 39 students; 2002, 104 students; 2003, 49 students). Out of these, 84% were male and 16% were female (162 males and 29 females in total; 2001, 34 males, 5 females; 2002, 86 males, 17 females; 2003, 42 males, 7 females). Here we present analysis of their responses to four questions.

How was your team organized? We collected valid answers to this question only in 2002 and 2003. In 2001, the question was phrased with open-ended responses rather than multiple choice answers, and many students misunderstood the question (for example, some answered "well"). So we changed the format of the question to multiple-choice in 2002. Based on the 2002 and 2003 data, many students responded that their teams were organized at after school programs (2002, 46%; 2003, 39%). About one fifth of teams were organized by one of the team members (2002, 18%; 2003, 22%). In both years, 14% of the students responded that their teams were organized by community youth groups or organizations. However, in 2002, 13% of the student participants reported that their teams were organized in class as part of their normal school day (as opposed to 2% in 2003. But in 2002, about half of the teams were Japanese teams and the event

Table 4. Return rates on surveys, 2001-2003

	2001		2002		2003	
	students	mentors	students	mentors	students	mentors
total responses	39 (48%)	16 (94%)	104 (57%)	16 (29%)	49 (21%)	27 (53%)
breakdown by gender						
male	34 (87%)	13 (81%)	86 (83%)	24 (89%)	42 (86%)	10 (63%)
female	5 (13%)	3 (19%)	17 (16%)	3 (11%)	7 (14%)	4 (25%)
breakdown by country						
Australia	11 (28%)	2 (13%)	17 (16%)	4 (15%)	0 (0%)	1 (6%)
Canada			5 (5%)	0 (0%)	0 (0%)	0 (0%)
China			0 (0%)	0 (0%)	0 (0%)	0 (0%)
Denmark			1 (1%)	1 (4%)	0 (0%)	
Finland			2 (2%)	1 (4%)	0 (0%)	0 (0%)
Germany	11 (28%)	6 (38%)	9 (9%)	2 (7%)	13 (27%)	1 (6%)
Iran					7 (14%)	2 (13%)
Italy					0 (0%)	0 (0%)
Japan			57 (55%)	15 (56%)	16 (33%)	3 (19%)
Korea			0 (0%)	0 (0%)		
Norway			2 (2%)	1 (4%)	1 (2%)	1 (6%)
Portugal					0 (0%)	0 (0%)
Singapore					9 (18%)	5 (31%)
Slovakia			2 (2%)	1 (4%)	3 (6%)	3 (19%)
Taiwan					0 (0%)	0 (0%)
Thailand			6 (6%)	2 (7%)		
UK	1 (3%)	1 (6%)			0 (0%)	0 (0%)
USA	16 (41%)	7 (44%)	3 (3%)	0 (0%)	0 (0%)	0 (0%)

Note that in 2003, 2 mentors (12%) did not answer the gender question.

was organized by the local city government. Because the local city government recruited teams from the city public district, many participating teams from the local area were organized at school.

How did you find out about RoboCupJunior? In 2001 and 2002, many students reported that they learned about RCJ from their school teachers (2001, 74%; 2002, 61%; 2003 24%). On the other hand, in 2003, the most popular informant was a local robotics society (35%), which was 0% in 2001 and 11% in 2002. This difference can be attributed to local influences in the host region.

What robot platform did your team use? The most popular robot platform used by RCJ teams is the Lego Mindstorms Robotics Invention Kit (2001, 72% of the student participants; 2002, 46%; 2003, 43%). This could be because Lego Mindstorms is also the most widely available robot platform around the world. One interesting trend to point out is that in 2002, 41% of student participants reported that their teams used the Elekit SoccerRobo. This is because RCJ-2002 was held in the city in Japan where the Elekit company is headquartered. Another trend to point out is, in recent years, more and more

Table 5. Survey return rates for students and mentors, by year and country

	2001		2002		2003	
	students	mentors	students	mentors	students	mentors
Australia	48%	67%	77%	57%	0%	25%
Canada			56%	0%	0%	0%
China			0%	0%	0%	0%
Denmark			100%	50%		
Finland			100%	0%	0%	0%
Germany	79%	120%	56%	50%	22%	11%
Iran					70%	67%
Italy					0%	0%
Japan			57%	71%	53%	30%
Korea			0%	0%		
Norway			100%	100%	50%	100%
Portugal					0%	0%
Singapore					56%	125%
Slovakia			100%	100%	38%	100%
Taiwan					0%	0%
Thailand			60%	50%		
UK	33%	50%			0%	0%
USA	37%	100%	100%	0%	0%	0%

teams are adding components of their own. In 2001, no student participants reported that they added components not included in the original kit. But 16% students in 2002 and 31% in 2003 reported that they added components.

How much time did you spend preparing for RCJ? Students were asked to specify when they began preparation for the event, how often their team met and how long each meeting lasted. Most teams spend 1-3 months preparing (36% in 2001, 30% in 2002 and 39% in 2003); however responses ranging from 3-12 months are only slightly lower. Very few teams spend less than 1 month preparing. Most teams meet once per week, although this data is hard to tally, since many students wrote in the margins of the survey that they started meeting once a week, and then met more frequently as the event drew closer. Overwhelmingly, teams spend more than 90 minutes at each meeting. All of these trends regarding preparation time are very similar from one year to the next, not deviating for different regions. It is interesting to note the length of meeting time. Since class periods in schools are typically shorter than 90 minutes, this points out that it is hard to find sufficient preparation time for RCJ only through classroom work.

3.2   Mentors' Responses

A total of 59 mentors participated in the survey (2001, 16 mentors; 2002, 27 mentors; 2003, 16 mentors). Out of the 59 mentor survey participants, 80% were male and 17% were female[2]. Here we present analysis of their responses to four questions.

---

[2] 47 males and 10 females in total; 2001, 13 males, 3 females; 2002, 24 males, 3 females; 2003, 10 males, 4 females, 2 did not provide gender data.

What was your role in the team? Out of all the mentor participants, 33 are school teachers, 13 are parents of the student participants, 10 are community group organizers, 3 are science museum/center staff, and 6 are from some type of organization. Every year, about half of the mentors are teachers (2001, 41%; 2002, 67%; 2003, 50%). In 2001, more parents (50% of the respondents) and fewer teachers got involved in than other years. Since RCJ typically occurs in July, many schools around the world are on summer holiday, so finding teachers to participate can be problematic. This was highlighted at RCJ-2001, when the event was held in the USA, because the summer school holidays are long and students tend to go to camp or get jobs.

What type of school and community does your team come from? Many teams are affiliated with public schools (2001, 63%; 2002, 41%; 2003, 69%). This shows that educational robotics is not limited for those who go to private schools with high-end technologies. Most of them are from either urban or suburban areas (2001, 38% in urban and 38% in suburban; 2002, 44% in urban and 44% in suburban; 2003, 38% in urban and 6% in suburban). This could be because of a lack of RoboCupJunior-related events and/or activities in rural areas in general. Many RCJ local competitions tend to be held in large cities. This suggests that the organizers of RCJ events need to examine ways to extend local events to more rural areas in the future.

How did your team fund its effort? Mentors reported that about 70% or more teams received money from their schools, sponsors, local government, or/and fundraising activities (2001, 75%; 2002, 68%; 2003, 88%). In 2001, half the teams received funding from sponsors. 19% of the teams did fundraising activities and 13% received support from their school. 31% of teams had their parents pay for them. On the other hand, in 2002, more teams had their parents pay (41%) and were less successful for receiving sponsorship (33%). Also, teams received more funding from their schools (30%) and local government/board of education (11%). However, some teams had to have the team members and/or mentors pay to participate (teacher, 2 teams; members, 1 team). In 2003, half of the teams received sponsorship including RoboCupJunior travel support, and 25% of them receive support from their school. However, parental support was still one of the main resources for the teams.

On the other hand, more than 75% of the mentors did not receive any direct funding (2001, 75%; 2002, 89%; 2003, 75%) (i.e., payment for their time, e.g., as after-school teachers). Some of them were able to get paid through their schools or from grants, but it is obvious that the mentors need avenues for financial support. Yet this statistic is astounding — as overworked as most schoolteachers are, the vast majority of them are motivated enough by RCJ to volunteer their time and participate, sometimes even spending their own money. Despite of the financial hardships, most mentors indicate their intention to participate the International competition again (2001, 63%; 2002, 89%; 2003, 88%).

Do you use robotics in your school curriculum, and if yes, how? Out of 33 teacher-mentors, 31 teacher-participants provided information about their schools. All of them teach middle (11-14 year-old) and/or high (older than 15) school age students. Only four teach elementary (5-10 year old) school age students (some of them also teach

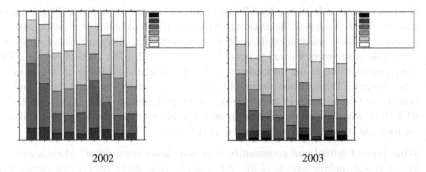

Fig. 3. Students' self-assessment

key: = math;   = physics;   = computer programming;   = mechanical engineering;
         = electronics;   = general science;   = communication skills;   = teamwork;   =
personal development (such as organization). NA means that the question was not answered.

older students). This suggests that educational robotics is used more with students who
are older than 10. The future questions will be finding out the reason why RCJ does
not attract elementary school teachers and how to make it more accessible for these
teachers; although we can speculate on two points. First, it is more difficult to travel
with younger children than with older students[3]; so it may be that, as a result, RCJ
is perceived as an event for older participants. Second, younger children typically have
trouble concentrating for extended periods of time. Given that the large majority of teams
spend over 90 minutes at each meeting time, it may be that many younger children do
not have the focus to stay involved for a sustained period of time, over a number of
months.

About half of the mentors teach technology. Other commonly taught subjects are
chemistry (6 teachers), general science (6 teachers), and physics (5 teachers). Despite
the fact that educational robotics can be used to illustrate a variety math concepts, only
three mentors are math teachers (all three participated in 2001).

Eighteen teachers responded that they use robotics in their curriculum and 13 teachers
do not use it in the curriculum but do use it in after school programs. However, 13
teachers out of the 18 teachers organized their RCJ teams after school, not within their
class room period. This suggests that we should investigate in the future what are the
obstacles for teachers using robotics in their curriculum to organize their teams as part
of classroom activities.

We are working on developing materials to help teachers take an integrated approach
to educational robotics and RCJ. Students should be encouraged to write lab reports,
journaling their efforts in engineering and programming. They can create posters and
oral presentations about their developments. As indicated above, robotics can be used
to demonstrate a wide variety of math skills; and some of our work involves creating
curriculum to do this.

---

[3] Less supervision and parental involvement is required.

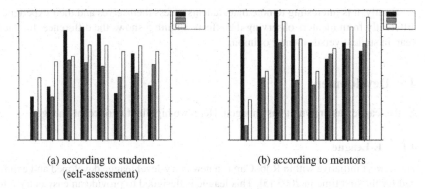

(a) according to students   (b) according to mentors
(self-assessment)

Fig. 4. Which aspects were positively affected by students' participation in RCJ?

3.3   Educational Value

A large part of the evaluation is dedicated to trying to identify what students are actually learning from participating in RCJ. The surveys administered in 2002 and 2003 include a series of statements (such as, "I am better in math because of working with robots.") to which respondents indicated agreement or disagreement on 5-point scale. Figure 3 shows how students responded.

In order to get more of a snapshot view, we took this data plus the responses to similar questions from 2001 and interpreted each response as to whether the students and mentors thought each aspect was positively affected by involvement in RoboCupJunior. The results are shown in figure 4.

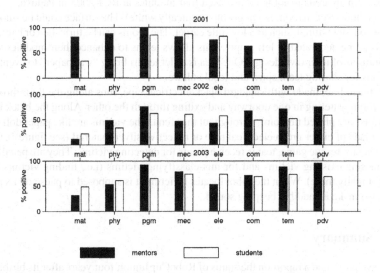

Fig. 5. Comparisons between students' self-assessment and mentors' observations

Finally, it is interesting to note that students assess themselves and their experience differently from mentors observing the effects. Figure 5 shows the difference, for each year, in student versus mentor opinion.

4   Development

As the league has grown, it has changed. Here, we highlight two recent changes.

4.1   E-League

The newest initiative within RoboCup is a new entry-level league developed and exhibited for the first time in 2003 [5]. This league is designed to provide an easy entry into RoboCup involving more undergraduate students, as a means for RoboCupJunior "graduates" to continue with the project between high school and graduate school (where most RoboCup senior league participants come from). The league is also intended to serve as an accessible entry-point for undergraduates who have not participated in RoboCupJunior. The league is inexpensive (compared to the other RoboCup senior leagues).

4.2   Rescue

In 2003, we designed and constructed a miniature, modular version of the NIST standard USAR test bed especially for RoboCupJunior [6]. The design features a varying number of "rooms", connected by hallways and ramps (see figure 1b). Two doorways are located at standard points in each room so that multiple rooms (modules) can be linked together easily. Modules can be stacked, to provide additional challenge; lighting conditions in lower rooms with a "roof" are different than in rooms with an open top. The number of modules in an arena is not fixed; we used four modules at RCJ-2003 in Padova.

The floor of each room is a light color (typically white). The surface could be smooth, like wood, or textured, such as a low-pile carpet. The rooms can be furnished or bare; the walls can be decorated or left empty. This allows teams to enhance their modules with decorations of their own design. One idea is to let teams bring "wallpaper" to events as a means for sharing team spirit and local culture.

A black line, made with standard black electrical tape, runs along the floor through each room, entering in one doorway and exiting through the other. Along the black line, "victims" are placed randomly throughout the arena. The victims are like paper doll cutouts, made of either green electrical tape or reflective silver material (see figure 1c). As in the senior rescue game, teams receive points for detecting victims. They are penalized for missing existing victims and for mis-classifying victims (i.e., finding victims that are not really there). When the robot locates a victim, it is supposed to pause on its path and also make an audible beeping sound.

5   Summary

We have presented a report on the status of RoboCupJunior, four years after its birth. We have provided statistical data on the demographics of participants, highlighting gender differences and a broadening range of internationalization. Further, we have offered results and analyses of evaluation surveys collected at RoboCupJunior international events since 2000. New developments were described, and we identified areas of improvement for the future.

One aspect of evaluating only the annual RCJ international event is that there is a concentration of teams from the local, host region. This presents a challenge from a research standpoint, since the cohort differs somewhat from year to year. However, it also helps highlight particular characteristics of these regions. Expanding our data collection to local and regional events world-wide will help identify broader effects that are (and are not) sensitive to cultural differences.

As RCJ expands worldwide, there is an increasing need to establish a better organizational foundation and structure for information dissemination and community-building. We have recently received significant support from the Association for Computing Machinery (ACM) to maintain and grow the initiative on an international basis. This support will help us improve the RCJ web-site and offer improved channels for information, communication and resources.

http://www.robocupjunior.org

References

1. Lund, H.H., Pagliarini, L.: Robot Soccer with LEGO Mindstorms. In: RoboCup-98: Robot Soccer World Cup II, Lecture Notes in Artificial Intelligence (LNAI). Volume 1604., Springer Verlag (1998)
2. Sklar, E., Johnson, J., Lund, H.H.: Children Learning from Team Robotics: RoboCup Junior 2000 Educational Research Report. Technical report, The Open University, Milton Keynes, UK (2000)
3. Sklar, E., Eguchi, A., Johnson, J.: RoboCupJunior: learning with educational robotics. In: Proceedings of RoboCup-2002: Robot Soccer World Cup VI. (2002)
4. Sklar, E.: RoboCupJunior 2002: League Description. In: Proceedings of RoboCup-2002: Robot Soccer World Cup VI. (2002)
5. Anderson, J., Baltes, J., Livingston, D., Sklar, E., Tower, J.: Toward an Undergraduate League for RoboCup. In: Proceedings of RoboCup-2003: Robot Soccer World Cup VII. (2003)
6. Sklar, E.: A long-term approach to improving human-robot interaction: RoboCupJunior Rescue. In: Proceedings of the International Conference on Robotics and Automation (ICRA-04). (2004)

Evolution of Computer Vision Subsystems in Robot Navigation and Image Classification Tasks

Sascha Lange and Martin Riedmiller

Neuroinformatics Group,
Institute for Computer Science and Institute for Cognitive Science,
University of Osnabrück, 49069 Osnabrück, Germany
{Sascha.Lange, Martin.Riedmiller}@uos.de

Abstract. Real-time decision making based on visual sensory information is a demanding task for mobile robots. Learning on high-dimensional, highly redundant image data imposes a real problem for most learning algorithms, especially those being based on neural networks. In this paper we investigate the utilization of evolutionary techniques in combination with supervised learning of feedforward nets to automatically construct and improve suitable, task-dependent preprocessing layers helping to reduce the complexity of the original learning problem. Given a number of basic, parameterized low-level computer vision algorithms, the proposed evolutionary algorithm automatically selects and appropriately sets up the parameters of exactly those operators best suited for the imposed supervised learning problem.

1 Introduction

A central problem within the field of robotics is to appropriately respond to sensory stimulation in real-time. At this, visual sensory information becomes more and more important. When applying techniques of machine learning to such tasks, one faces a number of recurring problems. In most cases it is simply not feasible to feed the whole image information directly into a neural network or into any other learning algorithm. Often a (hand-coded) computer vision subsystem is used to preprocess the images and extract useful visual features. Usually, the developer himself specifies such a restricted set of features and the structure of an intermediate data representation. By using the resulting low-dimensional feature vectors and the intermediate data representation, learning becomes possible.

Since there is still no all-purpose domain-independent computer vision system, adapting or redesigning the computer vision subsystem for a new problem is a time consuming and expensive task. But compared to finding a general-purpose computer vision subsystem and a transferable intermediate data representation it is generally easier and more reliable to build highly specialized, task-dependent preprocessing layers. Accordingly, we believe computer vision subsystems should concentrate on extracting only task-relevant features.

D. Nardi et al. (Eds.): RoboCup 2004, LNAI 3276, pp. 184–195, 2005.

What features are needed to be extracted in order to be fed to the learning algorithm, is highly dependent on the task itself. Therefore, when specifying the intermediate data representation and designing the computer vision subsystem, it is absolutely necessary to consider not only the input images but also the desired output signals.

To give an example (see experiment 2): Consider a set of some images containing a dice in front of a uniform background and others containing only the empty background. If the task were to detect the presence or absence of the dice in a particular image it would at least be necessary to look for a rectangular or quadratic area and return a boolean value. If the dice should be grabbed by a robot-arm the detected feature ("quadratic area") could still be the same. Additionally it would however be necessary to calculate its center of gravity or its bounding box. Finally, if the task was to read the dice it would be necessary to detect a completely different feature, namely the markers on it. However, in this task the position of the features is unimportant and does not need to be considered. Encoding their individual absence or presence is sufficient. It does not take much effort to come up with scenarios in which an extraction of even more features is needed.

This example clearly shows that different applications require different sets of features to be extracted. Searching the images for prominent or interesting features without considering the application of the extracted information can not give the optimal solution to a task. The approach presented in this paper, therefore is to treat the learning of both the preprocessing and the control subsystem as a whole, always considering both input and desired output.

The algorithm presented here, is able to directly solve a supervised learning problem on visual input without any further information or help from the outside. We have chosen to use a hybrid solution that comprises an evolutionary algorithm (outer loop) and the supervised training of feedforward nets (inner loop). The candidate solutions (individuals of the evolutionary process) are composed of a neural network realizing the decision making process and a specialized and highly task-dependent computer vision subsystem for preprocessing the image data. Candidate preprocessing subsystems are automatically constructed during the evolutionary process. The primitives forming the computer vision subsystem are more or less complex, highly parameterized, hand-coded programs, each realizing a feature detector, global operator or other low-level computer vision algorithm.

2   Related Work

A considerable number of methods has been developed to tackle the problem of feature selection. Statistical methods based on principle component analysis (PCA) (for example [5], [8], [9]) are able to remove redundant features with a minimal loss of information. One probelm of such methods is that they are often computationally expensive. Locally linear embedding (LLE) is a non-linear non-iterative unsupervised method that can be used for dimensionality reduction [15]. LLE can be extended to data with class information available [13], too.

Martin C. Martin has used genetic programming to construct visual feature extracting programs for a computer vision subsystem of a robot solving a simple navigation task [11]. Instead of considering the task as a whole, the subtask to be solved by the visual feature extractors and their correct output was explicitly specified.

Tony Belpaeme proposes a genetic algorithm for the evolutionary construction of visual feature extractors not needing any information about the target output of these extractors [2]. According to Belpaeme it would be possible to measure the fitness of the extractors as their performance in the task itself. But for performance reasons he proposes to use the entropy of the resulting feature vectors instead. Belpaeme does not provide any information about the generalization behavior of the algorithm or its performance in real tasks.

The Schema Learning System (SLS) proposed by Bruce A. Draper [6] is able to learn special-purpose recognition strategies under supervision using a library of executable procedures. The parameters of the computer vision algorithms in the library, however, have to be set by hand for every different task [7].

Bala et al. [1] use a genetic algorithm to search for a subset of user provided visual features in an object classification task. A classifier is constructed by inducing a decision tree using these feature subsets and a training set of labeled images.

In [3] Heinrich Braun and Joachim Weisbrod present a hybrid algorithm (ENZO) evolving the topology of feedforward nets. As ENZO is permitted to change the input layer, it is able to select features and reduce the dimensionality. The main fitness criterion is the performance of the evolved nets during a "local" training phase. The algorithm was later combined with reinforcement learning to solve complex learning problems [4].

3   Description of the Algorithm

3.1   Evolution of Candidate Solutions

The navigation and classification tasks examined have all been formulated as supervised learning problems: Given a training set

$$P = \{(\mathcal{B}_p; \boldsymbol{y}_p) \in [0, 255]^{3^{k \times l}} \times [0, 1]^m \mid p = 1, ..., n\}$$

of n training patterns $(\mathcal{B}_p; \boldsymbol{y}_p)$ consisting of images $\mathcal{B}_p$ and corresponding target vectors $\boldsymbol{y}_p$, the task is to find a control program minimizing an error term. We have chosen to use the "total sum of squares" (TSS)

$$E = \sum_{p=1}^{n} \sum_{j=1}^{m} (o_{pj} - y_{pj})^2$$

of the difference between the output $\boldsymbol{o}_p$ of the control program and the target $\boldsymbol{y}_p$. In all of these tasks, the input is a single multi-spectral image encoded in a $k \times l$-matrix $\mathcal{B}$. The entries of the matrix are color values given in the RGB color space.

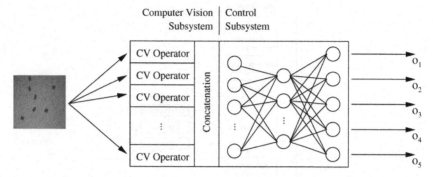

Fig. 1. The general structure of a candidate. The candidate is composed of the computer vision sublayer and a downstream neural network realizing the control subsystem. The computer vision subsystem consists of smaller building blocks (operators) that all access the image matrix directly. Their individual output is concatenated and returned as the output of the computer vision subsystem

In order to minimize the error term an evolutionary algorithm is proposed. The algorithm constructs and modifies complex candidates of a predefined architecture.

The Structure of the Individuals. The evolved candidates (individuals x_i of the population X_t) consist of two distinct subsystems. The layer processing the raw image matrices is a low-level computer vision subsystem. The control subsystem forming the second layer receives its input data from this computer vision subsystem. It performs the output generation (action selection) based on a more compact representation of the original data.

The subsystem realizing low-level vision tasks is a set of smaller subsystems that will be called "operators" throughout this text. These operators all realize a parameterized (by parameter vector p) function $h_p(\mathcal{B}) = v$ returning a q-dimensional vector v for every given image matrix $\mathcal{B}$. These operators could be simple filters like Gaussian blur, point-of-interest detectors like corner detectors or even complex programs like segmentation algorithms. The operators all reduce the input to a representation of the image much simpler and more compact than the image itself. This could be realized by (preferably) reducing the dimension or by choosing a "simpler" but still iconic encoding of prominent features (like edge or region maps) or by combining both of these methods.

These operators are ordered in parallel (Fig. 1) and totally independent of each other. They all access the image directly and do not consider the output of each other.

Generally the operators employed in this work need at least one parameter to appropriately adapt their behavior to the circumstances. Some operators may even change their behavior substantially according to different parameter settings. Each operator returns a vector of constant length that is concatenated to

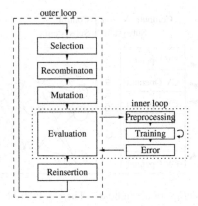

Fig. 2. The different steps of the algorithm. The outer cycle consists of the five steps of the evolutionary algorithm. The inner evaluation cycle evaluates each candidate by using its computer vision subsystems to obtain the preprocessed input data and by training the neural network several epochs on this data. The observed training and testing error are used to calculate the fitness

the output vectors of the other operators to form the final output of the image preprocessing layer. The length of the output vector is allowed to change only when the parameters are changed. Besides the parameters influencing the behavior of the operators, each operator has some boolean parameters to switch the output of a specific feature on or off.

In conclusion, the image processing layers of the candidates may differ in their composition of operators, in the setting of parameters controlling the internal behavior of the operators and in the selection of output values returned.

The control layer is formed by a simple feedforward net. The dimension of the input layer is determined by the size of the image processing layer's output vector, whereas the dimension of the output layer is determined by the dimension of the target vector provided with the training patterns. The internal topology of the net, however, is a free variable. Because the topology of a neural net has a major impact on its learning behavior and has to be chosen carefully it will be subject to evolutionary optimization too.

The Evolutionary Algorithm. Starting with a random initialized population, an (μ, λ)-evolutionary strategy with the addition of elitism is followed. The implemented algorithm comprises five different steps: Selection, recombination, mutation, evaluation and reinsertion (see Fig. 2).

During the selection phase, μ parents are chosen randomly for mating. The probability $P(\boldsymbol{x}_i)$ of selecting individual $\boldsymbol{x}_i$ as the k-th parent is proportional to its fitness:

$$P(\boldsymbol{x}_i) = \frac{f(\boldsymbol{x}_i)}{\sum_{\boldsymbol{x}_j \in X_t} f(\boldsymbol{x}_j)}.$$

During the recombination phase every two "neighboring" individuals x_1 and x_2 are recombined to form an offsprings x' by randomly choosing a subset of operators from the original operators of x_1 and x_2. The network topology is passed unchanged from the first individual.

Afterwards a mutation operator is applied to the resulting offsprings. A random number of entries of the computer vision operators' parameter vectors and the topology of the neural network are mutated by adding a zero-mean gaussian random variable. The chance of changing the entry p_i of the parameter vector p is set to 0.1 in all experiments.

Every newly formed individual is evaluated according to the fitness function described in the next section.

Finally, the fittest λ childrens replace the λ worst individuals of population X_t to receive the new population X_{t+1}. This is a combination of elitism and fitness based reinsertion.

It should be noted that although the evolution of good solutions typically takes several hours (when not parallelized), the basic operators and the resulting programs are able to analyze the input and calculate an output in real-time on present personal computers.

Training the Net to Measure the Fitness. The images $\mathcal{B}$ of the training patterns $(\mathcal{B}_i; y_i)$ are analyzed with each candidate's vision subsystem to form a set of training patterns $(h(\mathcal{B}_i); y_i)$ suitable for the input layer of its neural network. The net is trained on the resulting pattern set using resilient propagation (Rprop [14]) for a specific number of epochs. Afterwards the neural network is evaluated on the testing set. The TSS on the training and testing set are both normalized to yield a fitness value between 0 and 1 (bigger means better). The final fitness of the candidate is the weighted sum of these two values and some additional, less important and lower weighted components that may reward good runtime performance or smaller memory consumption.

In all experiments discussed in this paper, the fitness has been a weighted sum of training error, testing error (smaller gives higher fitness) and of three lower weighted factors penalizing higher numbers of hidden layer neurons and connections, bigger input layers and the runtime needed to process one image by the candidate's vision subsystem. These factors have shown to produce efficient candidates with rather small networks and a good generalization behavior.

3.2   Building Blocks of the Computer Vision Subsystem

From the huge number of well studied operators, detectors and algorithms, we have selected and adapted or implemented five algorithms to be used in the experiments. These operators are: a corner-detector employing the SUSAN principle [16], an operator returning a single image from the gauss pyramid, a histogram operator, returning n 2-dimensional UV histograms of n non-overlapping image partitions (inspired by [17]), a region-growing algorithm known as "color structure code" (CSC, [12]) and the operator with the most parameters, a segmentation algorithm based on the lower layers of the CVTK library [10].

Whereas two operators return iconic representations (of reduced dimensionality), the other operators return "meaningful" symbolic representations. For example the CVTK and CSC operators both return abstract descriptions of uniformly colored image regions containing information such as color, coordinates of the center of gravity, area, bounding rectangle and compactness of the regions found.

Each operator needs at least one parameter to be set properly; for example: the level of the gauss pyramid to return or the maximum color-distance being used by the CSC algorithm to decide, whether or not two neighboring pixels belong to the same region.

Since the detailed description of all algorithms and parameters would go beyond the scope of this paper, we will exemplarily describe in detail the CVTK operator.

This operator uses a given set S of samples $\boldsymbol{s} = (\boldsymbol{s}_{\mathrm{rgb}}, \boldsymbol{s}_c) = (r, g, b, c)$ of the discrete "classification" function $f : I^3 \mapsto 0, 1, .., N$ that assigns a class label c to every rgb-value $\boldsymbol{s}_{\mathrm{rgb}} \in I^3$. Implementation dependent, the values of the red, green and blue channel are from the set $I = \{0, 1, ...255\}$.

The algorithm uses a simple nearest-neighbor classification to segment the image. The label of each pixel b_{ij} of the image $\mathcal{B}$ is determined by finding the "closest" sample $\boldsymbol{s}_{\mathrm{closest}} = \arg\min_{\boldsymbol{s}_{\mathrm{rgb}}} d(\boldsymbol{s}_{\mathrm{rgb}}, b_{ij})$ and assigning its class label $\boldsymbol{s}_c$ iff the distance is smaller than a threshold t. If there is no sample having a distance smaller than t the class label of the background (0) is assigned to b_{ij}.

The distance d is the Euclidian distance between the color values of the pixel b_{ij} and the sample $\boldsymbol{s}$ after transforming both to one of four possible color spaces (RGB, YUV, UV, CIE L*a*b*). The color space can be chosen by setting a parameter of the operator.

Table 1. Parameters of the CVTK operator. Some parameters are influencing the segmentation process itself, whereas a second group of parameters selects what properties should be returned

	parameter	type	multitude	description
algorithm	max_classes	int	1	number of different color classes
	color_space	int	1	encodes the color space to use
	(r,g,b,c)	int^4	1..n	labeled color samples
	t	double	1	distance threshold
	min_size	double	1	minimum size of regions
output	regions_out	int	1	regions per color class to return
	order	boolean	1	order according to area
	area	boolean	1	return the area
	center	boolean	1	return the center of gravity
	class	boolean	1	return the class label
	compactness	boolean	1	return the compactness
	bounding_box	boolean	1	return corners of bounding box

After classifying the whole image pixel by pixel, neighboring pixels of the same color class are connected to regions. Internally, the algorithm uses a contour based region description to calculate region features like center of gravity, area, bounding rectangle or compactness. Finally the algorithm returns a list of n region descriptions. By setting boolean output parameters, the user can specify what features each description should contain. Table 1 lists all parameters of the CVTK operator.

4   Results

The algorithm has been tested on two classification tasks and on one robot-navigation experiment. We have not used any artificial images but only real world images captured by a digital camera under typical office-lighting conditions. There have been absolutely no efforts to reduce shadows or to ensure uniform lighting conditions. The image sets can be found at http://www-lehre.inf.uos.de/~slange/master/ .

Experiment 1: Subtraction of Colored Gaming Pieces. The images show different settings of up to eight red and blue gaming pieces on a white background. There are always at least as many blue as red pieces and never more than four pieces of the same color in the image frame. The task is to calculate the difference between the number of blue and red pieces and to label the images accordingly. There is an additional testing set that is used after finishing the evolution to judge the generalization performance of the whole evolutionary process – in contrast to the net's testing error that is calculated on the first testing set and used to guide the selection of the individuals. The training set contains 98 images, the testing set 27 images and the second testing set 9 images (320 × 240 pixel).

Experiment 2: Reading a Dice. In this setup a dice was filmed from a constant camera position. The images have to be classified into 6 classes according to the number of markers on the upper side of the dice. Again, there are three sets of equally sized (160 × 120 pixel) and correctly labeled images. The training set consists of 45 images, the testing set of 12 and the additional testing set contains 15 images. The second testing set also contains 5 images of a completely different and slightly bigger dice than the dice used during the evolution.

Experiment 3: Driving to the Ball. In this simple navigation task, a camera-equipped robot has to drive to a ball that is positioned somewhere in the surrounding area. There are no obstacles between the robot and the ball. The training and testing data is acquired by hand-steering the robot to the ball with a cableless joystick. The task is solved several times by a human. During this process the program stores a few hundred captured images together with the human provided control signal to form the training and testing sets. Afterwards

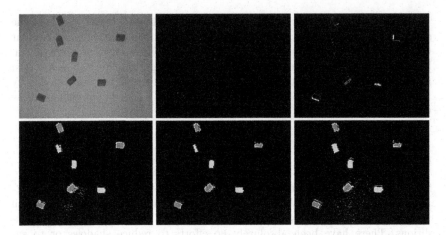

Fig. 3. From top to bottom, from left to right: Orinal image of the testing set of the subtraction task and the region image produced by the single cvtk operator of the best individual after $t = 0$, $t = 12$, $t = 25$, $t = 50$, und $t = 150$ epochs

a controller is evolved on this training data and finally tested in the real environment.

All experiments have been conducted with an initial population size of 60, $\mu = 20$ and $\lambda = 15$. The evolutionary process was always stopped after 200 epochs.

As a "proof of concept" the subtraction experiment has been solved first by allowing the candidate computer vision subsystems to contain only a single CVTK operator. This operator has a really huge parameter space and has to be set up carefully since the provided sample colors completely determine the result of the segmentation process. Due to changing lighting conditions and shadows the task of finding good sample colors is very difficult even for a human.

The progress made during the evolution in extracting the interesting regions is visualized in fig. 3. While the image processing is not substantially improved after the 50th epoch, the other settings keep improving.

Actually, in the very first run of the experiment, the best individual after 200 epochs adapted a quite intelligent and efficient strategy. Inspecting the parameter settings and generated intermediate representations, we have found the operator to not only robustly find the gaming pieces but also to return the smallest feature vector possible: The individual discards any information about the position, size and area and only considers the color of the regions found. The feature vector passed from the computer vision subsystem to the learning subsystem has only eight "boolean" entries – four for each of the two foreground color classes. This small representation is possible because the necessary minimal size of regions was set appropriately during evolution to filter out all false positive detections, that are consistently smaller in size than the correctly detected regions.

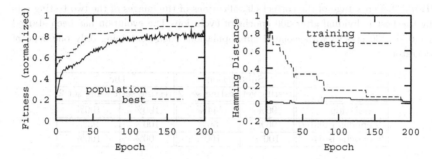

Fig. 4. Results of training a single CVTK operator in experiment 1. The fitness of the population and its best individual improve parallely (left). The Hamming Distance measured for the best individual continously decreases on the testing set (right). The Hamming Distance has beend divided by the number of patterns in order to be comparable

In all of the five repetitions we have performed with this setup, the best individual after 200 epochs always classified at least eight of the second, unseen testing set's nine images correctly. As can be seen in fig. 4 the training error of the best individual is very low right from the beginning. This is no suprise because a sufficiently big feedforward net is easily trained to memorize a relatively small training set. Obviously, it is the generalization error that has to be minimized during the evolutionary process.

Afterwards, the algorithm was tested in the first two experiments with all operators active. The classification results listed in tab. 2 show the resulting solutions to perform clearly above chance.

During the experiments, the algorithm was observed to have problems in the first task due to some operators dominating the entire population too early. This happens, because the population size – due to the computational cost – has been chosen to be relatively small and some operators may give notably better results than others during early stages of the optimization. If those "early starting" operators are not the operators that give the better results in the long run, the problem might occur.

This problem might be circumvented by simply training the different operators separately in the early stages of the evolutionary process. Table 2 shows that evolving the operators for the first 30 epochs separately solves the observed problems effectively.

Finally, the algorithm was tested successfully in the robot-navigation task[1]. Although the ball is nearly always reached, the robot drives very slowly at some positions. We believe this behavior results from inaccurate or contradictory control signals in the training pattern rather from an error in the evolved computer vision subsystem.

---

[1] Multimedia files of both the training and testing phase can be found at http://www-lehre.inf.uos.de/~slange/master/.

Table 2. Percentage of the correct classifications of the images of the two testing sets by the best individual after 200 epochs of evolution. The evolution has been started for 0, 10, 30 epochs with isolated subpopulations each containing only one type of operators

	Subtraction		Dice	
	testing set	testing set 2	testing set	testing set 2
no isolation	96%	44%	100%	100%
10 epochs isolated	96%	56%	100%	93%
30 epochs isolated	100%	100%	100%	100%

One interesting observation from the inspection of the vision systems constructed is that instead of detecting the whole area of the ball, some subsystems only search for an equatorial stripe. The center of the detected stripe always closely coincides with the center of the ball. The advantage of extracting this smaller region seems to be that problems due to highlights in the upper region of the ball and shadows in the lower half could be effectively circumvented. Compared to the subsystems which considered the whole area of the ball, the subsystems seem to be more robust against noisy pixels in the background and therefore had to sort out fewer false detections. In spite of this obvious difference in the preprocessing layer we were not able to detect any significant differences in the behavior of the robot.

5   Conclusion

We have introduced an elegant algorithm that is able to directly learn different tasks on visual input. In contrast to earlier work, it does not need any a priori knowledge neither about features to be extracted nor the layout of an intermediate representation. It is able to construct specialized, task-dependent computer vision subsystems that enable a learning algorithm to successfully learn the task. Moreover it finds good parameter settings for the employed operators even in huge parameter spaces. The generalization performance of the resulting strategies is clearly above chance. As yet, we have only used tasks having a "color-based" solutions. We plan to implement other operators and to try different and more difficult tasks in the future.

References

1. Bala, J. , DeJong K., Huang, J., Vafaie, H., Wechsler, H.: Hybrid Learning Using Genetic Algorithms and Decision Tress for Pattern Classification. 14th Int Joint Conf. on Artifical Intelligence (IJCAI), Canada (1995) 719-724
2. Belpaeme, T.: Evolution of Visual Feature Detectors. In: Proceedings of the First European Workshop on Evolutionary Computation in Image analysis and Signal Processing (EvoIASP99, Göteborg, Sweden), University of Birmingham School of Computer Science technical report. (1999)

3. Braun, H., Weisbrod, J.: Evolving feedforward neural networks. Proc. of the International Conference on Artificial Neural Nets and Genetic Algorithms (1993)
4. Braun, H., Ragg, T.: Evolutionary Optimization of Neural Networks for Reinforcement Learning Algorithms. In: ICML96, Workshop Proceedings on Evolutionary Computing and Machine Learning, Italy (1996) 38-45
5. McCabe, G.P.: Principal Variables. In: Technometrics vol. 26 (1984) 127-134
6. Draper., B.: Learning Object Recognition Strategies. Ph.D. dissertation, Univ. of Massachusetts, Dept. of Computer Science. Tech. report 93-50 (1993).
7. Draper, B.: Learning Control Strategies for Object Recognition. In: Visual Learning, Ikeuchi and Veloso (eds.). Oxford University Press (1996)
8. Jolliffe, I.T.: Principal Component Analysis. Springer-Verlag, Berlin Heidelberg New-York (1986)
9. Krzanowski, W.J.: Selection of Variables to Preserve Multivariate Data Structure, Using Principal Component Analysis. In: Applied Statistics- Journal of the Royal Statistical Society Series C vol. 36 (1987) 22-33
10. Lange, S.: Verfolgung von farblich markierten Objekten in 2 Dimensionen. B.Sc. thesis, Universität Osnabrück, Institut für Kognitionswissenschaft (2001)
11. Martin, C. M.: The Simulated Evolution of Robot Perception. Ph.D. dissertation, Carnegie Mellon University Pittsburgh (2001)
12. Priese, L., Rehrmann, V., Schian R., Lakmann, R.: Traffic Sign Recognition Based on Color Image Evaluation. In: Proc. Intelligent Vehicles Symposium (1993) 95-100
13. De Ridder, D., Kouropteva, O., Okun, O., Pietikäinen, M. and Duin, R.P.W.: Supervised locally linear embedding. In: Proc. Joint Int. Conf. ICANN/ICONIP 2003, Lecture Notes in Computer Science, vol. 2714, Springer Verlag, Berlin Heidelberg New York (2003) 333-341
14. Riedmiller, M., Braun, H.: A direct adaptive method for faster backpropagation learning: the Rprop algorithm. In: Proceedings of the ICNN, San Francisco (1993)
15. Roweis, S. T., Saul, L. K.: Nonlinear dimensionality reduction by locally linear embedding. In: Science, 290(5500) (2000) 2323-2326
16. Smith, S. M.: A new class of corner finder. In: Proc. 3rd British Machine Vision Conference (1992) 139-148
17. Steels, L., Kaplan, F.: AIBO's first words. The social learning of language and meaning. In: Gouzoules, H. (ed) Evolution of Communication, vol. 4, nr. 1, Amsterdam: John Benjamins Publishing Company (2001)

Towards Illumination Invariance
in the Legged League

Mohan Sridharan[1] and Peter Stone[2]

[1] Electrical and Computer Engineering,
The University of Texas at Austin
smohan@ece.utexas.edu
[2] Department of Computer Sciences,
The University of Texas at Austin
pstone@cs.utexas.edu
http://www.cs.utexas.edu/~pstone

Abstract. To date, RoboCup games have all been played under constant, bright lighting conditions. However, in order to meet the overall goal of RoboCup, robots will need to be able to seamlessly handle changing, natural light. One method for doing so is to be able to identify colors regardless of illumination: *color constancy*. Color constancy is a relatively recent, but increasingly important, topic in vision research. Most approaches so far have focussed on stationary cameras. In this paper we propose a methodology for color constancy on mobile robots. We describe a technique that we have used to solve a subset of the problem, in real-time, based on color space distributions and the KL-divergence measure. We fully implement our technique and present detailed empirical results in a robot soccer scenario.

Keywords: Illumination invariance, Color constancy, KL-divergence, mobile robots.

1   Introduction

Color constancy (or illumination invariance), though a major area of focus, continues to be a challenging problem in vision research. It represents the ability of a visual system to recognize an object's true color across a range of variations in factors extrinsic to the object (such as lighting conditions) [3]. In this paper, we consider the problem of color constancy on *mobile robots*.

In the RoboCup Legged League, teams of mobile robots, manufactured by Sony [2], coordinate their activities to play a game of soccer. To date, games have all been played on a small field ($4.4m \times 2.9m$) under constant, bright lighting conditions (Figure 1). However, the overall goal of the RoboCup initiative [1, 11] is, by the year 2050, to develop a team of humanoid robots that can beat the world champion human soccer team on a real outdoor soccer field. Color constancy is one of the important barriers to achieving this goal.

D. Nardi et al. (Eds.): RoboCup 2004, LNAI 3276, pp. 196–208, 2005.

In the past, color constancy has been studied primarily on static cameras with relatively loose computational limitations. On mobile robots, color constancy must be achieved in real time, under constantly changing camera positions, while sharing computational resources with other complex tasks such as localization, movement, decision-making etc. This paper contributes a color constancy method based on the KL-divergence measure that is efficient enough to work on mobile robots that must operate under frequent illumination changes. Our method is fully implemented and tested on a concrete complex robot control task.

The remainder of the paper is organized as follows. In Section 2 we present some basic information on the problem and identify a subset of the overall problem that we address in this paper. Section 3 provides a brief review of related approaches that have been employed to solve the color constancy problem. Section 4 describes the experimental setup, the basic algorithm involved, and the mathematical details of our comparison measure. Details on the experimental results are provided in Section 5 followed by the conclusions in Section 6.

2   Background Information

In this section, we present a brief description of our experimental platform and describe the specific problem addressed in this paper.

On the Sony Aibo ERS-7 robots [2], we perform the visual processing in two stages: color segmentation and object recognition. During the initial off-board training phase, we train a color cube C that maps a space of $128 \times 128 \times 128$ possible pixel values[1] to one of the 10 different colors that appear in its environment (pink, yellow, blue, orange, marker green, red, dark blue, white, field green, black – see Figure 1). C is trained using a Nearest Neighbor (NNr) (weighted average) approach based on a set of hand-labelled input images. Due to computational constraints, C is precomputed and then treated as a lookup table. The robot uses C during task execution to segment an image and then recognize objects of importance. For full details on this process see our technical report [17].

The actual pixel readings associated with a given object can vary significantly with changes in lighting conditions (both environmental and as a result of shadows) and there is significant overlap between some of the colors in the problem space. A color cube trained for one particular lighting condition can therefore be rendered ineffective by a reasonably small change (e.g., the difference between daytime and nighttime on the same playing field within a normal room with windows). In this paper, we propose an approach to solve this problem.

In our lab, the lighting on the soccer field varies significantly between the *bright* condition (≈ 1500 lux with all lights on) and the *dark* condition (≈ 350 lux with only the fluorescent ceiling lights on). One of the primary requirements

---

[1] We use half the normal resolution of 0-255 along each dimension to reduce storage space requirements.

to playing soccer is that of finding the ball and scoring a goal, which we define as the *find-ball-and-score-goal* task. If trained under the bright illumination condition, the robot is able to perform this task proficiently. But if the same robot is now made to function under the dark illumination condition (or any other illumination condition significantly different from the bright illumination condition), it is totally lost and cannot even recognize the ball. On the other hand, if the robot is equipped with a color cube trained under the dark illumination condition, it scores a goal in the same amount of time as in the bright illumination. Again, it is effectively blind when the lights are all turned on.

Our long-term goal is to enable the robot to perform this task in lighting conditions that may continuously vary between the two extremes (bright and dark). In this paper, we consider the subtask of enabling a robot to work in three illumination conditions: *bright, intermediate* and *dark* where *intermediate* refers to an illumination condition almost midway between the other two illumination conditions. Preliminary results indicate that solving this subtask may be sufficient for solving, or nearly solving, the long-term goal itself.

Fig. 1. An Image of the Aibo and the field

In order to work in all three illumination conditions, the robot must be able to:

1. Correctly classify its input images into one of the three illumination conditions;
2. Transition to an appropriate color cube based on the classification and use that for subsequent vision processing;
3. Perform all the necessary computation in real-time without having an adverse effect on its task performance.

We present an algorithm that meets all of these requirements in Section 4. First, we take a brief look at some of the previous techniques developed to achieve illumination invariance.

3   Related Approaches

Several approaches have been attempted to solve the problem of color constancy. Though they differ in the algorithmic details, most of them, to date, have focussed on static camera images. The Retinex Theory of Land [12] and the "Gray World" algorithm by Buchsbaum [5] are based on global or local image color averages, though these have later been shown to correlate poorly with the actual scene illuminant [4]. The *gamut mapping* algorithm, first proposed by Forsyth [9] and later modified by Finlayson [6, 7], using *median selection*, is based on a set of mappings that transform image colors (sensor values) under an unknown illuminant to the gamut of colors under a standard (canonical) illuminant. The probabilistic correlation framework, developed by Finlayson [8], operates by determining the likelihood that each of a possible set of illuminants is the scene illuminant.

The Bayesian decision theoretic approach, proposed by Brainard [3], combines all available image statistics and uses a maximum local mass (MLM) estimator to compute the posterior distributions for surfaces and illuminants in the scene for a given set of photosensor responses. Tsin [19] presents a Bayesian MAP (*maximum a posteriori*) approach to achieve color constancy for the task of outdoor object recognition with a static surveillance camera while Rosenberg [15] describes a method that develops models for sensor noise, canonical color and illumination, and determines the global scene illuminant parameters through an exhaustive search that uses KL-divergence as a metric. More recently, Lenser and Veloso [13] presented a tree-based state description/identification technique. They incorporate a time-series of average screen illuminance to distinguish between illumination conditions using the absolute value distance metric to determine the similarity between distributions. In this paper we explore an alternative similarity measure based on color space distributions.

In the domain of mobile robots, the problem of color constancy has often been avoided by using non-vision-based sensors such as laser range finders and sonar sensors [18]. Even when visual input is considered, the focus has been on recognizing just a couple of well-separated colors [10, 14]. There has been relatively little work on illumination invariance with a moving camera in the presence of shadows and artifacts caused by the rapid movement in complex problem spaces. Further, with few exceptions (e.g. [13]), the approaches that do exist for this problem cannot function in real-time with the limited processing power that we have at our disposal.

4   Approach

In this section, we introduce our experimental setup as well as our algorithmic framework.

4.1   Experimental Setup

We set out to see if it would be possible to distinguish between and adapt to the three different lighting conditions in our lab. Similar to our earlier work [16], we trained three different color cubes, one each for the *bright*, *intermediate*, and the *dark* illumination conditions.

We hypothesized that images from the same lighting conditions would have measurably similar distributions of pixels in color space. The original image is available in the YCbCr format, quantized into 256 bins: [0-255] along each dimension. In an attempt to reduce processing, but still retain the useful information, we transformed the image to the normalized RGB space, i.e. (r, g, b). By definition,

$$r = \frac{R+1}{R+G+B+3}, \quad g = \frac{G+1}{R+G+B+3}, \quad b = \frac{B+1}{R+G+B+3}$$

and $r + g + b = 1$. Thus any two of the three features are a sufficient statistic for the pixel values. For a set of training images captured at different positions on the field for each of the three illumination conditions, we then stored the distributions in the (r, g) space, quantized into 64 bins along each dimension. Once the distributions are obtained (one corresponding to each training image), the next question to address is the measure/metric to be used to compare any two given distributions.

4.2   Comparison Measure

In order to compare image distributions, we need a well-defined measure capable of detecting the correlation between color space image distributions under similar illumination conditions. We examined several such measures on sample images [16], we decided to use the KL-divergence measure [2].

KL-divergence is a popular measure for comparing distributions (especially discrete ones). Consider the case where we have a set of distributions in the 2D (r, g) space. Given two such distributions A and B (with $N = 64$, the number of bins along each dimension),

$$KL(A, B) = - \sum_{i=0}^{N-1} \sum_{j=0}^{N-1} (A_{i,j} \cdot ln\frac{B_{i,j}}{A_{i,j}})$$

The more similar two distributions are, the smaller is the KL-divergence between them. Since the KL-divergence measure contains a term that is a function of the log of the observed color distributions, it is reasonably robust to large peaks in the observed color distributions and is hence less affected by images with large amounts of a single color.

4.3   Algorithmic Framework

Once we had decided on the measure to be used for comparing image distributions, we came up with a concrete algorithm to enable the robot to recognize and adapt to the three different illumination conditions (*bright, intermediate,* and *dark*).

The robot starts off assuming that it is in the bright illumination condition. It is equipped with one color cube for each illumination condition and a set of (training) sample distributions generated from images captured at various

---

[2] Strictly speaking, KL-divergence is not a metric as it does not satisfy triangle inequality but it has been used successfully for comparing distributions.

positions on the field. In our experiments, we used 24 distributions for each illumination condition, though, as we shall show, we do not need so many to perform satisfactorily.

As new images are processed, one is periodically tested for membership in one of the three illumination class, using a fixed number of training samples from each class for comparison. The distribution obtained from a test image is compared with the training samples using the KL-divergence measure, to determine the training sample that is *most similar* to it. The test image is then assigned the same illumination class label as this training sample. If sufficient number of consecutive images are classified as being from an illumination condition, the robot transitions to another color cube (representing the new illumination condition) and uses that for subsequent operations. Parameters included in this process were as follows:

- num_{train}: the number of training samples from each class per comparison.
- t: the time interval between two successive tests.
- num_d, num_i, num_b: the number of consecutive dark, intermediate or bright classifications needed to switch to the corresponding color cube. We allow these parameters to differ.
- ang: a threshold camera tilt angle below which images are not considered for the purposes of changing color cubes. This parameter is introduced to account for the fact that the image appears dark when the robot is looking straight down into the ground (or the ball), which in most cases means that it is looking at a region enveloped in shadow.

5   Experimental Results

In this section, we describe the experiments that we used to determine the optimum values for the parameters. We then present the experiments that we ran to estimate the performance of the algorithm on the robot with respect to the goal-scoring task.

5.1   Estimation of Parameters

On the ERS-7 robot, with our current code base [17], generating a test distribution takes $\approx 25msec$ and the comparison with all 72 training distributions takes $\approx 130msec$, i.e., comparing the test sample with each training sample takes $\frac{130}{72} \approx 2msec$. Further, the normal vision processing takes $\approx 30\text{--}35msec$ per image, leading to a frame rate of 30 frames per second without any check for change in illumination. Thus, if we tested for illumination change on each input image, we would take $\approx 190msec$ per image giving us a very low frame rate of 5–6 frames per second, not including the processing required for action selection and execution. Doing so would lead to a significant loss of useful sensory information. As a result, we cannot afford to test each input image. Instead, we need to identify parameter values that do not significantly affect the normal operation of the robot while also ensuring that the illumination changes are recognized as soon as possible. With the parameter values that the robot ends up using we are still able to operate at around 25 frames per second.

num_{train}	Bright		Interm		Dark	
	Max	Min	Max	Min	Max	Min
24	$\frac{950}{1000}(0)$	$\frac{904}{1000}(0)$	$\frac{997}{1000}(0)$	$\frac{934}{1000}(0)$	$\frac{1000}{1000}(0)$	$\frac{1000}{1000}(0)$
12	$\frac{913}{1000}(0)$	$\frac{857}{1000}(1)$	$\frac{979}{1000}(0)$	$\frac{903}{1000}(0)$	$\frac{1000}{1000}(0)$	$\frac{999}{1000}(0)$
6	$\frac{874}{1000}(0)$	$\frac{712}{1000}(3)$	$\frac{964}{1000}(0)$	$\frac{850}{1000}(0)$	$\frac{1000}{1000}(0)$	$\frac{944}{1000}(0)$

Fig. 2. Classification accuracy and color cube transitions (all three illuminations)

Exp1: Classification Accuracy. The goal of this experiment was to determine how much the classification accuracy, using KL-divergence, depended on the parameters t and num_{train}. In order to do that the robot was placed at various positions on the field and only allowed to pan its head. The robot periodically sampled the images that it received from its camera and tested them to identify the illumination condition. The goal was to measure the real-time performance and see if there was any significant correlation between the performance of the robot and the associated position on the field. We chose six *test positions*, at each of which we measured the classification accuracy over a set of one thousand images. We first performed this experiment with $t = 1$ second, and with three different training sample sizes:

1. *Case1* : $num_{train} = 24$ (all training samples).
2. *Case2* : $num_{train} = 12$.
3. *Case3* : $num_{train} = 6$.

Figure 2 shows the results of the experiment when performed under each illumination, corresponding to cases 1, 2 and 3. We also measured the number of times the robot transitioned between color cubes for each given test condition. Since each test was conducted completely under one of the three illumination conditions, ideally there would be no transitions. The values in parentheses therefore represent the number of (incorrect) color cube transitions that would occur if we chose to use the corresponding values of the parameters during normal operation. Observe that in the dark lighting conditions there were practically no misclassifications at any of the *test positions*.

The transition to the *bright* illumination takes place after two consecutive test images are classified as being under the bright illumination condition while for the other two illumination conditions (intermediate, dark), this threshold is set at four and six respectively (i.e., $num_d = 6$, $num_i = 4$ and $num_b = 2$). The transition parameters were weighted in this manner because it was noted during experimentation that lighting inconsistencies (such as shadows) during both training and testing led to significantly more noise in the bright conditions.

From a close examination of the raw data, we report several observations:

- In the dark illumination, the robot did not err even in the case where $num_{train} = 6$. This makes sense considering the fact that the dark illumination condition is much different from the bright illumination condition (350lux vs 1500lux) and the presence of shadows or obstructions only makes the image *darker*.

num_{train}	Bright	Shifts	Intermediate	Shifts	Dark	Shifts
24	$\frac{432}{500}$	0	$\frac{456}{500}$	0	$\frac{480}{500}$	0
12	$\frac{398}{500}$	1	$\frac{421}{500}$	1	$\frac{482}{500}$	0
6	$\frac{360}{500}$	2	$\frac{395}{500}$	1	$\frac{490}{500}$	0

Fig. 3. Accuracy and transitions under each illumination under all three sampling cases

- In the bright and intermediate illumination conditions, the number of training samples made a difference in the performance of the robot; at each position, the performance worsens as num_{train} is decreased.
- The transitions (shifts) that occurred in the bright illumination condition were due to shadows or obstructions and this caused the robot to move from the bright to the intermediate illumination condition.
- In the rare case that the robot changed to the incorrect color cube, the error was sustained only for a few test frames before the robot recovered.

Next we varied t and repeated the experiments performed previously. Here, no significant change was noticed in the classification accuracy. With $t = 0.5$ seconds or 0.25 seconds instead of 1 second there was no significant change in the classification accuracy. However it did increase the processing performed. We quantify this effect in subsequent experiments.

Exp2: Task Execution. The *find-ball-and-score-goal* task was incorporated in this experiment to estimate the effect of the color constancy algorithm on the robot's task performance under constant illumination. The robot was placed at the center of the field facing one of the goals (say g_1) while the ball was placed at the center of the penalty box around the other goal (g_2) (see Figure 1). The robot had to find the orange ball and then score on g_2. We performed this experiment with the robot trying to score on either goal (blue/yellow) over one half of the total number of trials. In this process, it used the colored markers around the field to localize itself [17].

The robot performed the check for change in the illumination condition with $t = 1$ and $ang = -10$ (i.e. consider cases where the tilt angle of the camera is greater than 10°) and its accuracy was tested under all three sampling conditions used in experiment 1, i.e., we tested for $num_{train} = 24$, 12, and 6. We set the tilt angle threshold to ensure that when the robot is staring down at the ground or the ball, the shadows do not cause the robot to make a wrong transition. Figure 3 displays the classification accuracy and the number of transitions that occurred during task execution under all three cases. Since testing was done under each illumination condition separately, the *shifts* column represents the number of incorrect transitions that occurred.

From the results, we deduced that the value of num_{train} does not make a big change in the dark illumination case; the misclassifications that did occur in the dark illumination case happened when the robot fell down and was staring at the white border or field lines. But this was not the case under the bright

t (sec)	$num_{train}=24$	$num_{train}=12$	$num_{train}=6$
1.0	6.8±0.3(0)	6.8±0.4(0)	6.8±0.4(0)
0.5	6.9±0.3(0)	7.0±0.6(0)	7.0±0.5(0)
0.25	8.8±0.6(2)	9.1±1.8(0)	8.2±1.9(0)
0.125	51.8±31.7(4)	18.3±2.5(1)	11.7±6.2(0)
0.0	75.0±36.9(6)	52.8±13.1(3)	13.8±3.0(0)

Fig. 4. Time taken (in seconds) to *find-and-walk-to-ball* under bright illumination

and intermediate illuminations; with the decrease in num_{train}, the robot ended up making more errors (and wrong transitions between color cubes). The errors that occurred were mostly due to shadows when the robot was running into the goal. But the robot always recovered within a few test frames (fewer than 5).

Under this tilt angle setting, the (incorrect) color cube transitions were only one-off from the actual illumination condition, i.e., there were no incorrect transitions from *bright* to *dark* or vice versa. We could set higher tilt angle thresholds but then the robot is slow to identify changed illumination conditions which has a bad effect on its overall performance.

Exp3: Parameter Combinations. Next, we wanted to determine the parameter settings that would enable strong real-time performance. Specifically, we considered the parameters num_{train} and t.

To do so, we defined the *find-and-walk-to-ball* task. This task is identical to the *find-ball-and-score-goal* task, except that the robot only needs to find the ball and walk up to it, rather than actually score. This modification makes the measurements less dependent on the performance of other modules, such as kicking.

Under constant lighting conditions with a single color cube, the robot can *find-and-walk-to-ball* in 6.7(±0.6) seconds. The results for the bright illumination case, averaged over 10 trials, are in Figure 4. The values for the other two illuminations were not significantly different (as expected). We considered the cases where the robot did not find the ball after two minutes to be complete misses and omitted them from the results. The numbers in parentheses indicate the number of complete misses that occurred while collecting 10 values.

From Figure 4, we conclude that:

– The parameter values $t = 1$ second and $t = 0.5$ seconds (and to some extent $t = 0.25$ seconds), with all three sampling schemes, work fine on the robot in real-time without having an adverse effect on the normal game playing performance.
– With all the other testing frequencies there were instances, especially with $num_{train} = 24$, when the robot missed the ball during its scan due to the computation involved; by the time the robot had processed one frame that had the ball, its head was beyond the position were it could recognize the ball (the robot needs to see the ball continuously for around 3 frames before it accepts it as the ball [17]). In addition, when the testing was done on every

Lighting (start/after 1.5 seconds)	Time (seconds)
bright / intermediate	8.5 ±0.9
bright / dark	11.8 ±1.3
intermediate / bright	8.6 ±1.0
intermediate / dark	9.6 ±3.1
dark / intermediate	11.5 ±1.4
dark / bright	10.7 ±1.1

Fig. 5. Time taken to *find-and-walk-to-ball* under changing illumination

frame (or even once every 0.125 seconds), the robot's motion towards the ball was extremely jerky.

5.2 Changing Illumination

Once we had determined values for all the parameters, we were ready to test the robot on its task under changing illumination conditions. Based on the experiments described above, we chose the parameter values: $ang = -10^o$, $t = 1$ second, $num_d = 6$, $num_i = 4$, $num_b = 2$, and $num_{train} = 24$.

Real-Time Transitions. In these experiments, the robot was tested with the lighting conditions changing after a specific interval. The robot starts off in one illumination condition and after 1.5 seconds (the time it takes the robot to turn and see the ball), the illumination is changed by adjusting the intensity of all the lamps. The robot is then timed as it performs the *find-and-walk-to-ball* task. Recall that with a single color cube, the robot is unable to do so: when the illumination condition changes significantly, it is unable to see a ball that is right in front of its camera. Now, when the illumination conditions change, the robot seems lost for a couple of seconds while it recognizes the change and then functions as normal, scoring goals once again. The results are shown in Figure 5[3].

Stress Tests. To further explore the robustness of our approach, we report the results of two tests for which the current algorithm was not designed: intermediate lighting conditions (Test 1) and adversarial illumination changes (Test 2).

Test 1. The first experiment we performed involved reducing the intensity of the lamps in specific patterns. In our lab, we have four lamps mounted on stands along the shorter edges of the field, as shown in Figure 6.

During testing, we reduced the intensity of all the lamps such that the illumination on the field is in between the illumination conditions that the robot was explicitly trained for. To enable comparison of these results, we used the *find-and-walk-to-ball* task as in previous sections and recorded the time taken by the robot to perform the task. In Figure 7 we present the values corresponding to

---

[3] Video showing the robots performing under varying lighting conditions is available at http://www.cs.utexas.edu/~AustinVilla/legged/illumination

the case wherein the robot starts off in the *bright* illumination condition. About 1.5 seconds later, the lighting is changed such that it is between the *bright* and the *intermediate* illuminations (we also tested with the illumination changed to be midway between the *intermediate* and the *dark* conditions).

Test 2. Finally, we decided to test the robot under adversarial conditions. That is, we tried our best to confuse the robot completely by varying the illumination in the worst possible way for the robot. Here, we performed the *find-ball-and-score-goal* task.

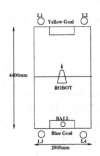

Fig. 6. A Line drawing of the field and the lamp arrangement

As soon as the robot recognized the ball and started walking towards it, we changed the illumination condition. That is, as soon as the robot transitioned to the correct color cube, as indicated by an LED on the robot, we would change the illumination to be one class away from the actual illumination condition. Due to the values of num_d, num_i and num_b, assuming that we start off under the bright illumination, the experiment would involve changing the illumination to correspond to the intermediate illumination after two seconds. We would turn the lamps off (dark illumination) after another four seconds, which would be followed by adjusting the intensity of the lamps to half the maximum value (intermediate illumination) after a further six seconds.

Then we would turn the lamps on at full intensity (bright illumination) after around four seconds. This cycle of change in illumination is performed repeatedly and Figure 8 depicts the corresponding average time (and standard deviation) taken to accomplish the *find-ball-and-score-goal* task.

Lighting	Time (seconds)
bet. bright and interm	12.27 ±0.5
bet. interm and dark	13.3 ±2.0

Fig. 7. Time taken (in seconds) to *find-and-walk-to-ball*

The fact that the robot can perform this task at all is due to the fact that it can occasionally recognize the ball even if it is not using the color cube corresponding to the current illumination condition. Even in the case where some of the lamps are selectively turned off (or their intensity is reduced), the robot transitions into an appropriate color cube and is still able to perform the task of scoring on the goal. The only way we could confuse the robot further would be to change between the bright and the dark illumination conditions in a similar manner. In that case, the robot does not make any progress to the ball at all.

An important point to note here is that in previous work [16] (where we had incorporated only two illumination conditions: bright and dark) we had problems when we tested the robot on illumination conditions that it was not trained for. There were problems especially while trying to score on the yellow goal in differentiating between yellow and orange (and between pink and orange). The robot would then walk away in an entirely wrong direction in an attempt to follow a spurious estimate of the ball. We had then hypothesized that adding a few more illumination conditions in between the two extreme ones might help alleviate some of the problems. We now see that with the added *intermediate* illumination condition, the robot does perform much better. In fact, in all the experiments mentioned above, the robot performed equal number of trials scoring/walking towards either goal (blue/yellow).

Also, in our earlier work [16] we used the previous version of the Sony robots: ERS210A. With very little modification in code (only to incorporate one more illumination condition), the strategy works fine to distinguish between three different illumination conditions. We find

Lighting	Time (seconds)
Adversarial	34.3 (±7.8)

Fig. 8. Time taken (in seconds) to *find-ball-and-score-goal*

that with this change, the robot is better able to work in illumination conditions corresponding to which the robot does not have training samples.

6   Conclusions/Future Work

In this paper, we have presented an approach that works in real-time to achieve color constancy on mobile robots in the RoboCup domain. The technique uses color space distributions, easy to train color cubes, and a simple and efficient comparison measure (KL-divergence) to determine and adapt to three discrete illumination conditions. Though we have solved only a subset of the problem, the results obtained seem to indicate that we do not need to consider a continuous spectrum of illuminations. When presented with illumination conditions that the robot is not trained for, there is little degradation of performance.

The problem of color constancy, on mobile robots or otherwise, is extremely challenging and is far from being solved. In the future, we shall first try to extend the approach to enable the robot to perform well under an even wider range of possible illumination conditions. One possible method would be to train a few discrete color cubes to represent significantly different illuminations (as we have done here) and then dynamically update the cubes for minor variations in illumination conditions. We shall also look into alternate stochastic approaches that may enable us to achieve illumination invariance without having to resort to training several color cubes. Ultimately, we aim to solve the daunting problem of developing efficient algorithms that enable a mobile robot to function under completely uncontrolled natural lighting conditions, with all its associated variations.

Acknowledgements

We would like to thank the members of the UT Austin Villa team for their efforts in developing the soccer-playing software mentioned in this paper. This research was supported in part by NSF CAREER award IIS-0237699.

References

1. The International RoboSoccer Competition. http://www.robocup.org.
2. The Sony Aibo robots. http://www.us.aibo.com.
3. D. H. Brainard and W. T. Freeman. Bayesian color constancy. *Journal of Optical Soceity of America A*, 14(7):1393–1411, 1997.
4. D. H. Brainard and B. A. Wandell. Analysis of the retinex theory of color vision. *Journal of Optical Soceity of America A*, 3(10):1651–1661, 1986.
5. G. Buchsbaum. A spatial processor model for object color perception. *Journal of Franklin Institute*, 310:1–26, 1980.
6. G. Finlayson. Color in perspective. *In IEEE Transactions of Pattern Analysis and Machine Intelligence*, 18(10):1034–1038, July 1996.
7. G. Finlayson and S. Hordley. Improving gamut mapping color constancy. *In IEEE Transactions on Image Processing*, 9(10), October 2000.
8. G. Finlayson, S. Hordley, and P. Hubel. Color by correlation: A simple, unifying framework for color constancy. *In IEEE Transactions on Pattern Analysis and Machine Intelligence*, 23(11), November 2001.
9. D. Forsyth. A novel algorithm for color constancy. *In International Journal of Computer Vision*, 5(1):5–36, 1990.
10. Jeff Hyams, Mark W. Powell, and Robin R. Murphy. Cooperative navigation of micro-rovers using color segmentation. *In Journal of Autonomous Robots*, 9(1):7–16, 2000.
11. Hiroaki Kitano, Minoru Asada, Yasuo Kuniyoshi, Itsuki Noda, and Eiichi Osawa. Robocup:the robot world cup initiative. proceedings of the first international conference on autonomous agents. In *Proceedings of the International Conference of Robotics and Automation*, pages 340–347, February 1997.
12. E. H. Land. The retinex theory of color constancy. *Scientific American*, pages 108–129, 1977.
13. S. Lenser and M. Veloso. Automatic detection and response to environmental change. In *Proceedings of the International Conference of Robotics and Automation*, May 2003.
14. B. W. Minten, R. R. Murphy, J. Hyams, and M. Micire. Low-order-complexity vision-based docking. *IEEE Transactions on Robotics and Automation*, 17(6):922–930, 2001.
15. C. Rosenberg, M. Hebert, and S. Thrun. Color constancy using kl-divergence. In *IEEE International Conference on Computer Vision*, 2001.
16. M. Sridharan and P. Stone. Towards illumination invariance on mobile robots. In *The First Canadian Conference on Computer and Robot Vision*, 2004.
17. Peter Stone, Kurt Dresner, Selim T. Erdoğan, Peggy Fidelman, Nicholas K. Jong, Nate Kohl, Gregory Kuhlmann, Ellie Lin, Mohan Sridharan, Daniel Stronger, and Gurushyam Hariharan. Ut austin villa 2003: A new robocup four-legged team, ai technical report 03-304. Technical report, Department of Computer Sciences, University of Texas at Austin, October 2003.
18. S. Thrun, D. Fox, W. Burgard, and F. Dellaert. Robust monte carlo localization for mobile robots. *Journal of Artificial Intelligence*, 2001.
19. Y. Tsin, R. T. Collins, V. Ramesh, and T. Kanade. Bayesian color constancy for outdoor object recognition. *In IEEE Pattern Recognition and Computer Vision*, December 2001.

Using Layered Color Precision for a Self-Calibrating Vision System*

Matthias Jüngel

Institut für Informatik, LFG Künstliche Intelligenz,
Humboldt-Universität zu Berlin, Unter den Linden 6, 10099 Berlin, Germany
http://www.aiboteamhumboldt.com

Abstract. This paper presents a vision system for robotic soccer which was tested on Sony's four legged robot Aibo. The input for the vision system are images of the camera and the sensor readings of the robot's head joints, the output are the positions of all recognized objects in relation to the robot. The object recognition is based on the colors of the objects and uses a color look-up table. The vision system creates the color look-up table on its own during a soccer game. Thus no pre-run calibration is needed and the robot can cope with inhomogeneous or changing light on the soccer field. It is shown, how different layers of color representation can be used to refine the results of color classification. However, the self-calibrated color look-up table is not as accurate as a hand-made. Together with the introduced object recognition which is very robust relating to the quality of the color table, the self-calibrating vision works very well. This robustness is achieved using the detection of edges on scan lines.

1 Introduction

The vision system that is described in this paper was implemented for the Sony four-legged league. This and the other RoboCup real robot leagues (middle-size, small-size) take place in a color coded environment. The robots play with an orange ball on a green ground and have to kick to yellow and sky-blue goals. At the corners of the field there are two-colored poles for localization. The environment for the games, as the size of the field and the colors of the objects, is well defined. However, the lighting conditions are not known in advance and might change during the competition and even during games.

This paper presents a vision system that uses a method for object recognition that is very robust relating to the quality of the color calibration. This is needed as the color calibration is not done by a human expert before the game, but during the game by the vision system itself.

The task of vision-based object recognition in color coded scenarios often is divided in several subtasks. The first step is the *segmentation* which assigns a

---

* The Deutsche Forschungsgemeinschaft supports this work through the priority program "Cooperating teams of mobile robots in dynamic environments".

D. Nardi et al. (Eds.): RoboCup 2004, LNAI 3276, pp. 209–220, 2005.

color class (orange, green, etc.) to each color (defined by three intensities) of the color space. How this can be done using thresholds or color look-up tables is described in [1].The thresholds and look-up tables usually are created by hand with the assistance of semi-automated tools. More sophisticated tools do an off-line self-calibration on sample images [5]. The general problem of all segmentation methods is, that the color calibration is only valid as long as the lighting conditions do not change. The effect of such a change is described in [6]. A method for autonomous dynamic color calibration is described in [2]. The drawback of this method is that it requires for special actions (walk and head motions) to be done by the robot in order to recalibrate. Sometimes there is a *color space transformation*, before the segmentation is done [4, 8]. In the *clustering* step connected regions of the same color class are identified [7, 1]. The final step is the *classification* which extracts objects and belonging properties like size and position from the set of clusters. This step is domain specific and not standardized. Besides the color based approach there are methods that concentrate on the boundaries of objects. A very robust method that finds image boundaries is described in [3] but this solution is too slow to work under real-time conditions on a robot with limited computational power.

This paper describes a vision system that is based on an analysis of the color of scan lines segments. How these segments are created and examined is described in section 2. The method of calibration-free color classification that is applied to the average colors of the segments is shown in section 3. Section 4 shows some of the experiments that were done to evaluate the vision system.

2   Segment Based Image Processing

The vision system subdivides the scan lines into several segments that are separated by edges. The segments are the starting point for all further image processing. Section 2.1 shows, how the scan lines are placed and how they are split to segments. The object recognition that is described in section 2.2 is based on the classification of the colors of the segments. For this classification an auto-calibrated color table is used (cf. section 3).

2.1   Distribution and Segmentation of Scan Lines

To detect objects in an image and measure their sizes, a set of scan lines is used. The scan lines are horizon-aligned and distributed such that the density of scan lines is higher in areas where more attention is needed to detect all objects. The horizon is used to align vertical scan lines which are perpendicular to the horizon. The lines run from 10 degrees above the horizon to the bottom end of the image and have a spacing of 3 degrees.

The segments of a scan line are found using a simple edge detection algorithm. The algorithm finds the first and the last pixel of an edge. An edge is a sequence of so called *edgels*. Edgels are the pixels of scan lines where there is a sharp

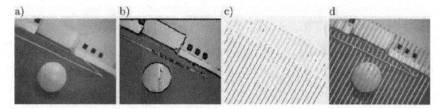

Fig. 1. Segments. a) The original image. b) For each vertical line of the image all edges on the line were detected. The color of the segments between the edges is the average color of the pixels of the respective segment. c) Same as b, but with horizon aligned scan lines that are subdivided into segments, each segment has the average color of all belonging pixels. d) The segments shown in c displayed with the color class that is assigned to the average color of each segment

variation of the intensity of at least one color channel. The segments are the parts between the edges of each scan line.

While the scan lines are scanned for edges, for each segment the first and the last point and the average intensity over all pixels for each color channel are determined. Fig. 1 shows such segments.

2.2   Finding Objects by Analyzing Segments of Scan Lines

Finding Points on Lines. On a RoboCup field there are different types of lines: edges between a goal and the field, edges between the border and the field, and edges between the field lines and the field.

The Border of the Field and Field Lines. To detect the border of the field and the field lines, all segments whose average intensities are classified as white are analyzed. The begin and the end point must be below the horizon, and the segment below must be green. To distinguish between the field lines and the field border, the expected size of a field line is examined. This size depends on the distance from the robot to the field line. Thus the distance to the end point of the white segment is calculated based on the rotation of the head and the position of the point in the image. If the length of the segment is more than fivefold the expected size of a field line, it is classified as a field border. To be classified as a field line, the segment's size must not be larger than twice the expected size of a field line. The figures 2b and 2c show how the vision system can distinguish between a field line and the field border.

Goals. To recognize the points at the bottom of the goals all sky-blue and yellow segments are analyzed. The end points must be below the horizon and the begin points must be above the horizon or lie on the image border. The length of the segment must be at least 5 pixels and the segment below must be green. Figure 2a shows such points.

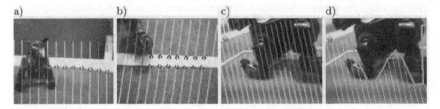

Fig. 2. Recognition of the border, the field lines, and obstacles. a + b) Points at the field border and at a field line. The border and the line have a similar size in the image and are around the same position in the image. They are distinguished based on the different direction of view of the head. c) The segments that are classified as green. d) Green Lines: The obstacles percept is the combination of adjacent green segments. Orange Line: The resulting Obstacles Model

Finding Obstacles. The obstacles percept is a set of lines on the ground that represents the free space around the robot. Each line is described by a *near point* and a *far point* on the ground, relative to the robot. The lines describe segments of green lines in the image projected to the ground. In addition, for each far point a marking describes whether the corresponding point in the image lies on the border of the image or not.

To generate this percept for each scan line the bottom most green segment is determined. If this green segment meets the bottom of the image, the begin and the end point of the segment are transformed to coordinates relative to the robot and written to the obstacles percept; else or if there is no green on that scan line, the point at the bottom of the line is transformed to coordinates relative to the robot and the near and the far point of the percept are identical.

Small gaps between two green segments of a scan line are ignored to assure that field lines are not misinterpreted as obstacles. In such a case two neighboring green segments are concatenated. The size limit for such gaps is $4 \cdot width_{fieldline}$ where $width_{fieldline}$ is the expected width of a field line in the image depending on the camera rotation and the position in the image.

Finding the Ball

Finding Points on the Border of the Ball. Usually there should be at most one orange segment per scan line. However, if there is a highlight or a shadow on the ball, this might result in a higher number of orange segments per scan line that represent the ball. In such a case all orange segments that lie on the same scan line and are part of a sequence of subsequent segments are grouped to one large segment. Then the begin and the end of the bottom most orange (grouped) segment on each scan line is added to the list of ball points if there is a green segment below.

Additional Scan Lines. If the set of ball points contains at least two points, three additional scan lines are calculated based on the smallest rectangle that

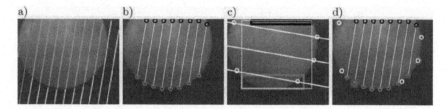

Fig. 3. Ball Recognition: a) The image of an ball and the color classified segments. b) White lines: The orange segments. Black circles: All end points of the orange segments, that are close to the border of the image. Gray circles: All meeting points of an orange and a green segment. c) White box: The bounding box of all black and gray points in b. White lines: Lines parallel to the horizon; the center line goes through the center of the white box. The spacing between the lines depends on the extension of the white box. White circles: Transitions from orange to green on the white scan lines. d) Red circles: The largest triangle amongst the gray and the white circles

contains all points of this set. The new scan lines are parallel to the horizon and evenly overlap the bounding rectangle. This distribution of the additional scan lines assures that more points on the border of the ball can be found on the left and the right side. These new scan lines are divided into segments in the same way as the long vertical scan lines. For each edge that separates an orange and a green segment a point is added to the set of ball points.

Calculation of the Circle. To calculate a circle that describes the ball first all points from the set of ball points that lie on the image border are removed. From the remaining points two points with the highest distance are chosen. Then the point with the highest distance to these points is selected. If these three points do not lie on one and the same straight line they are used to calculate the circle that describes the ball.

Finding Goals. Besides the points at the bottom of the goal, the bearings to the four sides of the goal are determined. Additionally the angle to the left and the right side of the larger part of the goal are determined.

The segments that were used to generate the points at the bottom of the goals are combined to clusters. Two segments of two different scan lines belong to the same cluster if there is no scan line without a goal colored segment between them.

Additional horizontal scan lines determine the horizontal extensions of the clusters. All clusters that have white below are rejected because goals are not outside the border. All clusters that have pink below are rejected because they probably are a part of a landmark. All other clusters are combined leading to a bounding box that includes all clusters. Such a bounding box, generated by combining all valid clusters, is accepted as a goal if it fulfills the conditions: 1: The bottom is below the horizon. 2: The top is above the horizon or intersects with the image border. 3: The width is at least 1.5 times the height of the bounding box or the left or the right intersects with the image border.

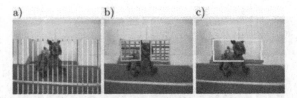

Fig. 4. Finding the goal. a) An image of the goal with color classified segments. b) The gray boxes show the vertical and horizontal extension of the clusters. The horizontal extension was measured using additional horizontal scan lines (the pink lines). c) Combination of the clusters. The red line marks the larger free part of the goal

For a robot that plays soccer the position of the angle and the distance of the goal are important for localization. To avoid to kick into the goalie the largest free part of the goal has to be determined. This is done by comparing the clusters that lead to the goal bounding box. If no cluster or the larger one intersects with the image border, the angles to the left and the right side of the larger cluster determine the best angle for a goal kick. If the smaller part intersects with the border, it can not be decided, whether this part extends outside the image and is the larger free part of the goal, or not. The vision system does not suggest a goal kick angle. Figure 4 shows how the goal is recognized.

3   Layered Color Precision

This section shows, how a color look-up table can be created and updated automatically during the robot plays soccer. The calibration system works on the segments of scan lines that are created for object recognition (cf. section 2). The output of the color calibration is the color look-up table that is used to classify the colors of the segments for object recognition.

3.1   Three Ways to Represent Color Classes

The vision system uses three layers of color precision. Each layer has its own method to assign colors to color classes.

The layer with the lowest precision, *layer 1*, represents the color classes $green_1$, $white_1$, and $black_1$. The colors $green_1$ and $white_1$ are represented by cuboids in the color space. The rest of the color space is $black_1$. Each cuboid is defined by 6 threshold (min and max for each color channel).

The next layer of color precision, *layer 2*, distinguishes between more colors. It uses a cuboid in the color space to define $green_2$ as a reference color. Relative to the reference color cuboids for the color classes $yellow_2$, $skyblue_2$, and $orange_2$ are defined. The colors $yellow_2$ and $orange_2$ can be overlapping in color space, which means that some colors can be classified as $yellow_2$ and $orange_2$.

The layer with the highest precision is *layer 3*, which uses a color look-up table to represent $white_3$, $green_3$, $yellow_3$, $skyblue_3$, and $orange_3$. This look-

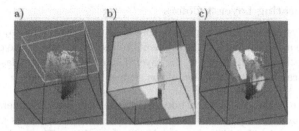

Fig. 5. Different layers of color precision. a) Layer1: Rough classification for green and white. The white and the green cuboid define the color classes. b) Layer2: Refined classification for green, other colors are classified based in the relation to green. The green cuboid is the reference color all other cuboids are defined in relation to the green cuboid. c) Layer3: Classification based on a color look-up table

Fig. 6. Color classified images based on the three different layers of color representation. a) The original image. b) Layer1: Rough classification for green and white. c) Layer2: Refined classification for green, other colors are classified based on the relation to green. d) Layer3: Classification based on a color look-up table

up table is used to classify the colors of the segments for object recognition as described in section 2.2.

Figure 5 shows all three ways to represent an assignment from colors to color classes. The color classification of an image is shown in figure 6 for each of the three forms of representation.

The output of the color calibration system is the *layer 3* color representation, the color look-up table. This is created based on segments that can be identified to belong to an object with a known color for sure. For example the average colors of all segments that belong to the ball are used to calibrate orange. To extract such segments the *layer 2*-based colors of the segments and knowledge about the environment is used. The reference color of the *layer 2* color representation is determined using the *layer 1*-based colors of the segments together with knowledge about the environment. The *layer 1* color representation is constructed based on statistics over the last images. The following subsections describe, how this process of *color-refinement* works in detail.

3.2   Calibrating Layer 3 Colors

The vision system identifies segments that belong to a certain object for sure and adds the average colors of the segments to the color class of the object. The segments are identified based on the layer 1 and layer 2 color classes and on spatial constraints for the position of the segments in the image.

White. White is the color of the border and the field lines. Segments that are $white_1$ are possibly a part of the field border. But in some cases parts of the ball or the ground or field lines can be classified as $white_1$. Thus only the $white_1$ segments that fulfill the following conditions are used to calibrate $white_3$: 1: The segment is not $green_2$. 2: The begin of the segment is below the horizon. 3: The length of the segment is at least 4 pixels. 4: The length of the segment is at least 5 times the expected size of a field line at the position of the center of the segment. (This filters noise) For each such segment the color look-up table is expanded such that its average color is classified as $white_3$.

Colors of the Goals. The color classification based on a reference color from *layer 2* can distinguish $orange_2$, $yellow_2$, and $skyblue_2$. But the color precision is not very high. The colors yellow and orange overlap and the rim of the field lines often is classified as yellow or sky-blue. If the white of the outer border is a little bit bluish it also might be classified as sky-blue. Thus only the segments that are $yellow_2$ or $skyblue_2$ and fulfill these conditions are identified as a part of goal: 1: The begin of the segment is above the horizon. 2: The end of the segment is below the horizon. 3: The length of the segment is at least 10. This filters all segments that are part of a ball or an other robot or caused by the rim of a scan line. Because of the last condition far distant and thus small goals can not be used for calibrating the goal colors. The average colors of all these goal segments are added to the corresponding color class in the *layer 3* color class representation.

Orange. To calibrate $orange_3$, segments of scan lines that lie on a ball are used. Not all segments that are $orange_2$ are a part of a ball. In some cases a goal or a small part of a red robot is classified as $orange_2$. Thus only the $orange_2$ segments that fulfill the following conditions are used to calibrate $orange_3$: 1: The begin of the segment is below the horizon. 2: There is a green segment below. 3: The length of the segment is at most 1.5 times the expected size of a ball at the position of the center of the segment. 4: The length of the segment is at least 0.2 times the expected size of a ball at the position of the center of the segment if the segment does not intersect with the image border. This method filters all segments that do not belong to a ball, and unfortunately even a lot of ball segments. However, in this way enough segments are determined that belong to a ball for sure. The average color of each such segment is added to the color look-up table for $orange_3$.

Green. To calibrate $green_3$ all segments that are classified as $green_1$ are examined. To be used for calibration of $green_3$ the segments have to fulfill the

Fig. 7. Calibration of green. a) The layer 1 color classes of the image. b) The frequent colors displayed bright. c) Black: All long scan lines with an average color classified as $green_1$ (cf. the green pixels in a) White: All long scan lines below the horizon and below the field border. Green: All long scan lines with a frequent color as average color (cf. the bright pixels in b). d) The scan lines that are used to calibrate green

following conditions: 1: The average color must be one of the most frequent colors of the current image. (As the green of the ground is the most frequent color in most images, this helps to filter all other objects) 2: The length must be at least 30 pixels. (This filters small segments that can be caused by other robots, the referee or unexpected objects on the field) 3: The field must be found above. (This ensures that the area outside the field is not used for calibration, in the case the robot looks over the image border) 6: The begin point must be below the horizon. (No ground segments are above the horizon) 5: The difference between the maximal and the minimal intensity of the v-channel must be less than 20. (If the border of the ball intersects with a scan line in a very small angle, it is possible that no edge is detected. In this case a segment is created that has a high difference between the minimal and the maximal intensity of the v-channel as green and orange differ in that channel very much.)

For each segment that fulfills these conditions the color look-up table is expanded such that $green_3$ includes the maximal and the minimal intensities for all 3 channels of the segment. Figure 7 shows which of these conditions are fulfilled by which scan lines in an example image.

3.3   Calibrating Layer 2 Colors

In layer 2 the colors are defined by cuboids in relation to the reference color $green_2$. Green is a refined version of the more rough $green_1$ from layer 1. The other colors in layer 2 do not directly depend on the classification from layer 1 but are approximated using $green_2$ as a reference color.

Green. To define the thresholds for the cuboid of the reference color $green_2$ all segments whose average color is classified as $green_1$ are used that fulfill the conditions for green segments that are used to calibrate $green_3$ (cf. 3.2). For each segment that fulfills these conditions the cuboid for the reference color $green_2$ is expanded such that it includes the maximal and the minimal intensities for all 3 channels of the segment. The other colors are defined in relation to this reference color.

Yellow. The yellow of the goal has more *redness* than the green of the field and less *blueness*. The brightness for yellow is not restricted. This leads to the thresholds $y_{min} = 0, y_{max} = 255 \mid u_{min} = 0, u_{max} = u_{min}^{green} \mid v_{min} = v_{max}^{green}, v_{max} = 255$ for *yellow$_2$* relative to the thresholds for *green$_2$*, where u_{min}^{green} is the lower threshold for the *blueness* of green and v_{max}^{green} is the upper threshold for the *redness* of green.

Orange. Orange has more red than yellow and in the YUV color space more blue than yellow. The brightness for yellow is not restricted. This leads to the thresholds $y_{min} = 0, y_{max} = 255 \mid u_{min} = 0, u_{max} = u_{average}^{green} \mid v_{min} = v_{max}^{green}, v_{max} = 255$ for *orange$_2$* relative to the thresholds for *green$_2$*. This means, that the yellow of the goal has more *redness* than the green of the field and less *blueness*. The threshold y_{min} means that sky-blue is not much darker than green.

Sky-blue. These are the thresholds for *skyblue$_2$* relative to the thresholds for *green$_2$*: $y_{min} = 0, y_{max} = 255 \mid u_{min} = u_{max}^{green}, u_{max} = 255 \mid v_{min} = 0, v_{max} = v_{average}^{green}$ This means that the sky-blue of the goal has more *blueness* than the green of the field and the same or less *redness*. The brightness for sky-blue is not restricted.

3.4   Calibrating Layer 1 Colors

To calibrate *green$_1$* and *white$_1$* for each channel the average intensity $\bar{i}_c$ of all scanned pixels over the last 10 images is calculated.

White. The minimal brightness for *white$_1$* is defined in relation to the average brightness of the last images. It turned out that for dark images all pixels that have more than twice the average brightness can be defined as white. For brighter images a minimal value for white that lies between the average brightness and the maximal possible brightness is useful. Thus the thresholds for the *white$_1$* cuboid are set to: $y_{min} = min\left(\bar{i}_y \cdot 2, \frac{(\bar{i}_y + 255 \cdot 2)}{3}\right)$, $y_{max} = 255 \mid u_{min} = 0, u_{max} = 255 \mid v_{min} = 0, v_{max} = 255$ Therefore all colors that are brighter than the threshold y_{min} are classified as *white$_1$*.

Green. The thresholds for *green$_1$* are set to: $y_{min} = \bar{i}_y \cdot 0.8, y_{max} = \bar{i}_y \cdot 1.2 \mid u_{min} = \bar{i}_u \cdot 0.9, u_{max} = \bar{i}_u \cdot 1.1 \mid v_{min} = \bar{i}_v \cdot 0.9, v_{max} = \bar{i}_v \cdot 1.1$ Thus all colors that are similar to the average color are classified as *green$_1$*. The tolerance for the brightness channel y is slightly larger than the tolerance for the color channels u and v.

3.5   Adaptation to Changing Lighting Conditions

The method for color calibration as described up to here only assigns new colors to color classes or changes the assignment of a color from one color class to another one. Therefore colors that are misclassified once will be misclassified for ever, except they are assigned to another color class. Thus for each entry of the color look-up table (layer 3 color representation) the time when it was last updated is stored. If an entry was not updated for a longer time, the assignment to the color class is deleted. This time depends on the color (green is updated more often than orange). All colors that are used to calibrate $green_3$ are also used to detect, if there was a sudden change of the lighting conditions. If $green_3$ is calibrated, the average value of the segments used for calibration in the current image is compared to the average values used for calibration in the last images. If there is a large difference in at least one color channel, all three layers of color representation are deleted. Thus the vision system reacts very quickly if there is a sudden change of the lighting conditions. In this case the robot has to see at least one object of each color for a few frames to recalibrate all colors.

4   Experimental Results

A crucial parameter for the auto calibration is the reference color. To test if the reference color does not contain colors different from green obstacles, an obstacle avoiding behavior was executed. The robot was able to pass the soccer field without touching any arbitrary positioned obstacle. There was no performance difference compared to a manually calibrated vision system. An easy way to test if the automatic color calibration is successful is to have a look at the resulting color classified images. These images should be similar to images that are the result of a color classification using a hand-made color table. A hand-made color

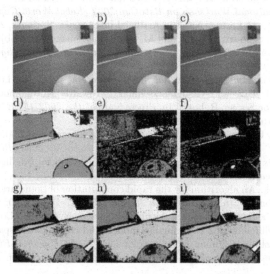

Fig. 8. a,b,c) Three images of the same scene, taken under different lighting conditions. d,e,f) The result of color classification based on a color table that was created by hand for image a g,h,i) The result of color classification based on a color table that was created automatically for each lighting condition

table that was created using sample images taken under certain lighting conditions usually is not usable under different lighting conditions (cf. Fig. 8a-f). The vision system described above was able to adapt the color table when the lighting conditions changed (cf. Fig. 8g-i). The system presented in this paper needs 20 ms to process an image of size 176x144 on an Aibo, which is equipped with a 400 MHz processor.

5   Conclusion

This paper presents a vision system for robotic soccer which needs no pre-run calibration. The independence of lighting conditions is reached by an auto-adaptation of color classes and an edge-based analysis of scan lines. The object recognition and the color calibration employ environmental constraints to determine size and position of the objects. The author would like to thank H.-D. Burkhard, Th. Röfer, and all members of the *Aibo Team Humboldt* for their support.

References

1. J. Bruce, T. Balch, and M. Veloso. Fast and inexpensive color image segmentation for interactive robots. In *Proceedings of the 2000 IEEE/RSJ International Conference on Intelligent Robots and Systems (IROS '00)*, volume 3, 2000.
2. N. B. Daniel Cameron. Knowledge-based autonomous dynamic colour calibration. In *7th International Workshop on RoboCup 2003 (Robot World Cup Soccer Games and Conferences)*, Lecture Notes in Artificial Intelligence. Springer, 2004. to appear.
3. R. Hanek, W. Schmitt, S. Buck, and M. Beetz. Fast image-based object localization in natural scenes. *IEEE/RSJ International Conference on Intelligent Robots and Systems (IROS) 2002*, pages 116–122, 2002.
4. A. K. Jain. *Fundamentals of Digital Image Processing*. Prentice-Hall, 1989.
5. G. Mayer, H. Utz, and G. Kraetzschmar. Towards Autonomous Vision Self-Calibration for Soccer Robots. In *Proceedings of the 2002 IEEE/RSJ Intl. Conference on Intelligent Robots and Systems*, 2002.
6. G. Mayer, H. Utz, and G. Kraetzschmar. Playing robot soccer under natural light: A case study. In *7th International Workshop on RoboCup 2003 (Robot World Cup Soccer Games and Conferences)*, Lecture Notes in Artificial Intelligence. Springer, 2004. to appear.
7. F. K. H. Quek. An algorithm for the rapid computation of boundaries of run length encoded regions. *Pattern Recognition Journal*, 33:1637–1649, 2000.
8. J.-C. Terrillon, H. Fukamachi, S. Akamatsu, and M. N. Shirazi. Comparative performance of different skin chrominance models and chrominance spaces for the automatic detection of human faces in color images. In *Fourth IEEE International Conference on Automatic Face and Gesture Recognition*, page 54ff, 2000.

Getting the Most from Your Color Camera in a Color-Coded World

Erio Grillo[1], Matteo Matteucci[2], and Domenico G. Sorrenti[3]

[1] Undergrad. Stud. (Informatica) at Universitá di Milano - Bicocca
[2] Politecnico di Milano, Dept. Elettronica e Informazione,
`matteucci@elet.polimi.it`
[3] Universitá di Milano - Bicocca, Dept. Informatica, Sist. e Comun.,
`sorrenti@disco.unimib.it`

Abstract. In this paper we present a proposal for setting camera parameters which we claim to give results better matched to applications in color-coded environments then the camera internal algorithms. Moreover it does not require online human intervention, i.e. is automated, and is faster than a human operator. This work applies to situations where the camera is used to extract information from a color-coded world. The experimental activity presented has been performed in the framework of Robocup mid-size rules, with the hypothesis of temporal constancy of light conditions; this work is the necessary first step toward dealing with slow changes, in the time domain, of light conditions.

1 Introduction

Color cameras are used in many application domains, in robotics especially, where they represent a relatively cheap, but powerful sensor. Whenever the operating environment of the camera, i.e. the working environment of the embodied agent, is such that the color of the objects carries the object semantic, then we say that the agent is immersed in a *color-coded world*. This work focuses on such situations, where the agent processing can color-discriminate the objects in the scene; the camera acquisition parameters should be matched to the color codes, to facilitate such discrimination.

Unfortunately, dealing with a color-coded world does not simplify things as much as one can expect; in fact, many aspects are not fixed and contribute to the difficulty of color-discriminating the world:

1. The actual values of the pixels of the objects can often be really different, even though the colors are appear as expected; in practice what is granted is just the human-usable string which represents the name of the color. A Robocup example: according to the rules *the playground has to be green*; this just means that we are happy when the color attribute of the playground is called *green* by most humans. On the other hand its color can range from the emerald-like playground we had at the European championship in Amsterdam 2000, to the pea green we had at the worlds in Padova 2004.

D. Nardi et al. (Eds.): RoboCup 2004, LNAI 3276, pp. 221–235, 2005.
© Springer-Verlag Berlin Heidelberg 2005

2. The actual light on the playground can be different in one or more aspects (e.g., intensity, color temperature) from site to site; of course this impacts on the apparent color of the objects

3. The spectral sensitivity curve (quantum efficiency diagram) of the camera changes with the device in use; of course this impacts on the apparent color of the objects, as seen by the color classification algorithms

4. The values of the parameters which affect the functioning of the many circuits inside the camera have also a large impact on the apparent color of the objects

5. This point applies only if the algorithms for the automatic setting of some parameters are active; the actual algorithm changes with the camera; we would not be happy to discover that our processing does not provide a reasonable output any more, if we change the camera model and/or brand

Cameras functioning usually depend on some parameters, which are in charge of controlling the image acquisition process. Examples of such parameters are: the gain of the amplifier which takes in input the output of the light collecting device and gives out the usable output of the camera, the color balancing "knobs", which allow to change the relative weight of the color channels in order to deal with different color temperatures of the light, etc. Even though the values of such parameters are not usually given the appropriate relevance, i.e. many users just leave them to their default, their role is really relevant in the image grabbing process. Slight changes of some parameters turns into very different images. As their values affect the best results that can be attained by the processing which is applied to the grabbed image, we would like image grabbing parameters set so to grab the near-best images, in terms of the noise acting on the color classification. This should be compared to current praxis, consisting in taking the default values and then working hard to discriminate the colors. Notice that the colors have largely been mixed-up during image grabbing.

It is important to recall that some parameters can be automatically set by the internal algorithms of the camera, while the others are usually left to the manual intervention (if any) of the user. The camera internal algorithms, however, cannot match the requisites of a specific application, mainly because of their generality with respect to the observed scene. As an example, a mid-size Robocup robot could require to discriminate between blue and cyan, but this capability is not part of the functionalities of a normal camera. Moreover, the internal algorithms do not perform well when the camera not in very good light conditions, e.g. when the camera has in view both bright and dark areas. To perform well here means to reach a quality of the image which a human operator could reach, by acting on the camera parameters. A last consideration concerns the internal algorithms, which are usually unknown; they are regarded as part of the source of revenues by the camera builder companies: we tried to gain knowledge of their functioning both from the literature and asking the manufacturer, without usable results. Therefore such algorithms are often turned off, at least to allow the user to understand what's actually going on in the camera, even though the user knows that doing so implies working far away from optimality.

It should be noted that the manual determination of near-optimal values is a time-consuming task and has to be performed by a human operator. This is especially true when working in not optimal light conditions, which is unfortunately when internal algorithms use to fail most. Moreover, the intervention of a human operator in setting these parameters introduces another problem, due to his/her inherently subjective perception of colors, which could make subsequent processing to fail. In order to leave out his/her subjectivity the operator should work basing on some quantitative index, therefore paving the way to an automatic use of the same index.

In this paper we propose to formalize the problem of selecting the acquisition parameters as an optimization task (see Section 2) and we solve such an optimization problem using the genetic meta-heuristic (see Section 3). The experimental activity, presented in Section 4, has been performed in the framework of Robocup mid-size rules, with time-constant light conditions. We consider this work as the necessary first step toward adaptation to slow time-changes of the light conditions. The overall aim of this work is to make the setup time short while minimizing the human intervention, in agreement with the short-setup issue in Robocup mid-size.

2   Setting Up the Problem

The Robocup mid-size league working environment is a so-called color-coded world: it can be described as follows, as ruled by current (2003) rules: the robots move on a flat and rectangular playground built with a green carpet; the two teams of four robots, like in real soccer, defend one goal and attack the other goal. Each of the two goals (one yellow, the other blue) lay on one of the two short sides of the playground. The poles of the goals are white, as the playground lines (side and touch lines, goal area lines, etc). The four corners feature a quarter circle arc white line in the playground and a pole with 3 bands of colors: the corner poles on the blue goal side have colored bands: yellow, blue, and yellow again, from the playground upward. The corners on the other side present the bands, but with the yellow and blue bands exchanged. More details, for the interested reader, can be found in the rule document [1].

The light on the playground is adjusted between 300lux and 900lux. Very slight spatial differences in the light intensity are allowed, e.g. shadows due to the goals, robots, etc. Starting from 2004 slight time changes of the light are allowed as it will be implied also by an imperfect (and cheaper) lightening system, complemented by natural light from outdoor. With such a wide range in light intensity typical issues regard highlights and shadows. With a strong light the red of the ball tends to get yellow in the un-avoidable highlights on the ball, the playground also gets filled with highlights where the green is so diluted to be easily perceived as cyan or even white. With too poor a light the blue gets undistinguishable from black and cyan, etc.

In order to describe the parameters involved in the image acquisition process, we considered those included in the IEEE1394 IIDC (DICAM, Digital Camera) version 1.30 standard [2]. IIDC is an open commercial standard on top of IEEE1394, the well-known Firewire serial bus, an affordable interface for high-speed devices, like cameras. IIDC is a standard for the so-called digital-video, i.e. uncompressed video on Firewire, and, because of the processing time constraints of applications, IIDC is the usual approach in Robotics since no un-compression is required. Many IIDC cameras are available on the market, from webcam-level to professional ones, CCD or CMOS based. When claimed to be compliant with IIDC, such cameras should export settings which are a subset of the settings defined in IIDC specifications. In the view of general usability in Robocup competitions, we carried out the work with low-cost webcam-level Firewire IIDC cameras; notice that this choice just applies to the experiments, i.e. the specific set of parameters on which to conduct the optimization, and does not affect the proposal per sé.

In the following we report a list of parameters in the IIDC standard (IIDC 1394-based Digital Camera Specification v. 1.30), with a very short explanation; some of them can be set to *automatically determined by the camera internal algorithms*. Of course the experimental work has been carried out just on the ones implemented in the actual camera we used, as described in Section 4.

1. EXPOSURE: the combination of the sensitivity of the sensor, the lens aperture and the shutter speed; most webcam-level cameras can actually change just the exposure time
2. IRIS: the lens aperture control; most webcam-level cameras do not have a mechanical iris; instead they use to act on the integration time of the incoming light, i.e. the exposure time
3. GAIN: the electronic amplification of the camera circuit gain control
4. SATURATION: the saturation of the COLOR, i.e. the degree to which a color is undiluted by white light; if a color is 100 percent saturated, it contains no white light; if a color has no saturation, it is a shade of gray
5. WHITE BALANCE - RED channel: the camera can automatically adjust the brightness of the red, green and blue components so that the brightest object in the image appears white, this is done by controlling the relative intensity of the R and B values (in RGB)
6. WHITE BALANCE - BLUE channel: see the point above
7. BRIGHTNESS: the black level of the picture, a pure offset of the intensity level of each pixel coming from A/D converter
8. GAMMA: this parameter defines the function between incoming light level and output picture level, see [3] for details
9. SHARPNESS: the camera can enhance, in a digital sense, the details of edges; this is done by applying a mathematical formula across the image; this is not of interest in the contest of this work
10. HUE: the phase of the color
11. SHUTTER: the opening time of the lens aperture; most webcam-level cameras do not have a mechanical shutter; instead they use to act on the integration time of the incoming light, i.e. the exposure time
12. FOCUS: the lens focus control, most webcam-level cameras just have a mechanical lens focus handle, which is not under software control

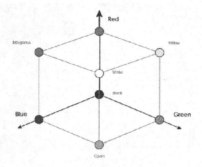

Fig. 1. The RGB cube and the 8 color codes relevant for the Robocup application

13. TEMPERATURE: the temperature inside the camera
14. TRIGGER: this is used to control the frame grabbing in applications which require accurate synchronization, e.g. with stroboscopic light
15. ZOOM the lens zoom control, not available on most webcam-level cameras
16. PAN movement: the motion around a vertical axis, not available on most webcam-level cameras
17. TILT movement: same as above, but referred to an horizontal axis, not available on most webcam-level cameras
18. OPTICAL FILTER CONTROL: changes the optical filter of a camera, not available on most webcam-level cameras

We propose a formalization of the task as an optimization problem, where the independent variables are the camera parameters; the dependent variable has to reflect the quality of the grabbed image, and, most important, the relationship between the independent and the dependent variables is not known. Under a pure theoretical point of view a complete model of image formation could be developed. Apart the complexity of such a model, the intrinsic limit of an approach which would try to create a complete model of the image formation is the difficulty and/or the un-accuracies in the determination of the values of the model parameters. As an example consider the many different physical incarnation of the playground, of the robot body, of the goal color, lights, etc. Therefore we consider not applicable a classical operating research approach, where the relationship *dependent* = $f(independent)$ has to be known.

Our formalization of the problem starts with the observation that the color-codes in Robocup are the vertexes of the so-called RGB cube, see Figure 1; this includes the extremal conditions of minimum light (black) and maximum light (white). Hence the task can be expressed as *how to adjust the camera parameters so that each image area of one color gets RGB values which are very near to the corresponding vertex of the RGB cube.* Here the vertexes are the human-defined color, which, in the following, we will call *color prototypes* for short.

Suppose we have an image where some areas have been collected and classified by a human operator and is taken as ground-truth, see Figure 2. We would

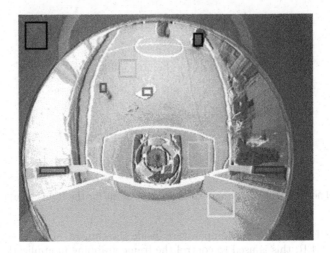

Fig. 2. An example of ground truth for color prototypes selected by the operator

like to push/stretch each color toward its corresponding *prototype* by acting on the acquisition parameters, ending up with pixel values (at least for the selected areas) which are easy-to-discriminate into the set of color prototypes. To perform this color space deformation in an automatic way, we first define an evaluation function, aiming at capturing the distance of the cloud of points of each color from its prototype and, after that, we seek the camera setting which minimize such a function. Of course the camera will not be moved during the whole optimization process.

Let GT^c be the set of ground-truth pixels of color c, i.e. $GT^c = \{p \mid$ declared by the user to be of color $c\}$. The prototype of color c is denoted by $prot^c$. The evaluation function f^c, i.e. computed with respect just to color c, is:

$$f^c = \frac{1}{|GT^c|} \sum_{\forall p \in GT^c} \|p(r, g, b) - prot^c(r, g, b)\| \tag{1}$$

where by $\|A - B\|$ we denoted a distance between two points A and B, in the RGB color space, and by $|Set|$ the cardinality of Set. Let $p(x)$, $x = r, g, b$ be the r or b or g coordinate of a pixel, then Equation 1, if we choose the classic Euclidean norm, translates into:

$$f^c = \sqrt{\sum_{x=r,g,b} [\overline{p}^c(x) - prot^c(x)]^2} \tag{2}$$

where:

$$\overline{p}^c(x) = \left(\frac{1}{|GT^c|} \cdot \sum_{\forall p \in GT^c} p(x) \right) \quad x = r, g, b. \tag{3}$$

If we consider $\bar{p}^c(x)$, $x = r, g, b$ as the r, g, b coordinates of the barycenter $\overline{P}^c$ of GT^c, Equation 2 can be written shortly as:

$$f^c = \|\overline{P}^c - prot^c\| \tag{4}$$

Put as above, the problem is a minimization problem, i.e. the smaller the evaluation function the better the solution. A relevant point is that this minimization has to be carried out for all the relevant colors at the same time. Fulfilling the constraint for just one color is an easy, but not useful, task.

In the case of mid-size Robocup league we have the 8 distinct color prototypes mentioned before; however the generality of the proposal is in its independency on the number of color-codes, on their specific location in the color space, and, to some extent, on the particular color space; this generality turns into applicability in other domains.

To account for the multi-objective aim of the work, which is to minimize the evaluation function for all the colors at the same time, we considered as the evaluation function average the average of the evaluation function of the single colors. So, let CC be the set of color codes, which in our example application are:

$$CC = \{white, black, yellow, blue, red, green, cyan, magenta\}$$

the overall evaluation function for a given set of parameters could be:

$$f_{ave}^{CC} = \frac{1}{|CC|} \sum_{\forall c \in CC} f^c, \tag{5}$$

however, this simple function turns out not to be capable to consistently reach good results because its formulation inherently allows a compensation of very good results for some colors with very bad ones for the others. Therefore, we propose to use a different function, a sum of quadratic (i.e. squared) terms (distances), which aims at weighting more large errors, so to drive the optimization toward a more homogeneous treatment of all the colors. Homogeneous here means forcing at the same time each color cloud towards each prototype.

$$f_{SSD}^{CC} = \sum_{\forall c \in CC} (f^c)^2. \tag{6}$$

3   How to Solve the (Optimization) Problem

The search space of the optimization problem introduced in the previous section is the space of the camera settings; the goal is to minimize an evaluation function computed on few color characteristics of the image obtained from the camera. However, it is not possible to face this optimization task with classical analytical or numerical optimization techniques due to several reasons:

- the absence of an analytical form for the relationship between the independent variables and the measured values,
- the evaluation function is noisy,
- the evaluation function is not linear nor continuous

Due to the previous reasons and given that we have no clue about the convexity of the problem, we propose to use a randomized algorithm to perform the optimization, in particular we suggest to use a genetic meta-heuristic.

Genetic algorithms [4] have proved to be a powerful search tool when the search space is large and multi-modal, and when it is not possible to write an analytical form for the error function in such a space. In these applications, genetic algorithms excel because they can simultaneously and thoroughly explore many different parts of a large space seeking a suitable solution. At first, completely random solutions are tried and evaluated according to a fitness function, and then the best ones are combined using specific operators. This gives the abil-

---

Algorithm 1 Sketch of Goldberg's Simple Genetic Algorithm

Begin Simple Genetic Algorithm
Create a Population P with N Random Individuals
for all $i \in P$ **do**
  Evaluate(i)
end for
repeat
  repeat
    Select i_1 and i_2 according to their Fitness
    with probability p_{cross}
      $i'_1, i'_2 \leftarrow$ **Cross**(i_1, i_2)
    otherwise {with probability $1 - p_{cross}$}
      $i'_1 \leftarrow i_1$ and $i'_2 \leftarrow i_2$
    end with probability
    with probability p_{mut}
      $i''_1 \leftarrow$ **Mut**(i'_1)
    otherwise {with probability $1 - p_{mut}$}
      $i''_1 \leftarrow i'_1$
    end with probability
    with probability p_{mut}
      $i''_2 \leftarrow$ **Mut**(i'_2)
    otherwise {with probability $1 - p_{mut}$}
      $i''_2 \leftarrow i'_2$
    end with probability
    Add Individuals i''_1 and i''_2 to New Population
  until (Created a new Population P')
  for all $i \in P'$ **do**
    Evaluate(i)
  end for
until (Stopping criterium met)
End Simple Genetic Algorithm

---

ity to adequately explore possible solutions while, at the same time, preserving from each solution the parts which work properly.

In Algorithm 1 we show the general scheme for the genetic algorithm we used. The initial population is randomly initialized choosing, for each individual, a random vector of camera parameters; each parameter range is sampled extracting a value out of a uniform distribution. After the definition of the first population of random solutions, each individual is evaluated by computing its fitness and ranked according to it. We evolve the population by selecting the individuals according to their fitness and stochastically applying to them the genetic operators *crossover* and *mutation*. Once a new offspring has been generated, the fitness of the new individuals is evaluated. This process continues until a stopping criterium is met, e.g. a maximum number of generations.

The basic genetic algorithm used in our implementation is the Simple Genetic Algorithm Goldberg describes in his book [4]. At each generation it creates an entirely new population of individuals by selecting from the previous population, and then mating them to produce the offspring for the new population. In all our experiments we use *elitism*, meaning that the best individual from each generation is carried over to the next generation. By using elitism we ensure the algorithm to be a monotonic any-time algorithm, meaning that the optimization process can be stopped at any point while getting always a reasonably good solution.

Solution are coded in genotypes by means of integer valued strings; genetic optimization is known to loose efficiency when individuals are coded by too long genotypes, hence we did not use binary coding. Each position, i.e. allele, represents a camera parameter. The alleles assume values in a limited range according to the camera specifications. We use uniform random initialization and classical uniform crossover; for the mutation we use a Gaussian probability mutation to select values in the neighbors of the actual value instead of a completely random ones. The fitness function described in Algorithm 2 is used to evaluate each individual according to the problem definition of Section 2. Experimental practice evidenced that changing the parameters takes effect after some time, so a couple of fake images are grabbed before each individual evaluation.

Algorithm 2 Evaluation of the fitness of an individual

Begin Evaluate (*Individual i*)
Set the camera parameters according to the genotype
Grab a couple of images to make the new parameter set effective
$fitness = 0$
for all $c \in CC$ **do**
　　Grab an image
　　Compute $\overline{P}^c$, the barycenter in the color space of pixels of color c
　　Compute the barycenter distance $f^c = \|\overline{P}^c - prot^c\|$ from the c color prototype
　　Set $fitness = fitness + (f^c)^2$
end for
return $fitness$
End Evaluate

---

4   Experimental Activity

We performed the experiments with one of the cameras in use on our robots; it is a quite widespread camera in the Robocup community. It is, as mentioned before, a webcam-level camera, with a single color CCD from Sony, progressive scan, capable to provide, via the Firewire interface, 640x480 pixel, 24bpp at 15 frames/s. The builder is OrangeMicro and the model IBot. After some tweaking we now have the camera reliably running onboard our goalkeeper. The camera exports a limited subset of the 18 parameters in the IIDC specification, the others being not software implemented or being related to physical feature which are not present on the camera, e.g. the mechanical iris. The set of parameters for our camera hence reduces to 8:

1. EXPOSURE, nominal range [0, 498]
2. IRIS, nominal range [0, ..., 4]
3. GAIN, nominal range [0, ..., 255]
4. SATURATION, nominal range [0, ..., 255]
5. WHITE BALANCE - RED channel, nominal range [0, ..., 255]
6. WHITE BALANCE - BLUE channel, nominal range [0, ..., 255]
7. BRIGHTNESS, nominal range is [0, ..., 511]
8. GAMMA, nominal range [0, 1]

It is possible to set EXPOSURE, WHITE BALANCE (both RED and BLUE), and BRIGHTNESS to take a value decided by the camera internal algorithm. Considering the nominal range of discrete values, the search space has a finite, but quite large number of points. The time required by the fitness evaluation is bounded by the image grabbing time: the camera gives out 15 frames/s, which means a period of about 67ms. Considering that more than one image transfer has to be awaited, in order to have the new settings operating, then the time required by a brute-force approach would be about:

$$BruteForceTime = 499 \cdot 5 \cdot 256 \cdot 256 \cdot 256 \cdot 256 \cdot 512 \cdot 2 \cdot 180 \; ms$$
$$= 65313835 \; years.$$

We call *current praxis* the situation where the *manual-only* parameters have been set to their default values and the others have been set so that the camera decides on them.

During the experimental activity we switched on all the lights, and obtained what we call an *intermediate* level of light intensity, i.e. in the range [260, 370] lux, which is quite homogeneous, in our opinion, although the light intensity is not as high as it should be in Robocup mid-size. It is worthwhile to notice that the conditions of this experiment are those where the *current praxis* is more likely to give good results, because of the homogeneity of the light intensity.

Figure 4 reports the evolution of the fitness for the best individual in the population compared to the fitness value of the image acquired with the *current praxis* parameters. As it can be seen in the plot, the steady-state is reached much before of the end of the process; the fitness value is lower than the one

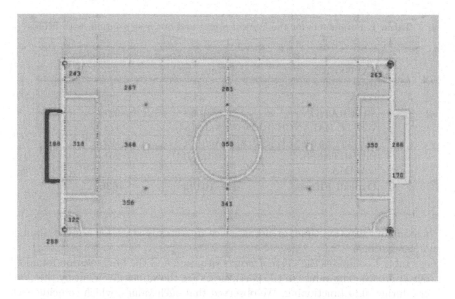

Fig. 3. Distribution of light in intermediate light conditions

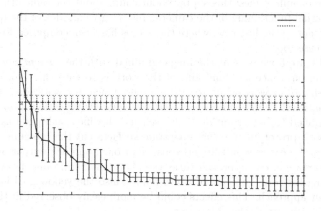

Fig. 4. An example of evolution of the fitness functions during the optimization

computed on an image grabbed with the *current praxis*, just after a very few generations. This experiment uses $p_{cross} = 0.7$, $p_{mut} = 0.2$, and a population of 20 individuals. These values have not been optimized and we feel that better results could be obtained by a more accurate selection of them. In our preliminary experiments, aiming at demonstrating the convergence of the proposal, the evolutionary process is terminated after a large number of generations; in the move to on-line applications we will define different stopping criteria such as

Table 1. Parameters for the *current praxis* and our best solution, with fitness

	current praxis	Evolved Parameters
EXPOSURE	Auto (498)	485
IRIS	4	1
GAIN	87	175
SATURATION	105	227
WHITE BALANCE [RED]	Auto (84)	128
WHITE BALANCE [BLUE]	Auto (91)	102
BRIGHTNESS	Auto (314)	201
GAMMA	1	0
Overall Fitness	**16193**	**4636**

maximum execution time, or permanence in a steady-state for a given number of generations.

Table 1 compares the *current praxis* situation and the best solution found by our approach. The values in the parentheses are the values given out by the camera during *auto* functioning. We observed that such values, which coincide with the default values of each parameter, do not change with the light conditions, whilst they should. Therefore we consider that the onboard software replies with just the default values; this supports our claim about the difficulties to discover the internal functioning of the camera. Just to give an idea of the quality of the results, notice the last row, where the fitness has been computed for the *current praxis* case too.

In Figure 5 we present the image grabbed with the *current praxis* and with our solution. Since the final aim of the work is to ease the color-classification processing, it is worthwhile to evaluate how much our work eases this processing. We therefore introduced a very simple color-classification scheme, which we then applied to both solutions. This scheme classifies each pixel as the color of the closest prototype, i.e. the separation surfaces cut the rgb-cube into 8 parts. As it can be easily seen the performance of our approach outperforms the current praxis: e.g. our approach turns the playground from mostly cyan to green, removing the large amount of white points; also the yellow goal is less reddish with our approach. This effects could be more easily observed in the rightmost column with the classified images.

In order to support the effectiveness of the approach we present hereafter a second experiment. In this experiment we reduced also the search space for some parameters, according to subjective considerations about the effect on the images obtained by changing (independently) each parameter, for instance in order to avoid removing colors form the image. When we made this experiment we were confident we had convincing reasons for such reductions of the search space.

In particular, we reduced the range to [200, 498] for EXPOSURE, to [80, 255] for GAIN, to [100, 255] for SATURATION, to [80, 255] for WHITE BALANCE (both RED and BLUE), to [100, 511] for BRIGHTNESS, and fixed GAMMA to 1 and IRIS to 4.

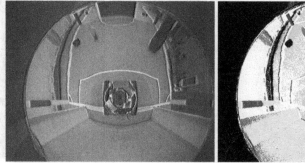

(a) RGB image, current praxis

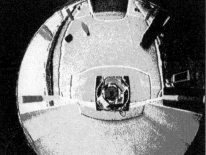

(b) Dumb-classified, current praxis

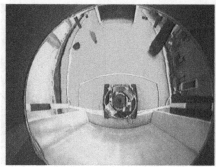

(c) RGB image, our approach

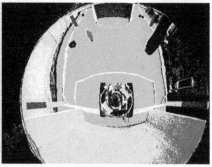

(d) Dumb-classified, our approach

Fig. 5. Images for *current praxis* (a, b), and our solution (c, d)

Table 2. Parameters for the *current praxis* and our solution, with fitness

	current praxis	Evolved Parameters
EXPOSURE	Auto (498)	490
IRIS	4	4
GAIN	87	217
SATURATION	105	242
WHITE BALANCE [RED]	Auto (84)	187
WHITE BALANCE [BLUE]	Auto (91)	126
BRIGHTNESS	Auto (314)	115
GAMMA	1	1
Overall Fitness	**21655**	**8143**

Table 2 compares the *current praxis* situation and the best solution found by our approach. Again our approach gets a lower fitness with respect to current praxis. However, the fitness is twice the fitness of the previous experiment (light conditions were very similar, and the observer is in the blue, darker, goal). According to us this is due to having reduced (improperly) the search space.

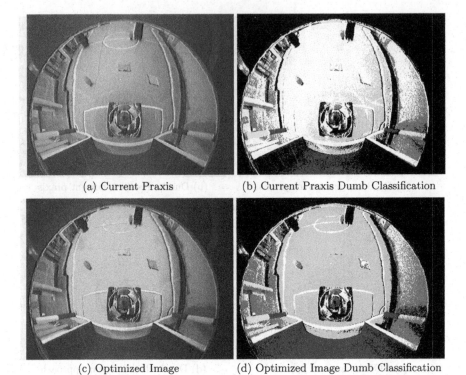

(a) Current Praxis   (b) Current Praxis Dumb Classification

(c) Optimized Image   (d) Optimized Image Dumb Classification

Fig. 6. Image grabbed with *current praxis* (a)(b), and our solution (c)(d)

This again supports the approach of blind optimization we took, avoiding any subjective interpretation of the parameters.

In Figure 6 we present the image grabbed with the *current praxis* and with our solution for this second experiment. As it can be easily seen, even with a sub-optimal solution, the performance of our approach is better than using the current praxis.

5   Conclusions

In this work we propose an approach to set the camera parameters affecting the image grabbing process, so to ease at the maximum extent the color classification task. The problem is cast to a minimization problem which we propose to optimize with a genetic meta-heuristic. An experimental activity is presented which validates the proposal.

This work is just a first step toward an automatic tool for tracking light changes. At the present stage it heavily depends on the training set to include patches from different light conditions, in order to be able to find a globally good parameter setting. We are now working on automatic acquisition of the ground-truth, which will allow to apply the approach to sequences of images, taken in different parts of the field.

Acknowledgements

This work has been partially supported by the project P.R.I.N. 2003 "Tecnologie software e modelli di conoscenza per progetto, sviluppo e sperimentazione di sistemi robotici multi-agente in ambienti reali", funded by M.I.U.R.

We thank the referees for their useful observations, which allowed us to improve the quality of the work.

References

1. Technical Committee of the Midsize League for 2003: Rules for Robocup Midsize Competitions, year 2003. available at http://wwwradig.in.tum.de/MSL-2003/ (2003)
2. 1394 Trade Association: IIDC 1394-based Digital Camera Specification Version 1.30. 1394 Trade Association, Regency Plaza Suite 350, 2350 Mission College Blvd., Santa Clara, California 95054, USA (2000) technical report id.: TA Document 1999023.
3. Poynton, C.A.: A technical introduction to digital video. Wiley (1996)
4. Goldberg, D.E.: Genetic Algorithms in Search, Optimization, and Machine Learning. Addison-Wesley Publishing Company, Inc., Reading, MA (1989)

Combining Exploration and Ad-Hoc Networking in RoboCup Rescue

Martijn N. Rooker and Andreas Birk

School of Engineering and Science, International University Bremen,
Campus Ring 1, D-28759 Bremen, Germany
m.rooker@iu-bremen.de

Abstract. In challenging environments where the risk of loss of a robot is high, robot teams are a natural choice. In many applications like for example rescue missions there are two crucial tasks for the robots. First, they have to efficiently and exhaustively explore the environment. Second, they must keep up a network connection to the base-station to transmit data to ensure timely arrival and secure storage of vital information. When using wireless media, it is necessary to use robots from the team as relay stations for this purpose. This paper deals with the problem to combine an efficient exploration of the environment with suited motions of the robots to keep data transmissions stable.

1   Introduction

At the International University Bremen (IUB), a team is working since 2001 in the domain of rescue robots [1, 2, 3](Figure1). Like in many other challenging domains for service robots, robot teams can be of huge benefit. As in many other applications, there is one basic chore that is of highest importance, namely to ensure the coverage of the entire environment by the robots. This task, commonly known as exploration, can obviously benefit from using multiple robots jointly working in a team. As the problem of exploration is a common one, there is already a significant amount of contributions using multiple robots as discussed in detail later on in section 2.

Another particular challenge is to ensure the transmission of data to an operators station. This is especially difficult when wireless media is used. Furthermore, it can not be assumed that the robots return to the spot where they are deployed, in contrary, their total loss during a mission is a likely risk. Therefore, they have to deliver all crucial information, like victims and hazards found or map-data[6], ideally on-line to an operators station, which is at a secured position. For this purpose, the robots either have to be in direct contact with the base-station or to use other robots as relays, i.e., to incorporate ad-hoc networking.

In this paper an approach is introduced that makes use of the Frontier-Based exploration algorithm [7, 8], which requires no explicit planning for the robots. This reactive approach for exploration has the disadvantage that all robots travel further and further into the unknown territory, hence loosing the contact to the

D. Nardi et al. (Eds.): RoboCup 2004, LNAI 3276, pp. 236–246, 2005.

Fig. 1. Two of the IUB rescue robots at the RoboCup 2003 competition

base-station. We extend the Frontier-Based exploration such that exploration takes place while the robots maintain a distributed network structure which keeps them in contact with the base-station. This *communicative exploration* algorithm is based on a utility function which weights the benefits of exploring unknown territory versus the goal of keeping communication intact. In our experiments, we show that the randomized algorithm yields results that are very close to the theoretical upper bound of coverage while maintaining communication.

The rest of this paper is structured as follows. Section 2 discusses related work. In section 3, the communicative exploration algorithm is introduced. Experiments and results are presented in section 4. Section 5 concludes the paper.

2   Related Work

Different approaches have been introduced in the field of exploration. Zelinsky [11] presents an algorithm where a quadtree data structure is used to model the environment and the distance transform methodology is used to calculate paths for the robots to execute. Makarenko et.al.[10] present an exploration strategy which balances coverage, accuracy and the speed of exploration. They also introduce a metric called localizability, that allows comparison of localization quality at different locations.

Another approach that is also used in this article is the Frontier-Based Exploration approach defined by Yamauchi [7, 8]. In this approach, a frontier is defined as regions on the boundary between open space and unexplored space. A robot moves to the nearest frontier, which is the nearest unknown area. By moving to the frontier, the robot explores new parts of the environment. This new explored region is added to the map that is created during the exploration. In the multi-robot approach different robots are moving over the frontier. Burgard et.al.[9] define a similar approach, but the difference is that the robots in their approach coordinate their behaviors, so that multiple robots will not move to the same position.

The other aspect that is tackled in this paper is Ad-Hoc Networking [12]. Ad-hoc networks are defined by Perkins as wireless, mobile networks that can be set up anywhere and anytime. The concept of ad-hoc networks is applied in different approaches in exploration. Howard [13] implements a deployment algorithm whereby the robots are placed over the environment in a way that they are able to explore the whole environment, but through the network are still able to communicate with each other. Nguyen [14] describes a system exploring a complex environment. The system consist out of four Pioneer robots that are used as autonomous mobile relays, to maintain communication between a lead robot and a remote operator. Some other examples where ad-hoc networking is applied are habitat monitoring [15], medical sciences [16] and childcare [17].

3   Adding Communication to Exploration

The *communicative exploration* algorithm builds upon Frontier-Based approach on the exploration side. The crucial aspect that has been added is the maintenance of communication. Before our new algorithm is introduced, the basic Frontier-Based approach is shortly re-visited in the next subsection.

3.1   Frontier-Based Exploration and Its Extension

Frontier-Based Exploration is introduced by Yamauchi in [7, 8]. He defines frontiers as regions on the boundary between explored and unexplored space. The idea behind this exploration approach is motivated as follows: *To gain as much new information about the world, move to the boundary between open space and uncharted territory.* When a robot moves to a frontier, it can look into the unexplored environment. By exploring the new environment, this data is added to a map that is maintained by the robot. Every time a robot explored new parts of the environment, the mapped region is expanded and the frontier moves over the environment. Through the moving of the frontiers, the robot increases its knowledge of the environment. The explored environment is represented by evidence grids [18]. On this grid a graph is created that is used to plan a path over the grid toward the nearest frontier.

Like the basic Frontier-based approach, the communicative exploration algorithm also avoids complex planning. We are interested in a reactive approach

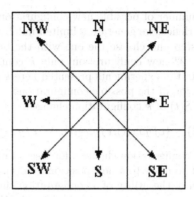

Fig. 2. The 9 new possible new positions of a robot

of exploration. To supplement the basic approach, a value is assigned to every movement of a robot, the so-called utility. The utility allows to penalize the loss of communication while rewarding the exploration of unknown space. Based on the utility, communicative exploration becomes an optimization problem.

3.2   The Utility of Robot Movements in Communicative Exploration

Communicative exploration proceeds in time-steps t. At every time-step t each robot has a position in the environment. The position of robot i at time t is denoted by $P[i](t) = (x_i(t), y_i(t))$. The set of positions of all n robots in the environment at time t is called a *configuration*, denoted as $cfg(t)$:

$$cfg(t) = \{(x_1(t), y_1(t)), (x_2(t), y_2(t)), \cdots, (x_n(t), y_n(t))\}$$

To calculate the Utility of a new configuration, the following algorithm is applied. For every robot a new configuration is calculated. In total there are 9 different possibilities for a new configuration for 1 robot, including that the robot stays at its position, as can be seen in figure 2.

So for n robots, there are 9^n different configurations possible. Formally, a configuration change at time t is defined as follows:

$$cfg\_c(t) = \{m_1(t), m_2(t), \cdots, m_n(t)\}$$

with $m_i(t)$ being the movement of robot i at time t, defined as:

$$m_i(t) \in M = \{N, NE, E, SE, S, SW, W, NW, R\}$$

As mentioned, there are 9^n different configurations for n robots. Some of these configurations are not possible, like for example when multiple robots move to the same position or if a robot moves into an obstacle.

The exponential number of possible new configurations makes it impossible to check all of them. Hence, we generate a limited number of random new configurations per time-step and choose the one with the best utility. So, instead of considering all the 9^n new configuration, only k configurations are considered, with $k << 9^n$. In the experiments presented here k is set to 50, giving an extremely fast evaluation of the possible configurations.

This creates a set $S(t)$ of k configuration changes:

$$S(t) = \{cfg\_c_1(t), cfg\_c_2(t), \cdots, cfg\_c_k(t)\}$$

wherein $cfg\_c_i(t)$ is a configuration change i for a set of robots. A configuration change causes a robot to move to a new position $P'[i](t)$.

For the calculation of the utility of a configuration change, the different options where a robot can move to have to be defined. When a robot wants to move to a new position, the following situations can occur:

- **Impossible position:** When one of the following situations occurs:
  - Two or more robots want to move to the same position.
  - A robot wants to move to a position that is occupied with an obstacle.
  These locations should be avoided, therefor a negative, repulsive value is assigned to those locations.
- **Loss of communication:** A robot wants to move to a location where there is no communication possible with the base-station, directly or indirectly. The process of checking if a robot is still in communication with the base-station is described in section 3.3. Also here it is the case that these locations need to be avoided, so once again a negative, repulsive value is assigned.
- **Frontier cell:** The location where the robot wants to move to is a location on the frontier. The frontier cells are the locations in the world that need to be explored, so a positive, attractive value is assigned to those locations.
- **Other:** The last option where a robot can move to is a location on the field that has already been explored. It could also mean that a robot maintains its position. This option is not optimal, but definitely not wrong as it avoids obstacles, so therefor a "neutral" value is assigned to these locations.

The following return values are defined for the different situation:

$$U(P'[i](t)) = \begin{cases} -100 & \text{if infeasible} \\ -10 & \text{if loss of communication} \\ 1 & \text{if frontier cell} \\ 0 & \text{other} \end{cases}$$

with $1 \le i \le n$.

The whole Utility of a new configuration is then calculated as follow:

$$U(cfg\_c_i(t)) = \sum_{i=1}^{n} U(P'[i](t))$$

The configuration change with the highest utility value is then selected. After all the robots have arrived at their new position, the whole process is repeated.

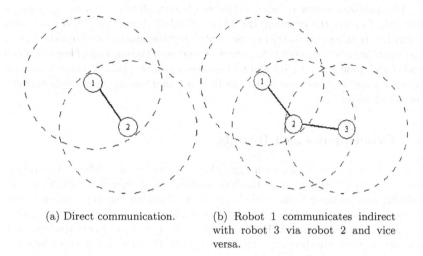

(a) Direct communication.   (b) Robot 1 communicates indirect with robot 3 via robot 2 and vice versa.

Fig. 3. Direct and indirect communication between robots

3.3   Detecting Communication

To properly determine the utility of a configuration, it is necessary that the robots check whether communication is maintained or not. A robot is able to directly communicate with another robot if it is within the communication range of that other robot. If a robot i is not within communication range of another robot j, the possibility still exist that robot i is able to communicate with robot j. This is illustrated in figure 3.

As the robots are moving around in the environment, connections between robots are broken and created. At a specific configuration cfg_i a certain amount of connections are possible. Through the locations of the robots and the connections between the robots a connection graph $CG = (V, E)$ is created, where $V = \#robots + \#base - station$ and E the set of communication connections between robots and base-station. If there is a path between robot i and robot j, than those robots can communicate with each other, i.e., we have a properly connected network.

If two robots are (indirectly) connected to each other, than there has to be a path between the two nodes in the graph. In our situation we are not only interested in connections between robots, but in connections between all the robots and the base-station. As the goal is to have to robots always in connection with the base-station, there always has to be a path between every robot and the base-station.

Of course, over a graph, different paths are possible. As it is ideal that the data is delivered as fast as possible to the base-station, the shortest communication path between a robot and the base-station has to be found. Shortest path in this situation is defined as the communication path with as few hops as possible.

This problem is now reduced to the well-known all-pairs shortest path problem [19]. To solve the problem, the *Floyd-Warshall Algorithm* [20] is used. This algorithm returns two matrices, one containing the amount of hops on a path and the other one containing the parent of a node. If the amount of hops between node i and node j is negative and node i does not have a parent, there is no path between these two nodes and thus there is no communication possible between these two nodes.

4   Experiments and Results

The communicative exploration algorithm is tested in simulation. The robots start at a fixed position near the base-station. From here the robots are calculating new configurations and slowly start exploring the environment while maintaining communication with the base-station. A snapshot of the start situation can be seen in figure 4. The red circles in the bottom center of the figure are the robots. They are starting at the base-station. The grey squares are obstacles.

As can be seen in figure 5, the robots spread out nicely, while exploring the environment and maintaining communication with the base-station.

From figure 5 it can be seen that the frontier between explored and unexplored space is constantly connected. Only if an obstacle intersects the frontier it is not connected at that point. Through this connected frontier, a continuous explored space is created.

For the experiments the following measurements are taken. As one of the goals of the experiments is to remain communication with the base-station, there is a hard limit to the maximum explored space. The robots can move furthest away from the base-station by forming a chain over which the communication is relayed.

For n robots an upper limit m_n for the space they can explore with working communication can hence be calculated as follows:

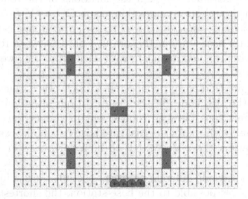

Fig. 4. The start positions of the robots at the center bottom. The five grey areas are obstacles

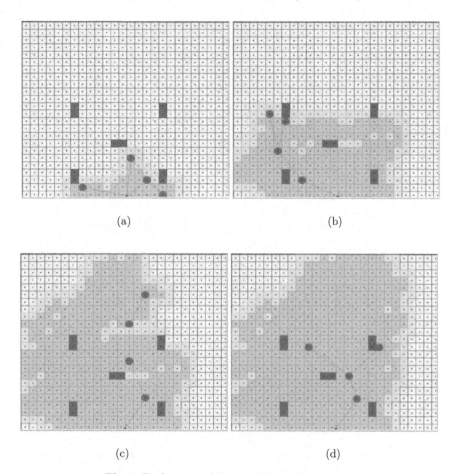

(a)   (b)

(c)   (d)

Fig. 5. Exploration of the environment by 4 robots

$$m_n = (\pi(r \cdot n)^2)/2$$

with r being the communication range of a robot.

This upper limit can be directly expressed in grid-cells:

$$mg_n = m_n/S(gc)$$

with mg_n being the maximum amount of grid-cells that can be explored with n robots and $S(gc)$ being the surface of a grid-cell.

This upper limit is roughly speaking the area of a half-cycle around the base-station with a radius of $r \cdot n$, i.e., the radius is achieved by chaining the robots at the limit of their communication ranges together. Note that this is an upper limit as there are several reasons why a smaller area is likely to be the real limit for the explorable area. Obstacles for example can easily limit the maximum range.

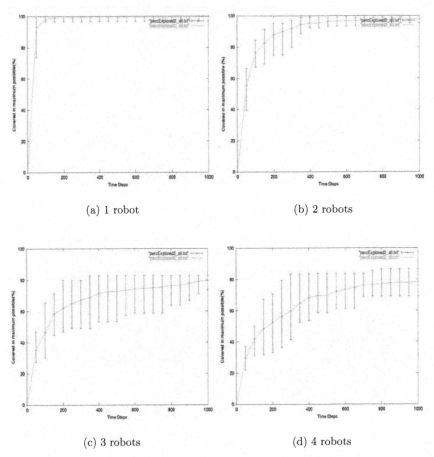

(a) 1 robot   (b) 2 robots

(c) 3 robots   (d) 4 robots

Fig. 6. Area coverage. With 1 and 2 robots the whole maximum possible area is covered. With more robots it takes longer before the maximum possible area is covered

Based on this upper limit, the percentage of explored space while maintaining communication can be calculated in experiments (figure 6). From the graphs it can be seen that in the case of 1 and 2 robots, almost constantly 100% of the upper limit is explored in all the runs. In the case of 3 or more robots, it appears that not the whole upper limit is achieved. A reason for this is the influence of the obstacles. Note that to make a guaranteed "optimal" exploration, i.e., to cover the largest possible area without communication loss, a proper motion-planning for the aggregate of all n robots would have to be done in *every* step, which is for complexity reasons infeasible. Our solution might "waste" a few cells that might have been reachable but that are not explored in the end. But this solution is highly efficient and yields large and continuous explored areas.

5   Conclusions

We presented the *communicative exploration* algorithm. It is an extension of the Frontier-Based approach for exploration [7][8]. Communicative exploration deals with the problem that in many real world applications, the robots have to be a part of a wireless network. In the original algorithm, all robots operate at the borderline to the unexplored space. They hence move further and further away from the point where they were deployed. But this point of deployment is typically an operator's station to which the robots should transmit their data. As they move away, all connections to the base-station get lost. Our extension of the algorithm ensures that each robot explored parts of the environment that are within the range of the robot's communication cell. Therefore, the transmission of data from all robots to the base-station can constantly be maintained.

References

1. Andreas Birk, Holger Kenn and Max Pfingsthorn. The IUB Rescue Robots: From Webcams to Lifesavers. *1st International Workshop on Advances in Service Robotics (ASER)'03*, 2003
2. Andreas Birk, Stefano Carpin and Holger Kenn. The IUB 2003 Rescue Robot Team. *RoboCup 2003: Robot Soccer World Cup VII*, Springer, Lecture Notes in Artificial Intelligence (LNAI), 2004.
3. Andreas Birk, Holger Kenn, Martijn Rooker, Akhil Agrawal, Horia Vlad Balan, Nina Burger, Christoph Burger-Scheidlin, Vinod Devanathan, Dumitru Erhan, Ioan Hepes, Aakash Jain, Premvir Jain, Benjamin Liebald, Gediminas Luksys, Marisano James, Andreas Pfeil, Max Pfingsthorn, Kristina Sojakova, Jormquan Suwanketnikom and Julian Wucherpfennig. The IUB 2002 Rescue Robot Team. *RoboCup-02: Robot Soccer World Cup VI*, Gal Kaminka, Pedro U. Lima and Raul Rojas, Springer, LNAI, 2002.
4. M. Micire, R. Murphy, J. Casper and J. Hymas. Potential tasks and research issues for mobile robots in robocup rescue. In Tucker Balch and Gerhard Kraetszchmar, editors *RoboCup 2000: Robot Soccer World Cup IV*, Lectures keywords in Artificial Intelligence 2000. Springer Verlag, 2001.
5. Rosalyn Graham Snyder. Robots assist in search and rescue efforts at wtc. *IEEE Robotics and Automation Magazine*, 8(4):26–28, December 2001.
6. Stefano Carpin, Holger Kenn and Andreas Birk. Autonomous mapping in the real robot rescue league. In *RoboCup 2003: Robot Soccer World Cup VII*, Lecture Notes in Artificial Intelligence (LNAI). Springer, 2004.
7. Brian Yamauchi. A Frontier-Based Approach for Autonomous Exploration. In *Proceedings of the 1997 IEEE International Symposium on Computational Intelligence in Robotics and Automation*, Monterey, CA, July 1997, pp. 146–151.
8. Brian Yamauchi. Frontier-Based Exploration using multiple Robots. In *Proceedings of the Second International Conference on Autonomous Agents (Agents '98)*. Minneapolis, MN, May 1998, pp. 46–53.
9. W. Burgard, M. Moors, D. Fox, R. Simmons and S. Thrun. Collaborative Multi-Robot Exploration. In *Proc. of the IEEE International Conference on Robotics & Automation (ICRA)*, 2000.

10. Alexei A. Makarenko, Stefan B. Williams, Frederic Bourgault and Hugh F. Durrant-Whyte. An Experiment in Integrated Exploration. Presented at *IEEE/RSJ Intl. Workshop on Intelligent Robots and Systems*, 2002.

11. Alexander Zelinsky. A Mobile Robot Exploration Algorithm. In *IEEE Transactions on Robotics and Automation*, Vol. 8, No. 6, December 92.

12. Charles E. Perkins. Ad Hoc Networking, Addison-Wesley, 2001

13. Andrew Howard, Maja J. Matarić and Gaurav S. Sukhatme. An Incremental Self-Deployment Algorithm for Mobile Sensor Networks. In *Autonomous Robots*, special issue on Intelligent Embedded System, G. Sukhatme, ed., 113-126, 2002

14. H. H. Nguyen, H. R. Everett, N. Manouk and A. Verma. Autonomous Mobile Communication Relays. In *SPIE Proc. 4715: Unmanned Ground Vehicle Technology IV*, Orlando, FL, April 1-5, 2002.

15. A. Cerpa, J. Elson, D. Estrin, L. Girod, M. Hamilton and J. Zhao. Habitat monitoring: Application driver for wireless communications technology. In *Proceedings of the 2001 ACM SIGCOMM Workshop on Data Communications in Latin America and the Caribbean, April 2001*, 2001.

16. Loren Schwiebert, Sandeep K. S. Gupta and Jennifer Weinmann. Research challenges in wireless networks of biomedical sensors. In *Mobile Computing and Networking*, pages 151-165, 2001.

17. Mani B. Srivastava, Richard R. Muntz and Miodrag Potkonjak. Smart kindergarten: sensor-based wireless networks for smart development problem-solving environments. In *Mobile Computing and Networking*, pages 132-138, 2001.

18. H. Moravec and A. Elfes. High resolution maps from wide angle sonar. In *Proceedings of the 1985 IEEE International Conference on Robotics and Automation*, (St. Louis, MO), pages 116-121, 1985.

19. http://www.nist.gov/dads/HTML/allpairsshrt.html

20. http://www.nist.gov/dads/HTML/floydWarshall.html

Robust Multi-robot Object Localization Using Fuzzy Logic

Juan Pedro Cánovas[1], Kevin LeBlanc[2], and Alessandro Saffiotti[2]

[1] Dept. of Information and Comm. Eng.,
University of Murcia, 30100 Murcia, Spain
`juanpe@dif.um.es`
[2] AASS Mobile Robotics Lab, Dpt. Technology,
Örebro University, 70182 Örebro, Sweden
{`klc, asaffio`}`@aass.oru.se`

Abstract. Cooperative localization of objects is an important challenge in multi-robot systems. We propose a new approach to this problem where we see each robot as an expert which shares unreliable information about object locations. The information provided by different robots is then combined using fuzzy logic techniques, in order to reach a *consensus* between the robots. This contrasts with most current probabilistic techniques, which average information from different robots in order to obtain a *tradeoff*, and can thus incur well-known problems when information is unreliable. In addition, our approach does not assume that the robots have accurate self-localization. Instead, uncertainty in the pose of the sensing robot is propagated to object position estimates. We present experimental results obtained on a team of Sony AIBO robots, where we share information about the location of the ball in the RoboCup domain.

1 Introduction

Cooperating robots can benefit in a number of ways from the exchange of information about perceived objects. For example, a robot which does not directly see an object can still get an estimate of its position. Also, an individual robot's estimate of an object position can be improved through information sharing. This occurs when some robots have more accurate and/or more reliable position estimates for that object, which can happen due a number of things. For example, a robot may be better localized, have a better view of the object, or have more effective sensors. In many situations, it is even possible for a group of robots, all having relatively poor estimates of an object's position, to obtain more accurate and reliable estimates through information sharing. However this requires that the information sharing be performed in an effective way.

The problem of cooperative object localization is the problem of fusing information from different sources in a way which produces agreement about object positions; the agreed upon positions should also be as close as possible to the

D. Nardi et al. (Eds.): RoboCup 2004, LNAI 3276, pp. 247–261, 2005.

real object positions. However, information fusion can result in degradation of information if it is not done carefully. For instance, if we combine a correct observation from a robot A with an incorrect observation from a robot B by simple averaging (or even weighted averaging), the result will be worse than what A would have established alone. This averaging problem occurs in many existing approaches (e.g. [11], [14], [17], [18]), as is mentioned in the next section.

Another limitation of many current approaches (e.g. [6], [18]) is that they assume that a robot has very little uncertainty about its own position. In reality, however, this assumption is often not valid. Ignoring self-localization uncertainty can severely restrict the applicability of these methods. For example, in the mid-sized and four-legged leagues of RoboCup [8], robots often have poor self-localization, due to the fact that the domain is highly dynamic, and also because of frequent undetected collisions.

In this paper, we propose a new approach to the cooperative localization problem, based on fuzzy logic. One distinctive point of our approach is that we see each robot as an expert, which provides unreliable information about object locations. We use fuzzy logic to fuse information provided by different sources in order to reach a *consensus* about object positions. This contrasts with many other methods, which yield a *compromise* between various data sources.

Moreover our method carefully takes into account different facets of uncertainty, including unreliability in perceptual information and uncertainty in self-localization. An important contribution of our method is that the uncertainty in a robot's own position is consistently propagated to its object position estimates. One of the obvious advantages of doing this is that high self localization accuracy is not required in order to get position estimates which reflect our knowledge. Our method provides estimates which are consistent with our observations, while taking into account the uncertainty present in both perception and in self-localization.

In the rest of this paper, we first describe some alternate approaches from the literature; then we describe our technique and discuss how we have implemented it on a team of Sony AIBO robots [5]; finally, we present experimental results based on the RoboCup domain (four-legged league).

2   Related Work

There are many existing approaches to cooperative object localization. One simple approach is to use a switching strategy. For example, Roth *et al* [13] use such a strategy for locating the ball in the four-legged league of RoboCup. In their implementation each robot maintains a local world model and a shared world model. A robot which has not seen the ball in some time selects the most probable location estimate from those available in the shared world model. Unfortunately it is fairly easy to select the wrong estimate, and even if the right estimate is selected, the accuracy is no greater than that achieved by a single robot.

Several approaches fuse information from multiple sources using variations on Kalman filtering techniques (e.g. [11], [14], [17]). Alternatively, Stroupe *et al* [18] represent observations as two-dimensional Gaussian distributions, and observations from different robots are merged by taking the average of the Gaussians. In case of disagreement, the merged position estimate will typically be weighted more heavily toward what the majority of robots believe. However, both these methods fuse observations using weighted averages of some sort, and fail to provide a robust solution in the presence of false positives and outliers. Observations which are incorrect, due to errors in perception or in self-localization, can make the fused estimate worse than that of some of the individual robots. In general, averaging produces a compromise between estimates. The method proposed in this paper, by contrast, seeks a consensus between them.

These methods have been improved upon through the use of gating strategies, which discard observations which are deemed invalid. Observations can be deemed invalid if, for example, they are very different from current position estimates. One could also discard observations which do not correspond to what the majority of robots believe. Marcelino *et al* [12] compare the method used by Stroupe *et al* [18] with a fusion algorithm described in [7], which uses a gating strategy to discard observations thought to be inconsistent with previous sensor readings. They show that the gating strategy improved performance and robustness. However simple gating strategies are often not enough. When a target object is unobserved for a certain time, or when a target object moves very rapidly, correct observations could be consistently disregarded since they may no longer correspond with the current belief about the state of the world. Typically, the confidence in the current belief should eventually decrease below some threshold, at which point valid observations would once again be accepted.

A more robust way to deal with outliers and false positives is to implement a *voting scheme*, which encourages belief in observations which are consistent with the majority opinion. Markov localization (e.g. [9]), which is widely used for both individual and cooperative object (and self) localization, implements such a voting scheme by maintaining a discrete, multi-modal probability distribution for position estimates. In the individual robot case, it encourages belief in positions which are consistent with previous and/or current sensor observations. The cooperative case extends this by also encouraging belief in positions which are consistent with information received from other robots. Markov localization is quite robust, though in general its accuracy is less than that of other common methods (e.g. Kalman filtering).

In [6] and [10], a hybrid method called Markov-Kalman localization (ML-EKF) is described. This method uses a grid-based version of Markov localization as a robust, low-resolution plausibility filter. It then uses an extended Kalman filter to compute precise object positions based on the observations which were deemed valid by the Markov process. This approach inherits the robustness of Markov localization and the precision of Kalman filtering. It provides an informed way of deciding which observations should be discarded, and yields increased robustness with respect to more arbitrary gating strategies. However it is still possible for false positives or outliers to affect the result, depending on how strict one is in tuning the plausibility filter. Moreover, this method assumes that a robot has very little uncertainty about its own position.

The method we propose in this paper can also be seen as a sort of voting scheme. However, there are two main differences with respect to other voting approaches. First, votes contain the full uncertainty in the agent's self-localization; second, this uncertainty can be multi-modal. One of the advantages of this careful treatment of uncetainty is that high self-localization accuracy is not required by our method.

A number of other methods for object localization using a single robot have been described in the literature, such as particle filters and multiple hypothesis tracking. However the cooperative aspects of these methods have not been thoroughly investigated. In [16], Schmitt *et al* describe an approach to cooperative perception using multiple hypothesis tracking; but careful tuning of the pruning parameters is required for this method to be effective.

3   Representing Location Information

3.1   Fuzzy Location Information

Location information may be affected by different types of uncertainty. Consider a robot that needs to grasp a given object. This task requires that the position of the object be known with a high degree of precision, as in the statement (a) "The object is at position $x = 81$". The statement (b) "The object is near the center of the table" is *vague*, since it does not give a crisp position. The statement (c) "The object is somewhere on the table" is *imprecise*, since it does not give a point position. The statement (d) "The object is either at position $x = 81$ or at position $x = 162$" is *ambiguous*, since it gives multiple options. And the statement (e) "The object was seen yesterday at position $x = 81$" is *unreliable*, as the object may no longer be there. An ideal uncertainty representation formalism should be able to represent all of these statements. Perhaps more importantly, it should account for the differences between these statements, by representing information at the level of detail at which it is available.

Fuzzy logic techniques are attractive in this respect [15]. We can represent information about the location of an object by a fuzzy subset μ of the set X of all possible positions [20, 21]. For any $x \in X$, we read the value of $\mu(x)$ as the degree of possibility that the object is located at x given the available information. Fig. 1 shows some examples, taken in one dimension for graphical clarity. Cases (a–e), correspond to the five items of information mentioned previously. Case (e) is especially interesting: in order to account for unreliability, we include in the distribution a uniform "bias" to indicate the possibility that the object could also be located somewhere else. Total ignorance, in particular, can be represented by the fuzzy set $\mu(x) = 1$ for all $x \in X$; that is, all locations are possible. Finally, case (f) shows a combination of the previous types of uncertainty.

Fuzzy positional information can be represented in a discretized format in a position grid: a tessellation of the space in which each cell is associated with a number in the range $[0, 1]$, representing the degree of possibility that the object

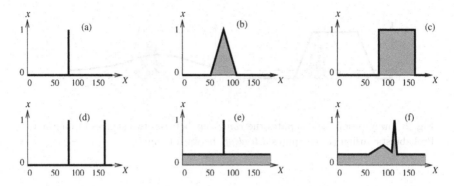

Fig. 1. Different types of fuzzy location information: (a) crisp, (b) vague, (c) imprecise, (d) ambiguous, (e) unreliable, (f) mixed

is in that cell. A common choice is to use a 2-dimensional grid of square cells with uniform size. A 3-D grid can be used if the orientation of the object is also relevant. In the approach proposed in this paper we use 2-D fuzzy position grids to represent our belief about the positions of objects, and a 3-D grid to represent our belief about the robot's own pose in the environment.

3.2   Fuzzy Information Fusion

An important component of a representation for uncertain information is how information coming from different sources can be *fused* together. Fusion can be used to combine the information provided by multiple robots, or by multiple sensors in the same robot.

If location information is represented by fuzzy sets, fusion can be performed by fuzzy intersection between these sets [1]. Let μ_1 and μ_2 be two fuzzy sets representing the information about the position of a given object, provided by sources 1 and 2, respectively. Then their combined information is given by the fuzzy set $\mu_{12} = \mu_1 \cap \mu_2$ defined by

$$\mu_{12}(x) = \mu_1(x) \otimes \mu_2(x), \tag{1}$$

where $\otimes$ is a t-norm.[1] Fig. 2 (a) illustrates fuzzy fusion. The result of the fusion of μ_1 and μ_2 is indicated by the shadowed area.

There are two facts about fuzzy fusion that should be noticed. First, only those locations which are regarded as possible by both sources are retained in the result of the fusion. Intuitively, the resulting fuzzy set μ_{12} represents the *consensus* between the two sources of information. This contrasts with standard

---

[1] T-norms are the general operators used to perform intersection of fuzzy sets [19]. The most common examples of t-norms are minimum, product, and the Łukasiewitz operator $\max(0, a + b - 1)$. In this work, we use the product t-norm.

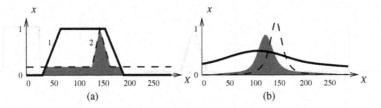

Fig. 2. Fuzzy fusion (a) computes the *consensus* between two sources of information. Probabilistic fusion (b) computes a *tradeoff* between them

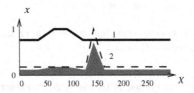

Fig. 3. Discounting unreliable information in fuzzy fusion. Information μ_1 is unreliable, as indicated by the high bias, and therefore only has a small impact on the result

probabilistic techniques, in which information fusion is typically performed by some sort of weighted average, representing a *tradeoff* between sources. Fig. 2 (b) shows how two pieces of information similar to the ones in Fig. 2 (a) might be fused in a probabilistic setting, by combining Gaussians. Notice that with fuzzy fusion the peak of the resulting distribution μ_{12} coincides with the peak of μ_2, since this is compatible with the peak of μ_1; it lies in between those peaks when using probabilistic fusion. As we mentioned earlier, averaging is often not the best solution when combining location information from multiple robots.

The second fact to note is that fuzzy fusion automatically discounts unreliable information. Consider the next example shown in Fig. 3. The information represented by μ_1 includes a high bias (0.8) to indicate that it is fairly unreliable, while the information represented by μ_2 only has a small bias (0.1). Correspondingly, the result of the fusion mostly reflects μ_2 and it is only marginally influenced by μ_1. In practice, this means that fuzzy fusion allows us to discard unreliable information, provided that this unreliability is correctly represented.

We sometimes need to extract a point estimate from the location information represented by a fuzzy set μ, e.g., to be used in other navigation modules. A common way to do this is by computing the center of gravity (CoG) of μ:

$$\hat{x} = \frac{\int_{x \in X} x\mu(x)\,dx}{\int_{x \in X} \mu(x)\,dx}. \tag{2}$$

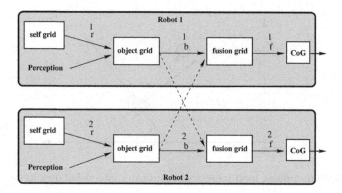

Fig. 4. Schema of our cooperative localization technique

4   Sharing Location Information

Fuzzy location information can be shared in order to get a common view of the environment. The key point here is to see each robot as a source of unreliable information. This information is represented and fused using the techniques described in the previous section. The diagram in Fig. 4 summarizes our fusion schema in the case of two robots. The extension to n robots is straightforward.

We use three fuzzy position grids in each robot. The "self grid" contains the information μ_r about the robot's self location. The "object grid" contains the information μ_o about the global location of the target object, derived from perceptual observations. The "fusion grid" contains the result of the fusion of location information computed by different robots. This grid is kept separate from the object grid in order to avoid circular dependencies. A final defuzzification step obtains a point estimate, if needed, using formula (2). Temporal aspects aside, the robots should end up with identical object position estimates.

4.1   From the Self Grid to the Object Grid

In order to simplify the exchange of location information, we assume that all robots use a common global reference system F_g. Each robot r, however, acquires perceptual data from its own point of view, and it estimates the positions of objects in its local reference frame F_r. In order to represent this information in the global frame F_g, we need to apply a coordinate transformation function $T_r^g : F_r \to F_g$.

Assume that the robot has observed an object at polar coordinates (ρ, ϕ) with respect to its own reference frame F_r — see Fig. 5. If we knew the robot's pose (x_r, y_r, θ_r) in the global frame F_g, then the computation of the global coordinates (x_o, y_o) of the object would be straightforward. In our case, however, the robot's pose is not known with certainty, but is represented by a fuzzy set μ_r in F_g. Accordingly, the object's global position is given by a fuzzy set $\mu_o(p)$ defined

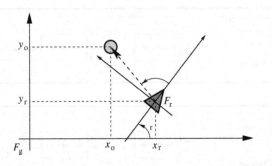

Fig. 5. Transforming local (observed) coordinates into global ones in the crisp case

as follows: for any $p = (x, y)$, the degree of possibility $\mu_o(p)$ that the object is located at the global position p is given by

$$\mu_o(p) = \sup\{\mu_r(q, \theta) \mid d(\overline{pq}) = \rho \text{ and } \angle(\overline{pq}) = \phi + \theta\}, \tag{3}$$

where $q = (x', y')$ is any 2-D position, and $d(\overline{pq})$ and $\angle(\overline{pq})$ respectively denote the length and the orientation of the segment $\overline{pq}$. Intuitively, this says that the object can be at location p as long as there is some possible pose (q, θ) for the robot such that, if observed from that pose, the location p would appear at distance ρ and angle ϕ.

It should be emphasized that the above transformation preserves the full self-localization uncertainty contained in the fuzzy set μ_r. In particular, if the robot is highly uncertain about its own position, then many poses (q, θ) will have a high value in the fuzzy set μ_r. Correspondingly, there will also be many possible positions for the object, which will then have a high value in μ_o. In this respect, our approach is different from most existing approaches, in which the global position of the object is computed by assuming a point estimate for the location of the robot, and uncertainty is added after the transformation. These approaches do not correctly propagate the uncertainty in the robot's pose, since they do not take into account the non-linearities in the local-to-global coordinate transformation. Moreover, they cannot deal with a multi-modal distribution in the robot's self-localization. Our transformation (3) addresses both of these issues.

In a discretized position grid, the transformation (3) can be computed by the following algorithm.

```
foreach cell p
    μ_o(p) := 1 - reliability(observation)
    foreach cell q such that dist(p, q) = ρ                    (4)
        μ_o(p) := max{μ_o(p), μ_r(q, (∠(p, q) - φ))}
end
```

In general, the μ_o distribution computed through this algorithm contains a "bias"; that is, its minimum value β is strictly positive. This indicates some

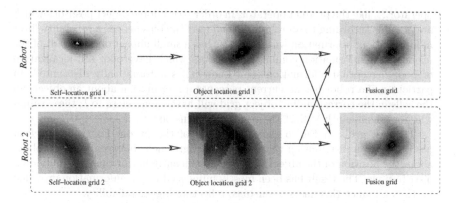

Fig. 6. Example of fusion of information between two robots

unreliability in the position of the observed object, as discussed in section 3.1, and illustrated in Fig. 1 (c). This unreliability can originate from two sources: (i) unreliability in perceptual information, represented by 'reliability(observation)' in the algorithm; and (ii) unreliability in the self-localization of the robot, represented by the bias in μ_r which is propagated to μ_o. This bias is instrumental in discounting unreliable information in our fuzzy fusion scheme.

The behavior of our transformation is illustrated in Fig. 6. The picture shows the individual self and object grids for two robots during information sharing, as well as the result of the fusion. Darker cells indicate higher degrees of possibility. The triangles represent the robot's estimates of their own positions. The circle in the object and fusion grids indicates the real position of the object. In this example, the object is a ball in a soccer field in the RoboCup domain. Both robots can see the ball.

The self-localization grid of Robot 1 shows that this robot has a fairly good estimate of its own position. The corresponding object grid indicates a larger uncertainty in the position of the ball. The reason for this is that the uncertainty in the ball location is affected by the uncertainty both in the (x, y) position of the robot and in its orientation (not shown in the self-grid for graphical clarity). The effect of the non-linearity in the coordinate transformation (3) is clearly visible. Robot 2 has just observed a landmark after a long time during which no observation was made, hence the distribution μ_r is concentrated around a circle at a given distance from the landmark. The low quality of the current self-location estimate for Robot 2 is reflected in the high bias, which appears as a uniform gray background in the grid. The corresponding object grid provides a very rough estimate of the location of the ball.

4.2   From the Object Grids to the Fusion Grid

Fusion of information from different robots is performed by fuzzy intersection of their distributions according to equation (1). We use a non-idempotent t-norm operator $\otimes$, like the product operator, in order to reinforce object positions which are possible according to all robots. Recall that the bias acts as a reliability filter in fuzzy fusion: information with high bias has a small impact on the result of the fusion. Thus information coming from robots which have poor self-localization or poor perceptual information (and know it) is automatically discounted. In particular, if a robot has no current perceptual information about an object, all the cells in the corresponding object grid will have values close to 1; these values will not (significantly) affect the result of the fusion.

The last grids in Fig. 6 show the result of the fusion. Fuzzy intersection has produced a distribution that reflects the agreement between the two robots about the position of the object, and it is significantly better than the individual distributions. The result has been mostly influenced by the information provided by Robot 1 since this has a lower bias, reflecting higher reliability.

The above schema assumes that the full object location grids are exchanged between robots. This may be an expensive operation in terms of time and communication bandwidth. As reported below, we have used some devices in our implementation in order to reduce complexity. With these devices, we were able to run information sharing among four Sony AIBO robots in real time, using the limited on-board computational resources of this platform, and using the limited bandwidth allowed by the rules of RoboCup.

5   Experiments

We have tested our method in the RoboCup domain using a team of Sony AIBO legged robots. This is a challenging, highly dynamic domain, characterized by: (i) imprecise sensor information, since the main sensor available to each robot is a color camera with limited resolution and a limited field of view; and (ii) high localization uncertainty, since legged locomotion and poor perception make the self-localization problem difficult.

5.1   Implementation

We have implemented the schema described in the previous section using, in each robot, a 3-D fuzzy position grid to represent self-localization, and a 2-D position grid to represent the ball position.[2]

The computation of the global ball position from observations is done according to an optimized version of algorithm (4). We exploit the fact that for every cell p in the ball grid we only need to look at the cells q in the self grid at a distance ρ from p. This is equivalent to saying that we only need to evaluate the cells in the circle with center p and radius ρ. To do so, we use an adaptation of Bresenham's algorithm for plotting 2D circles in a digital grid [2].

---

[2] The self-localization grid is actually implemented using a $2\frac{1}{2}$-D grid to allow for efficient real-time computation, as detailed in [3].

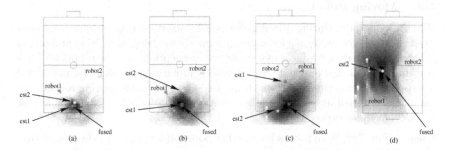

Fig. 7. Four examples of cooperative object localization in the RoboCup domain. The arrows show the individual position estimates produced by the two robots (est1, est2) and the fused estimate. The four cases are described in the text

The other expensive element of our schema is the transmission of the ball grids between robots. In order to preserve bandwidth, we convert each cell value to one byte and treat the grid as a gray-scale image. We then compress the grid using a simple run length encoding scheme. The ball grid is sent only if it contains information which is newer or of better quality than the one last sent.

5.2   Static Robots

A first set of experiments were performed to assess the quality of our fusion technique under static conditions. In these experiments, we used two robots standing at fixed positions and we put the ball at various known locations in order to evaluate the accuracy of object localization. The robots were alternatively looking at the landmarks in the field in order to self-localize, and at the ball in order to assess its position.

Two sample cases taken from these experiments are shown in Fig. 7 (a,b). The figure shows the fused object position grid resulting from the fusion of the two individual object grids (not shown). The three small circles represent the point estimates of the ball position according to each robot alone (marked 'est1' and 'est2') and as a result of the fusion (marked 'fused').

Case 1 (Fig. 7 a). Both robots can see the ball, and have accurate self-localization. Both the individual estimates and the fused estimate are very close to the real position of the ball (see quantitative data below).

Case 2 (Fig. 7 b). Both robots can see the ball, but while robot 1 is well localized, robot 2 is not. The result of the fusion in this case is almost identical to the information provided by robot 1, while the information provided by robot 2 is discounted since it is unreliable due to poor self-localization. This is similar to the case illustrated in Fig. 6.

5.3   Moving Robots

When robots move during a real game, self-localization may become very poor due to several factors: legged locomotion is poorly modeled; the tilting and rolling of the robot's body introduces inaccuracies in perception; and the robots are busy tracking the ball, and may note see the field landmarks very often. We performed a second set of experiments in which we used two constantly moving robots, and we placed the ball at various known positions. The following is a sample case taken from these experiments.

Case 3 (Fig. 7c). Both robots have rather poor self-localization. Because of this, the estimates of the ball position are very inaccurate and quite different from each other. When intersecting the corresponding fuzzy position grids, however, we obtain a fairly accurate fused estimate of the ball position. Note that this position does not lie between the two individual estimates, and hence could not have been obtained by averaging (e.g. using a Kalman filter).

5.4   Blind Robot

In a last set of experiments we tested the ability of robots which do not see the ball to use information provided by other robots. The following case illustrates a typical situation taken from these experiments.

Case 4 (Fig. 7d). The ball is behind robot 1, so only robot 2 can see it. In addition, the self-localization of both robots is rather poor. The fused information is similar to that provided by robot 2 alone. While this information suffers from poor self-localization, it is still made available to robot 1, who would otherwise have no knowledge of the ball position.

5.5   Results

In each experiment we have measured five quantities: the error in the self-location estimate for each robot, denoted by Δ_{self}^1 and Δ_{self}^2; the error in the ball position estimate produced by each robot individually, denoted by Δ_{ball}^1 and Δ_{ball}^2; and the error in the ball position estimate obtained by our fusion technique, denoted by Δ_{ball}^{fuzzy}. Errors were measured by comparing the center of gravity of the fuzzy location sets with the ground truth. The results for the above four cases are summarized in the following table. All errors are given in mm.[3]

Case	Δ_{self}^1	Δ_{self}^2	Δ_{ball}^1	Δ_{ball}^2	Δ_{ball}^{fuzzy}
1	110	230	78	103	85
2	186	439	117	609	106
3	n/a	n/a	940	502	180
4	312	566	n/a	842	862

---

[3] When the robots are moving (case 3) a source of ground truth of the robot positions was not available.

These results show that the estimates obtained by our fusion approach are, in practice, at least as good as the best individual estimate. When both robots gave good estimates (case 1) the fused estimate was as good as these. When one robot had a bad estimate due to localization errors (case 2), that robot's estimate was discounted in the fusion process, and the result was mostly influenced by the other robot's estimate. When both robots suffered from large localization errors (case 3), the fused estimate was at least as good as the individual estimates, and it was sometimes much better. When only one robot could see the object (case 4) the fused estimate was similar to the one from that robot.

The next table compares the error obtained by our fuzzy fusion technique with the error obtained by taking a simple average or a weighted average of the individual estimates (denoted by $\Delta_{\mathrm{ball}}^{\mathrm{avg}}$ and $\Delta_{\mathrm{ball}}^{\mathrm{wavg}}$, respectively). The weights in the latter case were given by the respective measure of reliability in self-localization. The errors refer to Case 3, above. As it can be seen, fuzzy fusion substantially outperformed averaging techniques in this situation.

$\Delta_{\mathrm{ball}}^{1}$	$\Delta_{\mathrm{ball}}^{2}$	$\Delta_{\mathrm{ball}}^{\mathrm{avg}}$	$\Delta_{\mathrm{ball}}^{\mathrm{wavg}}$	$\Delta_{\mathrm{ball}}^{\mathrm{fuzzy}}$
940	502	339	354	180

6   Conclusions

Our technique for multi-robot cooperative object localization has a number of advantages: it provides each robot with estimates which are, in practice, at least as good as the best ones available to any individual robot; it can effectively discount unreliable information; and there are no parameters to tune. The distinctive points of our approach are:

- The use of a sound technique to propagate the uncertainty in the self-localization of each robot to the uncertainty in object locations, and
- The use of an agreement seeking (fuzzy) operator instead of an averaging (e.g. probabilistic) operator to fuse the information provided by different robots.

We are currently using this technique in our RoboCup team [4]. Sharing ball information greatly improves the performance of robots in this domain. All robots can know the position of the ball if at least one member of the team sees it. Moreover, the fact that all the robots in the team have the same information about the ball position makes team coordination easier and more effective. The experiments reported here show that our technique produces useful results, even in the presence of high uncertainty. We are currently running a more systematic experimental evaluation of our approach, which includes empirical comparisons with other techniques.

Our future work is aimed at improving several aspects of our technique. First, our current approach is relatively demanding in terms of computation and bandwidth. Although we can run our algorithms in real time on the AIBO robots and using the limited bandwidth allowed by the RoboCup rules, it might be interesting to devise approximations of our technique which reduce the computational burden and/or the amount of information exchanged. Second, we would like to extend our technique to include sharing of information about multiple objects (e.g. opponent robots). Finally, we plan to investigate the extension of our approach to include cooperative self-localization, by merging the self-location grid of each robot with position grids from observing robots.

Acknowledgments

This work was supported by the EC Marie Curie Program, the Swedish CUGS (computer graduate school), the Consejería de Trabajo y Política Social (Región de Murcia) and the European Social Fund through the Seneca Foundation.

References

1. I. Bloch. Information combination operators for data fusion: A comparative review with classification. *IEEE Trans on Systems, Man, and Cybernetics*, A-26(1):52–67, 1996.
2. J.E. Bresenham. A linear algorithm for incremental display of circular arcs. *Communications of the ACM*, 20(2):100–106, 1977.
3. P. Buschka, A. Saffiotti, and Z. Wasik. Fuzzy landmark-based localization for a legged robot. In *Procs of the IEEE Int Conf on Intelligent Robots and Systems (IROS)*, pages 1205–1210, 2000.
4. Team Chaos. Team Chaos web site. http://www.aass.oru.se/Agora/RoboCup.
5. Sony Corporation. AIBO entertainment robot web site. http://www.aibo.com.
6. M. Dietl, J.-S. Gutmann, and B. Nebel. Cooperative sensing in dynamic environments. In *Procs of the IEEE Int Conf on Intelligent Robots and Systems (IROS)*, pages 1706–1713, 2001.
7. H. F. Durrant-Whyte. *Integration, Coordination, and Control of Multi-Sensor Robot Systems*. Kluver Academic Publishers, 1988.
8. The RoboCup Federation. RoboCup official web site. http://www.robocup.org.
9. D. Fox, W. Burgard, and S. Thrun. Markov localization for mobile robots in dynamic environments. *Artificial Intelligence Research Journal*, 11:391–427, 1999.
10. J.-S. Gutmann. Markov-Kalman localization for mobile robots. In *Procs of the Int Conf on Pattern Recognition (ICPR)*, pages 601–604, 2002.
11. R. Hanek, T. Schmitt, M. Klupsch, and S. Buck. From multiple images to a consistent view. In P. Stone, T. Balch, and Gerhard Kraetzschmar, editors, *Robocup 2000: Robot Soccer World Cup IV*, pages 169–178. Springer, 2001.
12. P. Marcelino, P. Nunes, P. Lima, and I. Ribeiro. Improving object localization through sensor fusion applied to soccer robots. In *Proc of the Scientific Meeting of the 3rd Robotics National Festival (Robotica 2003)*, Lisbon, Portugal, 2003.
13. M. Roth, D. Vail, and M. Veloso. A world model for multi-robot teams with communication. In *Procs of the IEEE Int Conf on Intelligent Robots and Systems (IROS)*, 2003.
14. S. I. Roumeliotis and G. A. Bekey. Synergetic localization for groups of mobile robots. In *Proc. IEEE Conf. on Decision and Control*, pages 3477–3482, 2000.

15. A. Saffiotti. The uses of fuzzy logic in autonomous robot navigation. *Soft Computing*, 1(4):180–197, 1997. Online at http://www.aass.oru.se/~asaffio/.
16. T. Schmitt, R. Hanek, S. Buck, and M. Beetz. Cooperative probabilistic state estimation for vision-based autonomous soccer robots. In A. Birk, S. Coradeschi, and S. Tadokoro, editors, *Robocup 2000: Robot Soccer World Cup IV*, pages 193–203. Springer, 2002.
17. S. Spors and N. Strobel. A multi-sensor object localization system. In *Int Workshop on Vision, Modeling and Visualization (VMV)*, pages 19–26, 2001.
18. A. Stroupe, M. Martin, and T. Balch. Merging gaussian distributions for object localization in multi-robot systems. In *Procs of the Int Symposium on Experimental Robotics (ISER)*. Springer, 2000.
19. S. Weber. A general concept of fuzzy connectives, negations and implications based on t-norms and t-conorms. *Fuzzy sets and systems*, 11:115–134, 1983.
20. L. A. Zadeh. Fuzzy sets. *Information and Control*, 8:338–353, 1965.
21. L. A. Zadeh. Fuzzy sets as a basis for a theory of possibility. *Fuzzy Sets and Systems*, 1:3–28, 1978.

Visual Robot Detection in RoboCup Using Neural Networks

Ulrich Kaufmann, Gerd Mayer, Gerhard Kraetzschmar, and Günther Palm

University of Ulm,
Department of Neural Information Processing,
D-89069 Ulm, Germany

Abstract. Robot recognition is a very important point for further improvements in game-play in RoboCup middle size league. In this paper we present a neural recognition method we developed to find robots using different visual information. Two algorithms are introduced to detect possible robot areas in an image and a subsequent recognition method with two combined multi-layer perceptrons is used to classify this areas regarding different features. The presented results indicate a very good overall performance of this approach.

1   Introduction

Due to the huge variety of robot shapes and designs in RoboCup middle size league vision based robot detection is a challenging task. A robot recognition method has to be very flexible to identify all the different robots but at the same time highly specific, in order not to misclassify similar objects outside of the playing field (e.g. black dressed children). Therefor many teams still recognize robots in their sensor data only as obstacles for collision avoidance.

At the same time, a good performance in RoboCup depends more and more on complex team behaviour. It is no longer sufficient, for a robot to localize itself on the playing field and behave autistically without taking care of other team robots. Is is rather necessary that the robots act as and interact within a team. Furthermore, recognizing and tracking of opponent robots becomes desirable to improve the team strategy.

Robot interaction is definitively only possible if the relative position of the partner is known exactly. Otherwise e.g. a pass may fail and an opponent robot can take possession of the ball. Regarding the bad experiences of past RoboCup competitions, it is also risky, if the robots solely base their decision on the shared and communicated absolute position on the field, because the communication may fail or the robot doesn't know its own position exactly (or even may be totally wrong). With respect to this, it is quite clear, that it is necessary for a robot to detect and recognize other players by himself without sharing position information explicitly. Whereas this might be bypassed with better selflocalization and a fault tolerant communication equipment, there are other tasks like for example to dribble around opponent robots and to plan a path without colliding

D. Nardi et al. (Eds.): RoboCup 2004, LNAI 3276, pp. 262–273, 2005.
© Springer-Verlag Berlin Heidelberg 2005

with any obstacle along this path. There is no way to do so without any kind of detection of opponent robots.

A visual robot detection method for the ROBOCUP environment has to weight carefully two opposing goals. It has to be reliable and computationally inexpensive at the same time. This requires on the one hand the use of good, significant features, on the other hand the computational complexity that is needed to calculate these features needs to be low. To detect the robots during a game, the whole recognition task lasting from recording the image to the last decision step mustn't take any longer than a few milliseconds.

In contrast to the classical object recognition problems, there are only little restrictions about the robots shape and spatial dimensions. Every team have their own, sometimes very different robots, ranging from large and heavy almost quadratic cubes to highly dynamic, small and fragile ones. So there is need for a highly flexible, yet fast method to find in a first step the possible robot positions within the image, because it is only computational maintainable to process a subset of each image. Another important point is, that the method has to be fast and easily adaptable to new and unknown robot shapes.

In this paper we present a robot detection system using neural networks which extracts (possibly multiple) robots from a recorded image. To be able to handle the special requirements of the robot recognition in ROBOCUP the algorithm is split up into three independent subtasks. The first task is finding the possible robot positions in the original image (i.e. defining one or multiple region of interests (ROI)). The next step is to extract the features from these ROIs for further classification. The final classification decision is then performed on the basis of the extracted features by two neural networks. The first two steps are always the same (means they are not adapted to specific robots or environments). Only the neural network may be adapted on the current situation (e.g. a completely other robot shape) if required. It may even be possible to do this adaption within shortest time e.g. on-site before competitions.

The paper is organized as follows: Section 2 first explains the robot detection task as a whole. After a small introduction into the ROBOCUP scenario in 2.1 the following subsections 2.2–2.5 explain the individual steps in detail. Experiments and results are presented in section 3. In section 4 this paper is discussed in the context of related work. Finally section 5 draws conclusions.

2   Method

In this section the individual steps of the presented robot recognition method are explained in more detail. The itemization of the different steps is as follows:

1. Detect the region of interest,
2. extract the features from the RIOs,
3. classification by two neural networks,
4. arbitration of the classification results.

Figure 1 illustrates the data flow up to the end result. The first step (1) shows the detection of the regions of interest. We present two alternative methods

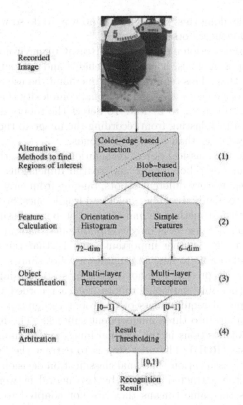

Fig. 1. Data-flow from image to recognition result

for finding them, one time a color-edge based one, the other time a color-blob based algorithm. Potential robot positions are searched here. For every possible area a data vector is calculated which includes orientation histograms and other, simpler features (2). These vectors are submitted to two artificial neural networks (3) and the results are then passed over to a final arbitration instance (4).

These are clearly separated tasks, so if there are robots that are not recognized well enough a customization of one of the networks (using the orientation histogram) can be applied. The second network only uses features described and predetermined by the ROBOCUP rules. As it is very important that the whole task is fast enough to process the images in "real time" (i.e. in an adequately short time) and is flexible enough to be adaptable to different opponents from game to game, in the following, each step is examined with respect to this.

2.1   The RoboCup Scenario

For those of the readers not familiar with the ROBOCUP scenario we first want to give you a short introduction. In the ROBOCUP middle size league four robots (three field player plus one goal keeper) playing soccer again four other ones.

All relevant objects on the field are strictly color coded: the ball is orange, the floor is green with white lines on it, the goals are yellow and blue, corner posts are blue and yellow colored and robots are mostly black with magenta or cyan color markers on it. During ROBOCUP championships there are also spectators around the field that can be seen from the robots.

There are also constraints about the robots size. Robots are allowed to have a maximal height of 80cm and a maximal width (resp. depth) of 50cm. Additional shape constraints and technical and practical limitations further restrict the possible robot appearances.

The game itself is highly dynamic. Some of the robots can drive up to 3 meters per second and accelerate the ball even higher. To play reasonably well within this environment at least 10–15 frames must be processed per second.

2.2   Region of Interest Detection

The first step is to direct the robots attention to possible regions within the recorded images. This is necessary because the feature calculation might be computational expensive and most time, large areas of the taken pictures are not of any interest. On the other hand, only a robot that stay within one of detected regions of interest can therefor be recognized later because all subsequent processing steps rely on this decision. So this attention control has to be as sound and complete as possible, to get all potential robot positions and not having to examine too much uninteresting areas. Two different algorithms are presented for this task each having its own quirks and peculiarities that score differentially in terms of speed and selectivity, so depending on the available computing power one of them may be used, or both may be mixed up in some way. Both methods rely on assertions made on the robots color and shape (as described in section 2.1) and therefore are rather specialized in the ROBOCUP scenario.

Histogram Method. The first approach examines all black areas within the picture. As it is currently save to assume, that each object on the playing field resides on the floor, we can calculate the size of the found regions easily. Regarding the already mention fact, that the robot size is restricted to a maximal size and that for now all robots in ROBOCUP middle size league are at least 30cm wide, all regions that do not achieve these restrictions are filtered out.

The blob detection is based on vertical and horizontal color histograms. To get the right position and size of an black blob we split the image in deliberate sub-images because one occurrence histogram of the black color for each direction may be not well-defined. By subdivision of the original picture to several areas it is easy to manage a detection of the interesting areas. Figure 2 shows the problem with histograms for several large areas.

The used image splitting is shown in Figure 3. It is assumed that all robots stand on the floor so the sub-images describe nearly the dimensions of the robots. Finally the histograms of the sub-images are subsequently searched for potential robots.

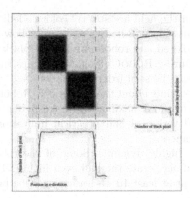

Fig. 2. Problems with the position-detection of color-blobs

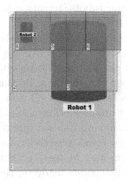

Fig. 3. Sub-images to check for black blobs

Color Edge Method. The second algorithm looks for black/green transitions within the picture starting at the bottom of the image, that indicate edges between (black) robots and the (green) floor. After calculating the real width of these lines and again filtering them with the minimal and maximal size constraints, the team color-makers above this lines are searched. This method already recognizes most robots reliably and selectively as long as they stand alone. Also, due to the high selectivity this method is much less robust against partially occlusions.

2.3 Feature Calculation

In the second step different features are calculated for the defined regions of interest. The different features describe different attributes of the robot. As the robot form cannot be predicted exactly, the features must be general enough to be applicable for different robot shapes, on the other hand specific enough to mask out uninteresting objects reliably. Beside that, the overall object detection is done with a combination of all used features as explained in section 2.4. The features used are the following:

- Size of the black/green transition lines.
- Percentages of black (robot) and cyan/magenta (team-marker) color.
- Entropy.
- Orientation histograms.

The features are calculated from the original picture and from a segmented image. The segmented image is created by assigning each real color to a color class describing one of the above mentioned (section 2.1) object classes on the playing field or to a catch-all class that is not mentioned further on (described in detail in [1]). The first three feature types mostly check attributes asserted by the rules (e.g. size, color, color-marker). However the orientation histograms contain indirect details on the shape of the robot in a rather flexible way, but are strongly dependent on the right selected region. If the window is too large, not only the orientation of the robot but also of the background is calculated and examined. Vice versa, if the window is too small, we may overlock important parts of the robot.

The size of the black/green transition line shows the visible robot size near the floor or the width of the ROI depending on the used attention control method.

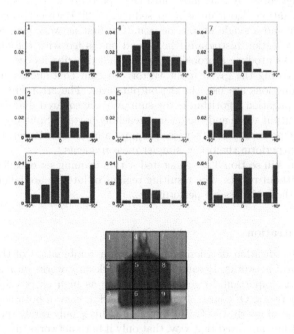

Fig. 4. Nine orientation histograms for one robot

The percentages of colors tells the team membership (team-maker) of a robot. A more general feature is the entropy. It is a indicator for the disorder within the area. A robot area is more disordered than a picture of the floor regarding the gray-scale values. Also a picture with the whole field is more disordered than a robot picture.

In Figure 4 the orientation histograms for one robot is shown. The histogram is made by accumulate the gradients in x and y direction detected by two Sobel filters on a grey image weighted by their quantity. The histogram is discretized into (in our case) eight chunks. Note that the histograms is calculated independently for nine sub-images, where the individual areas overlap of around 25%. This way, the orientation histograms are more specific for different prevailing edges within different parts of the image. In histogram number eight you can see e.g. the dominating vertical edge within the sub-image represented by the peak in orientation zero. In opposite, the horizontal bottom line of the robot is represented by another peak with orientation 90 degree in histogram number six. So the orientation histograms are a very flexible way to specify the robots shape within the region of interest.

2.4   Neuronal Networks Classification

Two neural networks do the actual classification in the whole robot recognition task. The networks are standard multi-layer perceptrons that are trained with a backpropagation algorithm as proposed e.g. in [2]. Both networks contain one hidden layer and a single output neuron. The first network gets only the data from the orientation histogram, the second one is fed with the other features. Both networks produce a probability value that describe its certainty of seeing a robot regarding the given input vector. To gain continuous output signals, sigmoidal functions are used in the output layer. The error function used for the backpropagation algorithm is the sum over the squared differences between the actual output value and the desired teaching pattern. Splitting the networks proved to be necessary, as otherwise the pure amount of orientation values suppress and outperform the other, simpler measurements.

The results in section 3 are generated with a training set of different pictures with and without robots. The resulting regions of interest are labeled manually to produce the best possible performance.

2.5   Arbitration

The final classification decision is made from a combination of the outputs of the two neural networks. Because every network only works on a subset of the features, it is important to get an assessment as high as possible from each individual network. Of course a positive feedback is easy, if both network deliver an assessment of nearly 100%. But in real life, this is only rarely the case. So the network outputs are rated that way, that only if both networks give a probability value bigger than 75%, it is assumed that a robot is found within the region of interest.

3   Experimental Results

In this section the results of the individual steps of the robot recognition task are described and discussed in detail. All tests are made using a training set of about 88 images. The size of the images is always PAL/4, i.e. 384×288 pixels. The images are taken with 3 different robots, containing 99 occurrences of them. The robots are recorded from different perspectives and from different distances. The teacher signal for training the neural network (i.e. the resulting bounding box around the found robot) is added by hand to assure the best possible performance. After the training phase the networks are able to calculate a classification from "robot" to "no robot" in the sense of a probability measure between zero and one.

The first two sections compare the different methods to calculate the regions of interest for the robots within the recorded images. After that, the overall performance is evaluated and finally the adaptation on new robots is discussed.

Because computational complexity and therefore the needed calculation time is very important in such a dynamic environment like the ROBOCUP, we measured the time needed for the individual processing steps on a 1.6 GHz Pentium4 mobile processor. These values are certainly highly implementation dependent, but may give an impression about how fast the whole object recognition task can be done.

3.1   Blob-Detection Method

The blob-detection method uses a more universal approach and detects all black areas within the picture. In our case, blob detection is simply performed by computing occurrence histograms in both vertical and horizontal direction for the color class of interest. Resulting blobs are then filtered with the size constraints found in the ROBOCUP rules. No further restrictions are taken into account apart from the size of the robots and the black color. As a result, this method detects more potential robot positions which implies more work in the subsequent two processing steps. On the other hand this method recognize all whole robots and more of the covered ones. The lower left image in Figure 5 shows, that overlapping objects have less influence on the detection performance than for the other described method. In the upper left image a false prediction can be seen (note the second white square in contrast to the upper right image).

This attention control method finds 93% (i.e. 92 of 99) of the robots within the images correctly. Additional 57 positions are found that do not contain a robot, but several other black objects. All robots which are not covered by other objects are detected.The accuracy of the robot detection is sometimes better compared to the other method as can be seen in the lower row of Figure 5.

On an average the method needs less then 1 ms to examine the whole picture and to detect all the possible robot positions. As a drawback of its flexibility, the following processing steps may take significantly more time than with the below mentioned attention control process due to the many false positives.

3.2   Black/Green Transition Detection Method

Using the attention control algorithm searching for black/green transitions within the image, the robots have to be "rule-conform" to be detected, otherwise they aren't selected or filtered out by the heuristics of this method. In the lower right image in Figure 5 you can see the consequence, if the robots size is determined by its bottom line only. If it is partially masked by another object, only parts of the robot may be used as region of interest or even filtered out because of the applied size assumptions.

This attention control method finds 92% of the robots within the images correctly. This means that 91 of the 99 robots are recognized that way, that a human expert marked them as sufficient. Additional 14 positions are found that do not contain a robot, but several other black objects. Again all robots which are not covered by other objects are detected. Missed robots are mostly far away and are hard to recognize in the image even for the human expert.

The advantage of this method is its speed and the low amount of wrong false classified areas, which again saves time in subsequent processing steps. On an average the method needs clearly less then 0.5 ms to examine the whole picture and to set the regions of interest.

3.3   Feature Calculation and Neural Processing

The time needed for the calculation of all features depends on the amount of found regions, as well as the implementation of the filters themselves. The used

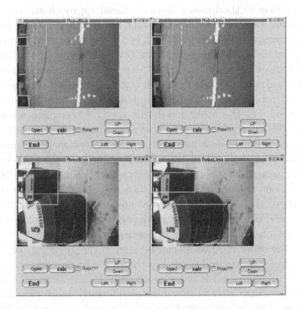

Fig. 5. Results using the different attention control algorithms

algorithms need around 174–223 ms depending on the size of the found region. Nevertheless the first attention control mechanism using the green/black transition needs significantly less overall processing time compared to the other, blob-based method because of the less false positives found. A preliminary, highly optimized version for the calculation of the orientation histogram (which consumes clearly the most time of the whole processing step) needs about 23 millisecond if applied to the whole (384×288 pixels) image.

The artificial neural networks are trained with around 200 feature vectors, about 40% of them contained data from real robots, the others are randomly chosen regions from the image. The final result of the neural networks again depend on the results delivered by the used attention control mechanism. Using the green/black transition method the overall correct robot recognition is about 95.3% regarding the delivered ROIs. When using the other (blob-based) algorithm, the result slightly decrease to 94.8%. This evaluation of the neural networks again is quite fast and uses clearly less than a millisecond.

3.4   Adaption and Retraining

If playing against robots with a totally different shape than that, used in the present training set, the network for the orientation histograms are likely to need adaption to the new situation. For this new training of the network, images of the new robots are needed. It is important to use images from different points of view and different distances. As the used learning algorithm is a supervised training method, the images have to be prepared so the precise robot positions are known. Now the network can be retrained for the orientation histogram. After a short time (around 1–2 minutes), the network is again ready to work.

Future work will focus on automating the training phase at the beginning of a game. Before a game-start only robots should be on the playing field, so every robot of the own team could take some pictures of the opponents which should fulfill the desired variety in orientation angle and distance. Herewith a reliable extraction should be possible and the learning of the new robot shape may be fully autonomous.

4   Related Work

Object detection is a well known problem in current literature. There are many approaches to find and classify objects within an image, e.g. from Kestler [3], Simon [4] or Fay [5] to name just a few that are developed and investigated within our department.

Within RoboCup the problems are rather less well defined then in their scenarios and real-time performance is not an absolute prerequisite for them, which may be the main reason that up to now there are only few workings are published about more complex object detection methods in RoboCup. Most of the participant within the RoboCup middle size league use a mostly color based approach, like e.g. in [6][7][8]. One interesting exception is presented from Zagal et. al. [9]. Although they still use color-blob information, they let the robot learn different parameters for the blob evaluation, like e.g. the width or the height of the blob using genetic algorithms. Therewith they are able to even train the robot to recognize multi-colored objects as used for the beacons on both sides of the playing field (as used within the Sony legged league, which is rather comparable to the middle size league).

One attempt to overcome the limitations of pure color based algorithms is presented from Treptow et. al. [10] in which they trained an algorithm called adaboost using small wavelet like feature detectors. They also attach importance to let the method work reliably and with virtual real-time performance. Another approach, that even don't need a training phase at all is presented from Hanek et. al. [11]. They use deformable models (snakes), which are fitted to known objects within the images by an iterative refining process based on local image statistics to find the ball.

5   Conclusions and Future Work

Considering all the mentioned boundary conditions, robot recognition in ROBO-CUP middle size league is a difficult task. We showed, that splitting the problem into several subtasks can made the problem controllable. The combination of relatively simple pre-processing steps in combination with a learned neural decision entity results in a fast and high-quality robot recognition system.

We think, that the overall results can be even increased with a temporal integration of the robots position as we use it already for our self-localization [12] and described by other teams [10][13]. So partially occluded robots can be detected even if the robot is not detected in every single image. Future work is also planned on the selected features. With highly optimized algorithms there is again computing power left over, that can be used to increase the classification rate.

We collected a huge amount of real test images during a workshop with the robot soccer team from Munich. So we will focus on doing a very detailed investigation, how this method behave for all the extreme situations that can happen in ROBOCUP, like e.g. occlusion, or robots at image boundaries. It's also of interest, how the neural networks behave, if they are confronted with opponent robots not yet in the training images data base.

Acknowledgment

The work described in this paper was partially funded by the DFG SPP-1125 in the project *Adaptivity and Learning in Teams of Cooperating Mobile Robots* and by the MirrorBot project, EU FET-IST program grant IST-2001-35282.

References

1. Mayer, G., Utz, H., Kraetzschmar, G.: Playing robot soccer under natural light: A case study. In: RoboCup 2003 International Symposium Padua (to appear). (2004)

2. Russell, S.J., Norvig, P.: Artificial Intelligence: A Modern Approach. Prentice Hall, Upper Saddle River, NJ (1995)
3. Kestler, H.A., Simon, S., Baune, A., Schwenker, F., Palm, G.: Object Classification Using Simple, Colour Based Visual Attention and a Hierarchical Neural Network for Neuro-Symbolic Integration. In Burgard, W., Christaller, T., Cremers, A., eds.: Advances in Artificial Intelligence. Springer (1999) 267–279
4. Simon, S., Kestler, H., Baune, A., Schwenker, F., Palm, G.: Object Classification with Simple Visual Attention and a Hierarchical Neural Network for Subsymbolic-Symbolic Integration. In: Proceedings of IEEE International Symposium on Computational Intelligence in Robotics and Automation. (1999) 244–249
5. Fay, R.: Hierarchische neuronale Netze zur Klassifikation von 3D-Objekten (in german). Master's thesis, University of Ulm, Department of Neural Information Processing (2002)
6. Jamzad, M., Sadjad, B., Mirrokni, V., Kazemi, M., Chitsaz, H., Heydarnoori, A., Hajiaghai, M., Chiniforooshan, E.: A fast vision system for middle size robots in robocup. In Birk, A., Coradeschi, S., Tadokoro, S., eds.: RoboCup 2001: Robot Soccer World Cup V. Volume 2377 / 2002 of Lecture Notes in Computer Science., Springer-Verlag Heidelberg (2003)
7. Simon, M., Behnke, S., Rojas, R.: Robust real time color tracking. In Stone, P., Balch, T., Kraetzschmar, G., eds.: RoboCup 2000: Robot Soccer. World Cup IV. Volume 2019 / 2001 of Lecture Notes in Computer Science., Springer-Verlag Heidelberg (2003)
8. Jonker, P., Caarls, J., Bokhove, W.: Fast and accurate robot vision for vision based motion. In Stone, P., Balch, T., Kraetzschmar, G., eds.: RoboCup 2000: Robot Soccer. World Cup IV. Volume 2019 / 2001 of Lecture Notes in Computer Science., Springer-Verlag Heidelberg (2003)
9. Zagal, J.C., del Solar, J.R., Guerrero, P., Palma, R.: Evolving visual object recognition for legged robots. In: RoboCup 2003 International Symposium Padua (to appear). (2004)
10. Treptow, A., Masselli, A., Zell, A.: Real-time object tracking for soccer-robots without color information. In: Proceedings of the European Conference on Mobile Robotics (ECMR 2003). (2003)
11. Hanek, R., Schmitt, T., Buck, S., Beetz, M.: Towards robocup without color labeling. In: RoboCup 2002: Robot Soccer World Cup VI. Volume 2752 / 2003 of Lecture Notes in Computer Science., Springer-Verlag Heidelberg (2003) 179–194
12. Utz, H., Neubeck, A., Mayer, G., Kraetzschmar, G.K.: Improving vision-based self-localization. In Kaminka, G.A., Lima, P.U., Rojas, R., eds.: RoboCup 2002: Robot Soccer World Cup VI. Volume 2752 / 2003 of Lecture Notes in Artificial Intelligence., Berlin, Heidelberg, Germany, Springer-Verlag (2003) 25–40
13. Schmitt, T., Hanek, R., Beetz, M., Buck, S.: Watch their moves: Applying probabilistic multiple object tracking to autonomous robot soccer. In: Eighteenth National Conference on Artificial Intelligence, Edmonton, Alberta, Canada (2002)

Extensions to Object Recognition
in the Four-Legged League

Christopher J. Seysener, Craig L. Murch, and Richard H. Middleton

School of Electrical Engineering & Computer Science,
The University of Newcastle, Callaghan 2308, Australia
{cseysener, cmurch, rick}@eecs.newcastle.edu.au
http://robots.newcastle.edu.au

Abstract. Humans process images with apparent ease, quickly filtering out useless information and identifying objects based on their shape and colour. However, the undertaking of visual processing and the implementation of object recognition systems on a robot can be a challenging task. While many algorithms exist for machine vision, fewer have been developed with the efficiency required to allow real-time operation on a processor limited platform. This paper focuses on several efficient algorithms designed to identify field landmarks and objects found in the controlled environment of the RoboCup Four-Legged League.

1 Introduction

Robot vision systems are often required to identify landmarks relevant to the robot's operation. In autonomous robot soccer, situations arise where the robot operates for extended periods without viewing critical landmarks on the field such as goal posts and other objects. This lack of information leads to the robot becoming disorientated and uncertain about its position.

In the past, methods used to alleviate this problem have mostly involved so-called "active localisation". Active localisation requires the robot to periodically interrupt its normal operation to better determine its location. For example, a robot may interrupt its gaze to search for a landmark before executing a critical manoeuvre (such as kicking the ball). However, because of the time critical nature of robot soccer, the use of active localisation can allow opposition robots to pounce on the ball while the localising robot wastes time determining its position and orientation. Identification of additional landmarks is therefore beneficial as it reduces the need for active localisation.

Many vision algorithms developed in recent years are inappropriate for limited processor systems and real-time applications. In this paper, we present some efficient algorithms to solve complex problems such as edge detection and data association so commonly seen in both theoretical and practical problems. While the solutions presented are in the context of the RoboCup Four-Legged League [1], it is expected that some of the ideas and algorithms could be useful in areas such as automation so commonly seen in industry.

D. Nardi et al. (Eds.): RoboCup 2004, LNAI 3276, pp. 274–285, 2005.

The objective of the methods described in this paper is to increase the number of landmarks on the field that the vision system can recognise. By locating field and sidelines in the image, it is possible to locate points such as the corners of the penalty box, the centre circle and the intersection of the centre line with the sideline (the "centre corner").

The accuracy of measurements to objects detected by the vision system is also of great importance. In the context of a soccer match and in many other areas, obstructions can occur which impede both the identification of objects and measurements to them. The practice of fitting circles to the ball provides a robust mechanism through which accurate heading and distance information can be determined, even in the presence of obstructions and despite the limited processing power available.

2 Field and Sideline Detection

As discussed in the introduction the attainment of additional information from which to localise is of paramount importance. One idea to increase the quantity of objects identified by the vision system is to make use of the field and sideline information. The following sections document the process and algorithm used in identifying and determining the equation of the field and sidelines within the image.

2.1 Line Detection

While existing techniques for edge detection, such as those of the German team, have focused on examining the camera image (in YUV colour space) for a shift in the Y component [2], this report focuses on the identification of field and sidelines using a classified image. A classified image is one which has, through the use of a look up table, been mapped to a set of colours identifiable by the robot. Using the classified image, the key to determining the presence of field and sidelines lies in the systems ability to discern field-green/boundary-white and boundary-white/field-green transitions. These transitions identify the boundary between the field and both field and side lines. For example, for a field line to be present, a field-green/boundary-white transition and a corresponding boundary-white/field-green transition must be found. These two transitions mark the change from field to field-line and back to field again, ruling out the possibility that the point was actually part of the sideline.

The algorithm searches a sparse grid structure over the entire image to discover the field and sideline transitions. While a complete scan of all pixels would have provided more accurate data, line transitions are typically several pixels wide and so are still identifiable when checking only the pixels located on gridlines. Under the current implementation, a grid with a spacing of 4 by 2 pixels is used in the identification of field and sideline points. While this grid is considered fine, testing has revealed that should additional processor time be required, the size of this grid may be increased without a serious negative effect on the identification of points in the image.

Fig. 1. Identification of field (blue) and side (red) line points

The points identified as field and sideline can be seen in Figure 1. The red line present in this image represents the horizon line, which is used to identify those parts of the image which should be scanned and those which can be ignored. This technique obviously reduces the area to be scanned and hence the processing load in many images. Further detail on the generation of the horizon line can be found in [2, 3].

2.2   Field and Sideline Identification

Having acquired the transitions in the image believed to be part of a field or sideline, the problem of associating points with individual lines offers a further challenge to be investigated. The sideline case is trivial, in that sidelines are always separated horizontally in the image. However, multiple field lines may appear stacked, a fact illustrated by Figure 1.

The data association problem for field lines was solved using a simple clustering algorithm such that if the distance of a point from any previously generated line is too great then a new line will be formed. Points are always processed from left to right. Once all points have been allocated to a line, the start and end points of each line are compared with other lines identified. This step determines whether the two lines actually represent a continuous field line broken in two due to a corner. Any pair of lines found to be continuous are merged. It is for this reason that the points are always processed from left to right: By placing an ordering on the initial line formation step, longer lines tend to generated and therefore less merging must be performed later.

For example, in Figure 1 the clustering algorithm results in three field lines. The first field line is obvious and represents the four points identified in the goalmouth, while the second and third lines are not so obvious and represent the two components of the penalty box. The presence of the corner forces the points to be split in two as in processing the points from left to right, the distance of the points from the end of the generated line eventually becomes too great forcing a new line to be formed. In this example, the two lines that form the corner due to the proximity of their ends will be determined to be part of the same field line and will be merged.

Upon successful discovery of a fixed number of points believed to be part of the same line, the principle of Least Squares Line Fitting can be used to determine the equation of the line in the image. The algorithm may also be used to filter outliers based on their distance from the generated line thus increasing the tolerance of the system to noisy input data.

3   Landmark Identification Using Field and Sidelines

This following sections document the way in which field and sideline data identified is used to locate landmarks on the field useful for robot localisation. The landmarks identified by the algorithms presented in the following sections are shown in Figure 2.

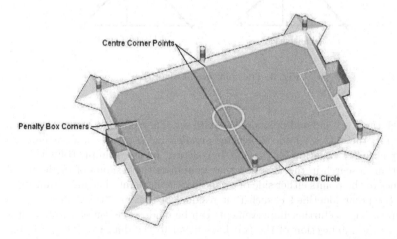

Fig. 2. Landmarks identified in the following sections

3.1   Penalty Box Corner Detection

While a number of approaches were trialled for the identification of penalty box corners, many of these proved too time consuming or error prone for use in a soccer game. The solution described below was successful in identifying corners both accurately and quickly, although it has the caveat of being unable to recognise the rare situation where multiple corners are present in a single line. This technique was inspired in part by the well known split and merge algorithm [4]. The first step is to identify the point furthest from the line fitted

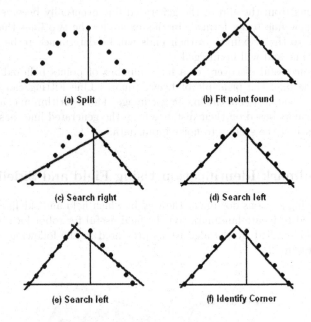

(a) Split   (b) Fit point found

(c) Search right   (d) Search left

(e) Search left   (f) Identify Corner

Fig. 3. The split and fit algorithm

through the start and end points of the data set. The Split and Merge algorithm then splits the data set into smaller and smaller line segments in an attempt to find an accurate equation for the line. In contrast, we use a simple Hill Climbing algorithm to search in the direction that minimises the residual of the fitting of two lines to the points either side of a candidate split point. The algorithm stops when the point identified is such that residual of the line fitting operations is minimised and no further improvements can be made. The corner point itself is taken as the intersection of the two lines fitted to the data set. This last step of taking the intersection point of the two lines minimises the risk of taking a point identified incorrectly in the original transition discovery phase due to misclassification or obstruction.

As can be seen in the example, the algorithm operates using a simple hill climbing process. The steps used can be described as follows:

a) Fit a line through the two extreme points of the dataset and find the point most distant to this line. This point defines the starting point of the search

b) Split the data set in two and calculate (using least squares) the residual of the two lines fitted either side of the centre point found

c) Take the first point to the right of the centre point and refit the two lines. Note that in this example the error of the system increases, indicating a move away from the corner

d) Take the first point to the left of the centre point and refit the two lines. Note that this decreases the error of the system and hence a move towards the corner
e) The algorithm continues to search in the direction that minimises the error of the system. In this case, the error has increased by searching left indicating that the corner point has already been passed.
f) Two lines are fitted through the point that minimises the error of the system and the intersection of these lines taken as the corner point. Despite the bad point in the example, the algorithm correctly identifies the true corner.

While the point identified in the first step is often the corner point, the additional steps have proved to give reliable and robust determination of the corner point even in situations where significant amounts of noise are present in the system. This algorithm has proved an extremely efficient solution to the problem of finding a corner in a system of ordered points, easily executed in real-time on the robot.

3.2 Centre Circle Detection

As the centre circle closely resembles a field line, it made sense that the detection of the centre circle should somehow be related to the detection of field lines within the image. This was achieved by observing that the centre circle provided a special case when present, in that from most angles three sets of field-green/border-white/field-green transitions occurred in the same column. The presence of the centre circle is therefore determined by the existence of three or more sets of field line transitions in the same column (referred to as a triple). An example of this situation can be seen in Figure 4.

Fig. 4. Centre circle

The parameters of the centre circle are determined from the sets of triples identified in the image. Using the sets of triples, the centre of the circle was defined by the point that bisected the line fitted through the two triples most separated in the image.

Although the discovery of the centre circle is only possible from certain angles upon the field, it does provide a foundation from which further work can be performed. Future work may include research into the fitting of circles or ellipses to determine accurately the presence and parameters of the circle.

3.3   Centre Corner Point Detection

Another area in which field and side line information is used, is in the identification of the two points where the centre field line intersects with the sideline. An example of the point considered the centre corner point can be seen in Figure 5.

Fig. 5. Centre corner point

The detection of these landmarks requires the detection of a field line intersecting with a sideline. This condition is satisfied at the goal mouth as well as at the centre corner point, so an additional constraint to reduce the number of false positives is that the goal must not be present in the image. On checking that the end of a field line is sufficiently close to the sideline, the intersection of the field and sideline is taken as the centre corner point.

4   Circle Fitting to the Ball

A commonly used method for object recognition in the Four-Legged League is to find a rectangular blob around all colour objects within the image. The size of the blob determines the distance to the object. While this is convenient to code and can be extremely efficient, it also results in an overall reduction of information and introduces error that makes the perceived location of objects less precise. This problem is most evident in the recognition of the ball due to its circular shape (other notable objects on the soccer field are roughly rectangular).

In ball recognition, it was decided that points located on the perimeter of the ball should be identified and a circle fitted through these points. It was hoped that this method would make the system more tolerant of obstructions in front of the ball, such as another robot or the edge of the image. For example, consider the situation shown in Figure 6(a) where the lower part of the ball is obstructed by the robots leg.

Fig. 6. (a) Obstruction caused by a robot. (b) Point determination after blob formation

The result is a blob smaller than the actual size of the ball and a resultant distance much greater than the actual distance to the ball. Fitting a circle using points on the perimeter helps overcome this problem, resulting in a better approximation for the radius of the ball.

4.1 Determining Fitted Points

In fitting a circle to the ball, it is necessary to know those points (coordinates in the image) which define the perimeter. The technique identified for locating perimeter points, involved the recovery of information lost during the identification of the ball. This algorithm performs a horizontal search from the edge of the circle blob towards the centre, recording the first pixel found to match the colour of the object (i.e. orange). In using this algorithm, the pixels searched are those shadowed by the blob identified as the ball, which are not orange. Specific knowledge of the point's location is used to exclude certain points from the set, such as points located on the edge of the image, which are not genuinely on the perimeter of the ball. Future work in this area may include the use of a Convex Hull to filter points incorrectly identified within the ball. Figure 6(b) shows the points determined by the algorithm presented above.

4.2   Least Squares Fitting to the Ball

Although techniques such as fitting a circle through three points were tested, the method of shape fitting using Least Squares Fitting (LSF) proved the most successful. Whilst this algorithm is slightly more computationally expensive, the algorithm generates the most accurate fit, especially in cases where obstruction of the ball occurs. It was also noted that as the algorithm uses a higher number of points the generated result is more tolerant of noisy data.

Whilst many different papers have been written about the Least Squares Fitting of circles through a set of points, the paper used as the primary reference for this research was that by Chernov and Lesort [6], which discusses the various strengths, weaknesses and efficiencies of several commonly used circle fitting algorithms. The concept of Least Squares Fitting of circles is based on minimising the mean square distances from the fitted circle to the data points used. Given n points $(x_i, y_i), 1 \leq i \leq n$, the objective function is defined by

$$F = \sum_{i=1}^{n} d_i^2 \tag{1}$$

where d_i is the Euclidian (geometric) distance from the circle to the data points. If the circle satisfies the equation

$$(x - a)^2 + (y - b)^2 = R^2 \tag{2}$$

where (a, b) is the centre and R is the radius, then

$$d_i = \sqrt{(x_i - a)^2 + (y_i - b)^2} - R \tag{3}$$

The minimisation of the objection function $min\{F\}$ for $a, b, R > 0$) is a non-linear problem with no closed form solution. For this reason, there is no algorithm for computing the minimum of F with all known algorithms being either iterative and computationally expensive or approximate. It was therefore important that the correct algorithm be chosen for implementation on the robot to minimise the load placed on both processor and memory.

Having reviewed the results in [6] it was decided that a combination of both geometric and algebraic methods be used. The Levenberg-Marquardt [5, 7, 8] geometric fit method was selected above others, such as the Landau and Spath algorithms, due to its efficiency. This algorithm is in essence a classical form of the Gauss-Newton method but with a modification known as the Levenberg-Marquardt correction. This algorithm has been shown to be stable, reliable and under certain conditions rapidly convergent. The fitting of circles with the Levenberg-Marquardt method is described in many papers.

It is commonly recognised that iterative algorithms for minimising non-linear functions, such as the geometric circle-fitting algorithm above, are sensitive to the choice of the initial guess. As a rule, an initial guess must be provided which is close enough to the minimum of F to ensure rapid convergence. It has been shown that if the set of data points is close to the fitting contour, then the convergence achieved is nearly quadratic. It was therefore required that in order to minimise the CPU utilisation on the robot, an initial guess should be provided by a fast and non-iterative procedure. After further research, it was decided that the fast and non-iterative approximation to the LSF, provided by algebraic fitting algorithms, seemed ideal for solving this problem.

Algebraic fitting algorithms are different from their geometric equivalents in that they attempt to minimise the sum of the squares of the algebraic errors rather than minimising the distances of the sum of the squares of the geometric distances. If the above equations are considered.

$$F = \sum_{i=1}^{n}[(x_i - a)^2 + (y_i - b)^2 - R^2]^2 = \sum_{i=1}^{n}(x_i^2 + y_i^2 + Bx_i + Cy_i + D)^2 \quad (4)$$

where $B = -2a$, $C = -2b$ and $D = a^2 + b^2 - R^2$. Differentiating the equation with respect to B, C and D yields the following system of linear equations

$$M_{xx}B + M_{xy}C + M_xD = -M_{xz} \quad (5)$$

$$M_{xy}B + M_{yy}C + M_yD = -M_{yz} \quad (6)$$

$$M_xB + M_yC + nD = -M_z \quad (7)$$

where M_{xx}, M_{xy} etc. define moments, for example $M_{xx} = \sum_{i=1}^{n} x_i^2$ and $M_{xy} = \sum_{i=1}^{n} x_iy_i$. The system can be solved using Cholesky decomposition (or any other method for solving a linear system of equations) to give B, C and D, which in turn gives parameters a, b and R that define the circle. The algorithm used on the robot varies slightly from this definition, in that it uses a gradient

Fig. 7. Circle fitting to the ball

weighted [9, 10] form of the algorithm. This algorithm is commonly considered
a standard for the computer vision industry [11, 12, 13] due to its statistical optimality. An example of a circle fitted on the robot can be seen in Figure 7.
The circle fitted in the image is resilient to noise caused by the obstruction (the
edge of the image), which would have previously affected both the distance and
elevation values of the resultant ball.

4.3   Results

A comparison of the distances returned by examining the height of the object, the
new circle fitting method and the actual distances are shown below in Figure 8.
The ball is approximately 75% obstructed by the robot in each measurement.

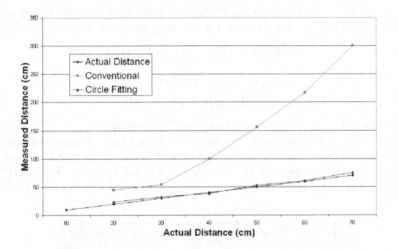

Fig. 8. Comparison of ball distances with 75% obstruction

5   Summary

The implementation of real-time object recognition systems for use on a robot
is a complex task. It was observed throughout this project that while complex
algorithms exist for image processing, many of these are not appropriate for use
on a robot.

This study focussed on the development of edge detection and data association techniques to identify field landmarks in a controlled environment. These
techniques were designed with efficiency as their key requirement so as to function effectively despite processor limitations. It has also been shown that a common method - circle fitting - can be scaled for use on a robot with a slow processor. Additionally, the circle fitting algorithm is able to operate in conjunction
with existing Four-Legged League vision systems

Hence while many conventional algorithms are unsuitable for use on the robot, appropriate algorithms have been developed or adapted for real-time visual processing, greatly assisting in the identification of key objects and in turn robot localisation.

Acknowledgements. We would like to thank all past and present members and contributors to the University of Newcastle Legged League RoboCup team.

References

1. RoboCup Four-Legged League web site. http://www.openr.org/robocup/index.html
2. Burkhard, H.-D., Düffert, U., Hoffmann, J., Jüngel, M., Lötzsch, M., Brunn, R., Kallnik, M., Kuntze, N., Kunz, M., Petters, S., Risler, M., v. Stryk, O., Koschmieder, N., Laue, T., Röfer, T., Spiess, Cesarz, A., Dahm, I., Hebbel, M., Nowak, W., Ziegler, J.: GermanTeam 2002, 2002. Only available online: http://www.tzi.de/kogrob/papers/GermanTeam2002.pdf
3. Bunting, J., Chalup, S., Freeston, M., McMahan, W., Middleton, R.H., Murch, C., Quinlan, M., Seysener, C., Shanks, G. Return of the NUbots! The 2003 NUbots Team Report, http://robots.newcastle.edu.au/publications/NUbotFinalReport2003.pdf
4. Tardós, J.: Data Association in SLAM, http://www.cas.kth.se/SLAM/Presentations/mingo-slides.pdf
5. Levenberg, K.: A Method for the Solution of Certain Non-linear Problems in Least Squares, Quart. Appl. Math. Vol. 2, 1944, 164-168.
6. Chernov, N., and Lesort, C.: Least Squares Fitting of Circles and Lines, http://arxiv.org/PS cache/cs/pdf/0301/0301001.pdf
7. Marquardt, D.: An Algorithm for Least Squares Estimation of Non-linear Parameters, SIAM J. Appl. Math., vol. 11, 1963, 431-441.
8. Shakarki, C.: Least-squares Fitting algorithms of the NIST algorithm testing system, J. Res. Nat. Inst. Stand. Techn., vol. 103, 1998, 633-641.
9. Turner, K.: Computer perception of curved objects using a television camera, Ph.D. Thesis, Dept. of Machine Intelligence, University of Edinburgh, 1974.
10. Sampson, P.D.: Fitting conic sections to very scattered data: an iterative refinement of the Bookstein algorithm, Comp. Graphics Image Proc., vol. 18, 1982, 97-108.
11. Taubin, G.: Estimation of Planar Curves, Surfaces and Nonplanar Space Curves Defined By Implicit Equations, With Applications To Edge And Range Image Segmentation, IEEE Transactions on Pattern Analysis and Machine Intelligence., vol. 13, 1991, 1115-1138.
12. Leedan, Y. and Meer, P.: Heteroscedastic regression in computer vision: Problems with bilinear constraint, Intern. J. Comp. Vision, vol. 37, 2000, 127-150.
13. Chojnacki, W., Brooks, M.J., and van den Hengel, A.: Rationalising the renormalization method of Kanatani, J. Math. Imaging & Vision., vol. 14, 2001, 21-38.

Predicting Opponent Actions by Observation

Agapito Ledezma, Ricardo Aler, Araceli Sanchis, and Daniel Borrajo

Universidad Carlos III de Madrid
Avda. de la Universidad, 30, 28911, Leganés (Madrid). Spain
{ledezma, aler, masm}@inf.uc3m.es, dborrajo@ia.uc3m.es

Abstract. In competitive domains, the knowledge about the opponent can give players a clear advantage. This idea lead us in the past to propose an approach to acquire models of opponents, based only on the observation of their input-output behavior. If opponent outputs could be accessed directly, a model can be constructed by feeding a machine learning method with traces of the opponent. However, that is not the case in the Robocup domain. To overcome this problem, in this paper we present a three phases approach to model low-level behavior of individual opponent agents. First, we build a classifier to label opponent actions based on observation. Second, our agent observes an opponent and labels its actions using the previous classifier. From these observations, a model is constructed to predict the opponent actions. Finally, the agent uses the model to anticipate opponent reactions. In this paper, we have presented a proof-of-principle of our approach, termed OMBO (Opponent Modeling Based on Observation), so that a striker agent can anticipate a goalie. Results show that scores are significantly higher using the acquired opponent's model of actions.

1 Introduction

In competitive domains, the knowledge about the opponent can give players a clear advantage. This idea lead us to propose an approach to acquire models of other agents (i.e. opponents) based only on the observation of their input-output behaviors, by means of a classification task [1]. A model of another agent was built by using a classifier that would take the same inputs as the opponent and would produce its predicted output. In a previous paper [2] we presented an extension of this approach to the RoboCup [3].

The behavior of a player in the robosoccer can be understood in terms of its inputs (sensors readings) and outputs (actions). Therefore, we can draw an analogy with a classification task in which each input sensor reading of the player will be represented as an attribute that can have as many values as the corresponding input parameter. Also, we can define a class for each possible output. Therefore, the task of acquiring the opponent model has been translated into a classification task.

In previous papers we have presented results for agents whose outputs are discrete [1], agents with continuous and discrete outputs [4], an implementation

D. Nardi et al. (Eds.): RoboCup 2004, LNAI 3276, pp. 286–296, 2005.

of the acquired model in order to test its accuracy [5], and we used the logs produced by another team's player to predict its actions using a hierarchical learning scheme [2]. In that work, we considered that we had direct access to the opponent's inputs and outputs. In this work we extend our previous approach in the simulated robosoccer domain by removing this assumption. To do so, we have used machine learning to create a module that is able to infer the opponent's actions by means of observation. Then, this module can be used to label opponent's actions and learn a model of the opponent.

The remainder of the paper is organized as follows. Section 2 presents a summary on our learning approach to modeling. Actual results are detailed in Section 3. In Section 4 discusses the related work. The paper concludes with some remarks and future work, Section 5.

2   Opponent Modeling Based on Observation (OMBO)

Our approach carries out the modeling task in two phases. First, we create a generic module that is able to label the last action (and its parameters) performed by any robosoccer opponent based on the observation of another agent (Action Labeling Module - ALM). We need this module given that in a game in the soccer simulator of the RoboCup, an agent does not have direct access to the other agents inputs and outputs (what the other agent is really perceiving through its sensor and the actions that it executes at each moment). This module can be used for labeling later any other agents actions. Second, we create a model of the other agent based on ALM data (Model Builder Module - MBM).

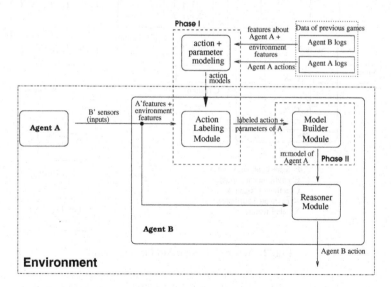

Fig. 1. Architecture for opponent modeling

Figure 1 shows a high level view of the general framework for opponent modeling.

2.1 Action Labeling Module (ALM)

In order to predict the behavior of the opponent (*Agent A*), it is necessary to obtain many instances of the form (Input Sensors, Output Actions), so that they can be used for learning. However, in a real match, *A*'s inputs and outputs are not directly accessible by the modeler agent (*Agent B*). Rather, *A*'s actions (outputs) must be inferred by *Agent B*, by watching it. For instance, if *Agent A* is besides the ball at time 1, and the ball is far away at time 2, it can be concluded that *A* kicked the ball. Noise can make this task more difficult.

The purpose of the ALM module is to classify *A*'s actions based on observations of *A* made by *B*. This can also be seen as a classification task. In this case, instances of the form (A's observations from B,A's actions) are required.

A general description of the action labeling module construction is shown in Figure 2.

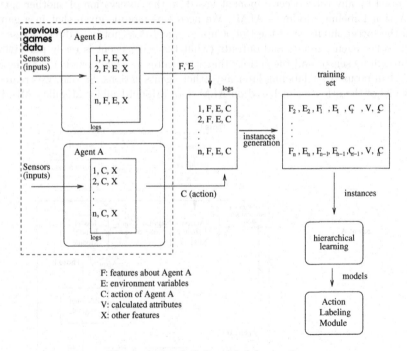

F: features about Agent A
E: environment variables
C: action of Agent A
V: calculated attributes
X: other features

Fig. 2. Action Labeling Module creation

The detailed steps for building the Action Labeling Module (ALM) are as follows:

1. The *Agent A* plays against *Agent B* in several games. At every instant, data about *A*, some environment variables calculated by *B*, and the actions of *A* are logged to produce a trace of *Agent A* behavior from *Agent B* point of view.

2. Each example *I* in the trace is made of three parts: a set of features about *Agent A*, *F*, some environment features, *E*, and the action of *Agent A*, *C*, in a given simulation step *t*, that is to say $I_t = F_t + E_t + C_t$. From this trace it is straightforward to obtain a set of examples *D* so that *Agent B* can infer by machine learning techniques the action carried out by *Agent A* using examples from two consecutive simulation steps. It is important to remark that, sometimes, the soccer server does not provide with information for two contiguous time steps. We ignore these situations.

3. Let *D* be the whole set of available examples from *Agent A* trace. Each example $d_i \in D$ is made of two parts: an n-dimensional vector representing the attributes $a(d_i)$ and a value $c(d_i)$ representing the class it belongs to. In more detail, $a(d_i) = F_t, E_t, F_{t-1}, E_{t-1}, C_{t-1}, V$ and $c(d_i) = C_t$. *V* represents attributes computed based on comparison of features differences between different time steps. We use 24 calculated attributes (e.g. *Agent A* position differences, ball position differences, etc).

4. When the actions $c(d_i)$ in *D* are a combination of discrete and continuous values (e.g. dash 100), we create a set of instances $\hat{D}$ with only the discrete part of the actions, and a set $\hat{D}_j$ for each parameter of the action using only the examples corresponding to the same action. That is, the name of the action and the parameter of the action will be learned separately. For instance, if the action executed by the player is "dash 100" only *dash* will be part of $\hat{D}$ and the value 100 will be in $\hat{D}_{dash}$ with all the instances whose class is dash.

5. The set $\hat{D}$ is used to obtain a model of the action names (i.e. classify the action that the *Agent A* carried out in a given simulation step). The $\hat{D}_j$ are used to generate the continuous values parameters associated to its corresponding action. We have called this way of learning the action and its parameter separately *hierarchical learning* (see Figure 3).

6. Once all classifiers have been built, in order to label the action carried out by *Agent A*, first the classifier that classifies the action is run in order to know which action it performed. Second, the associated classifier that predicts the value of the action parameter is executed. This set of classifiers constitutes the Action Labeling Module (ALM).

As kick, dash, turn, etc are generic actions, and the simulator executes in the same way the actions independently of the agent that executes them, this classifier will be independent of *Agent A*, and could be used to infer the actions of other agents as well. Also, the classifier has to be build just once.

2.2   Model Builder Module (MBM)

Our next goal is to learn a classifier to predict *Agent A*'s actions based on observations of *A* from *Agent B*'s point of view. It will be obtained from instances

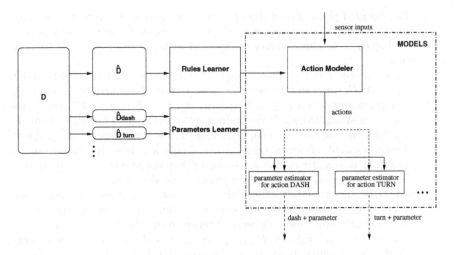

Fig. 3. Hierarchical Learning used by the ALM and MBM

(F_t, E_t, ALM_t) recorded during the match, where ALM_t is the action classified by ALM from observations at t and $t-1$. The aim of this classifier is to predict *Agent A* behavior.

More specifically, data consists of tuples $(I's)$ with features about *Agent A*, some environment variables, and the action that B predicted the *Agent A* has performed, together with its parameter (in this case, labeled by ALM). Instead of using a single time step, we have considered several of them in the same learning instance. Therefore, the learning tuples will have the form $(I_t, I_{t-1}, ..., I_{t-(w-1)})$, where w is the window size. Like in ALM construction, we used calculated attributes, V.

The detailed steps taken for obtaining the model of *Agent A* are as follows:

1. The ALM is incorporated into *Agent B*, so that it can label (infer) *Agent A*'s actions.
2. *Agent A* plays against *Agent B* in different situations. At every instant, *Agent B* obtains information about *Agent A* and the predicted action of *Agent A*, labeled by ALM as well as its parameters. All this information is logged to produce a trace of *Agent A* behavior.
3. Like in the ALM construction, every example I in the trace obtained is made of three parts: a set of features about *Agent A*, F, some environment variables, E, and the action of *Agent A*, C (labeled by ALM), at a given simulation step t $(I_t = F_t + E_t + C_t)$. In this case, we want to predict the action of *Agent A* in a given simulation step and we need information about it from some simulation steps behind. The number of previous simulation steps used to make an instance is denoted by w.
4. Let D be the whole set of instances available at a given simulation step. Each instance $d_i \in D$ is made of two parts: an n-dimensional vector representing

the attributes $a(d_i)$ and a value $c(d_i)$ representing the class it belongs to. In more detail, $a(d_i) = (F_t, E_t, F_{t-1}, E_{t-1} C_{t-1}, ..., F_{t-(w-1)}, E_{t-(w-1)}, C_{t-(w-1)}, V)$ and $c(d_i) = C_t$.

5. When the actions $c(d_i)$ in D are a combination of discrete and continuous values (e.g. dash 100), we create a set of instances $\hat{D}$ with only the discrete part of the actions, and a set $\hat{D}_j$ for each parameter of the action using only the examples corresponding to the same action. From here on, hierarchical learning is used, just like in ALM, to produce action and parameter classifiers. They can be used later to predict the actions of *Agent A*.

Although the Model Builder Module could be used on-line, during the match, the aim of the experiments in this paper is to show that the module can be built and is accurate enough. Therefore, in our present case, learning occurs off-line. On-line use is just a matter of the agent having some time for focusing its computer effort on learning. Learning could happen, for instance, during the half-time break, during dead times, or when the agent is not involved in a play. Also, incremental algorithms could be used, so as not to overload the agent process.

2.3   Using the Model

Predicting opponent actions is not enough. Predictions must be used somehow by the agent modeler to anticipate and react to the opponent. For instance, in an offensive situation, an agent may decide whether to shoot the ball to the goal or not, based on the predictions about the opponent. The task of deciding when to shoot has been addressed by other researchers. For instance, CMUnited-98 [6] carried out this decision by using three different strategies: based on the distance to the goal, based on the number of opponents between the ball and the goal, and based on a decision tree. In CMUnited-99 [7], Stone et al. [8] considered the opponent to be an ideal player, to perform this decision. Our approach in this paper is similar to Stone's work, except that we build an explicit model of the actual opponent.

It is also important to remark that, at this point in our research, how to use the model is programmed by hand. In the future we expect to develop techniques so that the agent can learn how to use the model.

3   Experiments and Results

This section describes the experimental sequence and results obtained in the process to determine whether the usefulness of using a learned knowledge generated by an *Agent B* by observing behavior of an *Agent A*, taken as a black box. To do so we have carried out three phases: first a player of a soccer simulator team (*Agent A*) plays against an *Agent B* to build the ALM; second, the model m of *Agent A* is built; and third, the model is used by B against A.

For the experimental evaluation of the approach presented in this paper, the player (*Agent A*) whose actions will be predicted is a member of the ORCA team [9], and the *Agent B* (modeler agent) is based on the CMUnited-99 code [7]. *Agent A* will act as a goalie and *Agent B* as a striker, that must make the decision of shooting to the goal or continue advancing towards it, depending on the predictions about the goalie.

The techniques to model *Agent A* actions have been, in both, ALM and model construction, C4.5 [10] and M5 [11]. C4.5 generates rules and M5 generates regression trees. The latter are also rules, whose then-part is a linear combination of the values of the input parameters. C4.5 has been used to predict the discrete actions (kick, dash, turn, etc.) and M5 to predict their continuous parameters (kick and dash power, turn angle, etc.). C4.5 and M5 were chosen because we intend to analyse the models obtained in the future, and decision trees regression trees obtained by them are quite understandable.

3.1   ALM Construction

As it is detailed in the previous section, the data used to generate the ALM, is a combination of the perception about *Agent A* from the point of view of *Agent B* and the actual action carried out by *Agent A*. Once the data has been generated, we use it to construct the set of classifiers that will be the ALM. Results are displayed in Table1.

There are three rows in Table 1. The first one displays the prediction of the action of *Agent A*, while the other two lines show the prediction of the numeric parameters of two actions: turn and dash. As *Agent A* is a goalie, for experimental purposes, we only considered relevant the parameters of these actions. Columns represent: the number of instances used for learning, the number of attributes, and the number of classes (continuous for the numeric attributes). In the last column, accuracy results are shown. For the numeric parameters, a correlation coefficient is displayed. These results have been obtained using a ten-fold cross validation.

ALM obtains a 70% accuracy for the discrete classes, which is a good result, considering that the simulator adds noise to the already noisy task of labeling the performed action. On the other hand, the results obtained for the parameters values are poor. Perhaps the techniques used to build the numerical model are not the most appropriate. Or perhaps, better results could probably be achieved by discretizing the continuous values. Usually, it is not necessary to predict the numeric value with high accuracy; only a rough estimation is enough to take advantage of the prediction. For instance, it would be enough to predict whether

Table 1. Results of ALM creation

Labeling task	Instances	Attributes	Classes	Accuracy
Action	5095	69	5	70.81%
Turn angle	913	69	Continuous	0.007 C.C.
Dash power	3711	69	Continuous	0.21 C.C.

C.C.: correlation coefficient

the goalie will turn left or right, rather than the exact angle. For the purposes of this paper, we will use only the main action to decide about when to shoot.

3.2   Model Construction

The ALM can now be used to label opponent actions and build a model, as explained in previous sections. C4.5 and M5 have been used again for this purpose. Results are shown in Table 2.

Table 2. Model creation results

Predicting	Instances	Attributes	Classes	Accuracy
Action	5352	73	6	81.13%
Turn angle	836	73	Continuous	0.67 C.C.
Dash power	4261	73	Continuous	0.41 C.C.

C.C.: correlation coefficient

Table 2 shows that the main action will be perform by the opponent can be predicted with high accuracy (more than a 80%).

3.3   Using the Model

Once the model m has been constructed and incorporated in-to the *Agent B* architecture, it can be used to predict the actions of the opponent in any situation. The task selected to test the model acquired is *when to shoot* [8]. When the striker player is approaching the goal, it must decide whether to shoot right there, or to continue with the ball towards the goal. In our case, *Agent B* (the striker) will make the decision based on a model of the opponent goalie.

When deciding to shoot, our agent (B) first selects a point inside the goal as a *shoot target*. In this case, a point at the sides of the goal. The agent then considers its own position and the opponent goalie position to select which point is the shoot target. Once the agent is near the goal, it uses the opponent goalie model to predict the goalie reaction and decides to shoot or not at a given simulation step. For example, if the goalie is predicted to remain quiet, the striker advances with the ball towards the goal.

In order to test the effectiveness of our modeling approach in a simulation soccer game, we ran 100 simulations in which only two players take part. The *Agent A* is an ORCA Team goalie and *Agent B* is a striker based on the CMUnited-99 architecture with ALM and the model of *Agent A*. For every simulation, the striker and ball were placed in 30 different positions in the field randomly chosen. This makes a total of 3000 goal opportunities. The goalie was placed near to the goal. The task of the striker is to score a goal while the goalie must avoid it.

To test the model utility, we compare a striker that uses the model with a striker that does not. In all situations, the striker leads the ball towards the goal until deciding when to shoot. The striker that does not use the model decides

Table 3. Simulation results

Striker	Simulations	Average of goals	Average of Shots Outside
without model	100	4.65	11.18
with model	100	5.88	10.47

when to shoot based only on the distance to the goal, while the striker that uses the model, considers this distance and the goalie predicted action.

The results of these simulations are shown in Table 3.

As results show, the average of goals using the model is higher than the average of goals without the model. They could be summarized as that every 30 shots, one extra goal will be scored if the model is used. Shots outside the goal are also reduced. We carried out a *t-test* to determine that these differences are significant at $\alpha = 0.05$, which they are. So, even with a simple way of using the model, we can have a significant impact on results by using the learned model.

4   Related Work

Our approach follows Webb feature-based modeling [12] that has been used for modeling user behavior. Webb's work can be seen as a reaction against previous work in student modeling, where they attempted to model the cognitive processing of students, which cannot be observed. Instead, Webb's work models the user in terms of inputs and outputs, that can be directly seen.

There are several work related to opponent modeling in the RoboCup soccer simulation. Most of them focus on the coaching problem (i.e. how the coach can give effective advice to the team). Druecker et al. [13] use neural networks to recognize team formation in order to select an appropriate counter-formation, that is communicated to the players. Another example of formation recognition is described in [14] that use a learning approach based on player positions. Recently, Riley and Veloso [15] presented an approach that generate plans based on opponent plans recognition and then communicates them to its teammates. In this case, the coach has a set of "a priori" opponent models. Being based on [16], Steffens [17] presents an opponent modeling framework in Multi-Agent Systems. In his work, he assumes that some features of the opponent that describe its behavior can be extracted and formalized using an extension of the coach language. In this way, when a team behavior is observed, it is matched with a set of "a priori" opponent models. The main difference with all these previous work is that in our work, we want to model opponent players in order to improve low level skills of the agent modeler rather than modeling the high level behavior of the whole team.

On the other hand, Stone et al. [8] propose a technique that uses opponent optimal actions based on an ideal world model to model the opponent future actions. This work was applied to improve the agent low level skills. Our work addresses a similar situation, but we construct a real model based on observation of an specific agent, while Stone's work does not directly construct a model, being his approach independent of the opponent.

In [18] Takahashi et al. present an approach that constructs a state transition model about the opponent (the "predictor"), that could be considered a model of the opponent, and uses reinforcement learning on this model. They also learn to change the robot's policy by matching the actual opponent's behaviour to several opponent models, previously acquired. The difference with our work is that we use machine learning techniques to build the opponent model and that opponent actions are explicitely labelled from observation.

5   Conclusion and Future Work

In this paper we have presented and tested an approach to modeling low-level behavior of individual opponent agents. Our approach follows three phases. First, we build a general classifier to label opponent actions based on observations. This classifier is constructed off-line once and for all future action labeling tasks. Second, our agent observes an opponent and labels its actions using the classifier. From these observations, a model is constructed to predict the opponent actions. This can be done on-line. Finally, our agent takes advantage of the predictions to anticipate the adversary. In this paper, we have given a proof-of-principle of our approach by modeling a goalie, so that our striker gets as close to the goal as possible, and shoots when the goalie is predicted to move. Our striker obtains a higher score by using the model against a fixed strategy.

In the future, we would like to do on-line learning, using perhaps the game breaks to learn the model. Moreover we intend to use the model for more complex behaviors like deciding whether to dribble, to shoot, or to pass. In this paper, how the agent uses the model has been programmed by hand. In future work, we would like to automate this phase as well, so that the agent can learn to improve its behavior by taking into account predictions offered by the model.

Acknowledgements

This work was partially supported by the Spanish MCyT under project TIC2002-04146-C05

References

1. Ricardo Aler, Daniel Borrajo, Inés Galván, and Agapito Ledezma. Learning models of other agents. In *Proceedings of the Agents-00/ECML-00 Workshop on Learning Agents,* pages 1–5, Barcelona, Spain, June 2000.
2. Christian Druecker, Christian Duddeck, Sebastian Huebner, Holger Neumann, Esko Schmidt, Ubbo Visser, and Hans-Georg Weland. Virtualweder: Using the online-coach to change team formations. Technical report, TZI-Center for Computing Technologies, University of Bremen, 2000.

3. H. Kitano, M. Tambe, P. Stone, M. Veloso, S. Coradeschi, E. Osawa, H. Matsubara, I. Noda, and M. Asada. The robocup synthetic agent challenge. In *Proceedings of the Fifteenth International Joint Conference on Artificial Intelligence (IJCAI97)*, pages 24–49, San Francisco, CA, 1997.

4. Agapito Ledezma, Ricardo Aler, Araceli Sanchis, and Daniel Borrajo. Predicting opponent actions in the robosoccer. In *Proceedings of the 2002 IEEE International Conference on Systems, Man and Cybernetics*, October 2002.

5. Agapito Ledezma, Antonio Berlanga, and Ricardo Aler. Automatic symbolic modelling of co-evolutionarily learned robot skills. In José Mira and Alberto Prieto, editors, *Connectionist Models of Neurons, Learning Processes, and Artificial Intelligence (IWANN, 2001)*, pages 799–806, 2001.

6. Agapito Ledezma, Antonio Berlanga, and Ricardo Aler. Extracting knowledge from reactive robot behaviour. In *Proceedings of the Agents-01/Workshop on Learning Agents*, pages 7–12, Montreal, Canada, 2001.

7. Andreas G. Nie, Angelika Honemann, Andres Pegam, Collin Rogowski, Leonhard Hennig, Marco Diedrich, Philipp Hugelmeyer, Sean Buttinger, and Timo Steffens. the osnabrueck robocup agents project. Technical report, Institute of Cognitive Science, Osnabrueck, 2001.

8. J. Ross Quinlan. *C4.5: Programs for Machine Learning*. Morgan Kaufmann, San Mateo, CA, 1993.

9. J. Ross Quinlan. Combining instance-based and model-based learning. In *Proceedings of the Tenth International Conference on Machine Learning*, pages 236–243, Amherst, MA, June 1993. Morgan Kaufmann.

10. Patrick Riley and Manuela Veloso. *Distributed Autonomous Robotic Systems*, volume 4, chapter On Behavior Classification in Adversarial Environments, pages 371–380. Springer-Verlag, 2000.

11. Patrick Riley and Manuela Veloso. Planning for distributed execution through use of probabilistic opponent models. In *Proceedings of the Sixth International Conference on AI Planning and Scheduling (AIPS-2002)*, 2002.

12. Patrick Riley, Manuela Veloso, and Gal Kaminka. Towards any-team coaching in adversarial domains. In *Proceedings of th First International Joint Conference on Autonomous Agents and Multi-Agent Systems (AAMAS)*, 2002.

13. Timo Steffens. Feature-based declarative opponent-modelling in multi-agent systems. Master's thesis, Institute of Cognitive Science Osnabrück, 2002.

14. Peter Stone, Patrick Riley, and Manuela Veloso. Defining and using ideal teammate and opponent agent models. In *Proceedings of the Twelfth Innovative Applications of AI Conference*, 2000.

15. Peter Stone, Manuela Veloso, and Patrick Riley. The CMUnited-98 champion simulator team. In Asada and Kitano, editors, *RoboCup-98: Robot Soccer World Cup II*, pages 61–76. Springer, 1999.

16. Peter Stone, Manuela Veloso, and Patrick Riley. The CMUnited-98 champion simulator team. Lecture Notes in Computer Science, 1604:61–??, 1999.

17. Yasutake Takahashi, Kazuhiro Edazawa, and Minoru Asada. Multi-module learning system for behavior acquisition in multi-agent environment. In *Proceedings of the 2002 IEEE/RSJ International Conference on Intelligent Robots and Systems*, pages 927–931, 2002.

18. Geoffrey Webb and M. Kuzmycz. Feature based modelling: A methodology for producing coherent, consistent, dynamically changing models of agents's competencies. *User Modeling and User Assisted Interaction*, 5(2):117–150, 1996.

A Model-Based Approach to Robot Joint Control

Daniel Stronger and Peter Stone

Department of Computer Sciences,
The University of Texas at Austin
{stronger, pstone}@cs.utexas.edu
http://www.cs.utexas.edu/~{stronger,pstone}

Abstract. Despite efforts to design precise motor controllers, robot joints do not always move exactly as desired. This paper introduces a general model-based method for improving the accuracy of joint control. First, a model that predicts the effects of joint requests is built based on empirical data. Then this model is approximately inverted to determine the control requests that will most closely lead to the desired movements. We implement and validate this approach on a popular, commercially available robot, the Sony Aibo ERS-210A.

Keywords: Sensor-Motor Control; Mobile Robots and Humanoids.

1 Introduction

Joint modeling is a useful tool for effective joint control. An accurate model of a robotic system can be used to predict the outcome of a particular combination of requests to the joints. However, there are many ways in which a joint may not behave exactly as desired. For example, the movement of a joint could lag behind the commands being sent to it, or it could have physical limitations that are not mirrored in the control software.

In this paper, we consider the case in which a joint is controlled by repeatedly specifying an angle that the joint then tries to move to (e.g. as is the case for PID control[1]). These are the *requested angles*, and over time they comprise a *requested angle trajectory*. Immediately after each request, one can record the *actual angle* of the joint. These angles make up the *actual angle trajectory*.

This paper presents a solution to a common problem in robot joint control, namely that the actual angle trajectory often differs significantly from the requested trajectory. We develop a model of a joint that predicts how the actual angle trajectory behaves as a function of the requested trajectory. Then, we use the model to alter the requested trajectory so that the resulting actual trajectory more effectively matches the *desired trajectory*, that is, the trajectory that we would ultimately like the joint to follow.

At a high level, our proposed approach is to:

---

[1] http://www.expertune.com/tutor.html

D. Nardi et al. (Eds.): RoboCup 2004, LNAI 3276, pp. 297–309, 2005.

1. Determine the various features of the joint that need to be taken into account by the model.
2. By experimenting with the joint and the various ways to combine these features, establish a mathematical model for the joint whose behavior mimics that of the joint when given the same input sequence.
3. Use the model to compute a series of requests that yields a close approximation to the desired trajectory. If the model is accurate, then these requests will cause the joint to behave as desired.

For expository purposes, we demonstrate the use of this technique on the Sony Aibo ERS-210A robot,[2] whose motors use PID control. However, this general methodology is potentially applicable to any situation in which robotic joints do not behave exactly as requested. We present empirical results comparing the direct approach of setting the requested angles equal to the desired angles against an approach in which the requested angles are set to a trajectory motivated by knowledge of the joint model.

The remainder of this paper is organized as follows. Section 2 relates our work to previous approaches. Section 3 introduces the Aibo robot and our model of its joint motion. Section 4 describes the process of inverting this model. Section 5 presents the empirical results validating our approach, and Section 6 concludes.

2   Related Work

One common approach to model-based joint control is to determine the complete physical dynamics of the system, which can in turn be used to construct a model-predictive control scheme. For example, Bilodeau and Papadopoulos empirically determine the dynamics of a hydraulic manipulator [1]. While this is a valuable technique, it is only applicable in situations where the physical parameters of the relevant joints can be ascertained. However, in many robotic systems, determining accurate values for these parameters can be very difficult or impossible. Another potential difficulty is that the low-level joint control policy is unattainable. This is the case for the Aibo; although we know that the joints are controlled by a PID mechanism, there is no information available about how the angle requests are converted to motor currents, nor about the motor specifications for the Aibo's joints. Our approach circumvents these problems by experimentally modeling the behavior of each joint as a function of the high-level angular requests that it receives. Furthermore, our approach extends beyond the construction and testing of a model to using the model to motivate more effective joint requests.

Although others have previously looked at the problem of correcting for joint errors, to our knowledge the approach proposed here has not been used. English and Maciejewski track the effects of joint failures in Euclidean space, but focus on joints locking up or breaking rather than correcting for routine inaccuracies [2]. An alternative approach to robot joint control is presented by Kapoor, Cetin, and

---

[2] http://www.aibo.com

Tesar [3]. They propose a system for choosing between multiple solutions to the inverse kinematics based on multiple criteria. Their approach differs from ours in that it is designed for robotic systems with more manipulator redundancy. Our approach is better suited to situations in which the inverse kinematics has a unique solution, but in which the joints do not behave exactly as requested.

3   Developing a Model

The Aibo robot has four legs that each have three joints known as the rotator, abductor, and knee. The robot architecture allows each joint to receive a request for an angle once every eight milliseconds. For all experiments reported in this paper, we request angles at that maximum frequency. The Aibo also reports the actual angles of the joints at this frequency.

Although we only have direct control over the angles of the three joints, it is often desirable to reason about the location of the robot's foot. The process of converting the foot's location in space into the corresponding joint angles for the leg is known as *inverse kinematics*. Since inverse kinematics converts a point in space to a combination of angles, it also converts a trajectory of points in space to an angle trajectory for each joint. We have previously solved the inverse kinematics problem for the legs of the ERS-210A robot [4].

As our primary test case for this project, we use a spatial trajectory for the foot that is derived from the walking routine of an entry in the 2003 RoboCup robot soccer competitions [5]. This trajectory is based on a half ellipse, and it is shown in Figure 1. The details of the trajectory and walking routine are given in [4]. The high-level motivation for this research is to enable more direct tuning of the legs' trajectories while walking.

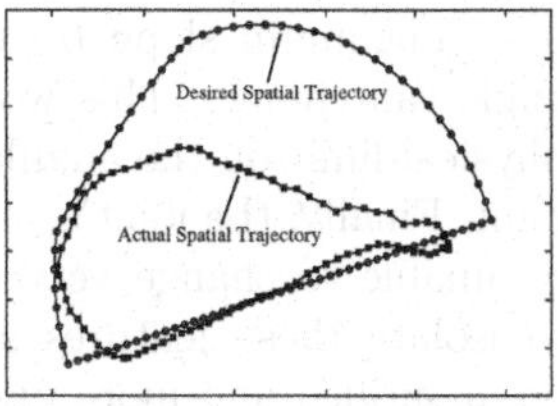

In an actual walking situation, the ground exerts significant and unpredictable forces on the Aibo's leg. In order to isolate the

Fig. 1. Desired and actual spatial trajectory of the robot's foot. Only two of the three spatial dimensions are depicted here

behavior of the joints unaffected by external forces, we perform all the experiments for this project with the robot held in the air, so that there is no interference with the leg's motion.

The most natural approach to trying to produce a desired trajectory for the foot is to convert the desired foot locations into joint angles by inverse kinematics, and then to set the requested joint angles equal to the resulting desired angles. To evaluate this method one can record the actual angle values of the leg's joints, convert them into an actual spatial trajectory with forward kinematics, and compare the desired spatial trajectory to the actual one. This comparison is shown in Figure 1.

The difference between these two trajectories is best understood in terms of the behaviors of the specific joints. We temporarily restrict our attention to only the rotator joint and determine the facets of the difference between its trajectory and the desired trajectory.

The goal of the joint model is to take as its input a sequence of requested angles and return a sequence of angles that mimics as closely as possible the angles that would actually be attained by the robot joint. We denote the sequence of requests by $R(t)$, where values of R are given in degrees, and the units for t is the amount of time between consecutive requests (eight milliseconds in our case). Furthermore, we restrict our attention to a period of time lasting the length of an Aibo step. We call this length of time t_{step}, which is 88 in our units (i.e. each step takes $88 \cdot 8 = 704$ milliseconds). During all of our experiments, we let the Aibo run through many steps continuously, so that an equilibrium is reached before measurements are taken.

To construct a model of how the joint responds to different requests, we observe the behavior of the joint as a result of various requested trajectories. Figure 2 shows how the joint responds to the sequence of requests that equal our desired angles for it. From this, we can start to infer the properties that our model needs to capture.

First, the actual angle lags behind the requests in time. Second, there appears to be a maximum slope by which the actual angle can change. This would amount to a physical limit on the angular velocity of the joint. Finally, the joint's velocity appears to be unable to change very quickly. In order to isolate these features so that they can be quantified and more precisely characterized, we perform a series of experiments on the joints.

For these experiments, we set $R(t)$ to a test trajectory given by

$$R(t) = \begin{cases} \theta_{test} & \text{if } t < \frac{t_{step}}{2} \\ 0 & \text{if } t \geq \frac{t_{step}}{2} \end{cases} \quad (1)$$

Since the sequence of requests is continuously repeated, this has the effect of alternating between requesting angles of 0 and θ_{test}. A graph of $R(t)$ with θ_{test} equal to 40 and the resulting actual angle is shown in Figure 3.

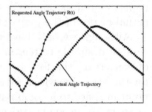

Fig. 2. This graph shows what happens when the angles requested of the rotator joint are set equal to our desired angle trajectory for that joint. These requests are compared to the actual angles that are reported by the robot at each time. Note that the graphs are cyclical, e.g. R(0) = R(88)

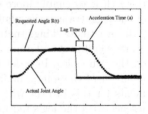

Fig. 3. Requested and actual angular trajectory of the rotator joint in test case with $\theta_{test} = 40$

This figure identifies a number of important facets that a model of the joint must have. First, when the requested angle suddenly changes by 40°, there is a period of time before the joint responds at all. We denote this *lag time* by l.

Then, after the lag time has elapsed, there is a period of time during which the joint accelerates toward its new set point. This *acceleration time* is denoted by a. After this, the joint's angular speed appears to plateau. We postulate that this maximum angular speed is the physical limit of the joint's speed, and we denote it by v_{max}.

At this point, two questions need to be answered. First, will a larger angle difference induce an angular speed greater than v_{max}, or is v_{max} really the joint's top speed? Second, is the acceleration time best modeled as a limit on the joint's angular acceleration, or as a constant acceleration time? These questions can both be answered by performing the same test but with θ_{test} equal to 110 degrees. The results of this test are shown in Figure 4.

In this situation, the joint has the same maximum angular speed as in Figure 3. This confirms that the joint cannot rotate faster than v_{max}, regardless of the difference between the requested and actual angles. Meanwhile, this test disproves the hypothesis that the acceleration time is due to a constant limit on the angular acceleration of the joint. This is because the joint takes the same amount of time to accelerate from angular velocity 0 to $-v_{max}$ in Figure 3 as it does to go from v_{max} to $-v_{max}$ in Figure 4. A constant accelera-

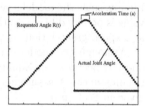

Fig. 4. Requested and actual angular trajectory of the rotator joint in test case with $\theta_{test} = 110$

tion limit hypothesis would predict that the second acceleration would take twice as long as the first one.

Although the joint's angular velocity is bounded by v_{max}, it is still the case that, within a certain range of differences between the requested and actual angle, the higher that difference is, the faster the joint will tend to rotate. This suggests the use of a function f defined as:

$$f(x) = \begin{cases} v_{max} & \text{if } x \geq \theta_0 \\ x \cdot \frac{v_{max}}{\theta_0} & \text{if } -\theta_0 < x < \theta_0 \\ -v_{max} & \text{if } x \leq -\theta_0 \end{cases} \qquad (2)$$

where θ_0 is a constant that denotes the size of the difference between the requested and actual angle that is needed for the joint to move at its maximum angular speed. Figure 5 depicts the function f.

In order to capture the effect of an acceleration time,

Fig. 5. The function f

we set our model's velocity to an average of a values of f. The model's values for the joint angle, which we denote by $M_R(t)$, are defined by:

$$M_R(t) = M_R(t-1) + \frac{1}{a} \sum_{i=l+1}^{l+a} f(R(t-i) - M_R(t-1)) \qquad (3)$$

Table 1. Model Parameters

Parameter	Description	Value
v_{max}	Maximum Angular Speed	2.5 degrees/t_u
θ_0	Angle Difference Threshold	7.0 degrees
l	Lag Time	4 t_u
a	Acceleration Time	6 t_u

where $R(t)$ is the sequence of requested angles. Since the most recent value of R that is included in this definition is $R(t - l - 1)$, the model captures the notion that any request takes a lag of l time steps before it begins to affect the joint. Finally, since the model's velocity is the average of a values of f, and f's absolute value is bounded by v_{max}, the model captures the fact that the joint never moves at a speed greater than v_{max}.

Although the model and its parameters were determined using only the request trajectories in Equation (1), it is highly accurate under a wide range of circumstances. Measurements of the fidelity of the model in our experiments are given in Section 5.

Table 1 summarizes the model's parameters and their values, as determined experimentally, for the Aibo. We use t_u to denote our eight millisecond unit of time.

It is worth noting that although the Aibo's joints are PID controlled, the features captured by the model are not predicted by that fact. PID control does not predict a maximum angular speed, and it does not explain the lag between the requests and their effects. This suggests that for the purposes of joint modeling, it is not particularly helpful to think of the Aibo joints as being PID controlled.

4   Inverting the Model

In order to compel the joint to move in our desired trajectory, which we call $D(t)$, we need to be able to convert it into a set of requests $I(t)$ such that when the values given by I are sent to the joint, the resulting behavior of the joint matches $D(t)$ as closely as possible. In terms of the model, given $D(t)$ we would like to find $I(t)$ such that $M_I(t) = D(t)$. This is the process of inverting the model.

4.1   Inverting the Model Explicitly

The first problem we encounter when trying to invert the model is that by the model, regardless of $I(t)$, the joint's angular speed is bounded by v_{max}. That is, $|M_I(t) - M_I(t-1)| \leq v_{max}$. If $D(t)$ violates this constraint, it is not in the range of the model, and there are no requests $I(t)$ such that $M_I(t) = D(t)$. In fact, many of the angular trajectories we get from inverse kinematics violate this constraint. It is theoretically impossible for the model to return these trajectories exactly.

Even when we know it is impossible to invert the model on a particular trajectory, one possibility is to try to construct an approximation to $D(t)$ that is in the range of the model and invert the model on that instead. However, doing so in general is complicated by the fact that f is not invertible, and in the range that it is invertible, there are points with infinitely many inverses.

We circumvent this problem by restricting our approximation of $D(t)$ to be a piecewise linear trajectory, which we call $P(t)$, with the property that the slope of each line segment is less than or equal to v_{max} in absolute value. Note that since $P(t)$ is a trajectory of joint angles, these segments represent an angle that varies linearly with respect to time (i.e. with a constant angular velocity). An example piecewise linear approximation is shown in Figure 6. One may be able to automate this approximation, even subject to the restriction on minimum and maximum angular velocity. However, for the purposes of this paper the approximations are constructed manually.

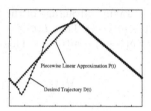

Fig. 6. Piecewise linear approximation. Note that the approximation diverges from the desired trajectory on the left side due to the slope restriction on the linear segments

Although $P(t)$ is not invertible in the model (due to instantaneous velocity changes), we can invert the mathematical model on the component line segments. Recombining these inverse line segments appropriately yields a series of requests, $R(t)$, such that the result of applying the model to it, $M_R(t)$, is a close approximation to $P(t)$, which is in turn a close approximation to $D(t)$.

4.2   Inverting Lines

Our proposed method relies on being able to determine the inverses of lines according to the model. That is, given a particular linear trajectory, what angles should be requested so that the given trajectory is actually achieved by the joint? The answer is a linear trajectory that has the same slope as the desired line but differs from it by a constant that depends on its slope. This is only possible when our line's slope corresponds to an angular velocity that is actually attainable, so we will restrict our attention to the case where the absolute value of the slope of the line is less than or equal to v_{max}.

Consider a linear series of requests, $L(t)$. We say that $L(t)$ has slope m, where m is $L(t) - L(t-1)$ for all t. We temporarily restrict our attention to the case where $|m| < v_{max}$ and furthermore assume without loss of generality that $m \geq 0$. We will determine the image of $L(t)$ in the model, $M_L(t)$. Since this will turn out to be a line of the same slope, it will enable us to compute the inverse of any line whose slope is in range.

We would like to be able to reason about the angular distance between points on the requested line and those on the resulting line according to the model, and specifically how that distance changes over time. Thus we denote the angular distance $L(t) - M_L(t)$ by $\delta(t)$. First, however, we must understand how $M_L(t)$ changes as a function of $\delta(t)$. This can be seen by plugging L into Equation (3) (for R):

$$M_L(t) = M_L(t-1) + \frac{1}{a} \sum_{i=l+1}^{l+a} f(L(t-i) - M_L(t-1)) \qquad (4)$$

Taking advantage of the fact that L is linear, we replace $L(t-i)$ with $L(t-1) - m(i-1)$. Thus $L(t-i) - M_L(t-1)$ is equal to $L(t-1) - M_L(t-1) - m(i-1)$, or $\delta(t-1) - m(i-1)$. This enables us to rewrite Equation (4) as:

$$M_L(t) = M_L(t-1) + \frac{1}{a} \sum_{i=l+1}^{l+a} f(\delta(t-1) - m(i-1)) \qquad (5)$$

This equation tells us how the the model's value for the angle varies as a function of the angular distance $\delta(t)$. In order to capture this relationship more concisely, we define the function S to be:

$$S(x) = \frac{1}{a} \sum_{i=l+1}^{l+a} f(x - m(i-1)) \qquad (6)$$

Thus we can characterize the relationship between M_L and δ as $M_L(t) = M_L(t-1) + S(\delta(t-1))$. Now that we understand how δ influences M_L, it is useful to analyze how $\delta(t)$ changes over time. We can isolate this effect by using the definition of $\delta(t)$ to replace $M_L(t)$ with $L(t) - \delta(t)$. This gives us:

$$L(t) - \delta(t) = L(t-1) - \delta(t-1) + S(\delta(t-1)) \qquad (7)$$

Then, since $L(t) = L(t-1) + m$, we can rearrange as:

$$\delta(t) = \delta(t-1) + m - S(\delta(t-1)) \qquad (8)$$

This equation indicates that as time progresses, $\delta(t)$ approaches an equilibrium where $S(\delta(t)) = m$. That is, $\delta(t)$ approaches a constant, which we denote by C_m, such that $S(C_m) = m$. Since $M_L(t) = L(t) - \delta(t)$, this means that $M_L(t)$ approaches $L(t) - C_m$.

Given a desired linear angular trajectory, $D_L(t)$, with slope m, C_m is the amount that must be added to D_L to get a sequence of requests such that the actual joint angles will approach our desired trajectory. Thus inverting each of our component line segments involves computing C_m for m equal to the slope of that segment. Unfortunately, calculating C_m in terms of m explicitly is not straightforward. Since S is defined as the average of a instances of the function f, we would need to determine which case in the definition of f is appropriate in each of those a instances. Nonetheless, we are able to approximate C_m quite accurately by approximating the sum in Equation (6) with an integral. The computation of C_m and an overview of its derivation are in the appendix.

In the case where $m = v_{max}$, C_m is not well defined, since $S(x) = v_{max}$ for any sufficiently large value of x. In fact, if $f(x - v_{max}(l + a)) \geq v_{max}$, then all of the values of f being averaged in the definition of S take on the value of v_{max}, and thus $S(x)$ equals v_{max}. This occurs when $x \geq \theta_0 + v_{max}(l + a)$. In this situation, $L(t)$ and $M_L(t)$ will both keep increasing at a rate of v_{max}, so the distance between them will not change. Thus we use the threshold value of $\theta_0 + (l + a)v_{max}$ for C_m and rely on switching between the inverses of the line segments at the right time to ensure that the actual angle trajectory stays close to the desired line.

4.3   Combining Inverted Line Segments

Our piecewise linear approximation, $P(t)$, is comprised of line segments $D_L(t)$ with slopes m. For any one of these line segments, we know that by requesting values of $D_L(t) + C_m$, the joint angle will closely approximate $D_L(t)$. For the joint to follow $P(t)$ to a close approximation the whole way through, we must transition between these lines at the appropriate times.

After the requests switch from one inverted linear trajectory to another, how long will it take for the joint to switch between the corresponding desired trajectories? According to the model, there is a lag time, l, before any effect of the change will be observed. After that, the joint will accelerate from one linear trajectory to the other over the course of an acceleration time of length a. Ideally then, the desired piecewise linear trajectory, $P(t)$, would transition between components in the middle of this acceleration period. In order to achieve this, we transition between inverted line segments $l + \frac{a}{2}$ time units before the transition between desired line segments. This completes the specification of our approach.

The whole process takes a desired angle trajectory, constructs a piecewise linear approximation to it, and formulates requests to the joint based on that approximation. These three trajectories and the resulting actual trajectory are shown in Figure 7. Notably, the actual trajectory runs very close to the piecewise linear approximation.

5   Experimental Results

The goal of this process is for the robot to move each joint so that it follows a desired trajectory as closely as possible. We can evaluate the success of the method by calculating the angular distance between the desired trajectory and the actual one (as depicted in Figure 7).

Although we have described our approach thus far using the Aibo's rotator joint as an example, we have implemented it on all three of the joints of an Aibo leg. Interestingly, we found the parameters of our model to be exactly the same for all three of the leg's joints. We analyze the method's success on all three joints.

We treat an angle trajectory as a t_{step}-dimensional vector, where t_{step} is the number of requests that comprise one Aibo step (88 in our case), and calculate the L_2 and L_∞ norms between the vectors corresponding to two trajectories. We consider distances between four different trajectories for each joint: Des, the desired trajectory, Dir, the actual angles under the direct method of setting the requests equal to the desired angles, Pwl, the piecewise linear approximation, and MB (Model-Based), the actual angle trajectory achieved by the our approach. Since these distances can vary from one Aibo step to the next, the numbers given in Table 2 are averages and standard deviations taken over 20 steps.

The actual angles achieved by our method come much closer to our desired angle trajectories than the ones obtained by setting the requested angles equal to the desired angles, as shown in the two bold rows. The very small distances between Pwl and MB indicate the strength of the fidelity of the model.

We also compare the attained spatial trajectory of the Aibo's foot to the desired spatial trajectory (see Figure 8). Here we measure the improvement in the Euclidean distance between the desired and attained foot trajectories. We calculate the distance between the desired and actual foot location at each time and apply the L_2 and L_∞ norms to these distances. The direct method yields an L_2 distance of 3.23 ± 0.01 cm and L_∞ of 4.61 ± 0.05 cm. Our model-based method gives us an L_2 of 1.21 ± 0.04 cm and an L_∞ of 2.34 ± 0.12 cm.

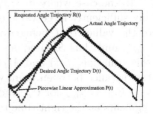

Fig. 7. This graph depicts the original desired angle trajectory, the piecewise linear approximation, the requests that are derived from that approximation, and the actual angle trajectory that results from this process

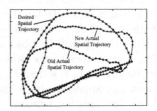

Fig. 8. This graph depicts the desired spatial trajectory compared to the effect of simply requesting the desired angles (old) and to the effect of the approach described in this paper (new). The spatial trajectories are computed by forward kinematics from the recorded actual joint angles

Finally, we compare our method to the following process. For any value of k, consider setting requested angles $R(t)$ equal to $D(t+k)$. That is, let the requests be exactly the same as the desired angles, but offset by a fixed amount of time. For each of the t_{step} possible values of k, we can compute the distance between the resulting actual angle trajectory and the desired one. The minimum of these distances with respect to k provides a measure of how much of our improvement can be attributed to modeling the lag in the joints. The distances returned by this approach were an L_2 of 1.55 ± 0.04 cm and an L_∞ of 3.57 ± 0.08 cm. These distances are smaller than those achieved by the direct method, but still

Table 2. Distances between angle trajectories

Comparison	Rotator	Abductor	Knee
$L_2(Des, Dir)$	31.0($\pm$0.2)	29.0($\pm$0.2)	20.1($\pm$0.1)
$L_\infty(Des, Dir)$	57.2($\pm$0.3)	59.5($\pm$0.5)	42.6($\pm$0.3)
$L_2(Des, MB)$	9.1($\pm$0.2)	10.4($\pm$0.1)	5.6($\pm$0.2)
$L_\infty(Des, MB)$	29.4($\pm$0.8)	24.5($\pm$0.7)	11.1($\pm$0.5)
$L_2(Pwl, MB)$	2.7($\pm$0.4)	2.7($\pm$0.3)	2.6($\pm$0.2)
$L_\infty(Pwl, MB)$	6.4($\pm$0.6)	6.0($\pm$0.4)	6.2($\pm$0.7)

significantly greater than the distances attained by our model-based method. This result indicates that the success of our approach is due to more than its ability to model lag in the joint.

6   Conclusion and Future Work

This paper demonstrates the development of a detailed joint model of a popular, commercially available robotic research platform. We show all the steps of the derivation of this model using a generally applicable methodology. We then approximately invert the model to determine control requests that cause the robot's joints to move in a desired trajectory. Using this approach, we successfully bring the robot's actual motions significantly closer to the desired motions than they were previously.

   The high-level motivation for this research is to enable direct tuning of the legs' trajectories while walking. In addition to applying the proposed approach towards that task, there are three main ways in which the work can be extended in future research. First, since the experiments reported in this paper were performed with the robot held in the air, we have a model of how the joints behave when the external torques being exerted on them are relatively small. An extension of this work would be to model how the joints respond to significant external torques, e.g. the torques that are exerted when a robot walks on the ground. Second, our approach could also be extended by implementing it on other platforms. Doing so will help elucidate the class of robotic problems on which these techniques are effective. Third, a possibility for future work is for the robot to learn the model of its joints automatically from experience based on its knowledge of the joint requests and actual angles over time. This would have the benefit that the robot could adjust its model over time to compensate for changing properties of the joints. While it would be challenging to learn a model of arbitrary functional form, we surmise that tuning the parameters of a model, such as l and a in our case, would be relatively straightforward. In this regard, one contribution of the research reported here is an identification of a class of functions that could be used as the space of models to explore during the learning process.

Acknowledgments

We would like to thank the members of the UT Austin Villa team for their efforts in developing the software used as a basis for the work reported in this paper. Thanks also to Chetan Kapoor and Ben Kuipers for helpful discussions. This research was supported in part by NSF CAREER award IIS-0237699.

Appendix

This appendix describes the computation of C_m from m and gives an overview of the derivation. As discussed in Section 4.2, C_m is the amount that must be added to a line of slope m to get its inverse, and we are considering the case where $m \epsilon [0, v_{max})$. The definition of C_m is that $m = S(C_m)$, where S is defined in Equation (6). The first step is to replace the sum in that definition with the corresponding integral, so that $m = S(C_m)$ becomes:

$$m = \frac{1}{a} \int_l^{l+a} f(C_m - mi)\, di \tag{9}$$

Next, since the definition of f is split into cases based on whether its argument is greater than $-\theta_0$, and θ_0, important thresholds in the analysis of Equation (9) are values of i for which $C_m - mi = -\theta_0$, and θ_0. We then divide our analysis into cases based on where these two thresholds fall with respect to our limits of integration (e.g. less than both, between them, or greater than both).

This results in three cases, each of which can be analyzed independently. Due to the particular parameters in our model, this analysis reduces to two cases. Finally, C_m is computed as follows. First, determine whether or not the following inequality holds:

$$m \left(\frac{\theta_0}{v_{max}} + \frac{a}{2} \right) < \theta_0 \tag{10}$$

If it does, C_m is given by the equation:

$$C_m = m \left(\frac{\theta_0}{v_{max}} + l + \frac{a}{2} \right) \tag{11}$$

If not, it is instead given by:

$$C_m = \theta_0 + m \left(l + a - \sqrt{\frac{2\theta_0}{m} \cdot \frac{a(v_{max} - m)}{v_{max}}} \right) \tag{12}$$

This computation of q relies on two numerical facts regarding the parameters of our model. These are:

$$\theta_0 > v_{max} \quad \text{and} \quad 4\theta_0 > v_{max}a \tag{13}$$

If these are not true, there may be more cases involved in the computation of C_m.

References

1. G. Bilodeau and E. Papadopoulos, "Modelling, identification and experimental validation of a hydraulic manipulator joint for control." in *Proceedings of the 1997 International Confrerence on Intelligent Robots and Systems (IROS '97)*, Victoria, BC, October 1998.

2. J. English and A. Maciejewski, "Measuring and reducing the euclidean-space measures of robotic joint failures," *IEEE Transactions on Robotics and Automation*, vol. 17, no. 1, pp. 20–28, Feb. 2000.

3. C. Kapoor, M. Cetin, and D. Tesar, "Performance based redundancy resolution with multiple criteria," in *Proceedings of 1998 ASME Design Engineering Technical Conference (DETC98)*, Atlanta, Georgia, September 1998.

4. P. Stone, K. Dresner, S. T. Erdoğan, P. Fidelman, N. K. Jong, N. Kohl, G. Kuhlmann, E. Lin, M. Sridharan, D. Stronger, and G. Hariharan, "UT Austin Villa 2003: A new RoboCup four-legged team," The University of Texas at Austin, Department of Computer Sciences, AI Laboratory, Tech. Rep. UT-AI-TR-03-304, 2003, at http://www.cs.utexas.edu/ftp/pub/AI-Lab/index/html/Abstracts.2003.html # 03-304.

5. P. Stone, T. Balch, and G. Kraetszchmar, Eds., *RoboCup-2000: Robot Soccer World Cup IV*, ser. Lecture Notes in Artificial Intelligence 2019. Berlin: Springer Verlag, 2001.

Evolutionary Gait-Optimization Using a Fitness Function Based on Proprioception*

Thomas Röfer

Center for Computing Technology (TZI), FB 3, Universität Bremen
roefer@tzi.de

Abstract. This paper presents a new approach to optimize gait parameter sets using evolutionary algorithms. It separates the crossover-step of the evolutionary algorithm into an interpolating step and an extrapolating step, which allows for solving optimization problems with a small population, which is an essential for robotics applications. In contrast to other approaches, odometry is used to assess the quality of a gait. Thereby, omni-directional gaits can be evolved. Some experiments with the Sony Aibo models ERS-210 and ERS-7 prove the performance of the approach including the fastest gait found so far for the Aibo ERS-210.

1 Introduction

In the Sony Four-Legged Robot League, two teams of four Sony Aibos each compete on a field of approximately 4.2 m × 2.7 m. For 2004, the two models allowed to be used are the ERS-210 and the ERS-7 (cf. Fig. 1a). As in all RoboCup leagues, speed is an essential factor in the game. Therefore, quite a lot of work has been done to optimize the gait of the Aibo using a variety of different approaches. In [3], a stationary state evolutionary algorithm was used to optimize the gait of the Aibo ERS-110, based on research done with Aibo's predecessor [4]. Thereby, a maximum speed of 100 mm/s was reached. On the ERS-210, Powell's method for multidimensional minimization was used to optimize the gait parameters [5]. The maximum speed reached was 270 mm/s. With an evolutionary hill climbing with line search approach, a parameter set resulting in a maximum speed of 296.5 mm/s was found [7]. In both latter works, the shapes of the foot trajectories were optimized. Using a variation of standard policy gradient reinforcement learning techniques, a maximum speed of 291 mm/s was reached, using a static, half-elliptical foot trajectory shape. In that paper it was also argued that the shapes of the foot trajectories sent to the robot significantly differ from the shapes actually performed, and therefore the adaptation of the shape of the locus is not necessarily important.

The typical training setup used in [4, 3, 5, 7, 6] was as follows: the robot walks from one side of the field the other one, then it turns on the spot and walks

---

* The Deutsche Forschungsgemeinschaft supports this work through the priority program "Cooperating teams of mobile robots in dynamic environments".

D. Nardi et al. (Eds.): RoboCup 2004, LNAI 3276, pp. 310–322, 2005.

a)   b)

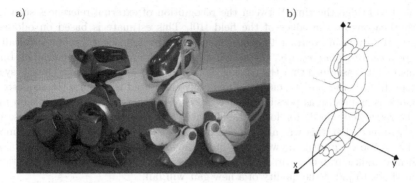

Fig. 1. a) The Aibos ERS-210 and ERS-7. b) A leg of an ERS-210 (modified from [6])

back to the first side, and so on. For orientation, it uses two beacons placed on both sides of the field (or color stripes in a pen in [4, 3]). The beacons are used to measure the speed of the walk as well as when to stop and to turn around. While in [4, 3, 5, 7] single robots are used, [6] trained three robots at once, thereby reducing the learning time by factor three. In [4, 3, 5, 7], the optimization was done on-board the robot. In contrast, [6] used an off-board computer for learning that controls the three robots used for evaluating different walk parameter sets.

2   Experimental Setup

The major drawback of the approaches described in the previous section is that they only optimize the gait for walking straight ahead with a more or less steady head. However, in most situations in actual RoboCup games, the robots have to perform different motions, especially when following the ball. They have to turn left and right while walking, move sideways and sometimes backwards. All these dynamic elements required for games cannot be learned when the robot only walks from beacon to beacon. Therefore, a more flexible scheme is required.

2.1   Measuring Robot Motion

So it must be possible to measure omni-directional robot motion. On the one hand, it is necessary to measure how good the motion performed matches the motion that was desired, on the other hand, the scheme must be simple enough to be performed at the actual competition site, because the carpet used there can differ from the carpet the robot was originally trained with in the laboratory.

Using a Self-Localization Method. All RoboCup teams have already implemented some kind of metric self-localization for their robots. They are either based on particle filters [10] or on extended Kalman filters [12]. However, both methods are based on an estimate of the actual motion of the robot, which is used to bridge the time between the recognition of external references such as the beacons or the edges of the field [10]. This estimate is based on odometry. However, in contrast to wheeled motion, the four-legged robots typically use *forward-odometry*, i. e., the actual motion of the robot is not measured, but instead it is assumed that the motion requested is actually performed by the system. In fact, the walking engine is manually calibrated to execute the requested motions as precise as possible. Hence, forward-odometry just means to sum up the requested speeds for forward, sideward, and rotational motion, because it is assumed that this will also be the resulting motion. Although this assumption is too optimistic anyway, it is quite obvious that it will certainly be false if new walk parameters are tested. Therefore, using traditional self-localization methods to judge the quality of a new gait will fail.

Using an External Sensing System. Another possibility would be to use an external sensing system to measure the actual motion of the robot. This can either be an overhead camera as used in the Small-Size League, or a laser scanner, as used in [11]. However, both methods are not suitable for training at the actual competition site, because there may be no possibility to mount a camera above the Legged League field, and the fields cannot be reserved long enough to be exclusively used for training with a laser scanner that must always be able to see the robot.

Using Proprioception. The solution is to use real odometry. The Sony Aibo robot is equipped with sensors that can detect whether a foot touches the ground or not. In addition, it can measure the actual angles of all joints. Thus using forward kinematics, the actual motion of the feet can be calculated, especially the part of it, in which they have contact to the ground. If this is done for at least two feet, the actual motion of the robot can be reconstructed, as long as the feet do not skid over the ground. It is assumed that skidding feet will result in an uneven walk, which can be detected by another proprioceptive system of the Aibo: its acceleration sensors in x, y, and z direction.

2.2   Odometry

Since gaits using the typical PWalk-style originally introduced by the team from the University of New South Wales [2] are state of the art, such a kind of walk is also used for the work presented in this paper (cf. Fig. 3). This is important, because with this kind of walk, only the contact sensors of the hind legs touch the ground, and therefore, only these two sensors can be used for calculating the motion of the robot. The current position of such a ground contact sensor can easily be determined based on the measurements of the angles of the three joints the Aibo has in each leg using forward kinematics. These angles are automatically determined by the Aibo every 8 ms. The three-dimensional positions (cf. Fig. 1b) for the left and right hind feet change over time with the motion of the legs and can be computed relative to some origin in the body of the robot. However, for odometry, only their lengthwise and sideward components are required. Therefore, p_t^{left} and p_t^{right} are only two-dimensional vectors.

a)   b)   c)   d)

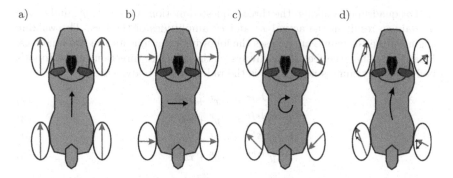

Fig. 2. Principle of treating legs as wheels [9]. Walking a) forwards, b) sideways. c) Turning. d) Turning while walking forward

Fig. 3. The Aibo ERS-210 walking with 311 mm/s

The overall offset a hind leg h has walked during a period of time $t_a \ldots t_b$ can be determined by summing up all the 2-D offsets between successive positions at which the foot had contact to the ground:

$$d^h_{t_a,t_b} = \begin{pmatrix} dx^h_{t_a,t_b} \\ dy^h_{t_a,t_b} \end{pmatrix} = \sum_{t=t_a+1}^{t_b} \begin{cases} p^h_t - p^h_{t-1} \,, & \text{if } g^h_t \wedge g^h_{t-1} \\ 0, & \text{, otherwise} \end{cases} \tag{1}$$

g just states whether the ground contact sensor of leg h was active at time t or not. Since each leg has sometimes contact to the ground and sometimes not while walking, it is important to always determine the distance with ground contact for a complete step phase. Otherwise, the distances measured by the left and the right feet are not comparable.

For quadruped walking, the three requested motion speeds x^r, y^r, and θ^r are overlaid to result in the actual x^h and y^h amplitudes of the legs. The walking engine of the GermanTeam [9] uses the following equations for the two hind legs. Please note that the feet move always in the opposite direction of the robot's body, so everything is negated (r is the radius to the body center):

$$x^{left} = -x^r + r\theta^r \tag{2}$$
$$y^{left} = -y^r + r\theta^r \tag{3}$$
$$x^{right} = -x^r - r\theta^r \tag{4}$$
$$y^{right} = -y^r + r\theta^r \tag{5}$$

For odometry, everything is the other way round. The foot motion is known, and the speed components have to be calculated. By transforming the motion equations, the measured walking speeds x^m, y^m, and θ^m can be calculated from the measured ground contact distances d^{left} and d^{right}. Please note that dx and dy measure only half of a step phase (while the corresponding foot touches the ground), so everything measured is multiplied by 2, i.e. all divisions by 2 are missing:

$$x^m_{t_a,t_b} = -\frac{dx^{right}_{t_a,t_b} + dx^{left}_{t_a,t_b}}{t_b - t_a} \tag{6}$$

$$y^m_{t_a,t_b} = -\frac{dy^{right}_{t_a,t_b} + dy^{left}_{t_a,t_b} + dx^{right}_{t_a,t_b} - dx^{left}_{t_a,t_b}}{t_b - t_a} \tag{7}$$

$$\theta^m_{t_a,t_b} = -\frac{dx^{right}_{t_a,t_b} - dx^{left}_{t_a,t_b}}{r(t_b - t_a)} \tag{8}$$

3   Evolutionary Algorithm

In evolutionary algorithms, a *population* of *individuals* is used to find a solution for a given problem. Each individual has a number of *genes* g_i that represent possible values for the variables of the problem. In each evolution step, the *fitness* f is determined for each individual, i.e., the current values of its genes are used to solve the problem and the solution is assessed by a so-called fitness function. Then, based on the current population and the fitness of its individuals, a new generation is created that replaces the current one. There are two major methods to generate individuals for the next population that can be combined in a lot of different ways: *mutation* and *crossover*. Mutation just adds noise to the genes of an individual and thereby creates a new one. In the crossover, the genes of two individuals are combined two a new one. There exist quite a lot of methods to select individuals for mutation or crossover [1].

In the work presented here, the genes represent the parameters of a gait. The fitness cannot directly be calculated, instead it has to be determined based on measurements, i.e. the gait parameter set has to be tested by letting the robot actually walk with these parameters. Thus, the fitness for a certain set of parameters has to be assumed to be noisy, so each time it will be measured it will be slightly different. This is especially true for learning omni-directional walking, because it is impossible to test all possible combinations of walking directions and rotation speeds. Therefore, the selection of individuals is based on a probabilistic approach. The ratio of an individual fitness to the overall fitness of the whole population is interpreted as the probability that the parameter set represented by this individual is the best one. So based on this interpretation, individuals are *drawn* from a population to form a new one. For each individual in the new population, two individuals are drawn from the previous one, and their genes are crossed. If a new population consists to more than 50% of a single individual from the previous population, every second individual is mutated.

The initial population is generated from a single individual (a hand-coded gait parameter set) by mutation. In addition, whenever the battery of the robot is empty, the best individual found so far is used to form the basis for the next evolution run.

Mutation. The mutation performed is rather simple. For each gene (i. e. gait parameter) a mutation radius is defined that has a reasonable size with respect to the possible range of that parameter. The genes are then mutated by adding random values within their mutation radius to them.

$$g_i^{new} = g_i^{old} random_{-r_i...r_i} \tag{9}$$

Crossover. As a new approach to evolutionary algorithms, there are two different methods for crossover: the *interpolation* and the *extrapolation*. They are used alternately, so one generation is created by interpolating between the individuals of the previous one, and the next one results from an extrapolation between the individuals of its predecessor. For the interpolating step, two individuals a and b are selected based on their fitness f and a new one is created by a random interpolation between their genes:

$$g_i^{new} = \beta g_i^a + (1 - \beta) g_i^b$$
$$\text{where } \alpha = random_{0...1} \tag{10}$$
$$\beta = \frac{\alpha f_a}{\alpha f_a + (1-\alpha)f_b}$$

While the idea of the interpolating step is to find a solution that is between two individuals, the extrapolating step shall find a solution that is outside. The difference between the genes of two individuals is used to extrapolate to one or the other side. The amount and direction of the extrapolation is again biased by the fitness of the two individuals selected:

$$g_i^{new} = \begin{cases} g_i^b + (1 - \beta)(g_i^a - g_i^b) \text{ , if } \beta < 1 \\ g_i^a + (\beta - 1)(g_i^b - g_i^a) \text{ , otherwise} \end{cases}$$
$$\text{where } \alpha = random_{0...1} \tag{11}$$
$$\beta = \frac{2\alpha f_a}{\alpha f_a + (1-\alpha)f_b}$$

That way, a solution can be found by alternately interpolating and extrapolating—without disturbing mutations. However, if a single individual is drawn too often, neither interpolation nor extrapolation can generate different sets of genes. Therefore, mutation is applied in such a case.

3.1   Fitness Function

The fitness function assessing the individual gait parameter sets is the sum of two values. On the one hand, it is determined, how good the motion performed matches the motion requested, and on the other hand, the measurements of the acceleration sensors are used to judge the smoothness of the gait.

Error Function. For each step, the motion actually performed is determined based on the calculations described in Section 2.2. However, instead of just using the odometry offset, a special kind of walk is forced by extending equation (1). In addition to determining the offset a foot moved on the ground, also the offset is calculated that the foot has moved while the *other foot* was on the ground. This second offset is negated and the average of both is used as a replacement for the original equation for the foot motion. This kind of fitness function enforces that when one foot is on the ground, the other one will swing in the opposite direction. Otherwise, the feet may waste time during their air phase by moving in the wrong direction.

$$d'^h_{t_a \ldots t_b} = \sum_{t=t_a+1}^{t_b} \frac{d^{h,h}_t - d^{h,other(h)}_t}{2}$$
$$\text{where } d^{h,h'}_t = \begin{cases} p^h_t - p^h_{t-1} \text{ , if } g^{h'}_t \wedge g^{h'}_{t-1} \\ 0 \qquad\quad \text{ , otherwise} \end{cases} \tag{12}$$

d' is used instead of d in the equations (6) to (8), resulting in the measured walking speeds x'^m, y'^m, and θ'^m. They are compared to the requested walking speeds x^r, y^r, and θ^r. These are determined by summing up all the requested speeds for the same whole step. The error w is determined as follows:

$$w_{t_a \ldots t_b} = |x'^m_{t_a \ldots t_b} - x^r_{t_a \ldots t_b}| + |y'^m_{t_a \ldots t_b} - y^r_{t_a \ldots t_b}| + \alpha |\theta'^m_{t_a \ldots t_b} - \theta^r_{t_a \ldots t_b}| \tag{13}$$

α is a weight that relates errors made in translation to the deviations in rotation. If more than one step is used for the assessment (which is normally the case), the average error is used.

Vibration. The acceleration sensors a^x, a^y, and a^z of the Aibo measure accelerations along the x, y, and z axes of the robot. The amount of vibration during a walk can easily be determined by calculating the standard deviation of the measurements:

$$v_{t_a \ldots t_b} = stdv(a^x_{t_a \ldots t_b}) + stdv(a^y_{t_a \ldots t_b}) + stdv(a^z_{t_a \ldots t_b}) \tag{14}$$

Fitness of an Individual. The overall fitness of an individual can then be computed as a sum of error and vibration. However, since small errors and few vibrations are desired, both values have to be negated. In addition, they have to be scaled to relate them to each other (by ϕ, β, and *gamma*):

$$f_{t_a \ldots t_b} = e^{\phi - \beta w_{t_a \ldots t_b} - \gamma v_{t_a \ldots t_b}} \tag{15}$$

4   Results

Experiments were conducted both with the ERS-210 and the ERS-7. As the ERS-210 was used as test bed in some recent work [5, 7, 6], a similar experiment is described here in detail. However, as another walking engine was used, the 23 parameters evolved are different. In fact, 26 parameters were evolved, but three of them are only relevant if the robot is also moving sideways or it is turning. In the final experiment conducted with the ERS-7, they were also used.

As in [6], a fixed half-ellipsoidal locus shape was used. The parameters are:

- The relative positions of the fore/hind feet as (x, y, z) offsets
- The height of the steps performed by the fore and hind legs
- The tilt of the locus (cf. Fig. 1b) of the fore and hind feet
- The timing of a step. The overall duration of a step is separated into a ground phase, a lift phase, an air phase, and a lowering phase. Since all phases sum up to 1, one of them can be calculated from the others and is left out. Please note that for the half-ellipsoidal shape of the locus used, the lifting and lowering phases are just waiting positions for the feet. As it will pointed out later, this does not mean that the feet actually stop during these phases, because the actual motion of the feet is different from the motion that they were requested for.
- The step size determines the amplitude of a step (here: only in x direction).
- The duration of a step (in seconds)
- The difference between the speed of the forefeet and the hind feed. In fact, this value modifies the amplitudes of the motion of the hind feet, so it is possible that the hind feet perform smaller steps that the front feet (here: only for x speed).
- A phase shift between the fore and the hind feet
- A body shift in x and y direction. This shifting is synchronous to the phases of walking, but also a phase shift can be defined. This allows the body to swing back and forth and left to right while walking.

4.1   ERS-210

The ERS-210 was trained to walk forward as fast as possible. As in [6], a population of 15 individuals was used. However, each parameter set was only tested for five seconds, the last three seconds of which were used to assess the performance of the resulting walk. The first two seconds were ignored to eliminate any influence of (switching from) the previous gait parameter set. The robot walked

Table 1. The parameters evolved for the ERS-210

Parameter	Initial Value	Best Value	Parameter	Initial Value	Best Value
Front locus			Rear locus		
z offset	76.85	58.97	z offset	108.72	111.54
y offset	78.1	67.24	y offset	76.95	58.09
x offset	55.88	63.26	x offset	-45.216	-35.96
step height	5.0	6.14	step height	24.0	27.27
tilt	-0.25	-0.48	tilt	0.05	0.12
ground phase	0.5	0.64	ground phase	0.5	0.34
lift phase	0.06	0.04	lift phase	0.06	0.19
lowering phase	0.06	0.12	lowering phase	0.06	-0.02
Step			Body shift		
size	76.0	89.77	x ratio	0	1.12
duration	0.64	0.512	y ratio	0	-0.10
Rear to Front offsets			phase offset	0	0.35
x speed ratio	1.1	1.0			
phase shift	0	-0.012			

from goal to goal and was turned manually. To avoid assessing the time when the robot was picked up, learning was interrupted when the back switch of the robot was pressed, and the assessment of the current individual was restarted when the button was released. To match the conditions in the related work, the robot's head did not move.

Learning was started with a variation of the hand-tuned gait of the German-Team 2003 [9]. The middle position of the hind feet was manually moved to the front to let the foot sensors have contact to the ground. The initial values of all parameters are given in Table 1. From this initial setup, the gait was learned automatically. After about three hours, the fastest gait found by the robot had a speed of 311 ± 3 mm/s (13.5 s for 4.2 m from goal line to goal line on a Sony Legged League field), which is also the fastest gait found so far for the Aibo ERS-210. In addition, the gait is quite smooth, i.e. the head of the robot does not shake significantly. Under [8] two videos can be downloaded showing the Aibo walking from the front and from the side. Six images from the latter video are depicted in Figure 3, showing a single step phase.

In addition, Figure 4 shows the trajectories of the feet and the knees of the robot. Please note that for the front legs (on the right in both subfigures), the robot is walking on its knees. Thus the feet never have contact to the ground. In [5, 7] it is analyzed, which shape of a foot trajectory is optimal. However, the resulting trajectories of the feet were never investigated in that work. Figure 4 shows that there is a severe difference between the shape of the locus that was desired (half-elliptical), and the resulting shape. While the knees follow the given trajectory quite well (in fact they never reach the maximum front point and overshoot the maximum back point), the feet have very significant deviations. The resulting trajectory of the hind feet can be explained by over- and

a)   b)

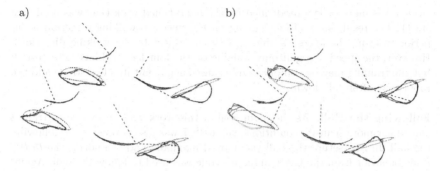

Fig. 4. The motion of the legs when moving with 311 mm/s (the orientation of the robot is the same as in Fig. 3). The dashed lines show a single snapshot of the positions of the legs. The dotted curves depict the trajectory of the feet and the knees as sent to the robot. The solid curves visualize the motion of the knees and the feet as measured by the joint sensors. a) Trajectories of a picked-up robot. b) Trajectories when walking on the ground. When ground contact was detected, the solid curve is drawn in gray

undershooting the given trajectory (it overshoots less when the robot is picked up, cf. Fig. 4a), but the control system seems to have real problems in following the desired loci of the forefeet.

4.2   ERS-7

All other experiments were conducted with the new Aibo model ERS-7, because it will be used in RoboCup 2004. However, the results are preliminary, because the experiments were only conducted with the beta version of the control software at a stage in which also the porting of the code from the ERS-210 to the ERS-7 was not completely finished. A major problem with the ERS-7 was that it switched itself off if there was too much load on one of its legs, and that this condition could easily be reached. These problems were only solved in the second release of the final version of the operating system, which came to late to repeat the experiments for this paper. So training the ERS-7 was a little bit complicated, because sometimes it had to be restarted after a few minutes. However, quite impressive results were achieved.

Walking Straight Ahead. The experiment conducted with the ERS-210 was repeated with the ERS-7. The robot is much more powerful, so it was able to reach 400±3 mm/s using a gait with 2.3 steps/s. The general appearance of the gait is quite similar to the gait of the ERS-210.

Walking Straight Backward. Then, the ERS-7 was trained to walk straight backward using the same experimental setup as before. The resulting gait performs less steps per second (1.9) and the robot is significantly slower (294 mm/s). In this walk, the robot still walks on the knees of the fore legs and the hind feet (which was more or less predefined by the hand-tuned walk that was used as a start). The result shows that walking on the knees in walking direction is superior to using the knees as hind legs (with respect to the walking direction). However, the speed of a gait only walking on the four feet still has to be tested, but the training may be complicated because such a gait it is less stable and the robot may often fall down.

Following the Ball. As the main goal of this work was to be able to train a gait under more realistic conditions, the ERS-7 was also trained while following the ball. In this setup, the ball was moved manually and the normal *go-to-ball* basic behavior from the GermanTeam's code was used to follow the ball. Again, each walking parameter set was switched every five seconds, and again, only the last three seconds were assessed. Learning was only interrupted when the motion requested by *go-to-ball* was too slow, because in such cases any kind of walk parameter set would get a good assessment. The ball was moved in a way that resulted in a good mixture between walking straight (less often) and turning for the ball (more often, due to the small field). The gait parameters were initialized with the ones from the fastest straight ahead gait of the ERS-7. After about of three hours of training, the Aibo had learned a gait with 3.1 steps/s, a maximum forward speed of 331 mm/s, and a rotation speed of 195°/s. The sideward and backward speeds were low, because the *go-to-ball* behavior seems not the use these motion directions. Due to the very high step frequency, the gait is extremely reactive.

5   Conclusions and Future Work

This paper presented a new approach to optimize gait parameter sets using evolutionary algorithms. On the one hand, it is the first method that is potentially suitable to learn appropriate parameter sets for any motion required in RoboCup, on the other hand, separating the crossover-step of the evolutionary algorithm into interpolation and extrapolating proved to be a good means to optimize the parameter sets. As a result, the fastest gait of the Aibo ERS-210 known so far has been found, outperforming the runner-up by 5%. In addition, preliminary experiments with the new Aibo ERS-7 also showed promising results.

However, as has already pointed out in [5], the more general a gait is, the less fast it will be. Therefore, several parameter sets have to be combined, each for a certain range of walking directions and rotations. This way, the gaits evolved for the ERS-7 were combined and used by the *Bremen Byters* at the RoboCup German Open 2004. They turned out to be the fastest gaits for forward motion, backward motion, and rotation in the whole competition, although other teams had also experimented with evolutionary gait optimization. In addition, the odometry calculation presented in this paper replaced the original forward odometry, because it scales better with different walking speeds.

However, since they were trained independently from each other, the transition between them is not guaranteed to be smooth, i. e. the robot may stumble. For instance, although they have a different step frequency, switching between the fast forward gait and the ball follower gait of the ERS-7 seems to work quite fine, while the transition between the ball follower (forward) gait and the backward gait is uneven. Therefore, several parameter sets for different directions/rotations have to be learned at once with interpolations between neighboring sets. Thus the transitions between the sets should be smooth, but it is still possible to have specialized sets for certain motions such as walking straight ahead.

Acknowledgments

The author thanks all members of the GermanTeam for providing the basis for this research, especially Uwe Düffert and Max Risler for writing the walking engine this work is based on.

References

1. J. E. Baker. Adaptive selection methods for genetic algorithms. In J. J. Grefenstette, editor, *Proceedings of an International Conference on Genetic Algorithms and their Application*, pages 101–111, Hillsdale, New Jersey, USA, 1985. Lawrence Erlbaum Associates.
2. B. Hengst, D. Ibbotson, S. B. Pham, , and C. Sammut. Omnidirectional motion for quadruped robots. In A. Birk, S. Coradeschi, and S. Tadokoro, editors, *RoboCup 2001: Robot Soccer World Cup V*, number 2377 in Lecture Notes in Artificial Intelligence, pages 368–373. Springer, 2001.
3. G. Hornby, S. Takamura J. Yokono, O. Hanagata, T. Yamamoto, and M. Fujita. Evolving robust gaits with Aibo. In *IEEE International Conference on Robotics and Automation 2000 (ICRA-2000)*, pages 3040–3045, 2000.
4. G. S. Hornby, M. Fujita, S. Takamura, T. Yamamoto, and O. Hanagata. Autonomous evolution of gaits with the sony quadruped robot. In W. Banzhaf, J. Daida, A. E. Eiben, M. H. Garzon, V. Honavar, M. Jakiela, and R. E. Smith, editors, *Proceedings of the Genetic and Evolutionary Computation Conference*, volume 2, Orlando, Florida, USA.
5. M. S. Kim and W. Uther. Automatic gait optimisation for quadruped robots. In J. Roberts and G. Wyeth, editors, *Proceedings of the 2003 Australasian Conference on Robotics and Automation*, Brisbane, Australia, 2003.
6. N. Kohl and P. Stone. Policy gradient reinforcement learning for fast quadrupedal locomotion. In *Proceedings of the IEEE International Conference on Robotics and Automation*, May 2004. to appear.
7. M. J. Quinland, S. K. Chalup, and R. H. Middleton. Techniques for improving vision and locomotion on the sony aibo robot. In J. Roberts and G. Wyeth, editors, *Proceedings of the 2003 Australasian Conference on Robotics and Automation*, Brisbane, Australia, 2003.
8. T. Röfer. Videos of Aibo ERS-210. Only available under "Images and Videos" of http://www.robocup.de/bremenbyters.

9. T. Röfer, H.-D. Burkhard, U. Düffert, J. Hoffmann, D. Göhring, M. Jüngel, M. Lötzsch, O. von Stryk, R. Brunn, M. Kallnik, M. Kunz, S. Petters, M. Risler, M. Stelzer, I. Dahm, M. Wachter, K. Engel, A. Osterhues, C. Schumann, and J. Ziegler. GermanTeam, 2003. Technical report, only availabe at http://www.robocup.de/germanteam.

10. T. Röfer and M. Jüngel. Vision-based fast and reactive Monte-Carlo localization. In *Proceedings of the IEEE International Conference on Robotics and Automation (ICRA-2003)*, pages 856–861, Taipei, Taiwan, 2003. IEEE.

11. T. Röfer and M. Jüngel. Fast and robust edge-based localization in the Sony four-legged robot league. In *International Workshop on RoboCup 2003 (Robot World Cup Soccer Games and Conferences)*, Lecture Notes in Artificial Intelligence, Padova, Italy, 2004. Springer. to appear.

12. C. Sammut, W. Uther, and B. Hengst. rUNSWift 2003. In *International Workshop on RoboCup 2003 (Robot World Cup Soccer Games and Conferences)*, Lecture Notes in Artificial Intelligence, Padova, Italy, 2004. Springer. to appear.

Optic Flow Based Skill Learning for a Humanoid to Trap, Approach to, and Pass a Ball

Masaki Ogino[1], Masaaki Kikuchi[1], Jun'ichiro Ooga[1], Masahiro Aono[1], and Minoru Asada[1,2]

[1]Dept. of Adaptive Machine Systems,
[2]HANDAI Frontier Research Center,
Graduate School of Engineering, Osaka University
{ogino, kikuchi, ooga, aono}@er.ams.eng.osaka-u.ac.jp
asada@ams.eng.osaka-u.ac.jp

Abstract. Generation of a sequence of behaviors is necessary for the RoboCup Humanoid league to realize not simply an individual robot performance but also cooperative ones between robots. A typical example task is passing a ball between two humanoids, and the issues are: (1) basic skill decomposition, (2) skill learning, and (3) planning to connect the learned skills. This paper presents three methods for basic skill learning (trapping, approaching to, and kicking a ball) based on optic flow information by which a robot obtains sensorimotor mapping to realize the desired skill, assuming that skill decomposition and planning are given in advance. First, optic flow information of the ball is used to predict the trapping point. Next, the flow information caused by the self-motion is classified into the representative vectors, each of which is connected to motor modules and their parameters. Finally, optical flow for the environment caused by kicking motion is used to predict the ball trajectory after kicking. The experimental results are shown and discussion is given with future issues.

1 Introduction

Recent progresses of humanoid robots such as ASIMO [3], QRIO [6], HOAP [8], and MORPH [2] have been attracting many people for their performances of human like behaviors. However, they are still limited to very few behaviors of individuals such as walking and so on. In order to extend the capability of humanoids, various kinds of behaviors with objects or other agents should be developed. RoboCup has been providing an excellent test-bed for such a task domain, that is, ball operation and cooperation with teammates (competition with opponents) [5]. Towards the final goal of RoboCupSoccer, the humanoid league has been held since 2002 in Fukuoka, and several technical challenges such as standing on one leg, walking, and PK have been attacked. However, the level of the performance is still far from the roadmap to the final goal [4]. Further, many teams developed humanoid behaviors based on the designers knowledge on the environment, and therefore seem to be brittle against the environmental

D. Nardi et al. (Eds.): RoboCup 2004, LNAI 3276, pp. 323–334, 2005.

changes. It is expected that a robot obtains the environmental model through the interactions with its environment.

Optical flow has been used to learn the sensorimotor mapping for obstacle avoidance planned by the learned forward model [7] or by finding obstacles that show different flows from the environments using reinforcement learning [9]. In addition, it is used for object recognition by active touching [1]. In these studies, the number of DoFs is much fewer than humanoids; therefore, it seems difficult to apply their methods to realize various kinds of humanoid behaviors. Especially, generation of a sequence of behaviors is very hard but necessary for the RoboCup Humanoid league to show not simply an individual robot performance but also cooperative ones between two robots. In the latter case, the following issues should be considered:

1. decomposition into basic skills,
2. basic skill learning, and
3. switching the learned skills to generate a sequence of behaviors.

A typical example task is passing a ball between two humanoids (face-to-face pass). Since attacking all of these issues together is so difficult, we focus on the second issue, and present three methods for basic skill learning (trapping, approaching to, and kicking a ball) based on optic flow information by which a robot obtains sensorimotor mapping to realize the desired skill, assuming that skill decomposition and planning are given in advance. First, optic flow information of the ball is used to predict the trapping point. Next, the flow information caused by the self-motion is classified into the representative vectors, each of which is connected to motor modules and their parameters. Finally, optical flow for the environment caused by kicking motion is used to predict the ball trajectory after kicking. The experimental results are shown and discussion is given with future issues.

2   Task, Robot, and Environment

2.1   Robots Used

Fig. 1 shows biped robots used in the experiments, HOAP-1, HOAP-2, and their on-board views. HOAP-1 is 480 [mm] in height and about 6 [kg] in weight. It has a one-link torso, two four-link arms, and two six-link legs. The other robot, HOAP-2, is a successor of HOAP-1. It is 510 [mm] in height and about 7 [kg] in weight. It has two more joints in neck and one more joint at waist. Both robots have four force sensing registors (FSRs) in their foots to detect reaction force from the floor and a CCD camera with a fish-eye lens or semi-fish-eye lens.

These robots detect objects in the environments by colors. In this experiment, a ball is colored orange, and the knees of the opponent robot are colored yellow. The centers of these colored regions in the images are recorded as the detected position.

Fig. 1. HOAP-1 with fish-eye lens and HOAP-2 with semi-fish-eye lens

2.2   Task and Assumptions

"Face-to-face pass" can be decomposed into a sequence of different behaviors:

- trapping a ball which is coming to the player,
- approaching to kick a trapped ball, and
- kicking a ball to the opponent.

All these basic behaviors need the appropriate relationship between motion parameters and the environment changes. For example, to trap a ball appropriately, the robots must estimate the arrival time and position of the coming ball. To approach to a kicking position, the robot should know the causal relationship between the walking parameters and the positional change of the objects in its image. Further, to kick a ball to the opponent, the robot must know the causal relationship between the kicking parameters and the direction the kicked ball will go.

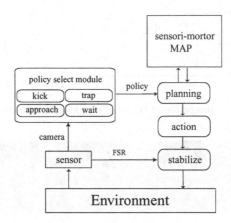

Fig. 2. A system overview

Moreover, basic skills to realize these behaviors should be activated at the appropriate situations. Here, the designer determines these situations to switch the behaviors, and we focus on the skill learning based on optic flow information. Fig. 2 shows an overview of our proposed system.

3   Skill Learning Based on Optic Flow Information

3.1   Ball Trapping

Fig. 4 shows the trapping motion by HOAP-1 acquired by the method described below. In order to realize such a motion, the robot has to predict the position and the arrival time of a ball from its optical flow captured in the robot view. For that purpose, we use a neural network which learns the causal relationship between the position and optical flow of the ball in visual image of a robot and the arrival position and time of the coming ball. This neural network is trained by the data in which a ball is thrown to a robot from the various positions. Fig. 3 shows several prediction results of the neural network after learning. Δx [pixel] and Δt [sec] indicates the errors of the arrival position and the time predicted at each point in the robot's view. Based on this neural network, the robots can activate the trapping motion module with the appropriate leg (right or left) at the appropriate timing (Fig. 4).

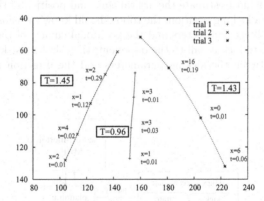

Fig. 3. The prediction of the position and time of a coming ball

Fig. 4. An experimantal result of a trapping skill

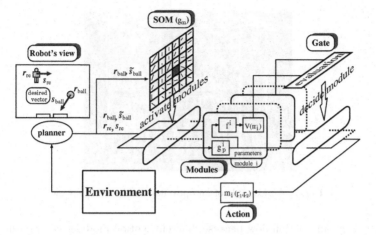

Fig. 5. An overview of the approaching skill

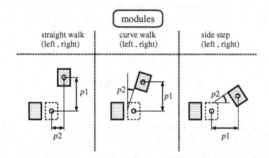

Fig. 6. Motion modules and parameters for approaching

3.2   Ball Approaching

Approaching to a ball is the most difficult task among the three skills because this task involves several motion modules each of which has parameters to be determined. These motions yield various types of image flows depending on the values of the parameters which change continuously. We make use of environmental image flow pattern during various motions to approach to the ball.

Let the motion flow vector Δr at the position r in the robot's view when a robot takes a motion, a. The relationships between them can be written,

$$\Delta r = f(r, a), \tag{1}$$
$$a = g(r, \Delta r). \tag{2}$$

The latter is useful to determine the motion parameters after planning the motion path way in the image. However, it is difficult to determine one motion to

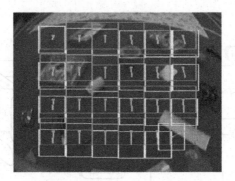

Fig. 7. An example of an optic flow in the robot's view

realize a certain motion flow because different motion modules can produce the same image flow by adjusting motion parameters. Therefore, we separate the description of the relationship between the motion and the image flow into the relationship between the motion module and the image flow, and the relationship between the motion parameters in each module and the image flow (Fig. 5), as follows.

$$m_i = g_m(\mathbf{r}, \Delta\mathbf{r}), \tag{3}$$

$$\mathbf{a}^i = (p_1^i, p_2^i)^T = g_p^i(\mathbf{r}, \Delta\mathbf{r}) \tag{4}$$

$$\Delta\mathbf{r} = f^i(\mathbf{r}, \mathbf{a}^i), \tag{5}$$

where m_i is the index of the i-th motion module and $\mathbf{a}_i = (p_{i1}, p_{i2})^T$ are the motion parameter vector of the i-th motion module. In this study, the motion modules related to this skill consists of 6 modules; *straight walk* (left and right), *curve walk* (left and right), and *side step* (left and right). Each of the modules has two parameters which have real values, as shown in Fig. 6.

Given the desired motion pathway in the robot's view, we can select appropriate module by g_m, and determine the motion parameters of the selected motion module by g_p^i based on the learned relationships among the modules, their parameters, and flows. If the desired image flow yields several motion modules, the preferred motion module is determined by value function.

Images are recorded every step and the image flow is calculated by block matching between the current image and the previous one. The templates for calculating flows are 24 blocks in one image as shown in Fig. 7.

g_m. All of the data sets of the flow and its positional vector in the image, $(\mathbf{r}, \Delta\mathbf{r})$, are classified by the self organizing map (SOM), which consists of 225 (15×15) representational vectors. After the organization, the indices of motion modules are attributed to each representational vector. Fig. 8 shows the classified image vector (the figure at the left side) and the distribution of each module in SOM. This SOM outputs the index of appropriate motion module so that the desired flow vector in the image is realized.

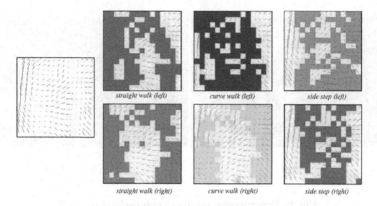

straight walk (left)   curve walk (left)   side step (left)

straight walk (right)   curve walk (right)   side step (right)

Fig. 8. Distribution of motion modules on the SOM of optic flows

f^i, g_p^i. The forward and inverse functions that correlate the relationship between the motion parameters in each module and the image flow, f^i, g_p^i, are realized by a simple neural network. The neural network in each module is trained so that it outputs the motion parameters when the flow vector and the positional vector in the image are input.

Plannning and Evaluation Function. In this study, the desired optic flow in the robot's view for the ball and the receiver, s_{ball}, s_{re}, are determined as a vector from the current position of a ball to the desired position (kicking position) in the robot's view, and as the horizontal vector from the current position to the vertical center line, respectively. The next desired optic flow of a ball to be realized, $\tilde{s}_{ball}$, is calculated based on these desired optic flows,

$$n_{step} = \|s_{ball}\|/\Delta r_{max}, \tag{6}$$

$$\tilde{s}_{ball} = s_{ball}/n_{step}, \tag{7}$$

where Δr_{max} is the maximum length of the experienced optical flow. This reference vector is input to the module selector, g_m, and the candidate modules which can output the reference vector are activated. The motion parameters of the selected module are determined by the function g_p^i,

$$a^i = g_p^i(r_{ball}, \tilde{s}_{ball}), \tag{8}$$

where r_{ball} is the current ball position in the robot's view. When the module selector outputs several candidates of modules, the evaluation function depending on the task, $V(m_i)$, determines the preferred module. In this study, our robots have to not only approach to a ball but also take an appropriate position to kick a ball to the other. For that, we set the evaluation function as follows,

$$selected\ module = \underset{i \in modules}{\arg\min}\Big[\|\tilde{s}_{ball} - f^i(r_{ball}, a^i)\| + k\|s_{re} - n_{step}f^i(r_{re}, a^i)\|\Big],$$

$$\tag{9}$$

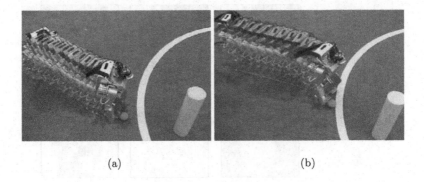

(a)	(b)

Fig. 9. Experimental results of approaching to a ball

where k is the constant value, and r_{re} is the current position of the receiver in the robot's view.

Fig. 9 shows experimental results of approaching to a ball. A robot successfully approaches to a ball so that the hypothetical opponent (a pole) comes in front of it.

3.3   Ball Kicking to the Opponent

It is necessary for our robots to kick a ball to the receiver very precisely because they cannot sidestep quickly. We correlate the parameter of kicking motion with the trace of the kicked ball in the robot's view so that they can kick to each other precisely. Fig. 10 shows a proposed controller for kicking.

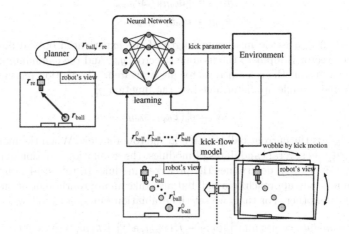

Fig. 10. The system for kicking skill

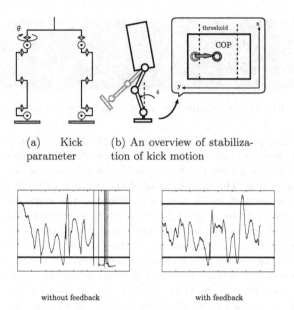

(a)   Kick parameter

(b) An overview of stabilization of kick motion

without feedback

with feedback

(c) The trajectories of CoP of the support leg during kicking motion

Fig. 11. The parameter and the stabilization of kicking

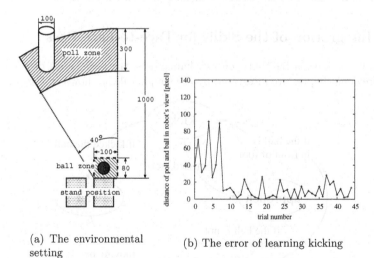

(a) The environmental setting

(b) The error of learning kicking

Fig. 12. The environmental setting and the learning curve for kicking

Fig. 13. An experimental result of kicking a ball to the pole

The kicking parameter is the hip joint angle shown in Fig. 11(a). The quick motion like kicking changes its dynamics depending on its motion parameter. The sensor feedback from the floor reaction force sensors is used for stabilizing the kicking motion. The displacement of the position of the center of pressure (CoP) in the support leg is used as feedback to the angle of the ankle joint of the support leg (see Fig. 11(b)). Fig. 11(c) shows the effectiveness of the stabilization of the kicking motion. The initial ball position and the parameter of the kicking motion affects sensitively the ball trace in the robot's view. To describe the relationship among them, we use a neural network, which is trained in the environment where the pole (10 [cm]) is put about 1 [m] in front of the robot (Fig. 13(a)). The trace of the ball (the effects of the self motion is subtracted) is recorded every 100 [msec], and the weights in the neural network are updated every one trial. Fig. 13(b) shows the time course of error distance between target pole position and kicked ball in the robot's view. It shows that the error is reduced rapidly within 20 [pixel], which is the same size of the width of the target pole. Fig. 13 shows the kicking performance of the robot.

4 Integration of the Skills for Face-to-Face Pass

To realize passing a ball between two humanoids, the basic skills described in the previous chapter are integrated by the simple rule as shown in Fig. 14.

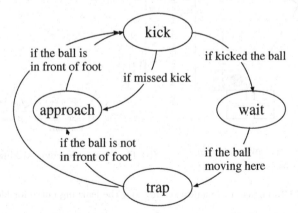

Fig. 14. The rule for integrating motion skills

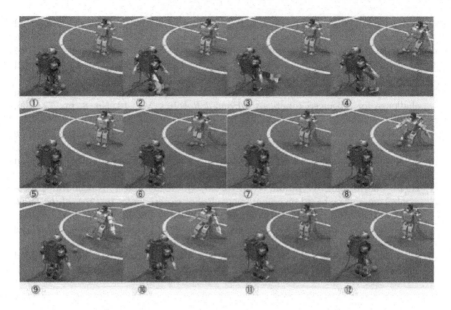

Fig. 15. An experimental result of passes between two humanoids

Fig. 15 shows the experimental result. Two humanoids with different body and different camera lens realize the appropriate motions for passing a ball to each other based on their own sensorimotor mapping. The passing lasts more than 3 times.

5   Conclusions

In this paper, acquiring basic skills for passing a ball between two humanoids is achieved. In each skill, optic flow information is correlated with the motion parameters. Through this correlation, a humanoid robot can obtain the sensorimotor mapping to realize the desired skills. The experimental results show that a simple neural network quickly learns and models well the relationship between optic flow information and motion parameters of each motion module. However, there remain the harder problems we skip in this paper. First is skill decomposition problem; it is concerning about how to determine what are the basic skills for the given task. Second is planning, that is how to organize each motion module to achieve the given task. In this paper, we assume skill decomposition and planning are given in advance. Combining the learning in each skill level with that in higher level is the next problem for us.

References

1. P. Fitzpatrick. First Contact: an Active Vision Approach to Segmentation, In *Proc. of IEEE/RSJ Int. Conf. on Intelligent Robots and Systems*, pp. 2161-2166, 2003.

2. T. Furuta, Y. Okumura, T. Tawara, and H. Kitano, 'morph': A Small-size Humanoid Platform for Behaviour Coordination Research, In *Proc. of the 2001 IEEE-RAS Int. Conf. on Humanoid Robots, pp. 165-171*, 2001.
3. M. Hirose, Y. Haikawa, T. Takenaka, and K. Hirai, Development of Humanoid Robot ASIMO, In *Proc. Int. Conf. on Intelligent Robots and Systems*, 2001.
4. H. Kitano and M. Asada, The RoboCup humanoid challenge as the millennium challenge for advanced robotics, *Advanced Robotics*, Vol. 13, No. 8, pp. 723-736, 2000.
5. H. Kitano, RoboCup-97: Robot Soccer World Cup I, *Springer, Lecture Note in Artificial Intelligence 1395*, 1998.
6. Y. Kuroki, T. Ishida, and J. Yamaguchi, A Small Biped Entertainment Robot, In *Proc. of IEEE-RAS Int. Conf. on Humanoid Robot, pp. 181-186*, 2001.
7. K. F. MacDorman, K. Tatani, Y. Miyazaki, M. Koeda and Y. Nakamura. Proto-symbol emergence based on embodiment: Robot experiments, In *Proc. of the IEEE Int. Conf. on Robotics and Automation, pp. 1968-1974*, 2001.
8. Y. Murase, Y. Yasukawa, K. Sakai, etc. Design of a Compact Humanoid Robot as a Platform. In *19th conf. of Robotics Society of Japan, pp. 789-790*, 2001.
9. T. Nakamura and M. Asada. Motion Sketch: Acquisition of Visual Motion Guided Behaviors. In *Proc. of Int. Joint Conf. on Artificial Intelligence, pp. 126-132*, 1995.

Learning to Kick the Ball Using Back to Reality

Juan Cristóbal Zagal and Javier Ruiz-del-Solar

Department of Electrical Engineering, Universidad de Chile,
Av. Tupper 2007, 6513027 Santiago, Chile
{jzagal, jruizd}@ing.uchile.cl
http://www.robocup.cl

Abstract. Kicking the ball with high power, short reaction time and accuracy are fundamental requirements for any soccer player. Human players acquire these fine low-level sensory motor coordination abilities trough extended training periods that might last for years. In RoboCup the problem has been addressed by engineering design and acceptable, probably sub-optimal, solutions have been found. To our knowledge the automatic development of these abilities has not been yet employed. Certainly no one is willing to damage a robot during an extended, and probably violent, evolutionary learning process in a real environment. In this work we present an approach for the automatic generation (from scratch) of ball-kick behaviors for legged robots. The approach relies on the use of UCHILSIM, a dynamically accurate simulator, and the *Back to Reality* paradigm to evolutionary robotics, a recently proposed method for narrowing the difference between simulation and reality during robot behavior execution. After eight hours of simulations successful ball-kick behaviors emerged, being directly transferable to the real robot.

1 Introduction

Kicking the ball with high power, short reaction time and accuracy are fundamental requirements for any soccer player. Until now all the four-legged teams use more or less the same method for kicking the ball, which works as follows: "First localize the robot with respect to the ball and targets, and then trigger some of the recorded ball kicks". We consider that this approach is quite restricted since first, it requires the ball to be placed relative to the robot into a discrete set of positions, the problem with this is that the robot should invest valuable time on repositioning itself; and second, there is only a discrete set of target directions to chose among. Human soccer players are able to kick the ball freely, i.e. from almost whatever relative position and with any target direction. Moreover within the four legged league we can notice a direct relation among the amount of ball-kicking alternatives of players and the team success in robot soccer.

However, some basic requirements are needed before starting to think on applying non-constrained ball-kick behaviors. First robots should be able to estimate with accuracy the relative ball position at the right instant of ball impact, and second, robots should be able to arbitrarily control their limbs with precision. We consider that these two requirements are currently fulfilled, at least by the best half portion of

D. Nardi et al. (Eds.): RoboCup 2004, LNAI 3276, pp. 335–346, 2005.
© Springer-Verlag Berlin Heidelberg 2005

the teams competing in the league; therefore it is a good time for further exploring this approach. Even more, this ability will be of fundamental importance when playing soccer with humanoid robots. Hopefully, in a near future we will see very effective fast reactive behaviors, such as a robot in motion kicking the ball while it is also in motion or kicking the ball while it is on air, as human players do.

Another important aspect that we aim to address here is learning. Human players acquire these fine low-level sensory motor coordination abilities trough extended training periods that might last for years. We believe that this type of reactive behavior can be derived from a tight interaction between the robot and the environment. There are several good examples in the literature of the successful generation of low level behaviors by learning from the real robot experience and practice [6][3][4]. However, learning to kick the ball with real legged robots seems to be an expensive and time-consuming task. We propose instead to use UCHILSIM, an accurate simulator, and the *Back to Reality* paradigm for combining virtual experiences with real ones. Our goal is to produce an optimal ball kick behavior for any given robot relative ball position and target direction that the robot might access without repositioning itself.

The remainder of this paper is organized as follows. In section 2 a short overview of RoboCup work related to this task is presented. In section 3 the general learning architecture that we have used is outlined. In section 4 the learning to kick the ball methodology is described. In section 5 the experiments are presented, and finally in section 6 the conclusions and projections of this work are presented.

2 Related Work

Kicking the ball is a key problem in RoboCup, however it has been addressed mainly at a high level within the RoboCup literature. Most of the work is concentrated on how to approach the ball and on how to choose the action to be triggered at the ball contact point. However some works are more related with low level behavior acquisition such as the work of Endo [1] in which a co-evolution of low level controller and morphology is proposed. The work of Hardt [2] surveys the role of motion dynamics in the design of control and stability of bipedal and quadrupeds robots. It is also related the work of Golubovic [5] which presents a software environment specially designed for gait generation and control design for AIBO robots. Certainly the work of Hornby [3] in the evolution of gaits for real legged robots is one important effort in order to produce low level behaviors from environmental interaction in a RoboCup directly related context. Beyond RoboCup the literature is quite rich in this subject, we recommend the reader to investigate further on the subject of evolutionary robotics [6].

3 The Learning Architecture

The experiments presented in this work are based on the recently proposed *Back to Reality* paradigm [7] to evolutionary robotics [6]. The main idea of this approach is to let the robots to behave both in their real environment and in a simulation of it. A fitness measure which tells the degree in which certain task is accomplished is computed after behavior execution. The difference in fitness obtained in simulation versus reality is minimized trough a learning process in which the simulator is adapted to reality. In other words the robot continuously generates an internal representation of their environment while adapting the simulator. In this way, as figure 1 shows, the simulation parameters are continuously tuned narrowing the reality-gap along the behavior adaptation process. The adaptation of the simulator is dependent on the particular behavior that is executed, thus the behavior defines the way in which the world is sampled or observed. During adaptation three learning process are taking place, first learning of the robot controller in the simulator, second learning of the robot controller in the reality, and finally learning of the simulator from real experiences. When the robot learns in reality, the controller is structurally coupled to the environment, while it learns in simulations the controller is structurally coupled with the simulator. Given these two couplings of the robot controller, it is necessary to conceive the robot as the union of the controller, the body and the simulator.

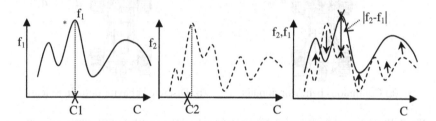

Fig. 1. How the reality-gap is narrowed by the Back to Reality paradigm (one iteration of the procedure is shown). The first/second curve shows the fitness curve f1/f2 over the controller parameter space in reality/simulation; the maximal value of fitness f1*/ f2* corresponds to the C1/ C2 controller. The last graph illustrates how the difference between both curves is narrowed during the evolutionary process; the fitness curve of the simulator, defined by the simulator parameters, converges towards the (real) fitness curve of reality, using the difference between the obtained real and virtual fitness as adaptation parameter

We will use UCHILSIM [8], a dynamically accurate simulator for performing the adaptation of the controller under a virtual environment. We use genetic algorithms for searching ball-kicks solutions over the space of controller parameters. A fitness measure f_1/f_2 will be used in order to measure the degree in which each individual solves the proposed task. Genetic algorithms will also be used for evolving the simulator parameters such that $|f_2-f_1|$ is minimized during the learning process.

4 Learning to Kick the Ball

We aim at learning the best ball-kick behavior that the AIBO robot is able to execute while introducing a minimum of designer bias. As figure 2 (a) shows, the desired behavior should operate over a continuum of robot accessible ball positions and target directions. We define a ball-kick behavior domain as a continuous three dimensional space composed by the dimensions of the starting ball distance d [14cm→18cm] measured from the robots neck, the robot relative ball angle [0°→180°] measured from the robots right side and the ball target direction defined by an angle [0°→180°] measured from the robots right side.

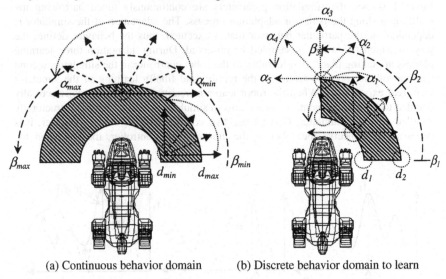

(a) Continuous behavior domain   (b) Discrete behavior domain to learn

Fig. 2. Illustration of the behavior domain defined by different robot relative ball positions and ball target angles. They are first shown (a) for the expected behavior to be generated and (b) for the discrete set of points that we consider for learning

In order to generate an optimal kick for each point (task) in this domain we will consider a discrete set of points for learning, as figure 2 (b) shows; two ball distances $\{d_1, d_2\}$, three relative angles $\{ _1, _2, _3\}$ and five target angles $\{ _1, _2, _3, _4, _5\}$, i.e. thirty points in total. We have selected only the right half portion of the range of since we expect solutions to be symmetric around = 90°. For each one of these thirty points, a specific ball-kick behavior will be generated by means of learning the set of parameters that define the corresponding sequence of robots limb displacements given rise to a ball-kick. The final continuous behavior for an arbitrary point, i.e. ball position $\{d, \}$ and target angle $\{ \}$, will be obtained by interpolating, in parameter space, the solutions which are obtained at the neighboring points. In all cases the idea is to maximize the distance traveled by the ball while maintaining high accuracy in the resulting shooting direction.

4.1 Parameterization of Ball Kick

The way in which the ball-kick is parameterized might be a strong source of designer bias, therefore care must be taken in order to provide flexibility. We will present a ball-kick parameterization which aims at serving as a compromise among simplicity for learning and flexibility for allowing several solutions. Under a physical perspective we should consider that a larger amount of energy might be transmitted to the ball when all the robot limbs are allowed to move. For example, the legs supporting the body can displace the torso and its resulting ground relative speed might be aggregated to the torso-relative speed of the limb extreme which finally kicks the ball. It is also important to provide freedom with respect to where on the ball surface the forces are applied. This freedom might allow the robot to discover, just as an example, that kicking the ball on its upper part is more efficient for gaining speed than doing that on their lower part (as it happens in the billiard game). Notice that the UCHILSIM simulator well suites for such high detail level.

The configuration sate of a limb is defined by the angular state of their three joins { , , }. The trajectory of a limb during a ball-kick might be described using hundreds of points in joint space, however we are analyzing only the case of fast reactive ball-kicks where no ball repositioning is required prior to ball-kick. For this case we estimate that it is sufficiently good to consider just four extreme points of displacements for each limb, which are (1) the starting limb configuration, (2) a limb configuration where energy is accumulated, (3) the limb configuration at the ball contact instant and (4) the ending limb configuration. In this case we consider the ending configuration to be the same as the starting limb configuration. Therefore for a given ball configuration and target angle a ball-kick correspond to the subsequent execution of the following four stages, which are depicted on figure 3:

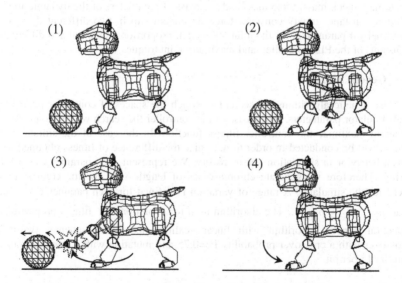

Fig. 3. Example of the subsequent configurations which are taken by one of the robot limbs during the execution of a ball-kick behavior. The limb starts from an equilibrium configuration (1), second it passes to a energy accumulation configuration (2), third it triggers the ball-kick (3) and finally the robot returns to the starting equilibrium configuration (4). Notice that for simplicity this illustration covers the action of just one limb, however the displacement of the four legs is considered in the presented experiments

1. The robot limbs are in their corresponding starting equilibrium configuration which we have selected to be the same as our walking equilibrium configuration.
2. A transition is taken from the equilibrium configuration of each limb to their corresponding energy accumulation configuration { $_{i1}$, $_{i1}$, $_{i1}$}, with i=[1,...,4].
3. Another transition is taken from the previous configuration to an energy liberation configuration (ball-kick instant configuration) { $_{i2}$, $_{i2}$, $_{i2}$}, with i=[1,...,4].
4. Finally the robot limbs are returned to their starting equilibrium configuration.

In this work 25 parameters should be obtained trough learning, they correspond to six parameters { $_{i1}$, $_{i1}$, $_{i1}$, $_{i2}$, $_{i2}$, $_{i2}$} for each one of the four robot legs, and one parameter which controls the speed of joints by incorporating certain amount of points in the trajectory which are passed as intermediate references to the joint servo motors. A larger amount of interpolation points makes slower the motion of legs.

4.2 Simulator Parameters

The robot simulator solves equations of motion derived from a Lagrange multiplier velocity based model [8]. We have selected a set of 12 parameters, which determine the simulator and robot model dynamics. These parameters include variables used for solving equations of rigid body dynamics and the PID controller constants used for modeling the leg joints. There are 4 parameters for the mass distribution in the robot: head mass, neck mass, torso mass and leg mass; 4 parameters of the dynamic model: friction constant, gravity constant, force dependent slip in two different directions; and finally 4 parameters for the joint leg model: proportional, integral and differential constants of the PID controller and maximum joint torque.

4.3 Genetic Search

We use genetic algorithms for searching trough the spaces of controller parameters and simulator parameters. In the case of the controller the search will be conducted in order to maximize the controller fitness function. In the case of the simulator the search will be conducted in order to minimize the difference of fitness obtained for a given behavior in simulation versus reality. We represent each parameter as a 8 bit string. Therefore we generate chromosomes of length 8x25 for the controller and 8x12 for the simulator. A range of variation is defined for each parameter p_i, such that p_i $p_{i\min}$, $p_{i\max}$. The algorithm used for evolution is a fitness-proportionate selection genetic algorithm with linear scaling, no elitism scheme, two-point crossover with a crossover probability Pc=0.75 and mutation with a mutation rate of Pm=0.015 per bit.

4.4 Definition of the Controller Fitness Function

For a given ball position and target angle we aim at finding a ball-kick which maximizes the ball displacement while maintaining high accuracy in the resulting direction. Deriving a good fitness function for this purpose might appear quite straight forward, however some considerations must be taken.

Equation 1 corresponds to the first fitness function that we have used during our experiments, it correspond to the measured Euclidean distance among the final ball position and the robot neck's position multiplied by the cosine of the error angle between the resulting direction vector and the target direction vector. This function is maximized when the error is zero.

$$fitness \quad d\cos(\ ) \tag{1}$$

However this function was shown to be useless for finding accurate ball-kick behaviors since genetic search first concentrates on finding individuals which are able to shoot the ball at large distances without having an accurate angle. The reason is that the cosine function is quite flat around zero. In order to solve this inconvenient we choose instead an exponential function as shown in equation 2. Using this function we observed that genetic search first concentrates on producing the right angle and then maximizes the distance.

$$fitness \quad de^{k^2} \tag{2}$$

4.5 Combining Solutions in Parameter Space

A different ball-kick behavior will be obtained for each one of the 30 points in the discrete behavior domain space. They will be obtained trough genetic search over the space of the 25 controller parameters using the back to reality approach. However the

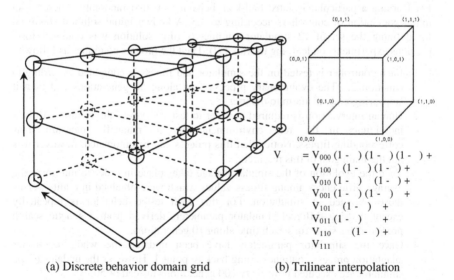

(a) Discrete behavior domain grid   (b) Trilinear interpolation

Fig. 4. Illustration of (a) the three dimensional behavior domain grid, each point in the grid represent the set of 25 parameters which are obtained trough genetic search, and (b) trilinear interpolation which is performed for obtaining the 25 parameters for an arbitrary point $\{d, \ ,\ \}$ in the behavior domain

desired behavior should perform the right ball-kick for an arbitrary point $\{d, \ , \ \}$ on the continuous behavior domain. For doing that we use trilinear interpolation over the obtained points. Figure 4 (a) shows an illustration of the three dimensional behavior domain grid which will be generated. Each corner in the grid contains the resulting 25 parameters of the best obtained behavior for the particular point. Figure 4 (b) shows the trilinear interpolation which is performed over each one of the 25 parameters in order to derive the corresponding interpolated value.

A requirement of this procedure in order to work is that neighboring grid solutions should produce a similar set of limb configurations. Otherwise the resulting interpolated ball-kick will not make sense. We measure this similarity as the Euclidean distance of the parameter vectors.

5 Experiments

In this section we present the experiments which were performed with the methodology presented in 4. First we present some examples of the evolution of fitness and resulting behaviors when learning a single isolated point in the behavior domain space. Finally we present different results on learning the whole discrete behavior domain grid and deriving a final overall continuous behavior.

5.1 Learning Isolated Points

For learning a particular isolated ball-kick behavior we first randomly generated an initial population of controllers according to 4.3. A known initial solution was used for defining the set of 12 simulator parameters (this solution was obtained form previous experiments on learning to walk [7]). Then the process continues as follows:

1. Each controller is tested on the simulator and fitness is computed according to equation 2. The evolutionary process runs along 50 generations and then it stops triggering an alarm to the user.
2. Human intervention is required in order to test a group of the resulting 10 best individuals in the real environment and to manually measure their corresponding fitness. Notice that this process can be automated, however it is not the main goal of this research.
3. Then the evolution of the simulator stage takes place in order to minimize the average differences among fitness values which were obtained in reality versus those obtained in simulation. For this each tested behavior is repeatedly executed using different simulator parameters derived from a genetic search process. This stage runs each time along 10 generations.
4. Once the simulator parameters have been updated the whole controller adaptation process continues going back to point 1. However the updates to the simulator are executed now every 100 generations instead of 50.

We ran this process for the 30 different points of the behavior domain grid that we aim at constructing. The above mentioned process was performed along 500 generations for each point. Figure 5 shows the evolution of fitness for the point $\{d=14cm, \ =90°, \ =90°\}$, i.e. the task was to kick the ball when it is placed at 14cm right in front of the robot neck with a frontal target direction. The vertical lines represent the interruption stages where the simulator was adapted using as feedback the fitness measures of the resulting 10 best behaviors on the real environment. It can be seen how updates in the simulator produce differences in the maximal and average fitness along the controller adaptation process. We observed that at the beginning of the adaptation process the behaviors obtained in the simulator where not directly transferable into reality, i.e. their corresponding fitness was quite different, however at the end of the process (beyond generation 300) the resulting behaviors where almost directly transferable to reality.

Similar observations were done in most of the remaining 29 points of the behavior domain, however learning was not successfully accomplished in some points like {$d=14cm$, =0°, =180°}. Certainly it is hard to imagine the robot performing a kick to the ball being placed at its right side with a left side target direction!. Some few neighbors to this point neither exhibit good results during the learning process. We consider that this is a design problem which can be easily solved by restricting the behavior domain. For the presented experiment we replaced the neighboring good solutions in those points where solution was not found (just 4 cases).

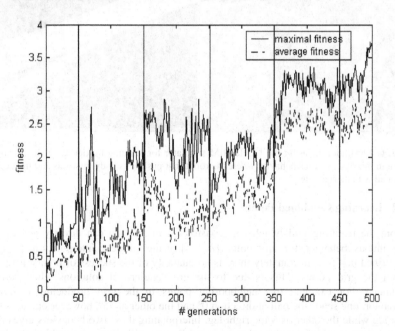

Fig. 5. Evolution of maximal and average fitness for the adaptation of the behavior domain point {$d=14cm$, =90°, =90°}, i.e. the task was to kick the ball when it is placed at 14cm right in front of the robot neck with a frontal target direction. The vertical lines represent the interruption stages where the simulator is adapted using the feedback obtained when measuring the fitness of resulting best behaviors on the real environment. It can be seen how updates in the simulator produce differences in the maximal and average fitness along the controller adaptation process

The resulting behaviors where quite interesting, even more if we consider that they were obtained from scratch. For example the robot learns that a way of shooting away the ball was to jump over it with its torso or with some legs as well. We obtained several behaviors were the concept of "accumulating energy" was heavily exploited. Some resulting ball-kick behaviors don't look very nice since the robot ends up in strange configurations, however they are really effective. Figure 6 shows a screenshot of UCHILSIM while learning a ball-kick behavior.

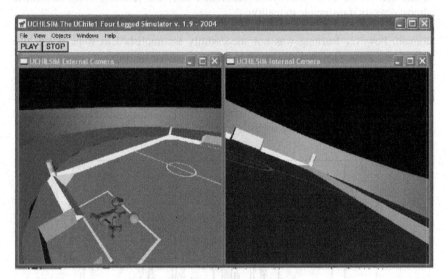

Fig. 6. Illustration of using the UCHILSIM simulator for learning to kick the ball. It can be seen in this case a curios behavior where the robot first pushes its body forwards and then kicks the ball with its right leg

5.2 Learning Combined Points

From the resulting grid which was constructed in 5.1 one can directly generate a continuous behavior for any point $\{d, , \}$ using trilinear interpolation as it is described in 4.5. Unfortunately there is no guaranty of the success of points different from the grid points. This is due to the great variety of solutions which where obtained. For example two neighboring points in the grid producing a similar behavior, one sends the ball at direction $0°$ and the other at $45°$, however one uses the left leg while the other uses the right leg. Interpolating these two behaviors gives rise to a behavior which let the robot to just fall over without even touching the ball. Under this scenario we visualize three alternatives: (1) to manually select similar solutions for the behavior grid disregarding at some point their optimality, (2) guide the evolutionary process by contaminating populations with manually selected individuals from neighboring points and (3) enforce similarity during the learning process by redefining the fitness function. We have produced a sufficiently good continuous overall behavior by using the second methodology. Starting from the point presented in 5.1 we contaminated their neighbor's populations with an 80% of individuals belonging to this successful point, and then we repeated the process over the complete domain. We really dislike this approach since it is quite time consuming and it goes against the philosophy of this learning approach, however it works well producing a continuous ball kick behavior over the entire domain.

6 Conclusions and Projections

It was presented a method for learning to kick the ball using virtual and real experience. The resulting behaviors are directly transferable from the simulator to the real robot. A few designer bias is introduced when (i) designing the ball-kick parameterization and (ii) when selecting some of the individuals for producing a continuum behavior. The resulting behavior allows kicking the ball with high power, short reaction time and accuracy over the entire domain. The employed method relies on the use of UCHILSIM, a dynamically realistic simulator, and the *Back to Reality* paradigm to evolutionary robotics. We remark that this method allowed us to learn from scratch a ball-kick which performs similarly as the most powerful ball-kick which is currently used in the league, i.e. when the robot leaves its body to fall over the ball (it was used for example by the German Team during the penalty definition against CMPack during RoboCup 2003). All independent ball kicks were obtained after just 8 hours of simulations. We should notice that the alternative is to let a student designing a ball-kick during a probably larger period of time. We are currently working on a method that automatically enforces similarity among behaviors in the grid. The fitness function contains a factor which depends on the Euclidean distance among the parameter vectors defining neighboring points. The key of this approach is to perform the search in parallel. That way allows us to maintain the 8 hours period of learning. Movies of some resulting behaviors, as well as a demo version of UCHILSIM are available at http://www.robocup.cl.

References

1. Endo, K., Yamasadki, F., Maeno, T., Kitano, H.:Co-evolution of Morphology and Controller for Biped Humanoid Robot. In: LNAI Proceedings of RoboCup 2002 Robot Soccer World Cup VI, Springer, pp. 327-339 (2002).
2. Hardt, M., Stryk, O.: The Role of Motion Dynamics in the Design, Control and Stability of Bipedal and Quadrupedal Robots. In: . In: LNAI Proceedings of RoboCup 2002 Robot Soccer World Cup VI, Springer, pp. 206-221 (2002).
3. Hornby, G.S., Fujita, M., Takamura, S., Yamamoto, T., Hanagata, O.: Autonomous Evolution of Gaits with the Sony Quadruped Robot. In: Proceedings of the Genetic and Evolutionary Computation Conference, Morgan Kaufmann, pp. 1297-1304 (1999).
4. Husbands, P., Harvey, I.: Evolution Versus Design: Controlling Autonomous Robots. In: Integrating Perception, Planning and Action: Proceedings of the Third Annual Conference on Artificial Intelligence, Simulation and Planning, IEEE Press, pp. 139-146 (1992).
5. Golubovic, D., Hu, H.:An Interactive Software Environment for Gait Generation and Control Design of Sony Legged Robots. In: LNAI Proceedings of RoboCup 2002 Robot Soccer World Cup VI, Springer, pp. 279-287 (2002).

6. Nolfi, S., Floreano, D.: Evolutionary Robotics – The Biology, Intelligence, and Technology of Self-Organizing Machines, In: Intelligent Robotics and Automation Agents. MIT Press (2000).
7. Zagal, J.C., Ruiz-del-Solar, J., Vallejos, P.: Back to Reality: Crossing the Reality Gap in Evolutionary Robotics. Proceedings of the IAV 2004, 5th IFAC Symposium on Intelligent Autonomous Vehicles, (in press), (2004).
8. Zagal, J.C., Ruiz-del-Solar, J.: UCHILSIM: A Dynamically and Visually Realistic Simulator for the RoboCup Four Legged League. Proceedings of the RoboCup 2004: Robot Soccer World Cup VIII, Springer (in this volume), (2004).

Cerebellar Augmented Joint Control for a Humanoid Robot

Damien Kee and Gordon Wyeth

School of Information Technology and Electrical Engineering,
University of Queensland, Australia

Abstract. The joints of a humanoid robot experience disturbances of markedly different magnitudes during the course of a walking gait. Consequently, simple feedback control techniques poorly track desired joint trajectories. This paper explores the addition of a control system inspired by the architecture of the cerebellum to improve system response. This system learns to compensate the changes in load that occur during a cycle of motion. The joint compensation scheme, called Trajectory Error Learning, augments the existing feedback control loop on a humanoid robot. The results from tests on the GuRoo platform show an improvement in system response for the system when augmented with the cerebellar compensator.

1   Introduction

Complex robots, with high degrees of freedom are becoming more common place in todays society. Robots with multiple limbs such as humanoids and octopeds are becoming more prominent in areas as varied as domestic robotics and all-terrain exploration. Such systems are difficult to model mathematically and hence analytical determination of feed forward dynamics for model based control can be both a complicated and time consuming process. In addition to this, these robots are progressively moving from a structured environment into regular society. Contact with the real world and human interaction further complicates the system loads.

Conversely, biological controllers do not use an accurate model of the system, rather incremental adjustment of control parameters is performed, based on the experience of the system. Initial response may be quite crude, but over time appropriate control parameters are learnt. Neural networks hold some promise in the field of trajectory control with the ability to learn system dynamic without explicit representation of a robots configuration.

This paper uses Trajectory Error Learning (TEL) [1], based on a CMAC neural network, to assist a conventional PI controller with trajectory tracking. The GuRoo humanoid robot with its high degree of freedom and non-linear dynamics forms a suitable platform to apply the system.

D. Nardi et al. (Eds.): RoboCup 2004, LNAI 3276, pp. 347–357, 2005.
© Springer-Verlag Berlin Heidelberg 2005

1.1   Previous Work

The use of a cerebellum models for motion control has been studied in the past. Infants of approximately 5 months of age display multiple accelerations and decelerations when moving an arm [2]. This series of sub-movements eventually guides the arm to the desired position. Over time, and with more experience, the child learns the required muscle movements to smoothly guide the arm. This shows that the human body is not born with a perfect plant model, but in fact learns it through experience.

Collins and Wyeth [3] used a CMAC to generate the required velocities needed for a mobile robot to move to a waypoint. Significant sensory delay was introduced that would cripple a traditional control system. The CMAC was able to learn the system dynamics, compensate for this delay and produce the required signals necessary to move to the waypoint with a smooth velocity profile.

Fagg et al [4] implemented a CMAC control system on a 2 degree of freedom arm, actuated by three opposing sets of muscles. The CMAC is responsible for the co-ordination of these three actuators to control the two joints. When the CMAC does not bring the arm to the required position, an additional external CMAC was engaged that produces short sharp bursts of motor activity until the target was reached. Once the desired position was reached, the trial was terminated and a new trial initiated. Over time, the external CMAC was made redundant as the original CMAC correctly learned the required muscle commands.

1.2   Paper Overview

Section 2 describes The GuRoo, the humanoid platform constructed at the University of Queensland, on which the research is applied. Section 3 outlines the CMAC neural network used as the basis for learning. Section 3 outlines the difficulty in using the current conventional control and described the application of Trajectory Error Learning (TEL). Section 5 describes the crouching experiment undertaken and presents results from before and after the implementation of the system. The final section draws conclusions from these results and discusses where these results may lead.

2   GuRoo Project

GuRoo is a fully autonomous humanoid robot (Figure 1) designed and built in the University of Queensland Robotics Laboratory [5]. The robot stands 1.2 m tall has a total mass of 34 kg, including on-board power and computation. GuRoo is currently capable of a number of demonstration tasks including balancing, walking, turning, crouching, shaking hands and waving.

The intended challenge for the robot is to play a game of soccer with or against human players or other humanoid robots. GuRoo has been designed to

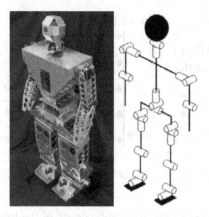

Fig. 1. The GuRoo humanoid robot with a schematic showing the degrees of freedom. In the cases where there are multiple degrees of freedom (for example, the hip) the joints are implemented through short sequential links rather than as spherical joints

Table 1. Type and axis of each DoF. "2 x" indicates a left and right side

Joint	Type	Axis	No.
Head/Neck	RC Servo	Pitch + Yaw	2
Shoulder	RC Servo	Pitch + Roll	2x2
Elbow	RC Servo	Pitch	2x2
Spine	DC Brushed	Pitch + Roll + Yaw	3
Hip	DC Brushed	Pitch + Roll + Yaw	2x3
Knee	DC Brushed	Pitch	2x1
Ankle	DC Brushed	Pitch + Roll	2x2
		TOTAL	23

mimic the human form and function to a degree, considering conflicting factors of function, power, weight, cost and manufacturability.

2.1   Electro-Mechanical Design

The robot has 23 joints in total. The legs and spine contain 15 joints that are required to produce significant mechanical power, most generally with large torques and relatively low speeds. The other 8 joints drive the head and neck assembly, and the arms with significantly less torque and speed requirements. Table 1 outlines the type and axis of actuation of each motor. Due the high power / low velocity nature of these joints, large gearboxes are used which contribute to the length of the actuators and hence the unnaturally wide legs. The centre of gravity of each leg lies outside the line of the hip rotation, and as such, the legs naturally swing inwards. The motors that drive the roll axis of the hip joints are each supplemented by a spring with a spring constant of 1 Nm/degree. These

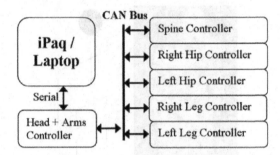

Fig. 2. Block diagram of the distributed control system

springs serve to counteract the natural tendency of the legs to collide, and help to generate the swaying motion that is critical to the success of the walking gait.

2.2   Distributed Control Network

A distributed control network controls the robot, with a central computing hub that sets the goals for the robot, processes the sensor information, and provides coordination targets for the joints. The joints have their own control processors that act in groups to maintain global stability, while also operating individually to provide local motor control. The distributed system is connected by a CAN network. In addition, the robot requires various sensor amplifiers and power conversion circuits.

2.3   Sensors

The GuRoo currently has encoders on each of the high powered DC motors, able to provide rotational position to a resolution of 0.001 of a degree. An inertial measurement unit consisting of 3 axis rate gyroscopes and 3 axis accelerometers has been obtained that is currently being integrated into the system. Provision has been made for the future inclusion of pressure sensors on the soles of the feet and a stereo vision system.

2.4   Software

The software consists of four main entities: the global movement generation code, the local motor control, the low-level code of the robot, and the simulator [6]. The software is organised to provide a standard interface to both the low-level code on the robot and the simulator. This means that the software developed in simulation can be simply re-compiled to operate on the real robot. Consequently, the robot needs a number of standard interface calls that are used for both the robot and the simulator.

3   CMAC Neural Network

The Cerebellar Model Articulated Controller or CMAC, was first described by Albus [7]. The CMAC network can be viewed as a number of lookup tables. Each table, or Association Unit (AU), has the dimensions equal to the number of input variables. Inputs to the system are quantized and scaled to create a global lookup address.

This address is mapped to a coarser address space in each AU where a weight is stored. The AUs are structured such that a single resolution change in one input signal will result in only one different weight chosen. The output signal is calculated by finding the sum of the weights of all AUs at this lookup address. As the output result is the sum of all association units weights, a greater number of association units results in a system that is better able to generalize the input space.

The input space is dominated by hyperplanes of plausible input combinations, with large empty spaces in each AU where real-life input combinations are not physically possible. Hashing techniques are used to reduce the memory requirements by mapping the global address space to a smaller, virtual, address space. The modulo function is the most simple way of achieving this. Hash collisions occur when two or more global address hash to the same virtual address. This is not necessarily fatal, as a large number of AUs will ensure the table weight in question to have a minor effect on the overall output.

Table weights are updated using the following rule:

$$\omega_{new} = \omega_{old} + \frac{\alpha}{\eta}(\theta_{des} - \theta_{act}) \tag{1}$$

where:

ω_{new} : New weight value
ω_{old} : Original weight value
α : Learning rate
η : Number of association units
θ_{des} : Desired joint position
θ_{act} : Actual joint position

As the output of the response of the network is the sum of the selected table weights, the change in weight between iterations is divided by the number of association units to ensure the learning rate has the same effect regardless of the number of AUs.

4   Trajectory Error Learning

Trajectory Error Learning (TEL) is a biologically inspired method of robot motion compensation where learning is driven from the difference between the intended trajectory of the robot and the actual trajectory measured by the feedback sensor (possibly after some sensory delay) [1]. This section illustrates how TEL can be applied to improve tracking performance in a humanoid robot.

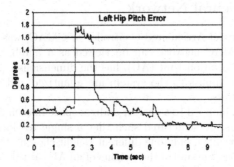

Fig. 3. Graph of the position error experienced in the left hip pitch joint over one complete walking cycle. The sudden increase in error after 2 seconds relates to single support phase of the gait

4.1   Joint Position Error

As can be seen in Figure 3, during the single support phase $(2 < t < 4)$, the joints are heavily loaded and experience significant position error. Conversely, during the swing phase each joint maintains positions adequately. It is these significant variations in load that prevent the PI control loop implemented on each of the GuRoos joints from obtaining a satisfactory response. Tests with gain scheduling techniques have, as yet, to provide any improvement in performance [8]. TEL uses a CMAC network to supply a compensating signal to eliminate this position error. As a typical walking gait of a humanoid is periodic in nature any errors experienced by the robot are also typically cyclic in nature: for example, the joint error during the support phase. By observing the gait phase, the CMAC learns which parts of the gait require compensation.

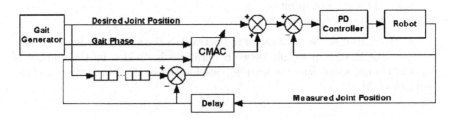

Fig. 4. System diagram. The Desired Joint Position is time delayed when calculating position error to account for sensor delay. Table weights are updated a set number of control loops after being used. This delay is equal to the sensory delay inherent in the system

4.2   System Implementation

The method of compensating the joint error is illustrated in Figure 4. The trajectory of the limb is expressed as a stream of desired joint positions which are generated by the gait generator. As the motion of the robot is periodic, the state of the trajectory can be expressed as the gait phase. The gait phase is implemented as a periodic counter, incrementing every control loop and resetting at the beginning of each motion cycle.

The desired joint position is augmented by the output of the CMAC, and passed to the feedback joint controller. The inputs to the CMAC consist of the gait phase and the measured joint position, where the measured joint position will be subject to some delay with respect to the desired joint position. In this form, the CMAC is used as a predictive modulator; seeking to eliminate the error it expects to see based on the errors that it has already seen at the same point in previous cycles. The sawtooth wave of the gait phase gives the CMAC the point in the cycle that it is currently compensating, while the measured joint position accounts for different disturbance conditions that may occur at that point in time.

The error in joint position is used to train the CMAC network. A history of previous desired position commands is kept to compensate for the sensory delay experienced in the system. This history buffer also ensures weight updates are performed with the correct time delayed error. The error signal used to train the CMAC is as follows:

$$\epsilon_k = \theta_{des(k-t)} - \theta_{act(t)} \tag{2}$$

where

ϵ_k = Error training signal(k)
$\theta_{des(k-t)}$ = Desired Joint Position(k - t)
$\theta_{act(k)}$ = Actual Joint Position(k)
t   = Sensory Delay

Thus weights are updated t control loops after they are used.

5   Crouching Experiment

The initial experiments that have been conducted using this method have been based on a slow crouching motion run over a period of 12 seconds. The pitch axis motors of the hip, knee and ankle joints follow a synchronised sinusoidal profile with a magnitude of 16, 35 and 22 degrees respectively to reach the bottom of the crouch position. This test exposes the joints to a range of dynamic and static loads, and can be repeated many times without moving the robot around the laboratory.

For this experiment, the following CMAC parameters were chosen. The number of receptive units and field width were chosen to provide the necessary discrimination, while also providing local generalisation. The hashing ratio was chosen to reduce memory requirements while still keeping a low probability of

Table 2. CMAC Parameters used for learning during the crouching experiment

CMAC Parameter	Value
Joint Position receptive units	9000
Gait Phase receptive units	1200
Field width (AU's)	50
Global Address Spaces	216204
Virtual Address Spaces	10001
Learning rate	0.001

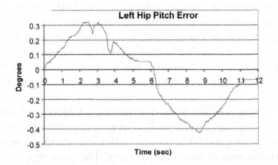

Fig. 5. Position error of the left hip pitch during a crouching motion. This signal is used to drive the learning in the CMAC network. Note the larger magnitude of error during the second half of the motion, as the robot accelerates itself upwards against gravity

hashing collisions. The learning rate was tuned to provide rapid learning of the compensation, without learning from noise. The measured joint positions were subject to a delay of 40 ms, which corresponds to a delay of 3 control cycles.

5.1   Existing Control

Each degree of freedom utilises a PI control loop on joint velocities which corresponds to PD control in position. Both the Proportional and Integral constants were determined by running a genetic algorithm, with a fitness function minimising trajectory error and maximising joint smoothness [8].

Without TEL, the hip pitch joint experiences the error in position seen in Figure 5. As the crouching motion is cyclic, these errors experienced do not change from cycle to cycle and are dependent on the current phase of gait, with larger errors present in the second half of the cycle as the robot accelerates itself upwards. This displays the inability of the existing control to provide a consistent response over the whole motion cycle.

The position error is roughly cyclic, as similar errors occur at similar points during the gait phase, where gait in the context of this experiment refers to phase of the crouch. When the TEL network is enabled, position error is quickly minimised. Figure 6 and Figure 7 show the position error of the left hip pitch and the compensating signal respectively. As the error signal reduces, the amount of learning also reduces, and the change in table weights decreases. The compen-

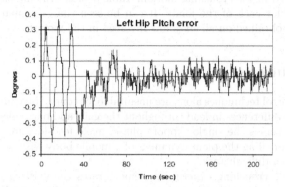

Fig. 6. Figure 6: Position error for the left hip pitch during the crouching motion. With the Joint Compensation system in place, the maximum error experienced by the joint is reduced by 75%

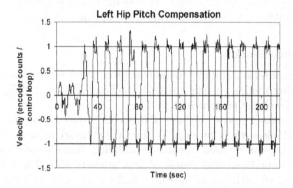

Fig. 7. Compensation signal of the Left Hip Pitch. As the error signal decreases, the rate of learning also decreases and the compensation signal takes on a constant periodic form

Table 3. Summary of peak error in position for all joints before and after learning

Joint	Peak Before	Peak After	Reduction
Left Hip	0.4°	0.1°	75%
Right Hip	0.4°	0.1°	75%
Left Knee	0.6°	0.16°	73%
Right Knee	0.6°	0.14°	77%
Left Ankle	0.4°	0.1°	75%
Right Ankle	0.4°	0.1°	75%

sating signal then becomes cyclic in nature and does not change until additional error develops. The error reduces from peak values of 0.4 degrees to the noise floor with peaks of 0.1 degrees. Similar results were obtained from all pitch movements involved in the crouching motion. Table 3 shows the increase in tracking performance for each of the joints. It can be seen that a similar performance increase was obtained on all six joints involved in the motion.

6   Conclusions

Feedback control techniques alone are unsuitable for control of a humanoid robot. The extreme differences in load throughout the gait, and particularly during the swing phase versus the single support phase, make feed-forward compensation necessary. Modelling the plant dynamics of a mobile body with so many degrees of freedom is a difficult task.

The simple crouching experiment demonstrates the existing control loop's inability to compensate for changes in load as a result of gravity. Using the error signal generated from the desired joint position and the actual joint position, the trajectory error, the cerebellar system is able to learn a response capable of decreasing the peak error by 75%. The experiments were conducted with a crouching motion on a real 23 degree of freedom humanoid and show marked reduction in position error of all joints with the implementation of the TEL system.

6.1   Further Work

In this implementation, suitable compensation of position error has been achieved for a crouching motion. It can obviously be trialled on a walking gait for improvement of walking performance.

The TEL system used as the basis of this work is suited to any control problem where a tracking error is present. Within a humanoid robot, there are many trajectories that can be used to enhance stability. Torso inclination, location of the Zero Moment Point and centre of foot pressure all follow a desired path. Deviations to this path can be measured and a trajectory error calculated. This error can be used to train a separate TEL structured CMAC to improve balance and walking gaits.

References

1. D. Collins: Cerebellar Modeling Techniques for Mobile Robot Control in a Delayed Sensory Environment, PhD Dissertation, University of Queensland, 2003
2. Barto, A: Learning to reach via corrective movements, Self-Learning Robots III Brainstyle Robotics, pp. 6/1, 1999
3. D. Collins, G. Wyeth: Fast and accurate mobile robot control using a cerebellar model in a sensory delayed environment, Intelligent Robots and Systems, pp. 233-238, 2000

4. A. Fagg, N. Sitkoff, A. Barto, J. Houk: A model of Cerebellar Learning for Control of Arm Movements Using Muscle Synergies, Computational Intelligence in Robotics and Automation, pp. 6-12, 1997
5. D. Kee, G. Wyeth, A. Hood, A. Drury: GuRoo: Autonomous Humanoid Platform for Walking Gait Research, Autonomous Minirobots for Research and Edutainment, 2003
6. S. McMillan: Computational Dynamics for Robotic Systems on Land and Underwater, PhD Dissertation, Ohio State University, 1995.
7. Albus J. S: A Theory of Cerebellar Function Mathematical Biosciences Vol : 10, pp. 25-61, 1971
8. J. Roberts, G. Wyeth, D. Kee: Generation of humanoid control parameters using a GA, ACRA2003, 2003

Dynamically Stable Walking and Kicking Gait Planning for Humanoid Soccer Robots

Changjiu Zhou, Pik Kong Yue, and Jun Ni

School of Electrical and Electronic Engineering,
Singapore Polytechnic, 500 Dover Road, Singapore 139651
{zhoucj, yue, JunNi}@sp.edu.sg
http://www.robo-erectus.org

Abstract. Humanoid dynamic walk and kick are two main technical challenges for the current Humanoid League. In this paper, we conduct a research aiming at generating dynamically stable walking and kicking gait for humanoid soccer robots with consideration of different constraints. Two methods are presented. One is synthesizing gait based on constraint equations, which has formulated gait synthesis as an optimization problem with consideration of some constraints, e.g. zero-moment point (ZMP) constraints for dynamically stable locomotion, internal forces constraints for smooth transition, geometric constraints for walking on an uneven floor and etc. The other is generating feasible gait based on human kicking motion capture data (HKMCD), which uses periodic joint motion corrections at selected joints to approximately match the desired ZMP trajectory. The effectiveness of the proposed dynamically stable gait planning approach for humanoid walking on a sloping surface and humanoid kicking on an even floor has been successfully tested on our newly developed Robo-Erectus humanoid soccer robots, which won second place in the RoboCup 2002 Humanoid Walk competition and got first place in the RoboCup 2003 Humanoid Free Performance competition.

1 Introduction

Humanoid soccer robot league is a new international initiative to foster robotics and AI technologies using soccer games [7]. The Humanoid league (HL) has different challenges from other leagues. The main distinction is that the dynamic stability of the robots needs to be well maintained while the robots are walking, running, kicking and performing other tasks. Furthermore, the humanoid soccer robot will have to coordinate perceptions and biped locomotion, and be robust enough to deal with challenges from other players. Hence, how to generate a dynamically stable gait for the humanoid soccer robots is an important research area for the HL, especially for the new technical challenge – Balancing Challenge which will be commencing in the coming RoboCup 2004.

The problem of gait planning for humanoid robots is fundamentally different from path planning for traditional fixed-base manipulator arms due to the inherent characteristics of legged locomotion – unilaterality and underactuation [3,11,12]. The humanoid locomotion gait planning methods can be classified into two main

D. Nardi et al. (Eds.): RoboCup 2004, LNAI 3276, pp. 358–369, 2005.
© Springer-Verlag Berlin Heidelberg 2005

categories [5]: one is online simplified model-based gait generation method; and the other is offline position based gait generation method. There are currently some ways for generating dynamically stable gaits, e.g., heuristic search approach, such as genetic algorithms (GAs) based gait synthesis [1]; problem optimisation method, such as optimal gradient method; model simplification with iteration [5], etc.

To have continuous and repeatable gait, the postures at the beginning and the end of each step have to be identical. This requires the selection of specific initial conditions, constraint functions and their associated gait parameters. However, finding repeatable gait when the constraint equations involve higher order differential equations remains an unsolvable problem. So, a natural way to solve this problem is to resort to numerical methods, e.g. Fourier series expansion and time polynomial functions. One advantage of this technique is that extra constraints can be easily included by adding the coefficients to the polynomials. Disadvantages, however, include the facts that the computing load is high for large bipedal systems and the selection of the polynomials may impose undesirable features to the joint profiles, e.g. oscillation. Moreover, the planning gait may not be human-like. To accomplish a human-like motion, it is quite natural to attempt using the Human Kicking Motion Captured Data (HKMCD) to drive the robot. However, some researches show that the HKMCD cannot be applied directly to humanoid robot due to kinematic and dynamic inconsistencies between the human subject and the humanoid, which usually require kinematic corrections while calculating the joint angle trajectory [2].

The rest of this paper is organized as follows. We will briefly present some basic constraints for the dynamically stable gait in Section 2. The humanoid soccer robot to be used for the experiment is introduced in Section 3. Most of the research on humanoids has been actively working to make a robot capable of dynamic walking on even floor. However, for the new technical challenges of humanoid league the robot is required to walk on uneven terrain such as sloping surfaces and stairs. So, we plan a dynamically stable gait for ascending a sloping surface in Section 4. In Section 5, we describe how to make use of human kicking motion capture data to drive the humanoid to perform penalty kick. Concluding remarks and some major technical challenges in this field are addressed in Section 5.

2 Robo-Erectus: A Fully-Autonomous Humanoid Soccer Robot

The Robo-Erectus (RE) project (www.robo-erectus.org) aims to develop a low-cost fully-autonomous humanoid platform so that educators and students are able to build humanoid robots quickly and cheaply, and to control the robots easily [16]. We have developed three generations humanoid soccer robots, namely RE40I, RE40II and RE40II (see Fig. 1). Our RE humanoid has participated in both the 1st and 2nd Humanoid League of RoboCup, won 2nd place in the Humanoid Walk competition at the RoboCup 2002 and got 1st place in the Humanoid Free Performance competition at the RoboCup 2003. The configuration of the hierarchical control system for the RE humanoid is shown in Fig. 2. We've also implemented reinforcement learning (see Fig. 3) to further improve the walking and kicking gait [14,15].

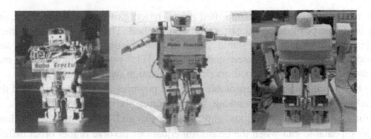

Fig. 1. RE40I at RoboCup 2002 (left), RE40II at the RoboCup 2003 (centre), and the newly developed RE40III (right)

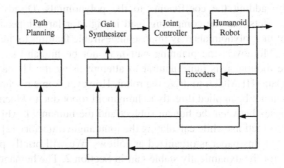

Fig. 2. Schematic diagram of the hierarchical control system for the RE humanoid robot

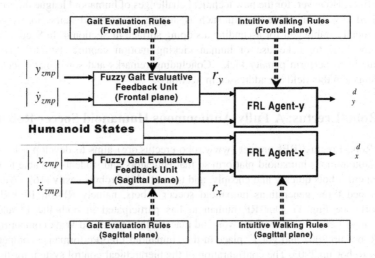

Fig. 3. Block diagram of the humanoid gait learning using two independent fuzzy reinforcement learning agents [15]

3 Dynamically Stable Gait

Since a humanoid robot tips over easily, it is important to consider stability in planning its gait. Many methods have been proposed for synthesizing walking patterns based on the concept of the zero moment point (ZMP) [11, 12]. The ZMP is defined as the point on the ground about which the sum of the moments of all the active forces equals zero. If the ZMP is within the convex hull (support polygon) of all contact points between the feet and the ground, the bipedal robot can walk dynamically.

Humanoid dynamics can be modelled using a multi-body system model consisting N chains involving the body parts, such as head, arms, trunk and pelvis. Each chain consists of n_i links ($i = 1, 2, \ldots, N$) interconnected with single DOF joints. The support-foot can only be controlled indirectly by ensuring the appropriate dynamics of the mechanism above the foot.

The humanoid robot is a highly redundant system with many extra degrees of freedom (DOF). Its gait consists of large number of unknown parameters. This allows us to formulate constraint equations for synthesizing gait. In this paper we formulate an optimization problem to determine the unknown parameters of the gait to achieve dynamic locomotion, i.e. to obtain a good match between the actual and the desired ZMP trajectories as follows

$$Minimize \quad \int_{t_i}^{t_f} \left\| P_{zmp}(t) - P_{zmp}^d(t) \right\|^2 dt \tag{1}$$

subject to the boundary conditions of both $p(t)$ and $\dot{p}(t)$ at time t_i and t_f, where P_{zmp} is the actual ZMP, and P_{zmp}^d is the desired ZMP position.

Due to the large number of the unknown parameters for the above optimization problem, we need to specify some constraints [2,4,8]. The following are some constraints that need to be considered.

Sabilization of the gait (ZMP constraint): The control objective of the humanoid dynamically stable gait can be described as

$$P_{zmp} - x_{zmp}, y_{zmp}, 0 \quad S \tag{2}$$

where ($x_{zmp}, y_{zmp}, 0$) is the coordinate of the ZMP with respect to O-XYZ. S is the support polygon.

Smooth transition constraint: The equation of motion of the centre of the humanoid can be described as

$$m_{cm} a_{cm} \quad f_L \quad f_R \quad m_{cm} g \tag{3}$$

Where m_{cm} and a_{cm} are the mass of the robot and the acceleration of the COM, respectively. f_R and f_L represent the ground reaction forces at the right and left foot. During single-support phase, the foot force can be obtained from (3) as one of f_R and

f_L will be zero. However, during double-support phase, only the total of the ground reaction forces is known. Hence, how to resolve the leg reaction forces appropriately to minimize the internal force has to be considered. This can ensure a smooth transition of the internal forces during placement and take-off [2].

Geometrical constraints: The swing limb has to be lifted off the ground at the beginning of the step cycle and has to be landed back at the end of it.

Maximum clearance of the swing limb: During swing phase, the foot of the swing limb has to stay clear off the ground to avoid accidental contact.

Repeatability of the gait: The requirement for the repeatable gait demands that the initial posture and velocities be identical to those at the end of the step.

Continuity of the gait: The horizontal displacements of the hip during the single and double support phases must be continuous.

Minimization of the effect of impact: During locomotion, when the swing limb contacts the ground (heel strike), impact occurs, which contacts sudden changes in the joint angular velocities. By keeping the velocity of the foot of the swing limb zero before impact, the sudden jump in the joint angular velocities can be eliminated.

4 Dynamic Walking Gait for Ascending Sloping Surfaces

For RE humanoid, we have developed a gait planner which is able to generate gait for the robot walking on both plat and uneven terrain, such as stairs and sloping surfaces. Note that, one of the major problems in controlling humanoids is the difficulty in choosing proper gait for a particular terrain.

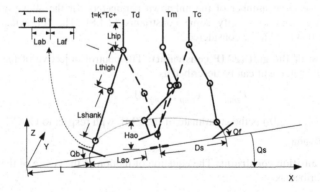

Fig. 4. The humanoid and gait parameters and coordinate on the slope

In this paper, we only consider how to generate a dynamically stable gait for humanoid ascending sloping surfaces, which will be a new technical challenge for

Humanoid League from 2004. To simply our analysis, we assume that the kth step begins with the heel of the right foot leaving the slop at t= $k*T_c$, and ends with the heel of the right foot making the first contact with the slope at t= $(k+1)*T_c$, as shown in Fig. 4.

We prescribe some joint trajectories, e.g. both foot trajectories, and then derive all joint trajectories by inverse kinematics with consideration of the constraints discussed in Section 3. These constraints can ensure that the gait is dynamically stable and also satisfies some geometric and periodical requirements. We also need some more constraints for the sloping surface. Eqs (4) to (7) show some constraints on ankle and hip joints.

$$x_a(t) \begin{cases} L*\cos Q_s \ k*D_s*\cos Q_s \ L_{an}*\sin Q_s, & t \ k*T_c \\ L*\cos Q_s \ k*D_s*\cos Q_s \ L_{an}*\sin(Q_s \ Q_b) \ L_{af}*(\cos Q_s \ \cos(Q_s \ Q_b)), & t \ k*T_c \ T_d \\ L*\cos Q_s \ k*D_s*\cos Q_s \ L_{ao}*\cos Q_s, & t \ k*T_c \ T_m \\ L*\cos Q_s \ (k \ 2)*D_s*\cos Q_s \ L_{an}*\sin(Q_s \ Q_f) \ L_{ab}*(\cos Q_s \ \cos(Q_s \ Q_b)), & t \ (k \ 1)*T_c \\ L*\cos Q_s \ (k \ 2)*D_s*\cos Q_s \ L_{an}*\sin Q_s, & t \ (k \ 1)*T_c \ T_d \end{cases} \quad (4)$$

$$z_a(t) \begin{cases} L_{an}*\cos Q_s \ L_{af}*\sin Q_s \ (L_{an}*\sin Q_s \ L_{af}*\cos Q_s)*tgQ_s \ x_a(t)*tgQ_s, & t \ k*T_c \\ L_{an}*\cos(Q_s \ Q_b) \ L_{af}*\sin(Q_s \ Q_b) \ (L_{an}*\sin(Q_s \ Q_b) \ L_{af}*\cos(Q_s \ Q_b))*tgQ_s \ x_a(t)*tgQ_s, & t \ k*T_c \ T_d \\ H_{ao} \ x_a(t)*tgQ_s, & t \ k*T_c \ T_m \\ L_{an}*\cos(Q_s \ Q_f) \ L_{ab}*\sin(Q_s \ Q_f) \ (L_{an}*\sin(Q_s \ Q_f) \ L_{ab}*\cos(Q_s \ Q_f))*tgQ_s \ x_a(t)*tgQ_s, & t \ (k \ 1)*T_c \\ L_{an}*\cos Q_s \ L_{ab}*\sin Q_s \ (L_{an}*\sin Q_s \ L_{ab}*\cos Q_s)*tgQ_s \ x_a(t)*tgQ_s, & t \ (k \ 1)*T_c \ T_d \end{cases} \quad (5)$$

$$x_h(t) \begin{cases} L*\cos Q_s \ k*D_s*\cos Q_s \ x_{ed}, & t \ k*T_c \\ L*\cos Q_s \ (k \ 1)*D_s*\cos Q_s \ x_{sd}, & t \ k*T_c \ T_d \\ L*\cos Q_s \ (k \ 1)*D_s*\cos Q_s \ x_{ed}, & t \ (k \ 1)*T_c \end{cases} \quad (6)$$

$$z_h(t) \begin{cases} H_{min} \ x_h(t)*tgQ_s, & t \ k*T_c \ 0.5*T_d \\ H_{max} \ x_h(t)*tgQ_s, & t \ k*T_c \ 0.5*(T_c \ T_d) \\ H_{min} \ x_h(t)*tgQ_s, & t \ (k \ 1)*T_c \ 0.5*T_d \end{cases} \quad (7)$$

From the above via points and the constraints, the walking gait on sloping surface can be generated using the spline interpolation. The gait is dynamically stable as the ZMP constraints (2) is applied.

Based on the planned gait for ascending the sloping surface, we have conducted both simulation and experimental studies.

Fig. 5(a) shows the horizontal displacements of the hip and both right and left ankle joints. It can be seen that the trajectories are smooth, i.e. the velocities are continuous for both single and double support phases. The horizontal displacement of the ZMP for different slope angles is shown in Fig. 5(b). One can see that the ZMP remains approximately at the centre of the stability region for the smaller slope angles, which ensures the largest stability margin and greatest stability. However, for bigger slope angle, e.g. 15 degrees, the dynamic stability cannot be guaranteed. In Figs. 5 (a) and (b), the angular velocities for both right ankle and right knee joints are given. It shown that the sudden changes in the angular velocities due to landing impact are reduced when the slope angle is smaller, e.g. less than 10 degrees. However, the landing impact is very obvious for the bigger slope angle.

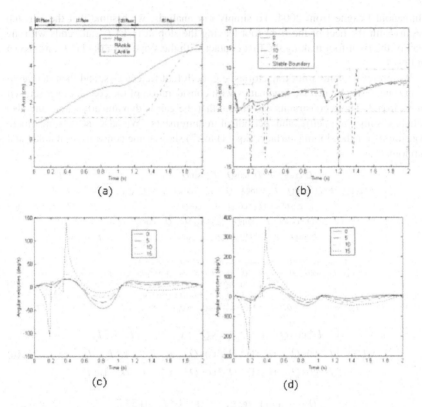

(a)   (b)

(c)   (d)

Fig. 5. (a) Movements of the hip and ankle joints: (b) Comparison of the ZMP trajectories for different slope angles (Q_s $0^o, 5^o, 10^o, 15^o$); (c) Right ankle joint angular velocities for different slope angles; (4) Right knee joint angular velocities for different slope angles

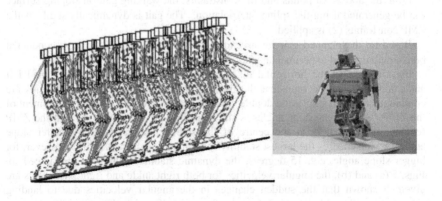

Fig. 6. RE40II ascending a sloping surface during the experiment

Fig. 6 shows the stick diagram of humanoid walking on the sloping surface.

By considering the humanoid performance constraints, a fuzzy reinforcement learning agent [14,15,16] has been used to further improve the gait. Fig. 7 shows humanoid walking sequence on flat platform (after learning). The turning sequence after learning is shown in Fig. 8. We are currently working on further improving the gait for ascending slope using reinforcement learning.

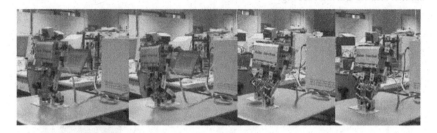

Fig. 7. The humanoid walking sequence (after learning)

Fig. 8. The Humanoid turning sequence (after learning)

5 Planning Dynamic Kicking Pattern Based on Human Kicking Motion Captured Data

To plan humanoid kicking trajectory, we have to consider some kicking constraints [16], e.g. the Maximum Kicking Range (MaxKR), the Effective Kicking Range (EKR), the Minimum Kicking Moment (MinKM) and so on. The challenge for the humanoid dynamic kick is that the ZMP must stay in the support polygon, which is the stance foot, during the kicking phase. In our previous research [16], by considering all the kicking parameters, an initial kicking pattern was generated using Kicking Pattern Generator developed by our Humanoid Robotics Group. However, the kick is clearly not a human-like one.

In this paper, we propose a new method to generate dynamically stable kicking pattern for humanoid soccer robot. To accomplish a human-like kick, it is quite natural to attempt using the Human Kicking Motion Captured Data (HKMCD) to drive the robot. However, by using human kicking data to prescribe the motion of the lower limbs, two immediate problems arise when using HKMCD directly. Firstly, a complex dynamic model is required. Secondly, the designer has no freedom to synthesize the joint profiles based on tangible gait characteristics such as walking speed, step length and step elevation. Please also note that there are kinematic and dynamic inconsistencies between the human subject and the humanoid, which usually require kinematic corrections while calculating the joint angle trajectory [2].

The adaptation of the HMKCD for the humanoid soccer robot uses mainly periodic joint motion corrections at selected joints to approximately match the desired ZMP trajectory. By optimisation with constraints, we can maximize the dynamic stability against sliding during kicking.

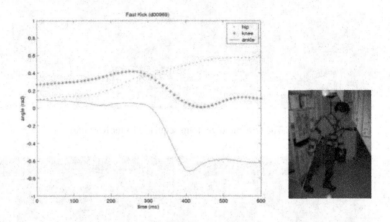

Fig. 9. Human kicking motion captured data (the experiment was conducted in Prof. Stefan Schaal's Computational Learning and Motor Control (CLMC) Lab at University of Southern California; and the data was captured by his PhD students Michael Mistry and Peyman Mohajerian)

The HMKCD was collected in the Computational Learning and Motor Control (CLMC) Lab at University of Southern California. The Sarcos SenSuit simultaneously measures 35 degrees of freedom (DOF) of the human body. It can be used for real-time capturing of full body motion, as an advanced human-computer interface, or to control sophisticated robotic equipment. The complete SenSuit is worn like an exoskeleton which, for most movements, does not restrict the motion while an array of lightweight Hall-sensors reliably records the relative positions of all limbs. For the arms, we collect shoulder, elbow and wrist DOF, for the legs, hip, knee and ankle data is recorded. In addition, the Sensuit measures head as well as waist motion. The experiment setup and result is shown in Fig. 9. At left, it is joint trajectories for the kicking leg recorded from a fast kick demonstration. At right, it is the human kicker who wore the Sarcos Sensuit.

By using optimization with ZMP constraints, we obtained the corrected joint trajectories which are shown in Fig. 10. The ZMP trajectories during humanoid kick are given in Fig. 11. It can be seen that by using the original HKMCD (left), the ZMP is out of the support polygon during the ending phase of the kick. From the ZMP constraints (2), it is clear that the original HKMCD cannot drive the robot to perform a stable kick. However, after correcting the HKMCD with ZMP constraints, the corrected ZMP trajectory is always inside the support polygon so that the dynamically stable kicking gait can be obtained. We have also conducted an experiment of humanoid kick using the corrected HKMCD. The kicking sequence is shown in Fig. 11. The striker's kicking posture is more or less "human-like".

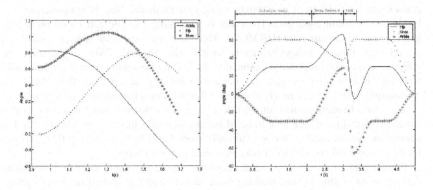

Fig. 10. Corrected joint trajectories for humanoid kick (At left, it is the joint trajectories for kicking phase. At right, it is the joint trajectories for a complete kicking cycle)

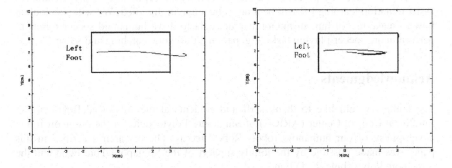

Fig. 11. Original and corrected ZMP trajectories (At left, the original ZMP trajectory is not always within the support polygon. At right, the corrected ZMP trajectory is always inside the support polygon so that the dynamic kicking gait is obtained)

Fig. 12. The humanoid kicking sequence (using the corrected human kicking motion capture data)

6 Concluding Remarks

Humanoid dynamic walk and kick are two main technical challenges for the current Humanoid League. In this paper, we conduct a research that aims at generating dynamically stable walking and kicking gait for humanoid soccer robots with consideration of different constraints using two methods. One is synthesizing gait based on constraint equations. The other is generating feasible gait based on human kicking motion capture data (HKMCD), which uses periodic joint motion corrections at selected joints to approximately match the desired ZMP trajectory. The effectiveness of the proposed dynamically stable gait planning approach for humanoid walking on a sloping surface and humanoid kicking on an even floor has been successfully tested on our newly developed Robo-Erectus humanoid soccer robots.

The Robo-Erectus project aims to develop a low-cost humanoid platform so that educators and students are able to build humanoid robots quickly and cheaply, and to control the robots easily. We are currently working to further develop this platform for educational robots, service robots and entertainment robots.

By using the proposed fuzzy reinforcement learning approach, we also demonstrate that the robot is able to start walking from an initial gait generated from perception-based information on human walking, and learn to further tune its walking and kicking behavior using reinforcement learning. Note that humans do not just learn a task by trial and error, rather they observe other people perform a similar task and then emulate them by *perceptions*. How to utilize perception-based information to assist imitation learning [9] will be a new challenge in this filed. We will also look at how to make use of human motion capture data to drive humanoid soccer robots to perform more soccer-playing tasks, e.g. passing, throwing, catching and so on.

Acknowledgments

The authors would like to thank staff and students at the Advanced Robotics and Intelligent Control Center (ARICC) of Singapore Polytechnic for their support in the development of our humanoid robots Robo-Erectus. The research described in this paper was made possible by the jointly support of the Singapore Tote Fund and the Singapore Polytechnic R&D Fund.

References

1. Cheng, M.-Y., Lin, C.-S.: Dynamic biped robot locomotion on less structured surface, Robotica 18 (2000) 163-170
2. Dasgupta, A., Nakamura, Y.: Making feasible walking motion of humanoid robots from human motion capture data, Proc. the 1999 IEEE Intl. Conf. on Robotics & Automation, Detroit, Michigan, May 1999, pp.1044-1049
3. Goswami, A.: Postural stability of biped robots and the foot-rotation indicator (FRI) point, Int. J. of Robotics Research 18(6) (1999) 523-533
4. Huang, Q., Yokoi, K. Kajita, S., Kaneko, K., Arai, H., Koyachi, N., Tani, K.: Planning walking patterns for a biped robot, IEEE Trans. on Robotics and Automation 17(3) (2001) 280-289

5. Kagami, S., Kitagawa, T., Nishiwaki, K., Sugihara, T., Inaba, M., Inoue, H.: A fast dynamically equilibrated walking trajectory generation method of humanoid robot, Autonomous Robots 12 (2002) 71-82
6. Kajita, S., Kanehiro, F., Kaneko, K., Fujiwara, K., Yokoi, K., Hirukawa, H.: A realtime pattern generator for biped walking, Proc. the 2002 IEEE Intl. Conf. on Robotics & Automation (ICRA2002), Washington DC, May 2002, pp.31-37.
7. Kitano H., Asada, H.: The RoboCup humanoid challenge as the millennium challenge for advanced robotics, Advanced Robotics 13(8) (2000) 723-736
8. Mu, X., Wu, Q.: Synthesis of a complete sagittal gait cycle for a five-link biped robot, Robotica 21 (2003) 581-587
9. Schaal, S.: Is imitation learning the route to humanoid robots? Trends in Cognitive Sciences 3(6) (1999) 233-242
10. Tang, Z., Zhou, C., Sun, Z.: Trajectory planning for smooth transition of a biped robot, 2003 IEEE International Conference on Robotics and Automation (ICRA2003), Taibei, Taiwan, September 2003, pp.2455-2460
11. Vukobratovic, M., Borovac, B., Surla, D., Stokic, D.: Biped Locomotion: Dynamics, Stability, Control and Application, Springer-Verlag, 1990
12. Vukobratovic, M., Borovac, B., Surdilivic, D.: Zero-moment point – proper interpretation and new applications, Proc. IEEE-RAS Intl. Conf. on Humanoid Robots, 2001, pp.239-244
13. Zheng, Y.F., Shen, J.: Gait synthesis for the SD-2 biped robot to climb sloped surface, IEEE Trans. on Robotics and Automation 6(1) (1990) 86-96
14. Zhou, C.: Robot learning with GA-based fuzzy reinforcement learning agents, Information Sciences 145 (2002) 45-68
15. Zhou, C., Meng, Q.: Dynamic balance of a biped robot using fuzzy reinforcement learning agents, Fuzzy Sets and Systems 134(1) (2003) 169-187
16. Zhou, C., Yue, P.K., Tang, Z., Sun, Z.Q.: Development of Robo-Erectus: A soccer-playing humanoid robot, Proc. IEEE-RAS RAS Intl. Conf. on Humanoid Robots, CD-ROM, 2003.

An Algorithm That Recognizes and Reproduces Distinct Types of Humanoid Motion Based on Periodically-Constrained Nonlinear PCA

Rawichote Chalodhorn, Karl MacDorman, and Minoru Asada

Department of Adaptive Machine Systems *and* Frontier Research Center,
Graduate School of Engineering, Osaka University,
2-1 Yamada-oka, Suita, Osaka 565-0871 Japan
kfm@ed.ams.eng.osaka-u.ac.jp
http://ed.ams.eng.osaka-u.ac.jp

Abstract. This paper proposes a new algorithm for the automatic segmentation of motion data from a humanoid soccer playing robot that allows feedforward neural networks to generalize and reproduce various kinematic patterns, including walking, turning, and sidestepping. Data from a 20 degree-of-freedom Fujitsu HOAP-1 robot is reduced to its intrinsic dimensionality, as determined by the ISOMAP procedure, by means of nonlinear principal component analysis (NLPCA). The proposed algorithm then automatically segments motion patterns by incrementally generating periodic temporally-constrained nonlinear PCA neural networks and assigning data points to these networks in a *conquer*-and-divide fashion, that is, each network's ability to learn the data influences the data's division among the networks. The learned networks abstract five out of six types of motion without any prior information about the number or type of motion patterns. The multiple decoding subnetworks that result can serve to generate abstract actions for playing soccer and other complex tasks.

1   Introduction

The development of robots that can learn to imitate human behavior as they participate in social activities is important both for understanding ourselves as a species and for transforming society through the introduction of new technologies. A mimesis loop [1] may be used to capture many aspects of this kind of imitative learning. This paper addresses one aspect of the mimesis loop: the abstraction of a robot's own kinematic motions from its proprioceptive experience.

Figure 1 roughly outlines how a mimesis loop might be realized in a soccer playing robot. Attentional mechanisms direct the robot's sensors toward the body parts of other players, and the robot maps successfully recognized body parts onto its own body schema. This paper introduces a method to abstract the robot's own kinematic patterns: our segmentation algorithm allocates proprioceptive data among periodic temporally-constrained nonlinear principal component neural networks (NLPCNNs) as they form appropriate generalizations.

D. Nardi et al. (Eds.): RoboCup 2004, LNAI 3276, pp. 370–380, 2005.
© Springer-Verlag Berlin Heidelberg 2005

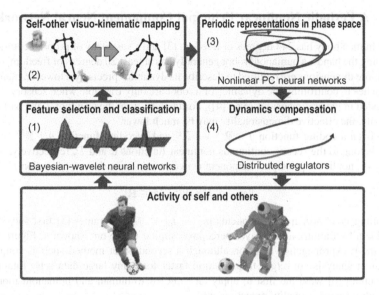

Fig. 1. Periodic nonlinear principal component networks may characterize motion patterns in a much larger system for recognizing, learning, and responding behavior

NLPCNNs, augmented with periodic and temporal constraints, provide an effective means of characterizing many typical human motions. These networks may be used to recognize, learn, and respond to behavior. A single network abstracts a particular type of periodic motion from joint angles and other proprioceptive data. A different network learns a different type of periodic motion until all the various kinds of motion have been learned. Networks can also learn transitions between motion patterns.

The robot can use NLPCNNs to recognize the activities of other players, if the mapping from their bodies to its own has already been derived by some other method. Since each network correspond to a particular type of motion in a proprioceptive phase space, it can act as a protosymbol. Thus, the robot would be able to recognize the behavior of others because it has grounded their behavior in terms of its own body.

Although periodic NLPCNNs may be used to generate motion patterns, the robot must continuously respond to unexpected perturbations. There are a number of approaches to this control problem that do not require an explicit model. For example, distributed regulators [2] could set up flow vectors around learned trajectories, thus, converting them into basins of attraction in a phase space of possible actions.

This paper is organized as follows. Section 2 extends an NLPCNN with periodic and temporal constraints. Section 3 presents a method of assigning observations to NLPCNNs to segment proprioceptive data. Section 4 reports experimental results using NLPCNNs to characterize the behavior of a Fujitsu HOAP-1 humanoid robot that has been developed to play RoboCup soccer.

2   A Periodic Nonlinear Principal Component Neural Network

The human body has 244 degrees of freedom [3] and a vast array of proprioceptors. Excluding the hands, a humanoid robot generally has at least 20 degrees of freedom — and far more dimensions are required to describe its dynamics precisely. However, many approaches to controlling the dynamics of a robot are only tractable when sensory data is encoded in fewer dimensions (e.g., [4]). Fortunately, from the standpoint of a particular activity, the effective dimensionality may be much lower.

Given a coding function $f : \mathbb{R}^N \mapsto \mathbb{R}^P$ and decoding function $g : \mathbb{R}^P \mapsto \mathbb{R}^N$ that belong to the sets of continuous nonlinear functions $\mathcal{C}$ and $\mathcal{D}$, respectively, where $P < N$, nonlinear principle component networks minimize the error function E

$$\|\boldsymbol{x} - g(f(\boldsymbol{x}))\|^2, \quad \boldsymbol{x} \in \mathbb{R}^N$$

resulting in P principal components $[y_1 \cdots y_p] = f(\boldsymbol{x})$. Kramer [5] first solved this problem by training a multilayer perceptron similar to the one shown in Figure 2 using the backpropagation of error, although a second order method such as conjugant gradient analysis converges to a solution faster for many large data sets. Tatani and Nakamura [6] were the first to apply an NLPCNN to human and humanoid motions, though for dimensionality reduction only.

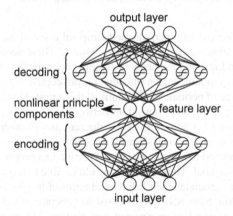

Fig. 2. Target values presented at the output layer of a nonlinear principal component neural network are identical to input values. Nonlinear units comprise the encoding and decoding layers, while either linear or nonlinear units comprise the feature and output layers

Nonlinear principal components analysis, unlike PCA (Karhunen-Loève expansion), which is a special case where $\mathcal{C}$ and $\mathcal{D}$ are linear, does not have a unique solution, and no known computational method is guaranteed to find any of the globally optimal solutions.

Nevertheless, for a 20-DoF humanoid robot, a hierarchically-constructed[1] NLPCNN has been shown to minimize error several times more than PCA when reducing to two-to-five dimensions [6].

2.1 The Periodicity Constraint

Because the coding function f of an NLPCNN is continuous, *(1)* projections to a curve or surface of lower dimensionality are suboptimal; *(2)* the curve or surface cannot intersect itself (e.g., be elliptical or annular); and *(3)* projections do not accurately represent discontinuities [7]. However, since the physical processes underlying motion data are continuous, discontinuities do not need to be modelled. Discontinuities caused by optimal projections can create instabilities for control algorithms (e.g., they allow points along the axis of symmetry of a parabola to be projected to either side of the parabola). Moreover, an NLPCNN with a circular node [8][9] at the feature layer can learn self-intersecting curves and surfaces.

Kirby and Miranda [10] constrained the activation values of a pair of nodes p and q in the feature layer of an NLPCNN to fall on the unit circle, thus acting as a single angular variable:

$$r = \sqrt{y_p^2 + y_q^2}, \ y_p \leftarrow y_p/r, \ y_q \leftarrow y_q/r$$

The delta values for backpropagation of the circular node-pair are calculated by the chain rule [10], resulting in the update rule

$$\delta_p \leftarrow (\delta_p y_q - \delta_q y_p)y_q/r^3, \ \delta_q \leftarrow (\delta_q y_p - \delta_p y_q)y_p/r^3$$

at the feature layer.

The hyperbolic tangent and other antisymmetric functions (i.e., $\varphi(x) = -\varphi(x)$) are generally preferred to the logistic function as the sigmoid in part because they are compatible with standard optimizations [11].[2] In addition, antisymmetric units can more easily be replaced with linear or circular units in the feature layer, since these units can produce negative activations. We propose using a slightly flatter antisymmetric function for the sigmoidal units with a similar response characteristic to $tanh$ (see Fig. 3). The advantage of this node is that it can be converted to a circular node-pair while still making use of its perviously learned weights.

2.2 The Temporal Constraint

Neither linear nor nonlinear principal components analysis represent the time, relative time, or order in which data are collected.[3] This information, when available, can be

---

[1] The joint encoder dimensionality of limbs is independently reduced, then the arms and the legs are paired and their dimensionality further reduced, and then finally the dimensionality of the entire body.

[2] These include mean cancellation, linear decorrelation using the K-L expansion, and covariance equalization.

[3] Although a temporal dimension could be added to an autoassociative network, one drawback for online learning is that this dimension would need to be continuously rescaled as more data is collected to keep it within the activation range of the nodes.

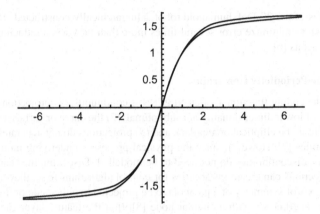

Fig. 3. The popular hyperbolic tangent activation function $y \leftarrow 1.7159 \tanh(\frac{2}{3}y)$ can be approximated by a pair of circular nodes where the activation of the second node y_q is fixed at $\sqrt{1.9443}$ and the activation of the first node is calculated accordingly $y_p \leftarrow 1.7159 y_p / \sqrt{y_p^2 + 1.9443}$

used to reduce the number of layers and free parameters (i.e., weights) in the network and thereby its risk of converging slowly or settling into a solution that is only locally optimal. Since the activations y_p and y_q of the circular node-pair in the feature layer in effect represent a single free parameter, the angle θ, if θ is known, we can train the encoding and decoding subnetworks separately by presenting $k \cos(\theta)$ and $k \sin(\theta)$ as target output values for the encoding subnetwork and as input values for the decoding network.[4] Once a single period of data has been collected, temporal values can be converted to angular values $\theta = 2\pi \frac{t_k - t_0}{t_n - t_0}$ for data collected at any arbitrary time t_k during a period, starting at t_0 and ending at t_n. A network may similarly learn transitions between periodic movements when using a linear or sigmoidal activation node in the feature layer because these open-curve transitions do not restrict us to using nodes capable of forming a closed curve.[5] NLPCNNs with a circular feature node remain useful to identify the period of a motion pattern, especially when the pattern is irregular and, thus, begins and ends at points that are somewhat far from each other.

3   Automatic Segmentation

We conceived of the automatic segmentation problem as the problem of uniquely assigning data points to nonlinear principal component neural networks. It is possible to partition the points without reference to the predictions of the networks.[6] However, for

---

[4] $k \approx 1.7$ for zero-mean data with variance equal to 1 based on principles discussed in [11].

[5] $y_{target} = 2k(\frac{t_k - t_0}{t_n - t_0} - \frac{1}{2})$, with $k \approx 1.4$.

[6] For example, data points may be partitioned at the point at which a trajectory most closely doubles back on itself, if the distance between the two paths is within a certain threshold and the paths then diverge beyond another threshold.

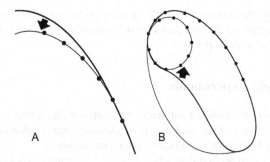

Fig. 4. The thick line shows the output of an NLPCNN and the thin line shows the underlying distribution. The dots are data points. **A.** Before learning converges, allowing the network to learn data points despite a high prediction error accelerates learning. **B.** However, after convergence, it leads to segmentation errors

our method each network's performance influences segmentation with more networks assigned to regions that are difficult to learn.

As the robot begins to move, the first network is assigned some minimal number of data points (e.g., joint-angle vectors), and its training begins with those points. This gets the network's learning started quickly and provides it with enough information to determine the orientation and curvature of the trajectory. If the average prediction error of the data points assigned to a network is below some threshold, the network is assigned additional data points until that threshold has been reached. At that point, data points will be assigned to another network, and a network will be created, if it does not already exist. To avoid instabilities, only a single data point may shift its assignment from one network to another after each training cycle.

$j \leftarrow 1, \; bucket \leftarrow 1, \; E \leftarrow 0$
$\forall x_i \{$
$\quad \texttt{train}(network_j, x_i)$
$\quad E_i = \|x_i - g(f(x_i))\|^2, \; E \leftarrow E + E_i$
$\quad \text{if} \, (\, bucket > B_{max} \, \vee$
$\quad (\, \texttt{learning?}(network_j) \, \wedge \, E/bucket > E_{max} \,) \, \vee$
$\quad E_i > E_{i+1} \,)$
$\quad j \leftarrow j + 1, \; bucket \leftarrow 1, \; E \leftarrow 0 \, \}$

Listing 1. Pseudocode for segmentation

Since a network is allowed to learn more data points as long as its average prediction error per point is low enough, it may learn most data points well but exhibit slack near peripheral or recently learned data points. At the start of learning, the network should be challenged to learn data points even when its prediction error is large (see Fig. 4A). As learning converges, however, the slack leads to segmentation errors (see Fig. 4B). Therefore, we alter the method of segmentation once the network nears convergence (as determined by Bayesian methods [12] or crossvalidation) so that a network may acquire neighboring points if its prediction error for those points is lower that the network currently assigned to those points.

4   Humanoid Experiments

This section shows the result of automatic segmentation and neural network learning. We assess the accuracy of the result based on a manual segmentation of the data points and an analysis of how they are allocated among the networks.

First, we recorded motion data while a HOAP-1 humanoid robot played soccer in accordance with a hard-coded program [13]. Each data point is constituted by a 20-dimensional vector of joint angles. A standard (noncircular) NLPCNN reduced the dimensionality of the data from 20 to 3, which was determined to be the intrinsic dimensionality of the data by the ISOMAP procedure [14] We then applied our algorithm to segment, generalize, and generate humanoid motion.

Our algorithm uniquely assigned the data points among a number of circularly-constrained NLPCNNs. Each of the networks learned a periodic motion pattern by conjugate gradients. Our algorithm successfully generalized five out of six primary motion patterns: walking forward, turning right or left, and side-stepping to the right or left. It failed to generalize as a single periodic trajectory the kicking motion, which has a highly irregular, self-intersecting shape. However, human subjects were also unable to determine the kicking trajectory from the data points.

Figure 5 shows that the automatic segmentation algorithm successfully employed circular NLPCNNs to separate and generalize five of the periodic motions. (The open-curve segmentation of transitions between periodic motions are omitted for clarity.) The periodic trajectories were generated by varying from 0 to 2π the angular parameter θ_i at the bottleneck layer of each of the circularly-constrained networks and mapping the result to the output layer for display. This demonstrates our method's capacity to generate periodic motions.

We calculated statistics based on running the automatic segmentation for 20 trails. The algorithm resulted in five decoding subnetworks for 45% of the trials, which is the most parsimonious solution. It resulted in six subnetworks for 50% of the trials, and seven for the remaining 5%. In results published elsewhere [15, 16], we developed a method resembling linear integration that consistently eliminated redundant networks for this data set. If the area between the predicted curves of two networks is sufficiently small, on network is removed and its data points are reassigned to the other network.[7]

Since the data was generated by the predefined behavior modules used by the Osaka University team in the 2003 RoboCup humanoid competition, each data point was already labeled and could be segmented into the five types of motion that had been successfully abstracted. To assess the accuracy of the automatic segmentation algorithm, we manually assigned the data points corresponding to each type of motion to five pe-

---

[7] The algorithms presented in [15] deviate from those presented here and in [17] in some minor details, the most significant being that learning and segmentation occur sequentially rather than simultaneously.

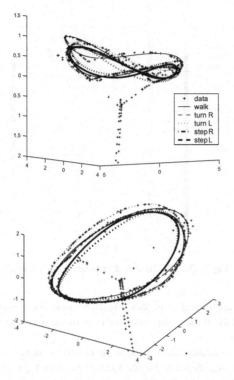

Fig. 5. Recognized motion patterns embedded in the dimensions of the first three nonlinear principal components of the raw proprioceptive data. The top and bottom plots differ only in the viewpoint used for visualization

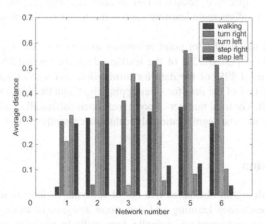

Fig. 6. The average distance between the prediction of a network trained on manually segmented data and each of the automatically generated networks

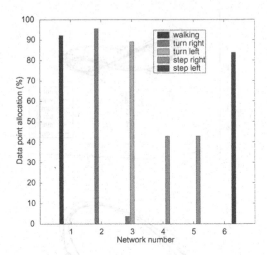

Fig. 7. The allocation of data points to each network

riodic temporally constrained NLPCNNs. Figure 6 shows the average distance between the prediction for each of these networks and each of the networks resulting from automatic segmentation.

The lowest bar indicates which pattern the networks, numbered 1 to 6 best match in terms of least average distance. Hence, the first network represents walking; the second represents turning right; the third turning left; the fourth and fifth sidestepping right; and the sixth sidestepping left. The fact that the fifth network is redundant, abstracting the same type of motion as the fourth, does not prevent the abstracted actions from supporting the mastery of soccer or some other task. Both networks can be used. The algorithm's capacity to reduce a vast amount of complex, raw data to just a few states may help reinforcement learning approaches to finesse the curse of dimensionality [18].

We counted the number of point belonging to each network. Figure 7 shows that the first network captured 92% of the walking data, the second 95% of the turning right data, the third 89% of the data for turning left and 3.6% for turning right, the fourth captured 43% of the data for sidestepping right, and the fifth 84% of the data for sidestepping left. The total number of point from each pattern allocated to the networks is not 100% because the segmentation algorithm successfully excluded most outliers.

5   Conclusion

Our proposed algorithm abstracted five out of six types of humanoid motion through a process that combines learning and data point assignment among multiple neural networks. The networks perform periodic, temporally-constrained nonlinear principal component analysis. The decoding subnetworks generate motion patterns that accurately correspond to the five motions without including outliers caused by nondeterministic perturbations in the data. During 45% of training episodes, the algorithm generated no redundant networks; a redundant network appeared in 50% of the training episodes, and two appeared in 5% of them. The fourth and fifth networks represent the same type of motion. Although they would symbolize a redundant state in the reinforcement learning paradigm, this does not prevent the learning of a complex task. In companion papers [15, 16], we propose a method that successfully removes redundant networks according to the proximity of their predictions. In future work, we will improve segmentation by competitively reassigning temporally-adjacent data points to the network that predicts the points with the least error.

References

1. Inamura, T., Toshima, I., Nakamura, Y.: Acquiring motion elements for bidirectional computation of motion recognition and generation. In Siciliano, B., Dario, P., eds.: Experimental Robotics VIII, Springer (2003) 372–381
2. Fujii, T.: A new approach to the LQ design from the viewpoint of the inverse regulator problem. IEEE Transactions on Automatic Control 32 (1987) 995–1004
3. Zatsiorsky, V.M.: Kinematics of Human Motion. Human Kinetics, Urbana Champaign (2002)
4. Okada, M., Tatani, K., Nakamura, Y.: Polynomial design of the nonlinear dynamics for the brain-like information processing of whole body motion. In: IEEE International Conference on Robotics and Automation. (2002) 1410–1415
5. Kramer, M.A.: Nonlinear principal component analysis using autoassociative neural networks. Journal of the American Institute of Chemical Engineers 37 (1991) 233–243
6. Tatani, K., Nakamura, Y.: Dimensionality reduction and reproduction with hierarchical NLPCA neural networks extracting common space of multiple humanoid motion patterns. In: Proceedings of the IEEE International Conference on Robotics and Automation, Taipei, Taiwan (2003) 1927–1932
7. Malthouse, E.C.: Limitations of nonlinear PCA as performed with generic neural networks. IEEE Transactions on Neural Networks 9 (1998) 165–173
8. Ridella, S., Rovetta, S., Zunino, R.: Adaptive internal representation in circular backpropagation networks. Neural Computing and Applications 3 (1995) 222–333
9. Ridella, S., Rovetta, S., Zunino, R.: Circular backpropagation networks for classification. IEEE Transaction on Neural Networks 8 (1997) 84–97
10. Kirby, M.J., Miranda, R.: Circular nodes in neural networks. Neural Computation 8 (1996) 390–402
11. LeCun, Y., Bottou, L., Orr, G.B., Müller, K.R.: Efficient BackProp. In Orr, G.B., Müller, K.R., eds.: Neural Networks: Tricks of the Trade, Springer (1998) 1–44
12. MacKay, D.J.: Probable networks and plausible predictions: A review of practical Bayesian methods for supervised neural networks. Network: Computation in Neural Systems 6 (1995) 469–505
13. Chalodhorn, R., Aono, M., Ooga, J., Ogino, M., Asada, M.: Osaka University "Senchans 2003". In Browning, B., Polani, D., Bonarini, A., Yoshida, K., eds.: RoboCup-2003: Robot Soccer World Cup VII, Springer Verlag (2003)
14. Tenenbaum, J.B., de Silva, V., Langford, J.C.: A global geometric framework for nonlinear dimensionality reduction. Science 290 (2000) 2319–2323

15. Chalodhorn, R., MacDorman, K., Asada, M.: Automatic extraction of abstract actions from humanoid motion data. In: IROS-2004: IEEE/RSJ International Conference on Intelligent Robots and Systems, Sendai, Japan (Submitted)
16. MacDorman, K., Chalodhorn, R., Ishiguro, H., Asada, M.: Protosymbols that integrate recognition and response. In: EpiRob-2004: Fourth International Workshop on Epigenetic Robotics, Genoa, Italy (2004)
17. MacDorman, K., Chalodhorn, R., Asada, M.: Periodic nonlinear principal component neural networks for humanoid motion segmentation, generalization, and generation. In: ICPR-2004: International Conference on Pattern Recognition, Cambridge, UK (2004)
18. Sutton, R.S., Barto, A.G.: Reinforcement Learning: An Introduction. MIT Press, Cambridge, Massachusetts (1998)

Three-Dimensional Smooth Trajectory Planning Using Realistic Simulation

Ehsan Azimi[1,2], Mostafa Ghobadi[1,2], Ehsan Tarkesh Esfahani[1,2],
Mehdi Keshmiri[1], and Alireza Fadaei Tehrani[1,2]

[1] Mechanical Engineering Department, Isfahan University of Technology (IUT)
[2] Robotic Center, Isfahan University of Technology
azimi@asme.org, {Ghobadi, e_tarkesh}@mech.iut.ac.ir,
mkeshmiri@istt.ir, mcjaft@cc.iut.ac.ir

Abstract. This paper presents a method for planning three-dimensional walking patterns for biped robots in order to obtain stable smooth dynamic motion and also maximum velocity during walking. To determine the rotational trajectory for each actuator, there are some particular key points gained from natural human walking whose value is defined at the beginning, end and some intermediate or specific points of a motion cycle. The constraint equation of the motion between the key points will be then formulated in such a way to be compatible with geometrical constraints. This is first done in sagittal and then developed to lateral plane of motion. In order to reduce frequent switching due to discrete equations which is inevitable using coulomb dry friction law and also to have better similarity with the natural contact, a new contact model for dynamic simulation of foot ground interaction has been developed which makes the cyclic discrete equations continuous and can be better solved with ODE solvers. Finally, the advantages of the trajectory described are illustrated by simulation results.

1 Introduction

Today, humanoid robots have attracted the attention of many researchers all over the world and many studies have been focused on this subject. When biped robots are viewed as general dynamical mechanisms, there are several problems such as limited foot-ground interaction, discretely changing dynamics, naturally unstable dynamics and multi-variable dynamics, which make using distinctive methods inevitable in controlling them and in planning trajectory for them [1].

Planning walking pattern is one of the most challenging parts. Recording human walking data aided Zerrugh to investigate human walking pattern [2]. Tad McGeer [3] developed the idea of using gravity as a main source of generating natural dynamic walking in the sagittal plane and Kuo [4] developed the stable walking in lateral motion.

Huang and his team [5] planned the walking pattern for a planner biped robot by interpolating third order spline between some key points of motion cycle. In the present study, we used this method to fulfill the trajectory for sagittal motion while also considering energy consumption optimization as well as geometrical consistency.

D. Nardi et al. (Eds.): RoboCup 2004, LNAI 3276, pp. 381–393, 2005.

Proposing a smooth pattern for lateral motion, we used only two parameters such that making changes in them could generate all possible trajectories.

Bipeds divide into two main categories: Static walkers, whose control is treated as a static control and that have very low speeds; and Dynamic walkers, which should be controlled dynamically and walk at relatively higher speeds. In the static case, the problem is solved as a static problem at each instant while in the dynamic case, differential equations should be solved at a time interval.

To ensure that an arbitrary trajectory will satisfy the stability of a biped, a criterion should be used. The most acceptable one for dynamic walkers is zero moment point (ZMP) criterion of stability but center of gravity (COG) should be evaluated for stability of static walkers. In some studies [6, 7, 8], a suitable ZMP is first assumed and then a trajectory is found that can maintain it. Others find the desired ZMP among several smooth trajectories. The second method is more useful (see section 2.2). Based on previous work [5], two parameters were used to compute all possible sagittal trajectories. We developed this method and added two parameters to cover all possible lateral trajectories. By executing a search algorithm for these trajectories and evaluating the ZMP criterion, stable walking trajectory is filtered out. Finally, the most stable trajectory was obtained by comparing different trajectories. The use of two parameters allowed us to overcome one of the biped problems. This will greatly reduce massive calculations that would be inevitable when dealing with several parameters.

As said above, one of the biped problems is the discontinuity of the state equations which is due to following reasons:

1. The dynamics of a bipedal walker changes as it transits from a single support to a double support and back to the single support.
2. Because of the multiple state friction equations such as Coulomb model, the dynamic equation of the contact is discrete in each phase.

To overcome this problem, we utilized a new contact model, which is more continuous and also has the advantage of being more similar to the real nature of the contact. In order to consider elasticity as a material characteristic of surfaces of contact, Young modulus of elasticity was used in calculating ground normal force on foot.

2   Three-Dimensional Walking Pattern Generation

For a biped robot to be able to walk with stability, two approaches are generally available. One is playing back offline trajectory and the other is applying online control techniques. The latter uses heuristic methods that lead to a motion in which specific trajectories, precision, and repeatability are not important factors and the motion can be different in each step. The former implies a pre-recorded joint trajectory to a robot based on some desired stability criteria, leading to a precise repeatable motion. The online method is useful in environments with unexpected surface terrains while the offline one must be selected in unchangeable situations especially while the massive calculations during motion are to be avoided. Since the ground condition is not assumed to change, the problem can be deled deeply and more accurate results will obtain. With these considerations in mind, the offline method was selected in the present study.

2.1 3D Kinematic of a Biped Motion

A biped with two legs, two arms and a trunk is considered here as shown in Fig. 1. Each leg consists of six degrees of freedom (DOF), two DOFs for hip, three for ankle, and one for knee. Each arm is composed of 3 DOFs. Physical properties for each part of the robot can be found in Table I.

The general motion of a biped consists of several sequential steps each composed of two phases: single support and double support. During the single support phase, one leg is on the ground and the other is in the swinging motion. The double support phase starts as soon as the swinging leg meets the ground and ends when the support leg leaves the ground. To be similar to human walking cycle, the period of this phase is considered to be 20% of the whole cycle. T_c and D_s are the duration and length of the whole cycle while T_d represents duration of double support phase in the cycle.

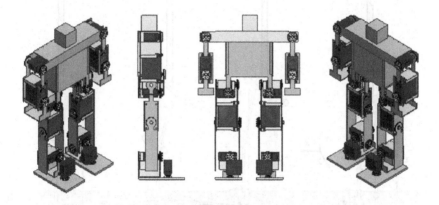

Fig. 1. Structure of the studied Biped

Table 1. Physical properties of a robot

Parameter	Foot				Shank		Thigh		Trunk	
Length (m)	L_{ab}	L_{an}	L_{af}	L_w	L_s	ab	L_t	ab	L_{tr}	H
	0.05	0.06	0.07	0.07	0.10	0.02	0.10	0.02	0.08	0.08
Mass (kg)	M_f				M_s		M_t		M_{tr}	
	0.15				0.3		0.4		1.2	

It can be shown that having ankle and hip trajectories is sufficient to formulate the trajectories of all leg joints uniquely [5]. With this assumption but unlike most previous works, we planned the motion not only in sagittal but also in lateral plane of motion. Let (x_a, y_a, z_a) and (x_h, y_h, z_h) denote the ankle and hip joint coordinates and $_a$, $_h$ to be their relative sagittal rotational angles and $_l$ to be the lateral rotational angle of the leg, All calculations are carried out for one leg but it can simply be repeated for the other leg. Considering each joint's location at some specific points and interpolating them by a particular function that both matches human motion and is compatible with geometrical and boundary conditions, we are able to plan walking patterns for the ankle and hip. The functions used here must behave smoothly. To satisfy this condition, the functions must be differentiable and their second derivative must be continuous. For this purpose, polynomial, spline, and sinusoidal functions are used here that normally satisfy these requirements.

Building upon pervious works with sagittal motion, we also consider lateral motion and use the same method as used in Huang et al. [5] to fulfill the 3D trajectory planning in a comprehensive manner.

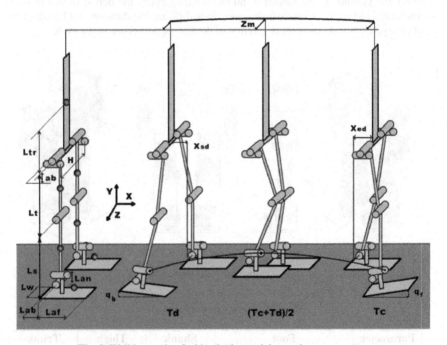

Fig. 2. Walking cycle of a biped robot and the motion parameters

Ankle Trajectory

A fourth degree polynomial is used when the foot is in the swinging motion while a sinusoidal function is used for the interval of foot-ground contact. The use of the latter function makes the foot trajectory compatible with geometrical constraints.

Applying transitional conditions (2), these two curves are joined smoothly as in Fig. 3. The resulting curves are illustrated in (1). In these trajectories, the foot angle is designed to vary in motion to better match human motion. If it is assumed to be always level with the ground, (like in most previous works), the biped speed will seriously reduce. T_m is the instant in which x_a and y_a reach L_{am} and h_{am}, respectively. $F(t)$ and $G(t)$ are 4th and 5th order polynomial functions, respectively, and p is used for adjusting swing acceleration.

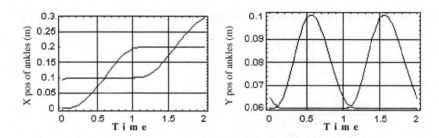

Fig. 3. The trajectory of ankle during walking simulated in *Mathematica*

$$0 \le t \le T_d:$$
$$x_a = L_{af}(1 - \cos\theta_a) + L_{an}\sin\theta_a$$
$$y_a = L_{af}\sin\theta_a - L_{an}(1-\cos\theta_a), \theta_a = q_b(\frac{t}{T_d})^P$$

$$T_d \le t \le T_m:$$
$$x_a = F_1(t), \; y_a = F_2(t), \; \theta_a = G(t)$$

$$T_m \le t \le T_c:$$
$$x_a = F_3(t), \; y_a = F_4(t), \; \theta_a = G(t)$$

$$T_c \le t \le T_c + T_d:$$
$$x_a = D_s + L_{ab}(1 - \cos\theta_a) + L_{an}\sin\theta_a$$
$$y_a = L_{ab}\sin\theta_a + L_{an}(1 - \cos\theta_a), \theta_a = q_f(\frac{T_c + T_d - t}{T_d})^P$$

$$T_c + T_d \le t \le 2T_c:$$
$$x_a = D_s \quad , y_a = 0 \quad , \theta_a = 0$$

$$(1)$$

Function $F_k(t)$ and $G_k(t)$ in the equation above are calculated using the constraints given below:

$$t = T_d$$
$$x = L_{af}(1 - \cos q_b) + L_{an}\sin q_b,$$
$$y = L_{af}\sin q_b + L_{an}(1 - \cos q_b), \; \theta_a = q_b$$
$$t = T_m$$
$$x = L_{am}, y = H_{am},$$

$$(2)$$

Hip Trajectory

Hip trajectory is interpolated with third–order splines assuming to have its highest position midway in the single support phase and its lowest midway in the double support phase. This makes the body to consume the energy in an optimal way as it has the minimum height while it has maximum velocity and vice versa. This situation is represented in (3) and in the corresponding curves in Fig. 4. Transitional conditions for hip equations are listed in Table II. In (3), x_{ed} and x_{sd} denote distances along the x axis from the hip to the ankle of the support foot and h_{min} and h_{max} designate the height of hip at minimum and maximum heights during its motion.

The equations of hip motion in sagittal plane in x and y positions are given by (3).

$$x \begin{cases} a_1 t^3 & b_1 t^2 & c_1 t & d_1 & \text{if} & 0 & t & T_d \\ a_2 t^3 & b_2 t^2 & c_2 t & d_2 & \text{if} & T_d & t & T_c \end{cases} \tag{3}$$

Where

$$a_1 \quad \frac{2(D_s(T_c \quad T_d) \quad T_c(x_{ed} \quad x_{sd}))}{T_c(T_c \quad T_d)T_d^2}$$

$$a_2 \quad \frac{2(D_s(T_c \quad T_d) \quad 2T_c(x_{ed} \quad x_{sd}))}{T_c(T_c \quad T_d)^2 T_d}$$

$$b_1 \quad \frac{3(D_s(T_c \quad T_d) \quad T_c(x_{ed} \quad x_{sd}))}{T_c(T_c \quad T_d)T_d}$$

$$b_2 \quad \frac{3(T_c \quad T_d)(D_s(T_c \quad T_d) \quad T_c(x_{ed} \quad x_{sd}))}{T_c(T_c \quad T_d)^2 T_d}$$

$$c_1 \quad \frac{D_s(T_c \quad T_d)^2 T_d \quad T_c T_d(T_c \quad 2T_d)(x_{ed} \quad x_{sd})}{T_c(T_c \quad T_d)T_d^2} \tag{4}$$

$$c_2 \quad \frac{(D_s(T_c \quad T_d)(T_c^2 \quad 4T_c T_d \quad T_d^2) \quad T_c(T_c^2 \quad 3T_c T_d \quad 2T_d^2)(x_{ed} \quad x_{sd})}{T_c(T_c \quad T_d)^2 T_d}$$

$$d_1 \quad x_{ed}$$

$$d_2 \quad \frac{2D_s(T_c \quad T_d)T_d^2 \quad T_d(T_c^2 x_{ed} \quad T_d^2 x_{ed} \quad 2T_c T_d x_{sd})}{(T_c \quad T_d)^2 T_d}$$

Table 2. The constraint applied to equation of hip motion

Time	X	Time	y	z
0	x_{ed}	$0.5T_d$	h_{min}	$z^*=0$
T_d	D_s-x_{ed}	$0.5(T_d+T_c)$	h_{max}	$z^*=z_m$
T_c+T_d	D_s+x_{ed}	$0.5T_d+T_c$	h_{min}	$z^*=0$

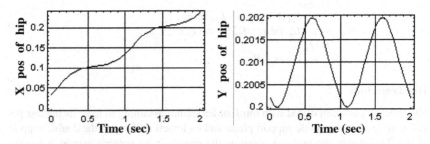

Fig. 4. The trajectory of hip joint during walking simulated in *Mathematica*

Assuming that the value of lateral displacement of the hip is zero midway in the double support and reaches its maximum midway in the single support phase, the lateral motion can be described by the lateral angle $_l$ (5). The motion is assumed to have sinusoidal form in the largest portion of the double support phase and tangents to

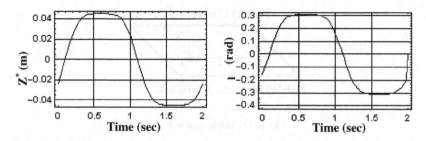

Fig. 5. The trajectory of hip joint in lateral plane during walking. Simulated in Mathematica

a forth-order polynomial mostly when the body is on one support (8). T_l is the duration of sinusoidal behavior, z_m is the maximum lateral deviation, and z* is the lateral position of the middle point.

Several lateral smooth trajectories can be obtained by varying these two parameters (z_m, T_l). The appropriate variation of lateral hip trajectory parameters are shown in Fig. 5. The Coefficient   ensures that z_m reaches its maximum in the mid single support phase.

$$\lambda \quad arctan(\frac{z_{sh}}{D}) \tag{5}$$

where, D is given by:

$$D \quad \sqrt{(x_{sh} \quad x_{sa})^2 \quad (y_{sh} \quad y_{sa})^2} \tag{6}$$

And Z_{sh} is substituted by either of:

$$z_{rh} \quad z^* \quad 0.5H \quad , \quad z_{lh} \quad z^* \quad 0.5H \tag{7}$$

depending on the supporting leg, right or left leg. Z^* is given by (8):

$$z^* \begin{cases} z_m sin(\dfrac{t \quad 0.5T_d}{T_l}) & (0 \quad t \quad \dfrac{T_d \quad T_l}{2}) \\ f_z(t) & (\dfrac{T_d \quad T_l}{2} \quad t \quad T_c \quad \dfrac{T_d \quad T_l}{2}) \\ z_m sin(\dfrac{t \quad T_c \quad 0.5T_d}{T_l}) & (T_c \quad \dfrac{T_d \quad T_l}{2} \quad t \quad T_c) \end{cases} \tag{8}$$

$$f_z(t) \quad a(t \quad 0.5(T_c \quad T_d))^4 \quad b(t \quad 0.5(T_c \quad T_d))^2 \quad c$$

Dynamic and Stability of a Biped Motion

There are several methods to ensure whether a particular trajectory is stable. As described above, we utilized the ZMP criterion which states that: if the ZMP is within the convex hull of all contact points, the stability of the robot is satisfied and the robot can walk [9]. In the biped, the stable region is the surface of the support foot (feet) (Fig. 6). The location of ZMP is given by (9).

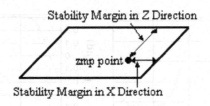

Fig. 6. The stable region of ZMP

$$
x_{zmp} \quad \frac{\sum\limits_{i=1}^{n} m_i(\ddot{y}_i \quad g)x_i \quad \sum\limits_{i=1}^{n} m_i \ddot{x}_i y_i \quad \sum\limits_{i=1}^{n} I_{iz} \ddot{\iota}_z}{\sum\limits_{i=1}^{n} m_i(\ddot{y}_i \quad g)}
$$

$$
z_{zmp} \quad \frac{\sum\limits_{i=1}^{n} m_i(\ddot{y}_i \quad g)z_i \quad \sum\limits_{i=1}^{n} m_i \ddot{z}_i y_i \quad \sum\limits_{i=1}^{n} I_{ix} \ddot{\iota}_x}{\sum\limits_{i=1}^{n} m_i(\ddot{y}_i \quad g)}
$$

(9)

The greater the distance between the boundaries of the stable region and the ZMP called stability margin, the higher the stability. So to find the safest trajectory, it is necessary to evaluate all possible trajectories planned by the method described above to satisfy this stability criterion. Then, the pattern in which the largest stability margin is gained will be selected as the optimum one.

For sagittal stability using previous works [5, 6, 7, and 10] and by defining different values for x_{ed} and x_{sd}, we were able to find the trajectory with maximum stability region through a search method.

Having developed this method for lateral motion, we used two lateral parameters T_l and z_m to make it possible to derive a highly stable, smooth hip motion without having to first design ZMP trajectory as other workers [8] had. Their method is associated with two disadvantages. First, they miss some suitable ZMP trajectories. Second, those trajectories not obtained through smooth functions may require too high an acceleration beyond the actuator's capacity so that the hip will fail to trace the desired trajectory.

3   Simulation Using Realistic Model

Modeling of contact is an important part of the modeling process. So using the equations that best describe the physics of contact and that also reduce the frequency of switching between equations is essential in avoiding highly frequently switching between differential equations which are usually hard to solve numerically.

3.1   New Continuous Contact Model Versus Traditional Discrete Types

It is well known that classical friction models such as Coulomb and Karnopp in which the relation between friction forces and the relative velocity between contact surfaces are discontinuous generate discontinuity in the biped robot model. This could result in difficulty in integrating equations of motion into one.

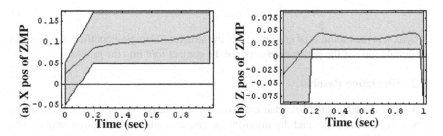

Fig. 7. ZMP trajectory in the stable region in x and z directions. *The gray surface is the stable region of ZMP*

Considering different types of friction laws, the Elasto-Plastic law[11], which renders both pre-sliding and stiction is selected.

To be able to apply this model to foot ground interaction, the foot is divided into some segments. The described model is employed to each segment by using (10) and (11) to calculate the reaction force in each segment (Fig. 8). The friction force in Elasto-Plastic Model is written as (10)

$$f_f \quad {}_0 z \quad {}_1 \dot{z} \quad {}_2 \, , \quad i \quad 0 \tag{10}$$

Details of this equation and description of the parameters are illustrated in the Appendix.

Finally, summing the effects of all segments, total surface reaction applied to the foot will be obtained. Using (10), the traditional two-state friction force will change to a single-state one and this avoids switching problems.

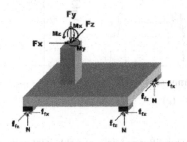

Fig. 8. The schematic of reaction force of contact model with four elements

For the normal reaction of the contact surface, Young's modulus-coefficient of restitution was used [12] to account for the elasticity of foot and ground material and is given by (11).

$$N \quad k \quad c^{\cdot} \tag{11}$$

where, k is the modulus of elasticity; c is the structural damping coefficient of contact material; and   is the penetration depth of the contact foot into the ground.

3.2 Simulation Results

To select the actuators, their characteristics were required. We employed the resulting trajectory in this paper and, by imposing inverse dynamics, the required torques were calculated (Fig. 9).

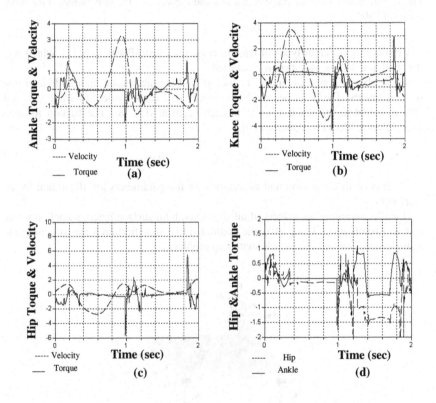

Fig. 9. Torque and velocity for different joints. a) Ankle joint in the sagittal plane b) Knee joint in sagittal plane c) Hip joint in the sagittal plane While d) Ankle and hip joint torque graphs in lateral direction.(*All torques and velocity measured in Nm and rad/s respectively*)

The graphical output of the simulation for two steps of walking of a robot is visualized (Fig10).

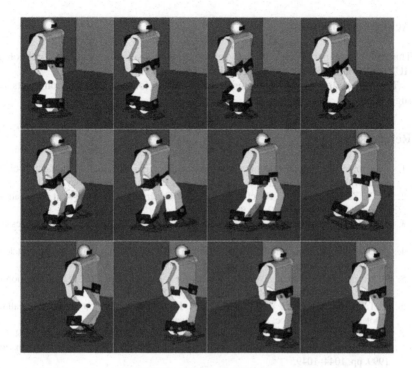

Fig. 10. The simulation result of 2 steps

4 Conclusion

In this paper, we described and developed a method for planning 3D smooth, stable walking pattern and, by using a new contact model in order to capture a better similarity with the real environment. The major contributions of the method are:

> Using sinusoidal functions, more useful trajectories for foot motion can be produced considering geometrical consistency;
>
> Energy consumption efficiency is achieved since height value changes inversely proportional to its speed;

1. Lateral trajectories are added to sagittal ones in order to simulate three dimensional motion;
2. ZMP criterion was developed for lateral motion so that the trajectory is stable in a general 3D motion;
3. Using a new contact model, the switching problem in differential equations is prevented which results in similarity with real natural contact.

Future study may consist of analyzing the effects of the upper body motion at very high speeds and online adaptation of trajectory to different ground conditions.

Acknowlededements

This study was partly supported by grant from Isfahan University of Technology (IUT) and Isfahan Science and Technology Town (ISST).

The authors also would like to thank the reviewers for their valuable comments and suggestion.

References

1. J. E. Pratt, "Exploiting Inherent Robustness and Natural Dynamic in the Control of Bipedal Walking Robot", Phd Thesis Massachusetts Institute of Technology, 2000.
2. M. Y. Zarrugh and C.W. Radcliffe, "Computer Generation of Human Gait Kinematics", Journal of Biomech, vol. 12, pp. 99–111, 1979.
3. T. McGeer, "Passive Walking With Knees", in Proc. of IEEE Int. Conf. Robotics and Automation, 1990, pp. 1640–1645.
4. A. D. Kou, "Stabilization of Lateral Motion in Passive Dynamic Walking", The Int Journal of Robotic Research, 1999, pp 917-930
5. Q. Huang et al., "Planning Walking Pattern for a Biped Robot", IEEE Int Transition on Robotic and Automation, Vol.17, N0.3, 2001, pp 280-289
6. A. Takanishi, M. Ishida, Y. Yamazaki, and I. Kato, "The realization of dynamic walking robot WL-10RD", in Proc. of Int. Conf. of Advanced Robotics, 1985, pp. 459–466.
7. A. Dasgupta and Y. Nakamura, "Making Feasible Walking Motion of Humanoid Robots from Human Motion Capture Data", in Proc. of IEEE Int. Conf. Robotics and Automation, 1999, pp. 1044–1049.
8. S.Kagami, T. Kitagawa, K.Nishiwaki, T. Sugihara, M. Inaba and H. Inoue", A Fast Dynamically Equilibrated Walking Trajectory Generation Method of Humanoid Robot", Autonomous Robots , Vol.12, No.1, Kluwer Academic Publishers, pp.71--82., Jan., 2002
9. Q. Huang, et al. "A high stability, smooth walking pattern for a biped robot", in Proc. of IEEE Int. Conf. Robotics and Automation, 1999, pp. 65–71.
10. Q. Huang, S. Sugano, and K. Tanie, "Stability compensation of a mobile manipulator by manipulator motion: Feasibility and planning", Adv. Robot., vol. 13, no. 1, pp. 25–40, 1999.
11. P. Dupont, B. Armstrong and V. Hayward, "Elasto-Plastic Friction Model: Contact Compliance and Stiction", in Proc. of American Control Conf, 2000, pp. 1072-1077.

Appendix: Elasto-Plastic Friction Law

In order to remove the discontinuity of the contact equation in Elasto-Plastic Model, z parameter is used to present the state of strain in frictional contact. The only difference which makes this theory more comprehensive (inclusive) than LuGre is an extra parameter which provides stiction and excludes pre-sliding. Using this model, the equation of friction may be written as:

$$f_f \quad {}_0 z \quad {}_1 \dot{z} \quad {}_2 v \tag{12}$$

where, σ_0 and σ_2 are Coulomb and viscous friction parameters and σ_1 provides damping for the tangential compliance, also $\eta(x,\dot{z})$ is used to achieved stiction. $\dot{z}$ is governed by

$$\dot{z} = \dot{x}(1 - \eta(x,\dot{z})\frac{\sigma_0}{f_{ss}(\dot{x})}\mathrm{sgn}(\dot{x})z)^i \tag{13}$$

where, i and $\dfrac{\sigma_0}{f_{ss}(\dot{x})}$ is always positive. $f_{ss}(\dot{x})$ is expressed by

$$\tag{14}$$

$$f_{ss}(\dot{x}) = f_c + (f_{ba} - f_c)e^{-\left|\frac{\dot{x}}{v_s}\right|}$$

where, f_c is obtained by Coulomb law. $\dot{x}$ is the relative velocity between the two contact surfaces and also f_{ba} is the breakaway force. Finally, v_s is the characteristic velocity.

$$\eta(z,\dot{x}) = \begin{cases} 0 & ,|z| \le z_{ba} \\ 0.5\sin\left(\dfrac{z - (0.5(z_{max} - z_{ba}))}{z_{max} - z_{ba}}\right) & , z_{ba} \le |z| \le z_{max}1 \\ 1 & ,|z| \ge z_{max} \end{cases} \tag{15}$$

$$\eta(z,\dot{x})$$

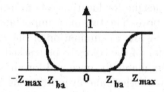

Fig. 11. Equations of $\eta(z,\dot{x})$ for $\mathrm{sgn}(\dot{x}) = \mathrm{sgn}(z)$

Plug and Play: Fast Automatic Geometry and Color Calibration for Cameras Tracking Robots

Anna Egorova, Mark Simon, Fabian Wiesel, Alexander Gloye, and Raúl Rojas

Freie Universität Berlin, Takustraße 9, 14195 Berlin, Germany
http://www.fu-fighters.de

Abstract. We have developed an automatic calibration method for a global camera system. Firstly, we show how to define automatically the color maps we use for tracking the robots' markers. The color maps store the parameters of each important color in a grid superimposed virtually on the field. Secondly, we show that the geometric distortion of the camera can be computed automatically by finding white lines on the field. The necessary geometric correction is adapted iteratively until the white lines in the image fit the white lines in the model. Our method simplifies and speeds up significantly the whole setup process at RoboCup competitions. We will use these techniques in RoboCup 2004.

1   Introduction

Tracking colored objects is an important industrial application and is used in the RoboCup small-size league for locating robots using a video camera which captures the field from above. The two most important problems which arise in this kind of object tracking are: a) elimination of the geometric distortion of the cameras, and b) calibration of the colors to be tracked. It is not possible to calibrate geometry and colors manually, once and for all, since lightning conditions change from one place to another, and even from one hour to the next.

In this paper we describe the techniques we have developed for fast calibration of the global camera(s) used in the RoboCup small-size league. The paper is organized as follows. First we comment on related work. Then we describe in detail our new semi-automatic color calibration method and compare its results to the hand-optimized ones. The next section deals with the calibration of the geometric transformation of the field and compare the automatic with the manual method. Finally, we describe our future plans.

2   Related Work and Motivation

Zrimec and Wyatt applied machine learning methods to the color calibration problem [10]. They recognize regions delimited by edges and classify them according to features such as average hue, saturation, intensity, and others. A computer vision module for Sony robots uses those features to locate field landmarks.

D. Nardi et al. (Eds.): RoboCup 2004, LNAI 3276, pp. 394–401, 2005.

Another approach for automatic calibration is to compute global histograms of images under different lightning conditions. Lookup tables for color segmentation are initialized in such a way as to make the new histogram equal to that found under controlled conditions [6]. In our case this approach would not work, since we do not use a single lookup table for the whole field. We recognize colors locally, using a color map which can change from one part of the field to the other. Some authors have tried decision trees in order to segment colors independently from light. However, they focus on object localization robustness and do not deal with the automation of the calibration [1].

Regarding the correction of the geometric distortion introduced by a camera, the canonical approach relies on determining first intrinsic camera parameters and then extrinsic parameters [2]. The intrinsic parameters can be measured in the laboratory. The extrinsic can be fitted by least squares, identifying points on the image with points in a geometric non-distorted model. We do not want to identify points on the field by clicking on them; the software should be able to automatically recognize the orientation of the camera and to select relevant points for matching them with the model.

Whitehead and Roth described an evolutionary optimization approach to camera auto-calibration [9]. Their method does not apply to our case, because they optimize the fundamental calibration matrix directly, without considering the radial camera distortion. Projective geometry methods alone solve one part of our problem, but not the whole problem. Some authors have studied real-time distortion correction for digital cameras, but without handling the projective transformation correction [4].

3   Semi Automatic Color Calibration

In [7] we described our use of color maps for robust color recognition of the robot's markers. They consist of a virtual grid superimposed on the field, one grid for each important color. At each grid node we store the RGB parameters of the color and the size of visible color markers for that part of the field, as seen by the camera. The grid must be initialized before starting to track objects, that is, we need an initialization step for each individual color. The color and marker size maps are further adapted during play, so that the tracking system updates the grid information whenever a change of illumination occurs.

3.1   Initializing the Color Maps

The global camera captures images in RGB format. This color space is not very useful when trying to achieve color constancy. Therefore, we operate in the HSV(hue/saturation/intensity value) color space. Fig. 1 shows the variability of the HSV components over the field as captured in our lab.

Fig. 2 shows the functional relationship between the intensity of the background pixels (green) and the color blobs on the field. The relationship is an affine function, whose parameters can be computed from the background intensity and from two probes of the desired color positioned on the field. The function

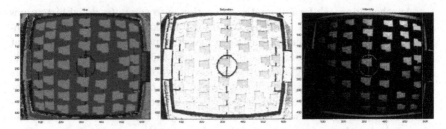

Fig. 1. The HSV components for the field of play with blue test blobs on it. The first image shows the hue component, the second the saturation, and the third the intensity. As can be seen, the intensity changes significantly throughout the field

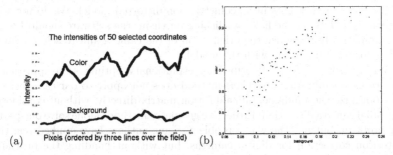

Fig. 2. The graph compares the intensity values of some random points on the green background and the color map. Illumination inhomogeneities produce different changes for different colors – scaling and translation factors are present: (a) shows the dependency of the intensities as a function of the field coordinates (three lines over the field); (b) shows the intensity of the color blobs as a function of the intensity of the background at the same points. The correlation is clearly visible

has two parameters: an additive constant and a scale factor. We compute both twice, from two probes, and average them for subsequent computations. More probes could also be taken, resulting in better averaging.

Given b_1 and b_2, the intensity of the background color at the point P_1 and P_2, and given c_1 and c_2, the intensities of the color C whose map we want to derive at the points P_1 and P_2, respectively, the translation factor t and the scale factor s are given by

$$t = \frac{(c_1 - b_1) + (c_2 - b_2)}{2} \qquad s = \frac{\left(\frac{c_1}{b_1}\right) + \left(\frac{c_2}{b_2}\right)}{2} \qquad (1)$$

When the background intensity b at a point P is known, the unknown intensity c of the color C at this point P is given by

$$c = bs + (b + t)$$

We also want to estimate the size of the robot markers (color blobs) around the field. The apparent size of the markers varies with lightning conditions and camera distortion. We assume that their size is proportional to the intensity of the blob. We estimate the marker size by interpolating the sizes of the samples, according to the intensity. One can also use the camera distortion to calculate the size of the robot markers according to it.

To reduce the influence of noise, a median color is computed for the marker and the background color at the picked point. The radius for the area around the blob is predefined with respect to the blob-object which is identified by this color (for example the orange ball is a little bit smaller than a team marker). The size of the median filter for the background color, for its color grid, is exactly the size of a grid tile [7].

3.2 Results and Comparison

In Table 1, the different forms of initializing the color map before starting a game are compared to the manual adjustment method, exemplarily for the ball. As shown, automatic initialization gives better results and smaller errors, compared to a uniform initialization of the color map. The most important improvement is the reduction of the maximum error, the relevant magnitude when trying not to lose track of an object in the field. The improvement in estimation of the ball size, as seen from the camera at different coordinates on the field, is also significant.

Table 1. Statistical results comparing initializations of color maps. The table shows the performance (relative to the hand-optimized color map) of a uniformly initialized color map and two different automatically initialized maps. The performance of one specific map is measured with the maximum and mean percentage deviation of computed values from those in the hand-optimized map

	Deviation from hand-optimized			
	Uniform		Automatic	
	maximum	mean	maximum	mean
hue	1.94%	0.85%	1.28%	0.61%
saturation	7%	1.4%	4.6%	1.7%
intensity	48%	9%	25%	9.4%
size	31 pixel	13 pixel	18 pixel	9.15 pixel
RGB distance	27.9%	5.2%	14.5%	5.7%
HSV distance	48%	8.7%	25.4%	9.73%

4   Automatic Geometric Calibration

The second camera setup problem we consider is the correction of the geometric distortion. Traditional methods require identifying pairs of points in the image and model space. One could artificially define those calibration points by using a carpet with marks, but that requires too much manual intervention. In this paper we show how to use the white lines on the field to determine the parameters of the transformation. The correlation between the extracted, and transformed lines, and the lines of the model, is a measure for the quality of the transformation. This measure enables us to find the best transformation with conventional optimization algorithms.

Our method is divided into the following steps: First, we extract the contours of field regions from the video image. Next, we find a simple initialization, which roughly matches the lines of the field to the model. Finally, we optimize the parameters of the transformation.

4.1   Extraction of Contours

The non-white regions on the field are found by applying our region growing algorithm as described in [3]. Discrimination between white and non-white pixels is based on their relative intensity with respect to the local background. First, we average the intensity of a larger region around a given pixel, which is assured to contain a high percentage of background pixels. Relative to that rough approximation, we reject foreground pixels in a smaller region and locally determine the intensity from the remaining pixels. The contours of the regions found are the borders of the field lines (see Fig. 3).

Fig. 3. F-180 Field Padova 2003, with artificial shadows with strong edges. Dashed lines show the field and out-of-field contours

4.2   Quality Measure

Assume that we have a hypothetical geometric transformation T and that we want to measure its error. Since we cannot determine the exact location of a contour point p in the model, we approximate its error $E(T, p)$ with the distance between the transformed point $T(p)$ and the nearest white line in the model. For a set of points P, we compute the quality error $\hat{E}$ as

$$\hat{E}(T, P) := \bar{E}(T, P) + \sigma\left(E(T, P)\right)$$

where $\bar{E}$ is the mean error, and σ its standard deviation.

4.3   Initialization of the Transformation

To minimize the complexity of the optimization step, a smart initialization of the transformation T is useful. We assume that the field has only limited distortion, radially and in the perspective. A linear transformation is a good initial approximation, except for the alignment of the field. Finding a matching pair of rectangles in model and field let us compute such a transformation. In our case a bounding box around region 1 (see Fig. 3) results in a good approximation to a rectangle corresponding to one half of the field in the model. The alignment can easily be determined by testing all four orientations of the bounding box as initialization for the linear transformation. Using the quality measure described above for the contour of region 1, the best match identifies the orientation of the field, since the region is not symmetric with respect to rotation. The association between the field sides in image and model is arbitrary. The transformation T can be initialized accordingly to the linear transformation T_{init}. We use a bi-quadratic transformation as our standard transformation (see Section 4.4), but other transformation models are possible.

4.4 Optimization

We use for both coordinates in model space a biquadratic interpolation in a single 3×3 grid to determine the coordinates in model-space, giving us 18 parameters to be optimized. The initialization can be analytically derived from T_{init}. However, our approach relies only on the convergence properties of the transformation function and not on the function itself.

Given the transformation T, gradient descent with linearly falling step-length suffices to optimize the placement of the vertices with sub-pixel accuracy. To achieve faster convergence adaptive selection of step-length and/or conjugate-gradient methods can be used. Gradient descent only requires a non-degenerate initialization with the approximate orientation. In our experiments, even the whole image as rectangle for the initialization converged against the global minimum.

Other transformation-functions can be optimized similarly. Depending on its convergence properties, a global optimization method may be necessary, for example simulated annealing.

4.5 Calibration Results

We applied our algorithm to a set of real video images, captured two different camera-systems (Basler 301fc, Sony XC-555). We presented the algorithm with two different fields, the field of the Padova World Cup 2003, built according to the F-180 2003 rules, and our lab-field, which has slightly different dimensions and several rough inaccuracies. The images had different brightness, and some were artificially modified to simulate shadows with strong edges (as can be seen in Fig. 3), which normally are not present on the field and strain region-extraction more than the soft shadows usually observed. Furthermore, some images were rotated up to 20°.

For a correct field model, our algorithm could adjust the geometric biquadratic transformation for all images presented, given the constraints on the distortion, without further adjustments to the algorithm after the initial selection of the Padova images. Subsequent attempts to improve manually the parameters of the transformation function resulted only in worse results, both with respect to the average and standard-deviation of the error.

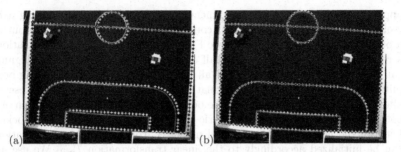

Fig. 4. The model matched on the rotated field (a) after the initialization and (b) after the optimization step

In order to speed up the algorithm, initially only a subset of the contour points is taken into account for the quality measure. The amount is gradually increased as the error decreases. We use an adaptive step-width, which slowly increases, unless the gradient step increases the error. Then the step-width is reduced. The step-width is independent of the norm of the gradient.

We started the optimization with a step-width of 1 pixel and only using every 20th contour point in the measure. The optimization step required on our set of images at most 6 seconds (5 seconds with conjugated gradient) to adopt the parameters in 1/100 pixel accuracy and in average 4 seconds (4 seconds) on an Athlon XP -2400+ (2GHz).

5   Future Work

The color map for a specific color is initialized by clicking on two markers during setup. This takes a few seconds. In the future, our vision system will automatically detect colors differing from the background, and will initialize a color map for them.

For the geometric calibration, we want to evaluate other transformation-functions, which are probably more accurate, due to their better modeling of the distortion.

The improvements reported in this paper have the objective of achieving true "plug & play" capability. In the future it should be possible just to place robots on the field and start playing immediately against another team, whose colors and marker orientation will be determined automatically. This would speed RoboCup competitions significantly.

References

1. Brusey, J., and Padgham, L., "Techniques for Obtaining Robust, Real-Time, Colour-Based Vision for Robotics", *Proceedings IJCAI'99 - International Joint Conference on Artificial Intelligence, The Third International Workshop on RoboCup - Stockholm*, 1999.

2. Forsyth, D. A., and Ponce, A., *Computer Vision: A Modern Approach*, Prentice Hall., 1st edition, 2002.
3. von Hundelshausen, F., and Rojas, R., "Tracking Regions", in D. Polani, B. Browning, A. Bonarini, K. Yoshida (Eds.): *RoboCup-2003 - Robot Soccer World Cup VII*, Springer-Verlag, 2004.
4. Gribbon, K. T., Johnston, C.T., Bailey, D.G., "A Real-time FPGA Implementation of a Barrel Distortion Correction Algorithm with Bilinear Interpolation", *Image and Vision Computing*, Palmerston North, New Zealand, pp. 408-413, November 26-28, 2003.
5. Jacobsen, "Geometric Calibration of Space Remote Sensing Cameras for Efficient Processing", *International Archives of Photogrammetry and Remote Sensing*, Vol.32, Part I, pp. 33-43.
6. Kulessa, T., and Hoch, M., "Efficient Color Segmentation under Varying Illumination Conditions", *Proceedings of the 10th IEEE Image and Multidimensional Digital Signal Processing Workshop*, July 12-16, 1998.
7. Simon, M., Behnke, S., Rojas, R.: "Robust Real Time Color Tracking" In: Stone, P., Balch, T., Kraetszchmar (eds): *RoboCup-2000: Robot Soccer World Cup IV*, pp. 239-248, Springer, 2001.
8. Rojas, R., Behnke, S., Liers, A., Knipping, L.: "FU-Fighters 2001 (Global Vision)", In: Birk, A., Coradeschi, S., Tadokoro, S. (eds): *RoboCup-01: Robot Soccer World Cup V*, Springer, 2001.
9. Whitehead, A., and Roth, G., "Evolutionary Based Autocalibration from the Fundamental Matrix", in S. Cagnoni, Stefano Cagnoni, Jens Gottlieb, Emma Hart, Martin Middendorf, Gnther R. Raidl (Eds.), *Applications of Evolutionary Computing – EvoWorkshops 2002, EvoCOP, EvoIASP, EvoSTIM/EvoPLAN*, Kinsale, Ireland, April 3-4, Springer-Verlag, 2002.
10. Zrimec, T., and Wyatt, A., "Learning to Recognize Objects - Toward Automatic Calibration of Color Vision for Sony Robots", *Workshop of the Nineteenth International Conference on Machine Learning (ICML-2002)*.

Real-Time Adaptive Colour Segmentation for the RoboCup Middle Size League

Claudia Gönner, Martin Rous, and Karl-Friedrich Kraiss

Chair of Technical Computer Science,
Technical University of Aachen (RWTH),
Ahornstr. 55, 52074 Aachen
{Goenner, Rous, Lraiss}@techinfo.rwth-aachen.de
www.techinfo.rwth-aachen.de

Abstract. In order to detect objects using colour information, the mapping from points in colour space to the most likely object must be known. This work proposes an adaptive colour calibration based on the Bayes Theorem and chrominance histograms. Furthermore the object's shape is considered resulting in a more robust classification. A randomised hough transform is employed for the ball. The lines of the goals and flagposts are extracted by an orthogonal regression. Shape detection corrects over- and undersegmentations of the colour segmentation, thus enabling an update of the chrominance histograms. The entire algorithm, including a segmentation and a recalibration step, is robust enough to be used during a RoboCup game and runs in real-time.

1   Introduction

A scenario for teams of mobile robots, in which optical sensors are very important, is RoboCup. In the middle size league this was forced by an environment, where objects like the orange ball can primarily be detected by optical sensors. The field is adapting towards the FIFA-rules step by step. The most important rule change in this category was the removal of the board. Illumination was decreased and variable lightening conditions, e.g. playing under sunlight, are subject of discussion.

At the moment RoboCup is ideal for applying colour segmentation algorithms. All objects have characteristic colours, e.g. the yellow and blue goals, permitting a classification in the colour space. To achieve frame rates between 10 and 30 Hz a fast scene segmentation is necessary. The main issue with colour segmentation, though, is calibrating the object colours.

Well known real-time colour segmentation methods are the definition of colour regions by thresholds [1, 2, 3], colour clustering [4], occupancy grids [5] and discrete probability maps [6]. Thresholds are either manually set [1] or trained by decision trees [2] and neural networks [3]. Probability maps may be approximated by Gaussian mixture-models [7] to save storage place.

To distinguish different colours by thresholds in an efficient manner, specialised colour spaces have been adapted to particular object colours [8]. In

D. Nardi et al. (Eds.): RoboCup 2004, LNAI 3276, pp. 402–409, 2005.

contrast to these threshold-based algorithms, probability maps make no assumptions about the outline of object colours. The number of colour regions does not need to be provided a priori as with k-means clustering or Gaussian mixture-models. Furthermore probability maps separate different objects based on the Bayes Theorem in a statistically sound manner and also consider the background. The occupancy grids of [5] are similar to probability maps, but do not model the background, use increment and decrement operations on the grids and accumulate a term considering the landmark confidence and area. Probability maps, however, distinguish a priori and object probabilities while computing the a posteriori probability.

In addition to colour information, the object shape may be utilised for object classification. Jonker, Caarls et al. [9] verify a ball hypothesis by two hough transforms, executed successively. Hanek, Schmitt et al. [10, 11] fit a deformable contour model to edges and propose an image processing that is independent of colour segmentation. Depending on the parametrisation of the contour model different shapes are detected. Their approach requires a rough initialisation of the adaptive contour model.

This work introduces a method to generate the mapping between single objects and their chrominance automatically. It is based on histograms which are combined according to the Bayes Theorem. Colour segmentation is combined with a shape detection to provide additional robustness and allow colour recalibration. The shape detection specialises on RoboCup and recognises the ball, the flagposts and the goals with their white posts. Whereas the adaptive colour segmentation is generally applicable, as long as a second segmentation is provided for the colour update. A prior version of the presented algorithm was published in [12], suggesting a supervised colour calibration.

2   Overview

Colour spaces that distinguish between luminance and chrominance are well suited for colour segmentation. Bright and dark objects of the same chrominance may be distinguished by their luminance. This work is based on the YUV space, which is supported by most video cameras and therefore does not require any conversion.

In order to compute the chrominance histograms of each object, an object detection is required that either does not utilise colour information or extrapolates incomplete results of a colour segmentation. The latter approach has been chosen here, sketched in Fig. 1. Image regions occupied by the objects of interest are marked using an initial table, which contains the mapping between the chrominance and the classified objects and is referred to as colour map. This map (Fig. 2) can be drawn by hand or recycled from a former tournament.

After applying the colour map, the object's contour and neighbourhood relations are checked to verify whether it really is a ball, a goal or a flagpost. On success histograms of the object colours are updated and a new colour map is calculated. To save computation time the histograms only need to be adjusted

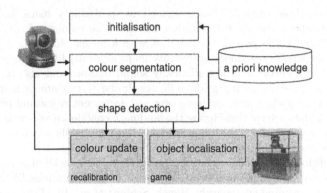

Fig. 1. Outline of the adaptive colour segmentation

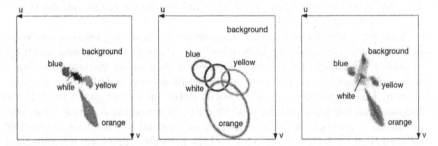

Fig. 2. *Left:* Colour map. *Middle:* Chrominance sets. *Right:* A posteriori probabilities

on change of the lighting conditions, i.e. the colour segmented area and the shape detected silhouette differ. Furthermore the colour map is recomputed after a fixed number of histogram updates. Colours that did not occur for a while are eliminated because the histograms are rescaled from time to time. To avoid complete deletion the histograms are scaled only if they contain more than a minimum number of entries.

Because of the iterative nature of the entire process a second map is added that specifies the maximum chrominance set (Fig. 2) for each object, including the background. In contrary to the colour map, these sets may overlap. They prevent divergence of the algorithm without loss of generality. Currently the maximum chrominance sets are provided a priori and kept static during a game.

3   Colour Training and Segmentation

Prior to colour segmenting an image, the chrominance histogram H_O for each object O in the object set $\mathcal{O}$ is computed. In this context, the background is also considered an element of $\mathcal{O}$. The relative frequencies in the histograms H_O can be interpreted as the a priori probabilities $P(u, v|O) = H_O(u, v)/|H_O|$ for occurrence of an (u, v)-pair under the assumption that object O is seen. The a posteriori probability, specifying how likely an (u, v)-pair belongs to object O, is given by the Bayes Theorem:

$$P(O|u,v) = \frac{P(u,v|O) * P(O)}{\sum\limits_{Q\in} P(u,v|Q) * P(Q)}. \tag{1}$$

McKenna, Raja et al. [7] note that the object probability $P(O)$ is related to the object size. However it is rather difficult to record images containing the objects in representative quantities. Therefore $P(O)$ is not computed, but set manually.

If a chrominance pair (u, v) has not been seen yet on object O an initial probability $P_{init}(u,v,O) = P_{init}(u,v|O) * P(O)$ is assumed, preventing that a chrominance pair gets a high a posteriori probability although it is barely contained in object O.

At each pixel the object with the highest a posteriori probability is classified. For an efficient classification the object belonging to a chrominance pair is stored in a look-up table, i.e. the colour map as visualised in Fig. 2. The figure also shows the probabilities of the classified objects, where darker values correspond to higher probabilities. In the centre, representing the achromatic area, white posts in the foreground overlap with white, black or grey background. The green floor is considered as background.

As long as the lighting condition remains constant the same colour map is applied and the complexity is identical with [13, 9]. On a lighting change the histograms and the colour map are updated. In order to eliminate old entries the histograms are scaled by a time-out factor prior to adding new images.

4   Ball Detection

Colour segmentation identifies image regions, which might contain interesting objects. Due to occlusions or bad segmentations, especially while the colour map is adapting, the object's segmentation might not be complete. Figure 3 shows a typical segmentation performed with an initial colour map. The upper half of the ball is not completely segmented. However, there is enough information to extrapolate a circle with the colour segmented contour points.

To detect partially occluded objects of known geometry we apply a randomised hough transform. Although the projection of a sphere on a plane results into an ellipse [11], we model the ball as a circle. This approximation is sufficient for verification and completion of the ball's shape.

A circle has three degrees of freedom: the centre (c_x, c_y) and the radius r, resulting in a 3D hough accumulator. If only a few round objects are inside the region of interest the search space is reduced with a random approach. The implemented randomised hough transform is based on the work of McLaughlin [14] and Xu, Oja et al. [15].

First, three points (x_i, y_i) on the contour are randomly selected. Considering the triangle, which is formed by these points, its centre of gravity coincides with

Fig. 3. *Left:* Initial colour segmentation. *Middle:* After recalibration. *Right:* Original image, in which the detected ball, flagpost and goal are surrounded by white lines

the circle's centre and all distances from the centre to these points have about the same length if the points lie on a true circle. If this test is passed the hypothised centre $(\hat{c}_x, \hat{c}_y)$ and its average distance $\hat{r}$ to all triples are inserted into the hough accumulator. It is advisable to store the occupied accumulator cells only because just a few cells are filled [14].

After the specified number of point triples has been drawn and added to the accumulator, the cells with the highest counts are verified. The best hypothesis is determined by the weighted sum of the ratio e_p between the circumference and the perimeter of the approximating polygon, the signal to noise ratio e_{SN}, and a radial error e_r. All measures are normalised to values between 0 and 1. Thereby the error e_r of the radius is modelled as Gaussian and a probability for the hypothised radius $P(\hat{r})$ is calculated from the mean μ_r and variance σ_r^2 of all point distances to the centre $(\hat{c}_x, \hat{c}_y)$:

$$e_r = P(\hat{r}) = \frac{1}{\sqrt{2\pi}\sigma_r} e^{\frac{(x - \mu_r)^2}{2\sigma_r^2}} \tag{2}$$

A hypothesis is valid if each measure is above a fixed threshold. If no circle is detected the procedure may be repeated. Partially occluded balls occasionally require such a repetition. Due to the randomised approach and noisy data it is recommended to verify more than one hypothesis at each cycle.

5   Detection of the Goal and Flagposts

Contrary to the ball, the shape of the goal and flagposts cannot be concluded from the colour segmented contour if a bad colour map is applied (Fig. 3). If so, the colour segmented area is ragged. Therefore object detection uses edges close to the colour regions. Also neighbourhood relations are considered. Sideways to a goal a white bar is expected. Due to partial occlusions a single collateral white region is sufficient. The flagposts are characterised by vertical yellow-blue or blue-yellow transitions.

As the robot's camera looks downward, the upper borders of the goal and flagposts are never visible. Thus only straight lines bordering the side and the bottom are checked. Lateral to the goal the inner and outer line of the white bar are determined. This allows construction of a histogram for white regions. The lower line of the flagpost is approximated by a horizontal line. All other lines are extracted by orthogonal regression [16] applied in a recursive manner. Thereby the point set is split along the regression line until the desired precision is reached or too few points are left. The line with the smallest distance to the colour segmented contour is chosen if more than one straight line fits.

6   Results and Discussion

At the moment colour recalibration supports the ball, the goals including the white posts and the flagposts. All other objects are modelled as background. By considering the objects' shape oversegmentations and undersegmentations are corrected. This enables learning new colours and updating the colour map. Figure 3 compares application of the initial and the calibrated colour map.

Due to the statistical approach, the system is able to distinguish between objects of similar colours, e.g. the ball and human skin (Fig. 4). If the colours of different objects overlap the probability $P(O)$ controls which object is preferred. As no restrictions are placed on the form of the clusters in colour space, an object may contain several non-connected clusters. Clusters belonging to different objects may interlock.

The invariance of the YUV colour space against changes in brightness and lighting is limited. Furthermore the colour map only contains values that have been seen before. Thus a recalibration is required upon new lighting conditions.

Fig. 4. *Left:* Colour segmentation of a ball, a bare foot and red posts. *Middle:* Circle detected in the left image. *Right:* Colour segmentation after training

Fig. 5. *Left:* Colour segmented ball after sudden occurrence of sunlight. *Middle:* Colour segmentation after recalibration. *Right:* Detected ball after recalibration

Both, Fig. 4 and 5 start with a segmentation based on the colour map trained for Fig. 3, but have been recorded in an office environment. In Fig. 4 the colour map specialises on the ball and discriminates skin and other red background objects. Figure 5 shows a ball under sunlight, which is recognised as an orange ball merely at the border regions, but still has a circular contour. After recalibration the entire ball is colour segmented again.

The system distinguishes between variably round objects and chooses the best. In Fig. 5 the ball is detected despite of its circular mirror on the floor. Partially occluded balls are extrapolated by the randomised hough transformation, see Fig. 6.

In general the detected circle is smaller than the true ball. This is caused by shadows, inter object reflections and a surface that is not lambertian. Depending on the lighting condition the ball's bottom is very dark and overlaps with achromatic colors (Fig. 6). The ball's boundary is sim-

Fig. 6. *Left:* Segmentation. *Right:* Extrapolated circle

ilar to yellow. Furthermore the ball's colours compete with skin and other background objects.

The circle's centre and radius tend to vary a bit from iteration to iteration. This is caused by randomly selecting point triples and a contour which is not perfectly circular. Increasing the number of samples, which are drawn when building the hough accumulator, results in marginal improvements. All circles in this paper are detected based on 25 samples.

The goal and the flagposts are modelled as quadrilaterals, whose lines are detected separately from another. Depending on the quality of edge detection and colour segmentation lines may be chosen that lie close by or on the objects, but not at the true border. The presented approach does not distinguish between the two lower lines of a goal looked at from the side.

However, too big goals and flagposts are critical merely at the beginning of the training process. Marking incorrect objects in an image influences the colour map if and only if the a posteriori probabilities of all other objects including the background are lower. These a posteriori probabilities primarily depend on the relative frequency in the histograms. As long as the histograms contain few entries, wrong entries result in higher frequencies as later on, when the correct maxima have many hits. Thus colour calibration is quite sensible to oversegmentations at the beginning. After some images have been trained, wrong segmentations have minimal effects. Old erroneous entries are decreased by the time-out factor and finally erased. To avoid divergence only potential object colours are filled into the histograms as specified a priori by the maximum chrominance sets.

The entire algorithm is implemented in C++ using the computer vision library   -Lib [17]. As runtime varies due to image content, 163 RoboCup images of 379×262 pixels are analysed on a Pentium III processor with 933 MHz, which is build into our robot. Both, the histograms and the colour map, are updated at each cycle. During a game recalibration only needs to be executed on lighting change. Object detection requires at least 24.28 msec, at most 55.68 msec and on the average 33.84 msec. Recalibration takes from 51.23 to 79.40 msec and averaged 59.23 msec, when the chrominance histograms and colour map are computed with 64 entries in each dimension. In our experiments we did not notice a quality difference between colour maps with 64×64, 128×128 and 256×256 entries.

References

1. Bruce, J., T.Balch, Veloso, M.: Fast and inexpensive color image segmentation for interactive robots. In: IROS'00. (2000) 2061–2066
2. Brusey, J., Padgham, L.: Techniques for obtaining robust, real-time, colour-based vision for robotics. In: RoboCup 1999. LNAI 1856, Springer (2000) 243–256
3. Amorosco, C., Chella, A., Morreale, V., Storniolo, P.: A segmentation system for soccer robot based on neural networks. In: RoboCup 1999. LNAI 1856 (2000) 136–147
4. Mayer, G., Utz, H., Kraetzschmar, G.: Toward autonomous vision self-calibration for soccer robots. In: IROS'02. (2002) 214–219
5. Cameron, D., Barnes, N.: Knowledge-based autonomous dynamic color calibration. In: Robocup 2003, Padua, Italy (2003)
6. Swain, M., Ballard, D.: Color indexing. Intl. J. of Computer Vision **7** (1991) 11–32
7. Raja, Y., McKenna, S., Gong, S.: Tracking and segmenting people in varying lighting conditions using color. In: 3rd Int. Conf. on Face and Gesture Recognition, Nara, Japan (1998) 228–233
8. Dahm, I., Deutsch, S., Hebbel, M., Osterhues, A.: Robust color classification for robot soccer. In: Robocup 2003, Padua, Italy (2003)
9. Jonker, P., Caarls, J., Bokhove, W.: Fast and accurate robot vision for vision based motion. In: RoboCup 2000. LNAI 2019, Springer (2001) 149–158
10. Hanek, R., Schmitt, T., Buck, S., Beetz, M.: Fast image-based object localization in natural scenes. In: IROS'02. (2002) 116–122
11. Hanek, R., Schmitt, T., Buck, S., Beetz, M.: Towards robocup without color labeling. In: Robocup 2002, Fukuoka, Japan (2002)
12. Gönner, C., Rous, M., Kraiss, K.F.: Robuste farbbasierte Bildsegmentierung für mobile Roboter. In: Autonome Mobile Systeme, Karlsruhe, Germany, Springer (2003) 64–74
13. Bandlow, T., Klupsch, M., Hanek, R., Schmitt, T.: Fast image segmentation, object recognition and localization in a robocup scenario. In: RoboCup 1999. LNAI 1856, Springer (2000)
14. McLaughlin, R.: Randomized hough transform: Improved ellipse detection with comparison. Pattern Rocognition Letters **19** (1998) 299–305
15. Xu, L., Oja, E., Kultanen, P.: A new curve detection method: Randomized hough transform (RHT). Pattern Rocognition Letters **11** (1990) 331–338
16. Duda, R., Hart, P.: Pattern Classification and Scene Analysis. John Wiley and Sons (1973)
17. http://ltilib.sourceforge.net   -Lib: C++ computer vision library

Visual Tracking and Localization of a Small Domestic Robot

Raymond Sheh and Geoff West

Department of Computing, Curtin University of Technology,
Perth, WA, 6102, Australia
{shehrk, geoff}@cs.curtin.edu.au

Abstract. We investigate the application of a Monte Carlo localization filter to the problem of combining local and global observations of a small, off-the-shelf quadruped domestic robot, in a simulated *Smart House* environment, for the purpose of robust tracking and localization. A Sony Aibo ERS-210A robot forms part of this project, with the ultimate aim of providing additional monitoring, human-system interaction and companionship to the occupants.

1   Introduction

This paper investigates using these two forms of sensors for the purpose of localizing a small quadruped robot in the presence of obstacles in a *Smart House*, as an aid to elderly and disabled people. Such a robot may be guided by the *Smart House* system, giving it the ability to interact with the occupants and visit areas not otherwise covered by *Smart House* cameras. For example, the robot could be used to disambiguate a person dropping out of view to retrieve an object from a person falling over out of view of the *Smart House* cameras. A combination of local (robot-based) and global sensing may be used to localize a robot in an environment. In domestic environments, low-cost and minimal interference with the existing environment are desirable, ruling out laser rangefinders, sonar rings and omnidirectional cameras. However, cheap surveillance cameras are available that can be mounted on the robot or fixed to the house.

This investigation makes use of the Sony Aibo ERS-210A Entertainment Robot, shown in figure 1(a). This robot possesses a low quality 176×144 pixel camera, a 385MHz MIPS processor, wireless ethernet, quadruped omnidirectional locomotion and approximately one hour of autonomous powered operation per battery charge, making it well suited for the *Smart House* project. However, its undulatory walking motion leads to poor odometry.

The system infrastructure used to operate and interface with the Aibo is derived from the code released by the UNSW/NICTA rUNSWift 2003 RoboCup

D. Nardi et al. (Eds.): RoboCup 2004, LNAI 3276, pp. 410–417, 2005.

team [1] [1]. This code provides a framework for controlling the robot's movements and acquiring and processing sensory information.

This paper describes how a partial model of the environment and information from robot odometry, the onboard camera and ceiling cameras can be combined to track and guide the robot through the environment. Preprocessed images from the robot camera are transmitted to the host PC and used to locate the robot relative to known carpet patterns. The ceiling cameras are also used to track potential robot positions. Both sensing techniques are complementary. Ceiling cameras often suffer from occlusion and clutter whilst information from local vision is often aliased or noisy. The process of sensor fusion makes use of Monte Carlo localization (MCL) to deal with incomplete or ambiguous data from one or both sources gracefully.

2   Environment

Four surveillance quality color overhead cameras are placed in the corners of the *Smart House* environment such that their fields of view overlap in the unobstructed center area as in figure 1(b,c). These cameras feed four PCs for analysis. The location of the robot on the floor is determined using a motion-based background subtraction technique [4] followed by a camera specific image to real world coordinate mapping[2]. The accuracy of the image to floor mapping is approximately 10cm at worst.

All of the joints in the legs and neck of the robot are equipped with angle encoders with sub-degree accuracy. However, mechanism slop and flexing of the robot structure reduce the accuracy of camera positioning relative to the ground plane to about 1° in angle and 5mm in height. These errors as well as the inability of the robot to determine which paws are in contact with the ground at any one time, means the accuracy drops further as the robot walks. The angle encoder values allow odometry to be obtained from the walk engine. Unfortunately, asymmetry in the robot's weight distribution, and variable slip in the paws, introduce significant errors into this odometry.

The rUNSWift software can color segment the images from the robot's onboard camera in realtime using a static 3D color lookup table. The resulting images, called *CPlanes*, can then be fed to the PC host via the robot's wireless ethernet interface at close to realtime speeds for further processing. Additionally, code exists to use the angle encoder values and known height to locate points observed with the camera in 3D space relative to the robot. However, positioning inaccuracies as discussed previously, combined with significant barrel and chromatic distortion severely limit the accuracy of this technique.

---

[1] The first author was a member of the 2003 rUNSWift RoboCup team. Components of the work described in this paper extend work undertaken by the first author and other members of the 2003 rUNSWift RoboCup team.

[2] The crosses in figure 1 are normally used to calibrate the overhead cameras.

Fig. 1. The Sony ERS-210A entertainment robot (a) and views from two of the *Smart House* overhead cameras (b),(c). Note the large overlapping region and the white crosses marked on the carpet. Arrows indicate the position of the robot

3   Monte Carlo Localization

The traditional approach to tracking is to use Kalman filtering that assumes that sensor measurements have Gaussian distributions. However, at any point in time the robot can only be represented in one location (with some variance). The MCL filter [2] allows multiple robot locations to be tracked because it does not commit to one location and instead stores many candidate locations.

MCL utilizes modified grid-based Markov localization [2] to represent the robot's possible location (with heading) or state $l = [x, y, \theta]$ as a probability map in a discretized state space. This representation is modified using a motion model and observation updates in order to track the robot's location over time. At any time, analysis of the distribution of locations allows the most probable location to be determined. The process is Markovian because the current distribution of possible locations is solely based on the distribution at the previous time step, the current observations and the robot's action. For efficiency, the grid is probabilistically sampled, each sample called a particle within a set of particles $S = \mathbf{s_i}|i = 1, ..., N$. Each particle $\mathbf{s_i}$ is a possible location l_i of the robot with an associated probability or weighting p_i.

Initially the distribution of particles can be random over the whole space of possible values of l but will soon converge to form a tight distribution after a number of iterations. In addition to S, two other parameters are required. First, a probability distribution $p(l|l', a)$: given the action a what is the probability that the robot goes to l from l'. Second, a probability distribution $p(s|l)$: given the location l what is the probability of a sensor derived location occurring. At each iteration, the steps in the algorithm are:

1. Generate N new samples probabilistically sampled from S. For each new sample, the location is called l'_i.
2. Generate l_i from $p(l|l', a)$ where a is the action or command given to the robot. Set the probability p for each l_i to be $1/N$.
3. Generate a new value of p_i for each new particle from $p(s|l)$ i.e. $p(l|s) \leftarrow \alpha P(s|l)$ where α is used to ensure $\Sigma_{i=1}^{N} p_i = 1$.
4. If required, determine the best estimate of location from the distribution S.

The main issues are defining the distributions $p(s|l)$ (based on observations) and $p(l|l', a)$ (based on odometry), and the number of particles N. In addition, in this work, the algorithm requires some approximations because of speed considerations.

4   Sensing

Preprocessing of data from the individual overhead cameras involves four major steps. Firstly, segmentation of the candidate robot pixels is performed by background subtraction using a mixture of Gaussians background model for each pixel and color channel [7]. Foreground objects are extracted as long as they move and are regarded as candidate robot pixels. Secondly, these pixels are projected onto the ground plane via a camera-specific transformation to form candidate robot points (assuming the robot is on the ground) [4]. Note that after projection, the points do not necessarily form a single connected grouping but will form clusters of points (figure 2).

Thirdly, these candidate points are filtered based on geometric properties to eliminate those that are unlikely to be part of the robot region. The filter applied to these points uses the kernel matrix of figure 2, tuned to give a high response when convolved with a cluster of points of a size that approximately matches the size of the robot. Finally the output is thresholded to segment out the candidate robot regions. The varying density of the candidate robot clusters is reduced by using morphological dilation [5] to directly yield a measure of $p(l|s)$.

A variant of the *NightOwl* system [6] developed by rUNSWift is used to extract localization information from the onboard camera. In order to determine a measure of $p(l|s)$ this algorithm matches known patterns on the ground and consists of three steps.

Firstly, patterns of interest on the gray-blue carpet are detected using a noise-reducing edge detector applied to the color segmented *CPlane* [6]. These patterns consist of white crosses on a 1m grid as in figure 1. These points are then projected onto the ground plane. Points projected further than 75cm from the robot are discarded due to potential inaccuracies. This process is illustrated in figure 3.

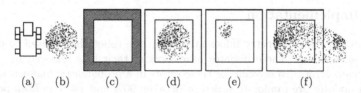

(a)   (b)   (c)   (d)   (e)   (f)

Fig. 2. The robot (a) from overhead appears as (b). The kernel (c) has a similarly sized positive inner region and negative outer region of a width equal to the minimum desired distance between detected robot clusters and other points. Three situations arise: the kernel matches a robot-sized cluster (d), a cluster that is too small (e) and too large (f)

(a)   (b)   (c)   (d)

Fig. 3. Robot at (a) observing a feature through its camera (b), transmitting a color segmented *CPlane* to the host PC (c) and projecting onto the ground plane in robot-relative space (d). Arcs in (d) illustrate points 50cm and 100cm from the robot respectively. Note that the robot is in the same position as in figure 1

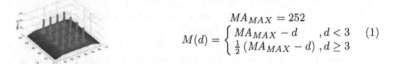

$$MA_{MAX} = 252$$
$$M(d) = \begin{cases} MA_{MAX} - d & ,d < 3 \\ \frac{1}{2}(MA_{MAX} - d) & ,d \geq 3 \end{cases} \quad (1)$$

Fig. 4. Matching array MA in heightfield and equation formats. $M(d)$ is the value of an element of MA based on its distance d from the nearest matchable feature

The matching of these observed, projected points with the known environment model is performed on an as-needed basis given a query state l for which an estimation of $p(l|s)$ is needed. Each edge point is transformed by this state to real-world co-ordinates. These points are multiplied with a matching array MA (figure 4) which serves as a precomputed model of detectable environmental features. Locations in the array representing detectable features (crosses) have a high value and the remaining locations are assigned successively lower values based on their Manhattan distance from the nearest feature, through a non-linear mapping (equation 1) to reduce the effect of outliers on the matching.

5   Implementation

The PC receives odometry information from the robot that is used to update particle positions. These updates have most of their bias removed but retain significant errors. To account for these errors in a probabilistic manner, the position updates are randomly scaled to between 90% and 100% in both position and heading. This motion model may be demonstrated by ignoring sensors and starting from a known state (figure 5). As the robot moves, the density of particles S disperses to account for odometric errors. In this application, 3000 particles were used to balance accuracy and ease of tracking against computational load.

The intermediate probability measure of each particle p_i is updated based on the observations from the cameras. The combination of each observational

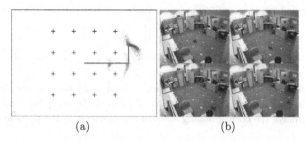

(a)   (b)

Fig. 5. The MCL filter tracking a non-sensing (dead reckoning only) robot (a), initialized to the robot's actual state. (b) shows successive actual robot positions

probability $p(l|s_i)$ from each source i to form a consolidated observational probability measure $p(l|s)$ is performed via additive heuristic rules formulated to reduce the bias towards areas of the environment that may be observed by many cameras. The result is summed with a probability measure $p(l|s_o)$ obtained from the *NightOwl* subsystem evaluated at l. These summations are possible since the measurements may be regarded as independent. Note that the overhead cameras do not contribute heading information θ.

To extract a single most likely robot location, an iterative algorithm based on Expectation-Maximization [3] is used. The average Euclidian distance in (x, y) is found between the last iteration's average position and each particle. A weight of 9 is assigned to particles within this distance and 1 to all other particles to reduce the effect of outliers. A weighted average position of all particles is then computed and the process repeated. This is efficient since the algorithm tends to converge within four iterations. The vector average of headings for particles close to this average position is then used to determine the most likely robot heading.

6   Experimental Results

When used alone, the data from the overhead cameras localized the robot even in the presence of some ambiguity, as long as the robot moved. For example, in figure 6 the robot has stopped long enough to become part of the background. After it begins to move again, the background "hole" left by the robot temporarily appears as another candidate robot (figure 6(b)). The filter continues to track the correct state once the hole becomes incorporated into the background.

When the robot is initially detected, the cluster of particles around the robot's state have uniformly distributed headings. As the robot moves, only those with headings matching the robot's heading will receive motion updates that cause their future predictions to coincide with subsequent observations. This technique is able to reliably achieve an angular accuracy of around 5° and spatial accuracy of around 20cm after only 1m of motion using only one overhead camera. When

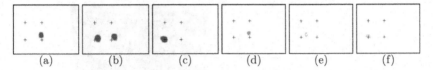

Fig. 6. Example of the particle filter coping with an ambiguous observation. Sequence (a) to (c) displays the input to the particle filter from the overhead camera, sequence (d) to (f) displays the corresponding state of the particles in the particle filter

there is significant clutter and noise, multiple cameras improve accuracy and speed of convergence.

The onboard camera is not able to globally localize the robot due to ambiguity i.e. observing a single cross provides four potential locations and with twelve crosses, 48 potential locations exist. Naturally, this ambiguity is reduced for more unique, natural and less repeating carpet patterns and features. Despite this, it is still possible to, over time, extract global localization from this data by tracking each of these potential locations. As the robot moves, some of these points may be eliminated as they coincide with environmental limits.

The combination of onboard localization information with that provided by the overhead cameras greatly improved the speed and accuracy with which the robot's heading was determined. If a carpet feature was visible whilst the robot's heading was unknown, the filter would often snap to two or three potential headings (due to similar features being visible in several directions). A small amount of movement, often only 20cm, would then eliminate all but one of these headings.

In the following example, only one overhead camera was used. The filter was able to track the robot (figure 7(a)) up to when it moved behind the barrier. The particles began to spread out as the robot was no longer visible (figure 7(b)). In a second test, the onboard camera observed the cross in front of it and was able to maintain localization and thus the cluster was tight even when the robot was clearly still occluded (figure 7(d)).

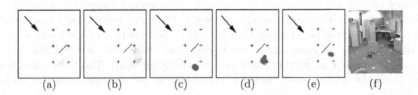

Fig. 7. Example of the robot moving behind an occlusion. In (a) and (b), no crosses are visible to the robot and the particles diverge. In (c) to (e), the robot continues to view crosses and maintain localization. Overhead camera vision (from the opposite corner of the room to the experiment camera) when the robot was occluded appears in (f)

7   Conclusion

This paper has demonstrated that the Monte Carlo filter can successfully track a robot using both local and global sensing and to cope when one of these is absent. It can accommodate ambiguous as well as erroneous observations. A number of heuristics have been used to speed up the algorithm such that real time performance is achieved. The *NightOwl* algorithm has been demonstrated to usefully contribute localization information in a real-world environment and proved to be very effective as a sensor despite the highly ambiguous information it provides because of the repeating pattern of crosses. Whilst *NightOwl* alone may not be effective for global localization in this specific application it allows the system to deal with inaccurate or missing global information.

Acknowledgements

We would like to thank the 2003 UNSW/NICTA rUNSWift RoboCup team, whose code infrastructure formed the basis for the robot and robot-PC communications code, and in particular Bernhard Hengst who initially proposed the *NightOwl* algorithm. We would also like to thank Sebastian Lühr and Patrick Peursum whose code infrastructure formed the basis for the overhead camera and PC-PC communications code. Finally we would like to thank Svetha Venkatesh for her useful advice and input through the course of this project.

References

1. Jin Chen, Eric Chung, Ross Edwards, Eileen Mak, Raymond Sheh, Nicodemus Sutanto, Terry Tam, Alex Tang, and Nathan Wong. rUNSWift, UNSW RoboCup2003 Sony Legged League Team. Honours thesis, School of Computer Science and Engineering, The University of New South Wales, 2003.
2. F. Dellaert, D. Fox, W. Burgard, and S. Thrun. Monte carlo localization for mobile robots.
3. A. P. Dempster, N. M. Laird, and D. B. Rubin. Maximum Likelihood from Incomplete Data via the EM Algorithm. *Journal of the Royal Statistical Society. Series B (Methodological)*, 39(1):1–38, 1977.
4. Patrick Peursum, Svetha Venkatesh, Geoff A.W. West, and Hung H. Bui. Object labelling from human action recognition. In *IEEE Conference on Pervasive Computing and Communications*, pages 399–406, March 2003.
5. Robjert J. Schalkoff. *Digital Image Processing and Computer Vision*. John Wiley and Sons, 1989.
6. Raymond Sheh and Bernhard Hengst. Nightowl: Self-localisation by matching edges. Technical Report UNSW-CSE-TR-0406, School of Computer Science and Engineering, University of New South Wales, 2004.
7. C. Stauffer and W. E. L. Grimson. Learning patterns of activity using real-time tracking. *IEEE Transactions on Pattern Analysis and Machine Intelligence*, 22(8):747–757, August 2000.

A Vision Based System for Goal-Directed Obstacle Avoidance

Jan Hoffmann, Matthias Jüngel, and Martin Lötzsch

Institut für Informatik, LFG Künstliche Intelligenz,
Humboldt-Universität zu Berlin,
Unter den Linden 6, 10099 Berlin, Germany
http://www.aiboteamhumboldt.com

Abstract. We present a complete system for obstacle avoidance for a mobile robot. It was used in the RoboCup 2003 obstacle avoidance challenge in the Sony Four Legged League. The system enables the robot to detect unknown obstacles and reliably avoid them while advancing toward a target. It uses monocular vision data with a limited field of view. Obstacles are detected on a level surface of known color(s). A radial model is constructed from the detected obstacles giving the robot a representation of its surroundings that integrates both current and recent vision information. Sectors of the model currently outside the current field of view of the robot are updated using odometry. Ways of using this model to achieve accurate and fast obstacle avoidance in a dynamic environment are presented and evaluated. The system proved highly successful by winning the obstacle avoidance challenge and was also used in the RoboCup championship games.

1 Introduction

Obstacle avoidance is an important problem for any mobile robot. While being a well studied field, it remains a challenging task to build a robust obstacle avoidance system for a robot using vision data.

Obstacle avoidance is often achieved by direct sensing of the environment. Panoramic sensors such as omni-vision cameras and laser range finders are commonly used in the RoboCup domain [1, 9]. Using these sensors, a full panoramic view is always available which greatly simplifies the task. In detecting obstacles from vision data, heuristics can be employed such as the "background-texture constraint" and the "ground-plane constraint" used in the vision system of the robot Polly [3]. In the case of the RoboCup world, this means that free space is associated with green (i.e. floor color), whereas non-green colored pixels are associated with obstacles (see introduction of [6] for an overview of panoramic vision systems).

In the Sony League, the robot is equipped with a camera with a rather limited field of view. As a basis for obstacle avoidance, a radial model of the robot's environment is maintained. In this model, current vision data is integrated with recent vision data. The approach to obstacle detection and obstacle modeling

D. Nardi et al. (Eds.): RoboCup 2004, LNAI 3276, pp. 418–425, 2005.

used by the GermanTeam in the RoboCup challenge turned out to be similar
the concept of "visual sonar" recently presented by [6]. Both bear a strong re-
semblance to the "polar histogram" used in [2]. Our work extends [6] and shows
how such a model can be used to achieve goal-directed obstacle avoidance. It
proved highly robust and performed extremely well in dynamic game situations
and the obstacle avoidance challenge.

Other approaches such as using potential fields [5] were not considered because the robot's environment changes rapidly which makes it hard to maintain
a more complex world model.

2   Obstacle Avoidance System

The following sections will describe obstacle detection, obstacle modeling, and
obstacle avoidance behavior. A Sony Aibo ERS210(A) robot was used in the
experiments. The robot has a 400 MHz MIPS processor and a camera delivering
YUV image with a resolution of 172x144 (8 bits per channel). A Monte Carlo
localization was used [8]; other modules not covered here such as walking engine,
etc. are described in more detail in the GermanTeam 2003 team description and
team report [7].

2.1   Obstacle Detection

Image processing yields what we call a *percept*. A percept contains information
retrieved from the camera image about detected objects or features later used
in the modeling modules. A percept only represents the information that was
extracted from the current image. No long-term knowledge is stored in a percept.

The *obstacles percept* is a set of lines on the ground that represents the free
space in front of the robot in the direction the robot is currently pointing its
camera. Each line is described by a *near point* and a *far point* on the ground,
relative to the robot. The lines in the percept describe segments of ground colored
lines in the image projected to the ground. For each far point, information about
whether or not the point was on the image border is also stored.

To generate this percept, the image is being scanned along a grid of lines
arranged perpendicular to the horizon. The grid lines have a spacing of 4°. They
are subdivided into segments using a simple threshold edge detection algorithm.
The average color of each segment is assigned to a color class based on a color
look-up table. This color table is usually created manually (algorithms that automate this process and allow for real-time adaptation exist [4]).For each scan
line the bottom most ground colored segment is determined. If this ground colored segment meets the bottom of the image, the starting point and the end
point of the segment are transformed from the image coordinate system into the
robot coordinate system and stored in the obstacles percept; if no pixel of the
ground color was detected in a scan line, the point at the bottom of the line is
transformed and the near point and the far point of the percept become identical.

Small gaps between two ground colored segments of a scan line are ignored to
assure robustness against sensor noise and to assure that field lines are not inter-

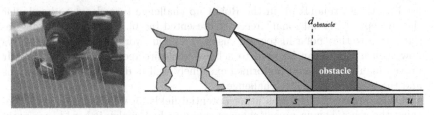

Fig. 1. Obstacle detection and diagram to illustrate what can be deduced from what is seen. Green lines in image: The obstacles percept as the conjunction of green segments close to the robot. Diagram: The robot detects some free space in front of it (s) and some space that is obscured by the obstacle (t). The obstacle model is updated according to the diagram (in this case the distance in the sector is set to $d_{obstacle}$ unless the distance value stored lies in r)

preted as obstacles. In such a case two neighboring segments are concatenated. The size limit for such gaps is 4 times the width of a field line in the image. This width is a function of the position of the field line in the camera image and the current direction of view of the camera. Figure 1 shows how different parts of scan lines are used to generate obstacle percepts and also illustrates how information about obstacles in the robot's field of view can be deduced from the *obstacle percept*.

2.2   Obstacle Model

The obstacle model described here is tailored to the task of local obstacle avoidance in a dynamic environment. Local obstacle avoidance is achieved using the obstacle model's analysis functions described below. The assumption was made that some high level controller performs path planning to guide the robot globally. Certain global set ups will cause the described algorithm to fail. This, however, is tolerable and it is a different type of problem that needs to be dealt with by higher levels of action planning. We therefore concentrate on a method to reliably steer the robot clear of obstacles while changing its course as little as possible.

In the model, a radial representation of the robot's surroundings is stored in a "visual sonar" [6]. The model is inspired by the sensor data produced by panoramic sensors such as 360° laser range finders and omni-vision cameras. In this model, free space in a certain direction θ is stored. θ is divided into n discrete sectors ("micro sectors").

If new vision information is received, the corresponding sectors are updated. Sectors that are not in the visual field are updated using odometry, enabling the robot to "remember" what it has recently seen. If a sector has not been updated by vision for a time period greater than t_{reset}, the range stored in the sector is reset to "unknown".

Micro sectors are 5° wide. Due to imperfect image processing the model is often patchy, e.g. an obstacle is detected partially and some sectors can be updated while others may not receive new information. Instead of using the

a)   b)   c)   d)

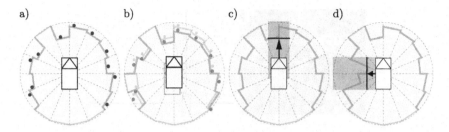

Fig. 2. Illustration of the obstacle model. The actual number of sectors is greater than shown here, it was reduced for illustration purposes. Fig. 3 shows the actual obstacle model used. a) The robot is at the center; dashed lines show sectors; solid orange lines (dark) show the free space around the robot; light gray lines are used if there is no information about free space in a sector; small circles denote *representatives*. b) illustrates how the model is updated using odometry when the robot is moving. Updated representatives are shown as dark circles dots. c) and d) illustration of analysis function used when determining the free space in front of the robot and to its side

model as such, analysis functions that compute information from the model are used. These functions produce high level output such as "how much free space is (in the corridor) in front of the robot" which is then used by the robot's behavior layers. These functions usually analyze a number of micro sectors. The sector with the smallest free space associated to it corresponds to the greatest danger for the robot (i.e. the closest object). In most analysis functions this sector is the most important overruling all other sectors analyzed. In the above example, the sector in the corridor containing the smallest free space is used to calculate the free space in front of the robot. Using analysis functions makes using the model robust against errors introduced by imperfect sensor information. It also offers intuitive ways to access the data stored in the model from the control levels of the robot.

In addition to the free space, for each sector a vector pointing to where the obstacle was last detected (in that sector) is stored. This is called a *representative* for that sector. Storing it is necessary for updating the model using odometry. Fig. 2 illustrates the obstacle model. The following paragraphs will explain in more detail how the model is updated and what analysis function are.

Update Using Vision Data. The image is analyzed as described in 2.1. Obstacle percepts are used to update the obstacle model. The detected free space for each of the vertical scan lines is first associated to the sectors of the obstacle model. Then the percept is compared to the free range stored for a sector; fig. 1 illustrates one of the many possible cases for updating the information stored in a sector θ.

If the distance in a sector was updated using vision information, the obstacle percept is also stored in the representative of that sector. The necessity to store this information is explained in the following paragraphs.

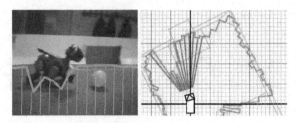

Fig. 3. *Left.* Camera image with superimposed obstacle percepts and obstacle model (projected onto the floor plane) *Right.* Actual obstacle model

Update Using Odometry. Sectors that are not in the visual field of the robot (or where image processing did not yield usable information) are updated using odometry. The representative of a sector is moved (translated and rotated) according to the robot's movement. The updated representative is then remapped to the - possibly new - sector. It is then treated like an obstacle detected by vision and the free space is re-calculated. In case more than one representatives are moved into one sector, the representative closest is used for calculating the free space (see Fig.2 b. for an example). If a representative is removed from a sector and no other representative ends up in that sector, the free space of that sector is reset to infinity). The model quality deteriorates when representatives are mapped to the same sector and other sectors are left empty. While this did not lead to any performance problems in our experiments, [6] shows how these gaps can easily be closed using linear interpolation between formerly adjacent sectors.

Analysis Functions. As explained above, the model is accessed by means of analysis functions. The micro sectors used to construct the model are of such small dimensions that they are not of any use for the robot's behavior control module. The way we model robot behavior, more abstract information is needed, such as "There is an obstacle in the direction I'm moving in at distance x" or "In the front left hemisphere there is more free space than in the front right." Of interest is usually the obstacle closest to a the robot in a given area relative to the robot. In the following paragraphs, some analysis functions that were used for obstacle avoidance and in RoboCup games are described. Other possible function exist for different kind of applications which are not covered here.

Macro Sector $sect(\theta, \Delta\theta)$. This function is used to find out how much free space there is in a (macro) sector in direction θ an of width $\Delta\theta$. Each sector within the macro sector is analyzed and the function returns the smallest distance in that macro sector. This can be used to construct a simple obstacle avoidance behavior. The free space in two segments ("front-left", $-22, 5° \pm 22, 5°$ and "front-right", $+22, 5° \pm 22, 5°$) is compared and the behavior lets the robot turn in the direction where there is more free space.

Corridor $corr(\theta, \Delta d)$. If the robot is to pass through a narrow opening, e.g. between two opponent robots, the free space not in a (macro) sector but in a *corridor* of a certain width is of interest. Usually, a corridor of about twice the width of the robot is considered safe for passing.

Free Space for Turning $corr(\theta = \pm 90°, \Delta d = \text{length of robot})$. When turning, the robot is in danger of running into obstacles that are to its left or right and thereby currently invisible. These areas can be checked for obstacles using this function. If obstacles are found in the model, the turning motion is canceled. (Note that this is a special case of the corridor function described above).

Next Free Angle $f(\theta)$. This functions was used in RoboCup games to determine which direction the robot should shoot the ball. The robot would only shoot the ball in the direction of the goal if no obstacles were in the way. Otherwise the robot would turn towards the "next free angle" and perform the shot.

2.3   Obstacle Avoidance

Goal-directed obstacle avoidance as used in the challenge. Obstacle avoidance is achieved by the following control mechanisms:

A. **Controlling the robot's forward speed.** The robot's forward speed is linearly proportional to the free space in the corridor in front of the robot.

B. **Turning towards where there is more free space.** If the free space in the corridor in front of the robot is less than a threshold value, the robot will turn towards where there is more free space (i.e. away from obstacles).

C. **Turning towards the goal.** The robot turns toward the goal only if the space in front of it is greater than a threshold value.

D. **Override turning toward goal.** If there is an obstacle next to it that it would run into while turning, turning is omitted and the robot will continue to walk straight.

When approaching an obstacle, *B.* causes the robot to turn away from it just enough to not run into the obstacle. *C.* and *D.* cause the robot to cling to a close obstacle, thereby allowing the robot to effectively circumvent it.

Obstacle Avoidance in RoboCup Games. In the championship games, a similar obstacle avoidance system was used. It worked in conjunction with a force field approach to allow for various control systems to run in parallel. The obstacle model itself was used for shot selection. When the robot was close to the ball, the model was used to check if there were obstacles in the intended direction of the shot. If there were obstacles in the way, it would try to shot the ball in a different direction.

Scanning Motion of the Head. In the challenge, the robot performed a scanning motion with its head. This gives the robot effective knowledge about its vicinity (as opposed to just its field of view), allowing it to better decide where it should head. The scanning motion and the obstacle avoidance behavior were fine tuned to allow for a wide scan area while making sure that the area in front of the robot was scanned frequently enough for it to not run into obstacles.

Rank	Team	# Colli- sions	Time [s]
1.	GermanTeam	0	35.7
2.	UT Austin	0	63.3
3.	AR AIBO	0	104.4
4.	UTS Unleashed	0	108.7
5.	ASURA	1	87.2
6.	rUNSWift	1	100.0
7.	Baby Tigers	2	141.5
8.	Team Sweden	2	179.9
9.	NUbots	1	(not reached)
10.	UW Huskies	1	(not reached)

Fig. 4. Image extracted from a video of the RoboCup World Cup 2003 obstacle avoidance challenge and table of results

In the actual RoboCup games, the camera of the robot is needed to look at the ball most of the time. Therefore, very little dedicated scanning motions were possible giving the robot a slightly worse model of its surroundings.

3   Application and Performance

RoboCup 2003 Technical Challenge. In the obstacle avoidance challenge, a robot had to walk as quickly as possible from one goal to the other without running into any of the other 7 robots placed on the field. The other robots did not move and were placed at the same position for all contestants. The algorithm used was only slightly altered from the one used in the actual, dynamic game situations. As can be seen from the results, the system used enabled the robot to move quickly and safely across the field. Avoidance is highly accurate: on its path, the robot came very close to obstacles (as close as 2 cm to touching the obstacles) but did not touch any of them. Very little time is lost for scanning the environment (as the obstacle model is updated continuously while the robot's head is scanning the surroundings) enabling the robot to move at a high speed without stopping. The system used in the challenge was not optimized for speed and only utilized about 70% of the robot's top speed. Furthermore, some minor glitches in the behavior code caused the robot to not move as fast as possible.

RoboCup 2003 Championship Games. The obstacle model was used for obstacle avoidance and for shot selection in the games. An improvement in game play was noticeable when obstacle avoidance was used. In several instances during the games, situations in which the robot would otherwise have run into an opponent it was able to steer around it.

4   Conclusion

The presented system enables the robot to reliably circumvent obstacles and reach its goal quickly. The system was developed for use in highly dynamic environments and limits itself to local obstacle avoidance. In our search for the simplest, most robust solution to the problem, maintaining a model of the obstacles was a necessity to achieve high performance, i.e. to alter the path of the robot as little as possible while assuring the avoidance of obstacles currently visible *and* invisible to the robot. The control mechanisms make use of this model to achieve the desired robot behavior. In the RoboCup 2003 obstacle avoidance challenge, the robot reached the goal almost twice as fast as the runner up without hitting any obstacles. The system was not used to its full potential and a further increase in speed has since been achieved. An improvement in game play in the RoboCup championship games was observed although this is very hard to quantify as it depended largely on the opponent.

Acknowledgments

The project is funded by Deutsche Forschungsgemeinschaft, Schwerpunktprogramm 1125. Program code is part of the GermanTeam code release and is available for download at http://www.robocup.de/germanteam.

References

1. R. Benosman and S. B. K. (editors). *Panoramic Vision: Sensors, Theory, and Applications*. Springer, 2001.
2. J. Borenstein and Y. Koren. The Vector Field Histogram Fast Obstacle Avoidance For Mobile Robots. In *IEEE Transactions on Robotics and Automation*, 1991.
3. I. Horswill. Polly: A Vision-Based Artificial Agent. In *Proceedings of the 11th National Conference on Artificial Intelligence (AAAI-93)*, 1993.
4. M. Jüngel, J. Hoffmann, and M. Lötzsch. A real-time auto-adjusting vision system for robotic soccer. In *7th International Workshop on RoboCup 2003 (Robot World Cup Soccer Games and Conferences)*, Lecture Notes in Artificial Intelligence. Springer, 2004.
5. O. Khatib. Real-Time Obstacle Avoidance for Manipulators and Mobile Robots. *The International Journal of Robotics Research*, 5(1), 1986.
6. S. Lenser and M. Veloso. Visual Sonar: Fast Obstacle Avoidance Using Monocular Vision. In *Proceedings of IROS'03*, 2003.
7. T. Röfer, I. Dahm, U. Düffert, J. Hoffmann, M. Jüngel, M. Kallnik, M. Lötzsch, M. Risler, M. Stelzer, and J. Ziegler. GermanTeam 2003. In *7th International Workshop on RoboCup 2003 (Robot World Cup Soccer Games and Conferences)*, Lecture Notes in Artificial Intelligence. Springer, 2004. to appear. more detailed in http://www.robocup.de/ germanteam/ GT2003.pdf.
8. T. Röfer and M. Jüngel. Vision-Based Fast and Reactive Monte-Carlo Localization. *IEEE International Conference on Robotics and Automation*, 2003.
9. T. Weigel, A. Kleiner, F. Diesch, M. Dietl, J.-S. Gutmann, B. Nebel, P. Stiegeler, and B. Szerbakowski. CS Freiburg 2001. In *RoboCup 2001 International Symposium*, Lecture Notes in Artificial Intelligence. Springer, 2003.

Object Tracking Using Multiple Neuromorphic Vision Sensors

Vlatko Bečanović, Ramin Hosseiny, and Giacomo Indiveri[1]

Fraunhofer Institute of Autonomous Intelligent Systems,
Schloss Birlinghoven,
53754 Sankt Augustin, Germany
{becanovic, hosseiny}@ais.fraunhofer.de

Abstract. In this paper we show how a combination of multiple neuromorphic vision sensors can achieve the same higher level visual processing tasks as carried out by a conventional vision system. We process the multiple neuromorphic sensory signals with a standard auto-regression method in order to fuse the sensory signals and to achieve higher level vision processing tasks at a very high update rate. We also argue why this result is of great relevance for the application domain of reactive and lightweight mobile robotics, at the hands of a soccer robot, where the fastest sensory-motor feedback loop is imperative for a successful participation in a RoboCup soccer competition.

Keywords: Neuromorphic vision sensors, analog VLSI, reactive robot control, sensor fusion, RoboCup.

1 Introduction

In our lab aVLSI technology is exploited in fast moving mobile robotics, e.g. RoboCup, where soccer-playing robots perform at high speed. The robot that is used in our experiments is a mid-sized league robot of roughly 45 by 45 cm with the weight of 17 kg. It is equipped with infra-red distance sensors in order to have fast and reliable obstacle avoidance, odometry together with an augmenting gyroscope in order to reduce the error in the odometry measurements, and contact sensitive bumper sensors. The robot uses a differential drive for movement, a pneumatic kicker for shooting and two small movable helper arms to prevent the ball from rolling away. The most important sensory inputs are streamed in via FireWire bus [1] from a digital color camera. The conventional part of vision processing is software based and consumes the most of the calculation resources on-board the robot [2].

One of the most difficult tasks in the RoboCup environment is to pass the ball from one player to another. This requires first of all that the robot can control the ball, that is, be in possession of the ball so that it can be kicked in any direction and this

---

[1] Institute of Neuroinformatics, Winterthurerstrasse 190, CH-8057 Zurich, Switzerland. E-mail: giacomo@ini.phys.ethz.ch

D. Nardi et al. (Eds.): RoboCup 2004, LNAI 3276, pp. 426–433, 2005.

while the robot is in motion. The ball needs to be close to the robot in order to be successfully controlled. This can be achieved by carefully controlling the velocity and position of the robot relative to the ball. The closer the ball the lower the relative velocity must be in order for it not to bounce off due to its lower momentum. In order to solve this very demanding problem the robot has to know where the ball is located at each instant, which requires a fast read-out and processing of the sensory information.

This paper is structured as follows: in section 2 a description of our robot platform is given. The neuromorphic vision sensors used in the experiments are presented in section 3. In sections 4 and 5 we investigate how the vision system can be aided with a set of neuromorphic vision sensors. Here, we present data collected during experimental runs with one of our robots. We show that this data is suitable for further higher level processing. In the conclusions we point out the importance of the results that were achieved.

2 Our Robot Platform

Our soccer playing robot has actuators in the form of motors to drive the robot and to turn a panning camera. A valve is used to kick the ball pneumatically and small robot arms attached to the left and right side of the robot keeps the ball in front of the kicker plate. Besides the optical sensors; camera and neuromorphic vision sensors, it has four infrared distance sensors, a contact sensitive bumper strip with rubber shield and odometry at the two actuated wheels of the robot. This is augmented by a gyroscope for fast turning movements. All of these peripheral devices are controlled by three 16 bit micro controllers [3]. They are interconnected with a bus interface (CAN), which is a standard in German automobile industry. A notebook PC operates the main behavior program and the operating system can be either Windows or LINUX. The cyclic update rate is 30 Hz (~33 ms) which is governed by the frame rate of the digital camera.

For the experiments we increased the observation rate for the neuromorphic sensors to the maximum effective sampling rate of the micro-controller module that is used which is ~2 kHz (0.5 ms). In the various experiments the signal is down-sampled to 153 Hz in the first experiments and up to 520 Hz in the more complex experiment done at the end.

The robot vision system does color blob tracking of multiple objects and delivers information from tracked objects such as position of geometrical center, bounding box and pixel area. In our experiments only the position of the geometrical center of the tracked object will be used to train the system. Other parameters like pixel area are only used indirectly, in order to prepare data for the training phase of the system by removing noisy information from distant objects and other artifacts. The vision software used for the experiments is a free software developed at the Carnegie Mellon University and used be many robot teams in RoboCup tournaments [2].

3 Neuromorphic Vision Sensors

Neuromorphic vision chips process images directly at the focal plane level. Typically each pixel in a neuromorphic sensor contains local circuitry that performs, in real time, different types of spatio-temporal computations on the continuous analog brightness signal. Data reduction is thus performed, as they transmit only the result of the vision processing off-chip, without having to transmit the raw visual data to further processing stages. Standard CCD cameras, or conventional CMOS imagers merely measure the brightness at the pixel level, eventually adjusting their gain to the average brightness level of the whole scene. The analog VLSI sensors used in our experiments are made using standard 1.6 and 0.8 micron CMOS technologies. They are small 2x2 mm devices that dissipate approximately 100mW each. Specifically, they a 1D tracking chip [5], a 1D correlation-based velocity sensor [6], a single 1D chip comprising both tracking and correlation-based velocity measurements, and, a gradient based 2D optical flow chip [7] (cf. Fig. 1). The 2D optical flow chip is the most complex and computes the optical flow on its focal plane providing two analog output voltages. The correlation-based velocity sensor delivers the mean right or left velocity computed throughout the whole 1D array in two separate output channels, and the 1D tracker sensor provides an analog output voltage that indicates the position of the highest contrast moving target present in its field of view.

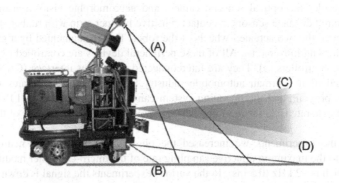

Fig. 1. Four aVLSI sensors mounted on the robot with their respective fields of view: The 2D optical flow sensor (A) is pointing straight towards the ground and also the absolute tracker (B) is pointing towards the ground. The absolute tracker (B) is mounted at a somewhat lower angle and with its pixel array vertically aligned. The 1D velocity tracker (C) and the 1D integrating tracker (D) are directed as a divergent stereo pair and with their respective pixel arrays horizontally aligned

4 Experiment

The purpose of the experiment is to investigate the plausibility of neuromorphic vision sensors to aid higher level vision processing tasks, in particular color blob tracking, which is a standard real-time vision processing application that is commonly used on mobile robots. The test consists of two stages; firstly to investigate if the sensors can be made sensitive to a moving primary colored object, and secondly, to validate this against a somewhat cluttered background. The first stage is performed to investigate the precision of the prediction from the fused sensory readings. The second stage is performed to investigate if there is enough discrimination against background patterns, that is, to investigate the robustness of the object tracking task when the robot is moving. If both stages are successful, this would imply that a set of neuromorphic vision sensors, sensitive to different types of motion, could aid a standard camera based digital vision system in a local domain of the scene.

The experiment consists of data collection from the neuromorphic vision sensors and the digital vision system of our soccer robot. The RoboCup soccer playing robot is fully autonomous and is operated by a behavior based program that was used by our team at the last world championships in Padua Italy [8],[9]. The test field is prepared with white lines that are located in a dense non-uniform grid and with an average spacing of about one meter. On the field there is a red soccer football.

Three experiments were performed, two stationary experiments followd by a moving robot experiment at the end [10]. In the stationary epxeriments the ball is moved according to certain patterns that ensure an even distribution of events when projected onto the focal plane of the digital vision system. In the moving robot experiment the robot will constantly try to approach the red ball in different maneuvers. During this time the robot will frequently pass lines on the floor which will influence the tracking task of the red ball. Optimally, the system should recognize what sensory input belongs to white lines and what input belongs to the red ball.

5 Experimental Results

The first step here consists of two stationary robot experiments treated in section 5.1, and the second step, which is a moving robot experiment is treated in sec. 5.2. The data is evaluated by comparing the results from a standard dynamical prediction model. A root mean square error is calculated relative to the reference signal from the standard vision system. The prediction model used for the two stationary robot experiments is a multivariable ARX model of 4'th order. The model, which is part of the Matlab™ system identification toolbox is performing parametric auto-regression that is based on a polynomial least squares fit [11]. For the dynamic experiments the best overall model was chosen in the range of up to a 15 ARX coefficients (15'th order ARX model).

5.1 Stationary Robot Experiment

In the first experiment the robot is not moving and the camera and neuromorphic vision sensors detect a single moving red RoboCup soccer football. The ball was moved so that it passed the robot along horizontal paths. The fields of view of the neuromorphic vision sensors were divided into four zones that were partially overlapping, and, within the field of view of the standard vision system. During the experiment the ball was thrown 25 times back and forth in each zone, but in random order, so that the data set would be easily split into a training and testing set of equal size. By this procedure the distribution would be close to uniformly distributed in the spatial domain and normally distributed in the temporal domain. The prediction efficiency is given in Table 1. For example, the horizontal x-channel over-all RMS error is about 13 %, which for the horizontal camera resolution of 320 pixels would mean an error of 40 pixels, which corresponds well to the fact that the resolution of the neuromorphic sensors is between 10 and 24 pixels.

In the second experiment, that is performed with a non moving robot and the same boundary conditions as the first experiment, the ball was moved so that it passed straight towards the robot hitting it and bouncing off, where the ball with its significantly lower momentum got deflected in an elastic collision. During the experiment the ball was thrown 25 times back and forth in different zones, but in rando`m order and at the same point of impact, so that the data set would be easily split into a training and testing set of equal size. The results here indicate similar efficiency as for the first stationary robot experiment for estimating the horizontal trajectories of the red ball, but with a better efficiency in the estimation of the vertical component (cf. Table 1). An example from the stationary robot data set used in this experiment is given in Figs. 2 and 3, where the predicted result for the horizontal and vertical blob position is plotted with a solid line and the "ground truth" reference signal is plotted with a dotted line.

Table 1. First and second stationary robot experiment – test data: The overall RMS error for the x-value and y-value of the centroid of the pixel blob delivered by the standard vision system (SVS). RMS errors of sensors are calculated only in their trig-points, thus the lower and irregular sample size. The RMS error is calculated as the difference between the object position given by the vision reference and the one predicted with the 4'th order ARX model

Stationary robot Data Set I:			
	0.1295	0.1920	38044
	0.1101	0.2069	569
	0.06250	0.1449	4647
	0.2405	0.2505	126
	0.1089	0.2304	112
Stationary robot Data Set II:			
	0.1386	0.1245	37974
	0.1586	0.1236	236
	0.1416	0.1172	1004
	0.1803	0.1210	387
	0.1316	0.1396	161

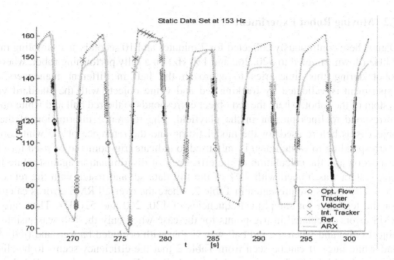

Fig. 2. An example from the stationary robot experiment for the red channel of the standard vision system. The predicted result for the *horizontal* blob position is plotted with a solid line and the "ground truth" reference signal is plotted with a dotted line. The activity of all the sensors is indicated as trig-points on top of the reference signal

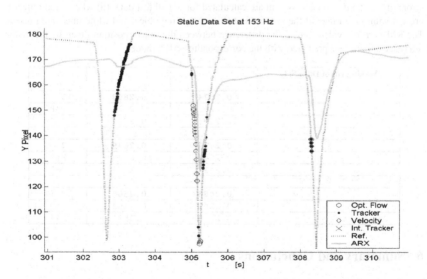

Fig. 3. An example from the stationary robot experiment for the red channel of the standard vision system. The predicted result for the *vertical* blob position is plotted with a solid line and the "ground truth" reference signal is plotted with a dotted line. The activity of all the sensors is indicated as trig-points on top of the reference signal

5.2 Moving Robot Experiment

Data is here continuously collected for 7 minutes and 10 seconds at a sampling rate of 2 kHz (down-sampled to 520, 260 and 130 Hz) on a fully performing robot, where the robot during this time tries to approach the ball in different maneuvers. The experiment is validated by tracking red and white objects with the standard vision system of the robot, where the red object corresponds to the red ball and white objects correspond to lines present in the playfield. The reference information of the red object is as before used for the model fitting and the reference of the white objects (corresponding to white lines) is only used to indicate trig-points to be used for visual inspection and the calculation of the efficiency of discrimination against white lines. The system was trained with 75% of the full data set and tested with the remaining 25%. The results are presented in Table 2, where the over-all RMS error is calculated for the test data for sampling frequencies of 130, 260 and 520 Hz. There are also RMS errors calculated in trig-points for the case when only the ball was visible (red object only) and when the red ball was visible with occluded background (red object and white line). It can be seen from Table 2 that the efficiency seems to be slightly improved at higher update rates and that the ball can be recognized in occluded scenes (with close to over-all efficiency).

Table 2. Moving robot experiment – test data: The overall RMS error for the x-value and y-value of the centroid of the pixel blob delivered by the standard vision system (SVS). RMS errors of the standard vision system are calculated for: (i) all test data, (ii) when a red object is present within the range of the sensors and (iii) when a red object and white line/s are present. The RMS error is calculated as the difference between the object position given by the vision reference and the one predicted with the corresponding ARX model

Moving robot Data Set:			
	0.2574	0.2808	13967
	0.2293	0.2331	758
	0.2195	0.2714	320
	0.2471	0.2679	27936
	0.2241	0.2328	829
	0.2113	0.2983	363
	0.2485	0.2568	55872
	0.2247	0.2163	829
	0.2116	0.2571	361

6 Summary and Conclusions

In our work we investigate if the output signals from a small number of neuromorphic vision sensors can perform the elementary vision processing task of object tracking. For our experiments we use a soccer playing robot as a test-platform, but are looking for a general application domain that can be used for all types of mobile robots, especially smaller robots with limited on-board resources. Those robots can benefit from neuromorphic vision systems, which provide high speed performance together with low power consumption and small size which is advantageous for reactive behavior based robotics [12], where sensors are influencing actuators in a direct way. In general it can be concluded that the results of the robot experiments presented indicate that optical analog VLSI sensors with low-dimensional outputs give a robust enough signal, and, that the visual processing tasks of object tracking and motion prediction can be solved with only a few neuromorphic vision sensors analyzing a local region of the visual scene.

Acknowledgments

The authors would like to thank Dr. Alan Stocker, from the Center for Neural Science, New York University, for providing the 2D optical flow sensor. Special thanks are due to Stefan Kubina, Adriana Arghir and Dr. Horst Günther of the Fraunhofer AIS for help regarding the set-up of the robot experiments. This work is funded by the Deutsche Forschungsgemeinschaft (DFG) in the context of the research program SPP-1125 "RoboCup" under grant number CH 74/8-2. This support and cooperation is gratefully acknowledged.

References

[1] http://www.1394ta.org/
[2] J. Bruce, T. Balch, M. Veloso, "Fast and inexpensive color image segmentation for interactive robots", *Proc. IEEE/RSJ Int. Conf. on Intelligent Robots and Systems (IROS)*, 2000, Vol. 3, 2061–2066.
[3] http://www.infineon.com/
[4] S. Kubina,"Konzeption, Entwicklung und Realisierung – Micro-Controller basierter Schnittstellen für mobile Roboter" *Diploma thesis at GMD Schloss Birlinghoven*, 2001. (in German).
[5] G. Indiveri, "Neuromorphic Analog VLSI Sensor for Visual Tracking: Circuits and Application Examples", IEEE Transactions on Circuits and Systems II, Analog and Digital Signal Processing, 46:(11) 1337-1347, 1999
[6] J. Kramer, R. Sarpeshkar, C. Koch, "Pulse-based analog VLSI velocity sensors", IEEE Transactions on Circuits and Systems II, Analog and Digital Signal Processing, 44:(2) 86-101, 1997
[7] A. Stocker, R. J. Douglas, "Computation of Smooth Optical Flow in a Feedback Connected Analog Network", Advances in Neural Information Processing Systems, 11:, Kearns, M.S. and Solla, S.A. and Cohn, D.A (Eds.), MIT Press, 1999
[8] A. Bredenfeld, H.-U. Kobialka, "Team Cooperation Using Dual Dynamics", *Balancing reactivity and social deliberation in multi-agent systems* (Hannebauer, Markus[Hrsg.]: Lecture notes in computer science, 2001), 111 – 124.
[9] A. Bredenfeld, G. Indiveri, "Robot Behavior Engineering using DD-Designer", *Proc. IEEE/RAS International Conference on Robotics and Automation (ICRA)*, 2001.
[10] R. Hosseiny, "Fusion of Neuromorphic Vision Sensors for a mobile robot", *Master thesis RWTH Aachen*, 2003.
[11] L. Ljung, System Identification: Theory for the User, Prentice Hall, 1999.
[12] R. Brooks, "A robust layered control system for a mobile robot", *IEEE Journal of Robotics and Automation*, Vol, RA-2, No. 1, 1986.

Interpolation Methods for Global Vision Systems

Jacky Baltes and John Anderson

Department of Computer Science,
University of Manitoba, Winnipeg, Canada
jacky@cs.umanitoba.ca
http://www.cs.umanitoba.ca/~jacky

Abstract. In 2004, the playing field size of the small sized league was
significantly increased, posing new challenges for all teams. This paper
describes extensions to our current video server software (Doraemon) to
deal with these new challenges. It shows that a camera with a side view is
a workable alternative to the more expensive approach of using multiple
cameras. The paper discusses the camera calibration method used in
Doraemon as well as an investigation into some common two–dimensional
interpolation methods, as well a novel average gradient method. It also
proves that (ignoring occluded parts of the playing field) it is possible
to construct a realistic top down view of the playing field with a camera
that only has a side view of the field.

1   Introduction

The small sized league (SSL) of the RoboCup competition is a very competitive
league which has introduced many innovations and modifications into the rules
in recent years. After a period of incremental changes to the field size (1997:
1.52m * 2.74m; 2001: 2.40m * 2.90m) the field size has been almost quadrupled
for the 2004 competition. At the moment the rules have not been finalized but
the proposed field size is 4.00m by 5.40m. Such a large change in the field size
obviously led to much discussion of the pros and cons of this change. The authors
of this paper hope that the larger field leads to more fundamental research in
the SSL league and consider it necessary to move towards games of 11 vs 11
players in the future.

The greatest impact of the larger field size is on the global vision system.
Whereas a single global vision camera with a wide angle lens mounted centrally
on top of the playing field is sufficient for the 2002 field (2.40m by 2.90m), this
is no longer true for the 2004 field.

Teams have suggested several solutions to this problem: (a) buy a better
wide angle lens, (b) mount the camera at a higher position, and (c) mount
several cameras over the playing field. The authors believe that a new model–
based approach to global vision in the SSL is a viable alternative to these that
will be beneficial to research. It is clear that few challenges remained for vision
processing with a single overhead camera. However, there are still many open

D. Nardi et al. (Eds.): RoboCup 2004, LNAI 3276, pp. 434–442, 2005.

problems when dealing with a side view of the field, not the least of which is to be able to compensate for occlusion of objects.

DORAEMON, the video server used by the Little Black Devils (LBD, named after the Winnipeg Rifles) from the University of Manitoba has always been an exception in the RoboCup competition. Instead of mounting the camera centrally overhead, the LBD have always used a side view of the playing field. The camera used by the LBD is an off-the-shelf camcorder *without a wide angle lens*, and DORAEMON is able to control our robots from any view of the playing field.

To be able to achieve this, Doraemon includes a sophisticated camera calibration routine based on the Tsai camera calibration. However, the larger playing field requires even better image processing and interpolation so that smaller geometric features can be detected and tracked successfully.

In this paper, the authors compare the accuracy and efficiency of several interpolation methods in the global vision robotic soccer domain. We also show a quasi-reconstructed view of the playing field obtained from a side angle.

2   Doraemon and the University of Manitoba Vision Kart

Figure 1 shows a picture of the University of Manitoba Vision Kart. The Vision Kart is a fully mobile global vision robotics setup. It contains a P4 2.4GHz small sized PC, a camera on a tripod, a side-mounted Infrared transmission system, and a wireless router. The Vision Kart can run off a car battery for about an hour and is an ideal platform for demonstrations and fund raising events for the LBD. Students usually use laptops with wireless cards to connect to the vision server and to control the robots.

Figure 2 shows the view of the camera in the vision kart overlooking the playing field. Clearly, the perspective distortion of this setup is much more extreme than in the case of an overhead camera. The top view shows the playing field with three robots. The bottom row shows a zoomed in view of the three robots.

Fig. 1. University of Manitoba Vision Kart. A fully mobile SSL platform

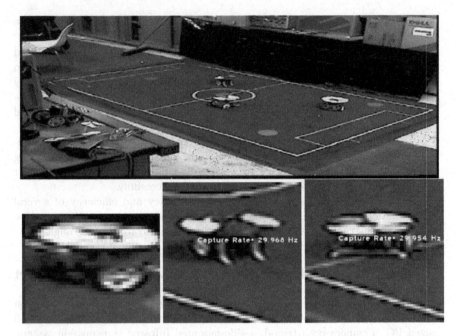

Fig. 2. Overview from the camera on the vision kart (top) as well as a zoomed-in views of three robots

From the views in Fig. 2, it is easy to see that colour features are hard to extract. For example, the small pink spot on the robot has been washed out to pure white because of the angle of the incoming light. We have therefore investigated alternative methods to use geometric features. For example, in [1] we describe a method that uses line segments of rectangular robots to determine the ID and orientation of robots. One problem of this approach is that most geometric features that may be used as feature points (e.g., area, aspect ratio, and angles) are greatly distorted by the low position of the camera.

3   Tsai Camera Calibration

DORAEMON's camera calibration uses the well-established Tsai camera calibration [2] which is popular among computer vision researchers. It is also suitable for global vision in robotic soccer since it can compute the calibration from a single image.

The Tsai camera calibration computes six external parameters (x, y and z of the camera position as well as angles of roll, pitch and yaw) and six internal parameters (focal length, center of the camera lens, uncertainty factor S_x, and κ_1, κ_2 radial lens distortion parameters) of a camera using a set of calibration points.

Calibration points are points in the image with known world coordinates. In practice, Tsai calibration requires at least 15 calibration points. DORAEMON uses a fast, robust and flexible method for extracting calibration points from the environment. A simple colored calibration carpet is used.

The user selects a colored rectangle and specifies the distance in the x and y direction between the centers of the rectangle. DORAEMON's calibration is iterative, so it can compensate for missing or misclassified calibration points. Even using a side view of the playing field, the calibration results in object errors of less than 1 cm.

4   Interpolation Algorithms

Interpolation has long been an active research area in computer vision, mathematics, and statistics. The problem can be stated as finding the value of a target function for points in between a supplied set of data points. Often the data points are assumed to be corrupted with noise, and smoothing must be performed first. This is often the case in computer vision applications such as global vision robotic soccer.

Figure 3 is a simple example of a one dimensional interpolation problem. Given the value of the target function at positions $x_1, x_2, \ldots, x_n$, find the value of the target function for position x_t.

By comparing the right and left half of the example interpolation problem, one can see that the right half is easier to interpolate. The data looks much more regular than the left side of the image.

4.1   Square Interpolation

The simplest interpolation algorithm is the square or pulse approximation, which results in a zero order interpolation of the target value as shown in Fig. 4.

This is identical to the one nearest neighbor algorithm (1-NN) algorithm where each sub-scanned pixel is assigned the value of its closest neighbor.

Although this algorithm is computationally very efficient, it does not lead to good results. For example, both the spot and the robot top have very jagged

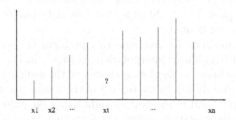

Fig. 3. One dimensional interpolation problem. Intuitively, one can see that the right side of the function is easier to interpolate than the left side

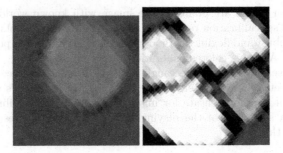

Fig. 4. Results of square pulse interpolation for the spots (left) and a robot (right)

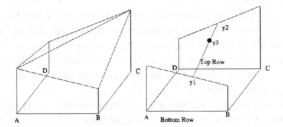

Fig. 5. Piecewise linear (left) and bi-linear interpolation of two dimensional data

edges. This jaggedness also enhances false colour information in the image. For example, the yellow edges on the robot (which resulted from errors in the colour processing of the camera) are emphasized.

4.2   Bilinear Interpolation

A slightly more complex interpolation method is the use of triangle instead of square approximation, also referred to as first order interpolation. In this case, the value of the target pixel is calculated along the line segment of its two neighbors.

The extension of this method to the two dimensional case (e.g., images) is slightly more complex. The problem is that four data points do *not* in general lie on an interpolation plane.

Two common methods to overcome this problem are: (a) piecewise linear interpolation, and (b) bilinear interpolation, as shown in Fig. 5.

In the piecewise linear interpolation approach, the sample surface is split into two planes (ABC and ACD). In the top left of the image, points are interpolated using the plane ABC and in the bottom right using the plane ACD.

Using bilinear interpolation, points are first linearly interpolated along the rows/columns and the resulting points are then used to linearly interpolate the value of the target point. This bilinear interpolation is commonly used because it is computationally efficient and leads often to better results in practice than

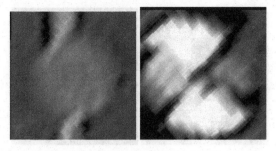

Fig. 6. Results of bi-linear interpolation for the spots (left) and a robot (right)

a square pulse interpolation. This can clearly be seen in the example pictures in Fig. 6. For example, the edges of spots are rounder and the shape of robot is also not as jagged as the square pulse interpolation.

4.3   Cubic B-Spline Interpolation

Another popular method of interpolation is to fit cubic b-splines to the data. In general, this method results in smoother interpolation of the data.

The main drawback is that it uses a larger neighborhood around the pixel to calculate the parameters of a cubic function. Because of computational constraints, the neighborhood is usually limited to a 4 by 4 pixel region.

The results of cubic b-spline interpolation are only slightly better than bi-linear interpolation, but larger interpolation neighborhoods are computationally much more expensive. Therefore, we restricted ourself to 2 by 2 neighborhood interpolation functions.

4.4   Average Gradient Method

Because of the noise in the image, it is often necessary to apply blurring to pixels (e.g., to get a better estimate of the colour of region of pixels). The most popular method for blurring is to replace a pixel with the average or median of the pixels in its 4, 9, or 16 neighborhood.

However, blurring an image results in a loss of contrast around the edges of an object. Since these are important cues to determine the geometrical features that we use in determining the position, orientation, and identity of our robots, blurring is not appropriate in our case.

It is also computationally more efficient if the blurring and interpolation steps are combined.

We therefore designed and implemented a gradient blurring interpolation. The average gradient along the rows and columns are computed. The closest point to the target point is selected and the average gradient is applied at the starting point.

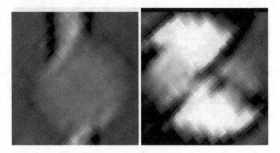

Fig. 7. Results of our average gradient interpolation for the spots (left) and a robot (right)

The resulting image of a spot and our robot are shown in Fig. 7. The result is similar to the bilinear interpolation and better than the square pulse interpolation.

5   Evaluation

One method of evaluating the scene recognition performance of a computer vision system is to reconstruct the scene from a different view point. Figure 8 shows the results of applying our interpolation method over the entire playing field. The resulting view is an overhead view isometric projection of the playing field using the original image as shown in Fig. 2.

The errors in the right edge of the playing field are due to errors in the camera calibration. In most cases, these errors are not significant, because the field is locally still consistent (i.e., the relative position of a robot and the ball is still correct), so robots can approach a ball to score, etc.

Although interpolating over the entire field is inefficient during a match, it nevertheless provides evidence of the accuracy that can be achieved with a solid camera calibration and reasonable interpolation methods.

Currently, no additional model knowledge has been used. The robots are assumed to be flat on the playing field. Although this is sufficient for detecting the position, orientation, and velocity of a robot, the quasi-overhead view reconstruction could be improved even further by correcting the image for the known height of robots that were found in the image.

We tested square pulse interpolation, bilinear interpolation, cubic b-splines interpolation as well as our own average gradient method on about 50 test images. Subjectively, we determined the accuracy of edges, area, and angles in the resulting image. In all cases, the performance of bilinear interpolation, cubic b-spline interpolation, and our own average gradient method was comparable and significantly better than that of the square pulse interpolation. The bi-linear interpolation is computationally the most efficient method, so we selected it as default interpolation method in DORAEMON.

Fig. 8. This is *not* an image taken by an overhead camera, but rather a reconstructed overhead view. The original image is shown in Fig. 2. The image has been corrected for perspective distortion and our average gradient interpolation was used

One drawback of the interpolation method is that DORAEMON is unable to maintain 60 fields per second when controlling a field with 11 objects in it. The capture rate drops to 30 frames per second. We believe that better optimized interpolation routines or better object tracking can overcome this problem.

6   Conclusion

This paper argues for a more model–based approach to computer vision to tackle the problems posed by the new field size in the SSL league. The goal of our research is to use a single camera without a wide angle lens that is able to cover the new larger playing field. As a first step, we built on the existing camera calibration routines in DORAEMON to geometrically interpolate pixels. This interpolation is used to extract important geometric features such as the area, aspect ratio, and angles in a shape. These features are then used to determine the position, orientation, and identity of objects on the playing field.

This paper described a comparison of several interpolation methods including square pulse interpolation, bi-linear interpolation, cubic b-spline interpolation, and a novel average gradient method. In our evaluation, bi-linear interpolation had good results, but ran faster than cubic b-spline and average gradient interpolation. Square pulse interpolation has the worst results.

To show the effectiveness of the reconstruction, we showed the result of calculating the interpolation of the entire playing field. The resulting image will hopefully convince other researchers that overhead cameras are not the only solution to the vision problem in the SSL.

442   J. Baltes and J. Anderson

References

1. Jacky Baltes. Doraemon: Object orientation and id without additional markers. In *2nd IFAC Conference on Mechatronic Systems*. American Automatic Control Council, December 2002.
2. Roger Y. Tsai. An efficient and accurate camera calibration technique for 3d machine vision. In *Proceedings of IEEE Conference on Computer Vision and Pattern Recognition*, pages 364–374, Miami Beach, FL, 1986.

A Method of Pseudo Stereo Vision from Images of Cameras Shutter Timing Adjusted

Hironobu Fujiyoshi[1], Shoichi Shimizu[1], Yasunori Nagasaka[2],
and Tomoichi Takahashi[3]

[1] Dept. of Computer Science, Chubu University, Japan
[2] Dept. of Electronic Engineering, Chubu University, Japan
[3] Dept. of Information Science, Meijo University, Japan
hf@cs.chubu.ac.jp, shiyou@vision.cs.chubu.ac.jp,
any@nn.solan.chubu.ac.jp, ttaka@ccmfs.meijo-u.ac.jp

Abstract. Multiple cameras have been used to get a view of a large area. In some cases, the cameras are placed so that their views are overlapped to get a more complete view. 3D information of the overlapping areas that are covered with two or three cameras can be obtained by stereo vision methods. By shifting the shutter timings of cameras and using our pseudo stereo vision method, we can output 3D information faster than 30 fps. In this paper, we propose a pseudo stereo vision method using three cameras with different shutter timings. Using three cameras, two types of shutter timings are discussed. In three different shutter timings, 90 points of 3D position for a sec are obtained because the proposed method can output 3D positions at every shutter timing of three cameras. In two different shutter timings, it is possible to calculate the 3D position at 60 fps with better accuracy.

1 Introduction

In a soccer robot match, it is important to calculate the position of an object as quickly as possible in order to control the robot by visual feedback. Also, it is necessary to calculate the 3D position of the ball, not 2D position on the soccer field, because some robots have an ability of striking a loop shot [1].

As an approach to implementing a high speed vision system, a 60 fps camera has been used in small-sized robot league [2, 3]. The system processes NTSC camera images at a 60 fps rate with double buffering. However, they can't calculate a 3D position because they use a single camera. Stereo vision using multiple cameras is needed for measuring the 3D position. They require cameras to synchronize with each other for tracking an object accurately and measuring its depth.

We have proposed a pseudo stereo vision method for calculating the 3D position of an object using two unsynchronized cameras [4]. The method can obtain the 3D position of a moving object at 60 fps making use of the time lags of the shutter timing between the two cameras. In this paper, we present two kinds

D. Nardi et al. (Eds.): RoboCup 2004, LNAI 3276, pp. 443–450, 2005.

of vision systems based on the pseudo stereo vision method using three normal cameras (that take pictures at 30 fps), which can output 3D positions at 60 fps or 90 fps by adjusting the shutter timing of each camera.

2   3D Position Measurement with Multiple Cameras

The stereo vision method which measures the 3D position of the object requires two images captured at the same time to reduce error in the measurement. Using a general stereo vision system, 3D positions can be obtained at 30 fps maximum using a normal 30 fps camera with fast vision algorithm described in [5].

Using two unsynchronized cameras for calculating the 3D positions of objects, we have proposed a pseudo stereo vision method taking advantage of time lag between the shutter timing of each camera [4]. To obtain a higher speed with better accuracy in 3D position, we investigate a vision system consisting of three cameras and a method for calculating the 3D position with two kinds of shutter timing of three cameras.

2.1   Shutter Timings of Three Cameras

Three combinations of cameras might be considered as shown in Figure 1 by adjusting the shutter timings of the cameras. One of them is the case of same shutter timings which are used in multiple baseline stereo as shown in Figure 1(a). We focus on two cases where the shutter timing of each camera is different as shown in Figure 1(b) and (c). In the case of type A, the shutter timing of each camera is shifted for 1/90 second. Since the 3D position is calculated at every shutter timing of each camera, the 3D position can be obtained at 90 fps. In the case of type B, the shutter timings of camera1 and camera3 are synchronized, and the 3D position is calculated using stereo vision. The shutter timing of camera2 is shifted for 1/60 sec from the shutter timing of camera1 and camera3. The 3D position can be obtained at 60 fps, and we can obtain better accuracy. The methods for estimating the 3D position of two kinds of shutter timing are described as follows.

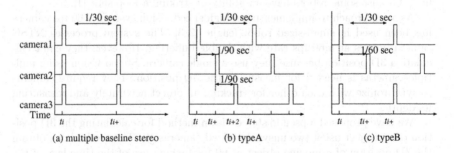

Fig. 1. Possible two combinations of shutter timing

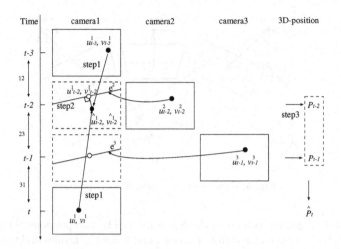

Fig. 2. Proposed calculation method of corresponding position

2.2 TYPE-A: Algorithm for Three Different Shutter Timings(90fps)

The 3D position in the last frame is estimated by using the time interval δ between the shutter timings of each camera and the results from the previous two frames. The procedure of 3D position measurement is as follows:

Step1. Calculation of 3D positions in the previous two frames
Step2. Linear prediction of the 3D position
Step3. Prediction using constraints from ray information

Calculation of 3D Positions in the Previous Two Frames

In order to obtain an accurate 3D position from the current frame by linear prediction, it is necessary to accurately calculate the previous 3D positions P_{t-1}, P_{t-2}. The algorithm of the 3D position calculation at $t-2$ is described as follows:

Step1. Using two observed points in the frame t and $t - 3$, a pseudo-corresponding point from camera1 on frame $t - 2$ $(u_{t-2}^{\hat{1}}, v_{t-2}^{\hat{1}})$ is interpolated by the following equation:

$$u_{t-2}^{\hat{1}} = \frac{\delta_{12}u_t^1 + (\delta_{23} + \delta_{31})u_{t-3}^1}{\delta_{12} + \delta_{23} + \delta_{31}}, \quad v_{t-2}^{\hat{1}} = \frac{\delta_{12}v_t^1 + (\delta_{23} + \delta_{31})v_{t-3}^1}{\delta_{12} + \delta_{23} + \delta_{31}} \quad (1)$$

Step2. Calculate the epipolar line e^2 on the image from camera1 using the corresponding point (u_{t-2}^2, v_{t-2}^2) from the image from camera2. Then, the nearest point $(u_{t-2}'^1, v_{t-2}'^1)$ from the interpolated point, calculated by step1, can be set as corresponding point for (u_{t-2}^2, v_{t-2}^2).

Step3. We can measure the 3D position P_{t-2} using triangulation as a crossing point of the two lines in 3D space.

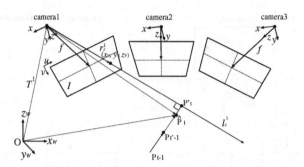

Fig. 3. Constraints in estimating 3D position

Step 1, 2, 3 are repeated in the frame $t - 1$. The two previous 3D positions, at $t - 1, t - 2$ are calculated from previous and following frames and the epipolar constraint. These two previous points, P_{t-1} and P_{t-2}, will be used and decrease the prediction error at the next step of linear prediction.

Linear Prediction of 3D Position
As shown in Figure 3, the predicted position $\hat{P}_t = [x_w, y_w, z_w]^T$ of the last frame t is calculated by the following equation using the already measured positions P_{t-1} and P_{t-2} in the previous two frames.

$$\hat{P}_t = P_{t-1} + \frac{\delta_{31}(P_{t-1} - P_{t-2})}{\delta_{12}} \tag{2}$$

Note that Equation (2) is based on the analyzed image of the last frame from camera1. For the position estimate, the Kalman filter [6, 7] and spline curve fitting have been proposed.

Prediction Using Constraints from Ray Information
In order to decrease the prediction error, the 3D position is calculated once more using the constraint of a viewing ray in 3D space obtained from the current image. Let $T^1 = [T_x, T_y, T_z]$ be the translation matrix from the origin of the world coordinate to the focus point of camera1, and $r_t^1 = [x_w, y_w, z_w]^T$ be the vector which denotes the direction of the viewing ray, l_t^1, passing through the position on the image coordinate (u_t^1, v_t^1) and the focus point of the camera. The viewing ray shown in Figure 3 can be expressed by

$$l_t^1 = k r_t^1 + T^1 \tag{3}$$

where k is a real number. Although Equation (2) gives a good 3D position prediction, the position may not exist on the viewing ray l_t^1 as shown in Figure 3 because of its prediction error. In order to solve this problem, the 3D position P'_t is calculated by the following equation as the nearest point on the ray l_t^1.

$$P'_t = \frac{(\hat{P}_t - T^1) \cdot r_t^1}{|r_t^1|^2} r_t^1 + T^1 \tag{4}$$

Finally, P'_t will be the 3D position of the object at the last frame. In case camera2 and camera3 are the latest frame, the image of the 3D position is calculated in the same way as mentioned above.

2.3   TYPE-B: Algorithm for Two Different Shutter Timings(60fps)

In order to estimate better 3D position by linear prediction, it is important to calculate 3D positions in the previous two frames. In the case of Figure 1(c), the shutter timings of two cameras(camera1 and camera3) are synchronized so that 3D position is calculated by stereo vision. Therefore, P'_t is estimated by linear prediction using P_{t-1} and P_{t-3}, which are obtained by stereo. Futhermore, 3D position of camera2 P'_t is calculated by constraint from ray information using the same algorithm as type A.

Therefore, 3D positions, which are calculated by stereo from two synchronized cameras and estimated by constraint from ray information of a single camera are obtained respectively. In this case, the total number of points can be obtained for a second is 60 points, which is less than type A (90 points).

3   Simulation Experiments

3.1   Recovery of Object Motions

We evaluated the proposed method by simulation of recovering the object's motion with uniform and non-uniform motion in 3D space (3,000 × 2,000 × 2,000 mm). In the simulation, we assumed that three cameras would be mounted at the height of 3,000 [mm].The proposed method is evaluated by following three motions.

- uniform motion(straight): An object moves to $(x, y, z) = (3,000, 1,200, 0)$ from $(x, y, z) = (0, 1,200, 2,000)$ at velocity of 3,000 mm/sec.
- uniform motion(spiral) An object moves in a spiral by radius of 620 mm at velocity of 3,000 mm/sec at center $(x, y) = (1,000, 1,000)$.
- non-uniform motion: An object falls from the height of 2,000mm, then an object describes a parabola (gravitational acceleration:g=9.8 m/s^2).

The trajectory of the object is projected to the virtual image planes of each camera. A 3D position is estimated by the proposed method described in section 2.1 using the projected point on the virtual image plane (u, v) of each camera.

3.2   Simulation Results

Table 1 shows averages of estimation error with simulation experiments. The unsynchronized method in Table 1 shows the result of stereo vision by corresponding points in time delay using two cameras, and the synchronized method shows the result of general stereo vision with no time delay. In the case of using two cameras, it is clear that the proposed method(type A) has a better result than the unsynchronized method, and its accuracy is close to the synchronized method.

Table 1. Average of absolute errors in 3D positions [mm]

method		fps	uniform		non-uniform
			straight	spiral	
2cameras	unsynchronized	60	23.2	21.4	16.2
	synchronized	60	0.03	0.14	0.12
3cameras	type A	60	1.1	2.2	1.8
	type A	90	1.1	2.0	1.7
	type B	60	0.2	0.5	1.5
	linear prediction	60	0.2	1.4	4.4

Linear prediction in Table 1 shows the result of linear prediction using the past two positions calculated by stereo vision. Comparing type B to linear prediction, it is clear that type B has better accuracy because constraint from ray information decreases the error generated by linear prediction. In the simulation experiment of non-uniform motion, type A has better accuracy compared to linear prediction even though the shutter timings of the three cameras are different. This is why the time interval of the shutter timing is small ($\delta = 1/90$ sec).

4   Experiments Using Real Cameras

We evaluated our method using real data in the same way as the simulation experiments.

4.1   Configuration of Vision System

Figure 4 shows the camera placement of our vision system that uses three cameras, camera1, camera2 and camera3. These cameras are mounted at a height of 2,800 mm, and each camera has a view of an area of 2,000 × 3,000 mm. Each camera is calibrated using corresponding points of world coordinates (x_w, y_w, z_w) and image coordinate (u, v) [8]. The shutter timing of each camera is controlled by a TV signal generator. Three frame grabbers for the three cameras are installed on a PC. Our hardware specifications are described as follows:

Process-1, process-2 and process-3 analyze images from camera1, camera2 and camera3 at every 1/30 second respectively. The analyzed results such as (u^i, v^i) and the time instant at which the analyzed image was captured are sent via UDP interface to process-4 that calculates the 3D positions of the object. There is negligible delay due to communications among processes because this work is done on the same computer.

4.2   Experiments

Figure 6 shows results of recovering the motion of a hand-thrown ball for about 1.5 sec. Figure 6(a) shows that the numbers of plotted points is 135. This indicates that the speed is the same as 90 fps camera. Figure 6(b) shows that the numbers of plotted points is 90, which is same as 60 fps camera.

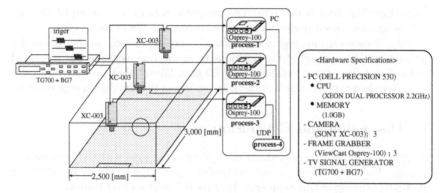

Fig. 4. Overview of our vision system

Fig. 5. Hardware specifications

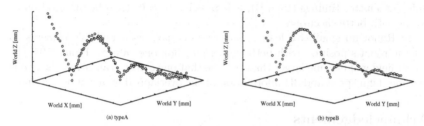

(a) typeA

(b) typeB

Fig. 6. Results of 3D position measurement

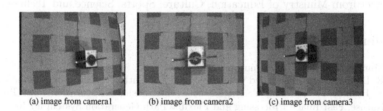

(a) image from camera1   (b) image from camera2   (c) image from camera3

Fig. 7. Captured images of turntable

Table 2. Average and variance of z values of 3D positions

	average[mm]	variance
type A	664.5	2143.7
type B	662.3	112.0

As an evaluation for the accuracy of estimated 3D positions, we used a turntable and a ball as shown in Figure 7. A ball attached on the edge of ruler(1,000 mm length) makes a uniform circular motion with a radius of 500 mm. The turntable is placed on a box at the height of 500 mm, and the ball's height from the floor is 660 mm. The turntable rotates at a speed of 45 rpm, and its rotation speed per second is $(45 \times 2\pi)/60 = 0.478$ radian.

Table 2 shows the average and variance of the 3D position on the z_w axis for both types. The average of the positions from the two methods was measured within 5 mm from the actual height of 660 mm. We see that variance of type B is smaller than type A, which is the same result as the simulation experiments.

5   Discussion and Conclusion

We proposed a pseudo stereo vision method using cameras with different shutter timings. The method can output 3D position at 60 fps or 90 fps by adjusting the shutter timing of three cameras. In three different shutter timings (type A), 90 points of 3D position for a sec are obtained because the proposed system can output 3D positions at every shutter timing of the three cameras. In two different shutter timings (type B), it is possible to calculate the 3D position at 60 fps with better accuracy.

In RoboCup small-size league, some teams have used multiple cameras to get the robot's position with better precision than one camera. From 2004, the soccer field will become larger than the size that one camera can cover. Using our method, high speed and 3D information of the overlapped area can be obtained.

Acknowledgements

This work was supported by the High-Tech Research Center Establishment Project from Ministry of Education, Culture, Sports, Science and Technology.

References

1. Muratec FC.http://www.muratec.net/robot/
2. R. D'Andrea, et al. Detailed vision documentation.http://robocup.mae.cornell.edu/
3. S. Hibino, Y. Kodama, Y. Nagasaka, T. Takahashi, K. Murakami and Tadashi Naruse. Fast Image Processing and Flexible Path Generation System for RoboCup Small Size League, RoboCup2002, pp.53-64, 2002.
4. S. Shimizu, H. Fujiyoshi, Y. Nagasaka and T. Takahashi. A Pseudo Stereo Vision Method for Unsynchronized Cameras. ACCV2004, vol.1, pp.575-580, 2004.
5. J. Bruce, T. Balch and M. Veloso. Fast and Inexpensive Color Image Segmentation for Interactive Robots, IROS-2000, vol.3, pp.2061-2066, 2000.
6. B. Browning, M. Bowling and M. Veloso. Improbability Filtering for Rejecting False Positives, In Proc. IEEE International Conference on Robotics and Automation, pp.120-200, 2002.
7. K. Horiuchhi, S. Kaneko and T. Honda. Object Motion Estimation based on Multiple Distributed Kalman Filters, In the IEICE, Vol.J79-D-II, Num.5, pp.840-850, 1996.
8. R. Y. Tsai. A versatile Camera Calibration Technique for High-Accuracy 3D Machine Vision Metrology Using Off-the-Shelf TV Cameras and Lenses, In IEEE Journal of Robotics and Automation, Vol.RA-3, Num.4, pp.323-344, 1987.

Automatic Distance Measurement and Material Characterization with Infrared Sensors*

Miguel Angel Garcia and Agusti Solanas

Intelligent Robotics and Computer Vision Group,
Department of Computer Science and Mathematics,
Rovira i Virgili University,
Av. Països Catalans 26, 43007 Tarragona, Spain
{magarcia, asolanas}@etse.urv.es

Abstract. This paper describes a new technique for determining the distance to a planar surface and, at the same time, obtaining a characterization of the surface's material through the use of conventional, low-cost infrared sensors. The proposed technique is advantageous over previous schemes in that it does not require additional range sensors, such as ultrasound devices, nor a priori knowledge about the materials that can be encountered. Experiments with an all-terrain mobile robot equipped with a ring of infrared sensors are presented.

1 Introduction

Infrared sensors are commonly utilized in mobile robotics as low-cost proximity sensors, basically for immediate collision avoidance. Their non-linear behavior and high dependence on the reflectivity of the sensed objects has prevented their application as range sensors, in favor of more sophisticated and costly devices, such as sonar (ultrasound) or laser systems.

Notwithstanding, since infrared sensors are inexpensive and readily available in most commercial mobile robots, some researchers have tried to overcome their limitations. A pioneering work is due to Novotny and Ferrier [4]. They applied the *Phong illumination model* [5] in order to compute the distance to a planar surface and, simultaneously, determine two model coefficients that represent the reflective properties of that surface. Unfortunately, this technique also requires a second ring of ultrasound sensors, each conveniently placed next to an infrared sensor, in order to obtain a parameter that is crucial to the solution of the problem: the minimum distance d between the robot and the surface being measured, Fig. 1.

In a similar direction, [2] presents a technique for computing distances to unknown planar surfaces by means of infrared sensors. In this case, a simpler illumination model based on the *photometry inverse square law* is utilized. This model characterizes the reflectivity of the surface being measured with a single coefficient. Once that coefficient has been established, the infrared sensor readings can be directly mapped to distances. However, that crucial coefficient must be estimated from the distance to the surface, which is also measured by an ultrasound sensor.

---

* This work has been partially supported by the Government of Spain under the CICYT project DPI2001-2094-C03-02.

D. Nardi et al. (Eds.): RoboCup 2004, LNAI 3276, pp. 451–458, 2005.

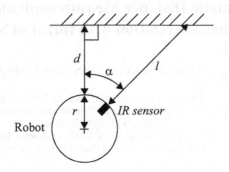

Fig. 1. Diagram of a robot emitting an infrared signal, adapted from [4]

Following a different approach, [1] presents a technique that applies infrared sensors for determining the shape of an unknown surface from among four basic shapes (plane, corner, edge and cylinder) independently of its position. When the shape is known, the distance can also be estimated. No ultrasound sensors are necessary. However, this technique requires an off-line, supervised training stage for generating an exhaustive collection of reference angular scans that describe the sensor output as a function of the incidence angle (α in Fig. 1) for all types of materials that can be encountered, the four considered shapes, and different distances (at regular increments of 2.5 cm). When a real scan is acquired, the system finds a matching reference scan within the database. Such an exhaustive training with a limited number of materials can be a serious limitation that may prevent the application of this technique to real situations in which there is not a priori knowledge of the kind of surfaces that can be encountered, especially in case of exploration tasks.

Based on the theory presented in [4], the present paper describes a new technique that exclusively utilizes low-cost infrared sensors for estimating the reflective coefficients of an unknown planar surface and then computing its distance to a robot, without any prior training or knowledge about the materials that can be encountered in the scene. The technique only requires two angular scans of the unknown surface from two positions from where the surface can be detected, such that the distance between both positions is known. In our experiments, this is done with a single infrared sensor mounted at the front of a robot. The robot rotates around its center to perform the first scan, then moves a predefined distance toward or against the surface and scans it again. The whole process can be performed with a single scan by using two infrared sensors mounted on concentric circles of different known radius.

This paper is organized as follows. The proposed technique is described in Section 2. Experimental results are discussed in Section 3. Finally, conclusions and further improvements are presented in Section 4.

2   Surface Characterization and Distance Measurement

The infrared sensors considered in this work are constituted by an infrared LED emitter and a photo-diode that measures the amount of emitted energy that is reflected by the surface of a nearby object.

Based on the Phong's illumination model and the photometry theory, [4] establishes that the energy absorbed by an infrared sensor mounted on a robot at distance r from its rotation center (Fig. 1) can be approximated by:

$$E_\alpha^d = (C_0\cos(\alpha) + C_1\cos(2\alpha))/\left[\frac{d}{\cos(\alpha)} + r\left(\frac{1}{\cos(\alpha)} - 1\right)\right]^2 \qquad (1)$$

where d is the minimum distance between the sensor and the surface, α is the incidence angle of light with respect to the surface and C_0 and C_1 are two coefficients that express the reflectivity properties of the surface being sensed. This energy corresponds to the actual sensor reading.

The proposed technique aims at determining the two reflective coefficients and the minimum distance d by means of two angular scans. Each scan consists of a sequence of N sensor readings obtained while the robot and, hence, the sensor, turn around the robot's rotation center. Since the robot rotates at a known angular speed and the sensor is read at a constant frequency, it is straightforward to determine the angular increment between any pair of consecutive readings. From that increment, the relative angle α between the current sensor orientation and the one in which the sensor is perpendicular to the surface and, hence, receives maximum energy, is computed. Let Φ be the maximum angle (e.g., $\Phi = \pi/6$). When the distance between the sensor and the surface is the sought minimum distance d, angle α is zero.

Let $\{E_\alpha^d\}$ and $\{E_\alpha^{d'}\}$, $\alpha \in [0, \Phi]$, be two angular scans of N readings obtained at two unknown minimum distances, d and d', $d < d'$, whose separation is known:

$$\Delta = d' - d \qquad (2)$$

According to the previous definition of α, E_0^d and $E_0^{d'}$ are the maximum energies detected by the sensor in both scans respectively. From (1), the minimum distance to the surface is expressed as [4]:

$$d = r(\cos(\alpha) - 1) + \cos(\alpha)\sqrt{(C_0\cos(\alpha) + C_1\cos(2\alpha))/E_\alpha^d} \qquad (3)$$

By combining (2) and (3):

$$\Delta = (1/\sqrt{E_\alpha^{d'}} - 1/\sqrt{E_\alpha^d})\cos(\alpha)\sqrt{C_0\cos(\alpha) + C_1\cos(2\alpha)} \qquad (4)$$

Hence:

$$C_0\cos(\alpha) + C_1\cos(2\alpha) = \Delta^2/(\cos(\alpha)(1/\sqrt{E_\alpha^{d'}} - 1/\sqrt{E_\alpha^d}))^2 \qquad (5)$$

Given an orientation α and its corresponding sensor readings, E_α^d and $E_\alpha^{d'}$, the previous expression defines a linear equation in which the two unknowns are the reflective coefficients C_0 and C_1. Since N readings are available, each with a different α, an over determined system of N linear equations over two unknowns is defined and solved by applying least-squares.

Once the two reflective coefficients are known, the minimum distance to the surface is simply found by applying (3) for $\alpha = 0$:

$$d = \sqrt{(C_0 + C_1)/E_0^d} \qquad (6)$$

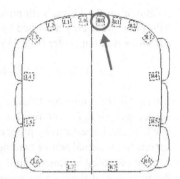

Fig. 2. K-Team's Koala robot equipped with a ring of 16 infrared sensors. The proposed technique is applied to the readings obtained with the highlighted frontal sensor

3 Experimental Results

The proposed technique has been tested on real surfaces of different colors and materials by using an all-terrain Koala mobile robot (Fig. 2). This robot is equipped with a ring of 16 infrared sensors distributed around the robot's chassis. In particular, each sensor is a TSL252 consisting of an infrared LED that emits at 880nm and a photo-diode that detects the reflected light. Each device generates an output voltage that is read by the robot through a 10-bit A/D converter.

This technique has been applied to the readings obtained by a single frontal sensor highlighted in Fig. 2(*right*). Once the robot is close to the surface to be measured, it performs a first angular scan by rotating around its center at constant angular speed while the sensor is read at a constant frequency. Once the sequence of readings is complete, the robot turns back to the orientation in which the maximum energy was found —this is the orientation in which α is considered to be zero. At this point the robot moves backwards (against the surface) a predefined distance Δ and performs a second scan similar to the initial one.

The experiments presented in this paper have been performed with a Δ equal to 2 cm. Other values are possible and do not significantly affect the results whenever the sensor is kept within the operational range in which it has sufficient accuracy. However, if Δ is too small, the discretization of the sensor readings will not allow to distinguish variations of light in case the distance to the surface is relatively large.

Fig. 3 shows two examples of angular scans respectively acquired at approximately 10 and 12 cm ($\Delta = 2$) away from a white thin cardboard, by following the operational procedure described above. The angular resolution between consecutive readings is 0.015 degrees. The maximum energies are $E_0^d = 935$ and $E_0^{d'} = 653$ respectively. These energies are not expressed in physical units but in sensor units (integers ranging between 0 and 1,023).

Therefore, the reflective coefficients, C_0 and C_1, do not really have the physical interpretation that the Phong's coefficients have (energy $\times$ distance^2). Instead, they are expressed in terms of sensor units: sensor-units $\times$ distance^2. This does not pose any problem in order to compute the minimum distance d in the same units as Δ (cm

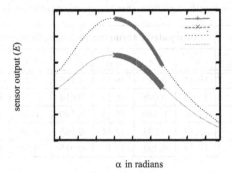

α in radians

Fig. 3. Two infrared scans acquired at positions separated by 2 cm ($\Delta = 2$), at minimum distance d approximately equal to 10 (d10) and 12 (d12) cm respectively from a white thin cardboard. The thickened curves (D10 and D12) are sections of the approximating parabolas

in our case), since, according to (6), both C_0 and C_1 are divided by E_0^d and, hence, the sensor units cancel each other.

In order to remove noise inherent to the sensor and compensate for discretization errors due to the A/D conversion, the readings E_α^d and $E_\alpha^{d'}$ obtained from the sensor are filtered out before applying them to (5). This is done by approximating a second-order polynomial (a parabola) to the sensed readings comprised between $-\Phi$ and Φ through least-squares fitting. Within this range of angles, the original curves can be approximated by a second order polynomial without significant error. This filtering stage is also beneficial in order to extrapolate the missing energy readings in case the sensor saturates in being too close to the measured surface. The thickened curves in Fig. 3 correspond to the sections of both parabolas (between 0 and Φ) which are utilized in the subsequent estimation process.

An over-determined linear system of N equations (5) is defined and solved by applying a least-squares system equation solver. In the example, the result is $C_0 = 17.7$ and $C_1 = -8.12$. According to these values and E_0^d and $E_0^{d'}$, the minimum distances after applying (6) are: $d = 10.14$ cm and $d' = 12.12$ cm.

In order to determine the influence of distance on the computation both of the reflective coefficients, C_0 and C_1, and the estimated minimum distance d, the same process described above has been carried out at different test distances within the sensor's operational range for various types of material. Each experiment has been run a number of times for every distance. Table 1 and Table 2 show the average results corresponding to white and black thin cardboard respectively. Fig. 4 shows the evolution of the computed distances with respect to the test distances for those two materials. At every test distance, the reflective coefficients have been computed through the procedure described above.

The previous results indicate that the distance estimation starts degrading beyond a certain point away from the measured surface. In our experiments, the distance error is below 1 mm for distances up to 14 cm from the surface. From 14 cm to 18 cm, the error is kept below 1 cm. Beyond this distance, the error progressively degrades. This degradation is inherent to the infrared technology being used and

Table 1. Average Results for White Thin Cardboard at Different Test Distances

Test d (cm)	Computed Parameters (average and standard)					
	$\bar{d}$	σ_d	$\overline{C_0}$	σ_{C_0}	$\overline{C_1}$	σ_{C_1}
10	10.05	0.119	16.35	1.189	-7.22	0.786
12	11.92	0.098	16.24	0.614	-7.17	0.517
14	13.96	0.149	15.28	0.540	-6.05	0.404
16	15.59	0.291	13.61	0.105	-5.06	0.081
18	17.23	0.291	13.15	0.025	-4.91	0.227

Table 2. Average Results for Black Thin Cardboard at Different Test Distances

Test d (cm)	Computed Parameters (average and standard)					
	$\bar{d}$	σ_d	$\overline{C_0}$	σ_{C_0}	$\overline{C_1}$	σ_{C_1}
10	10,04	0,140	13,53	0,671	-6,30	0,511
12	12,00	0,320	12,40	0,458	-5,14	0,172
14	13,87	0,497	11,53	0,270	-4,49	0,389
16	15,68	0,209	10,55	0,413	-3,72	0,567
18	17,16	0,358	10,38	0,468	-3,82	0,277

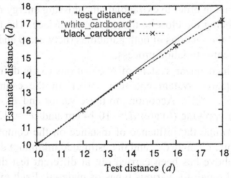

Fig. 4. Average estimated distances versus test distances for two types of material

agrees with the results reported in [4]. As a consequence, it is advisable to obtain the reflective coefficients C_0 and C_1 at distances below 14 cm (e.g., 10 cm) in order to have a reliable characterization of the measured surface.

Table 3 shows the average and standard deviation of the reflective coefficients obtained for five different materials at a test distance of 10 cm. Due to the small standard deviations associated with those coefficients, it is possible to perform the distance measurement process in a single run (two angular scans separated by Δ) without significant variation with respect to the average values.

Table 3. Average Reflective Coefficients of Different Materials Estimated at the Same Test Distance (10 cm)

Test Material	$\overline{C_0}$	σ_{C_0}	$\overline{C_1}$	σ_{C_1}
White thin cardboard (w_card)	16.8	1.39	-7.6	0.97
Black thin cardboard (b_card)	13.8	0.76	-6.5	0.56
Unfinished brown wood (u_wood)	14.4	1.42	-3.9	1.10
Rugged gray wall (r_wall)	11.9	1.3	-5.2	1.15
Shiny brown wood (s_wood)	7.37	0.38	-2.9	0.41

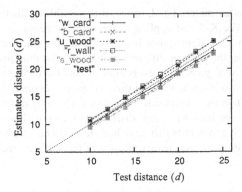

Fig. 5. Estimated distances versus test distances for five different materials by considering reflective coefficients measured at 10 cm

Once C_0 and C_1 have been computed, (6) makes it possible to estimate the current distance d between the sensor and the same surface from a single sensor reading E_0^d obtained with the sensor perpendicular to the surface. This condition can be guaranteed at every distance by making a small angular scan that allows to determine the maximum reading. Fig. 5 shows the estimated distances obtained by applying the aforementioned procedure to the five previous materials, by considering test distances between 10 and 24 cm.

The sensor response at distances above 24 cm is not accurate enough. In this work, both reflective coefficients were obtained from two angular scans taken at 10 and 12 cm respectively (Table 3). These results show that the distance error is kept within a small interval approximately between 0.8 mm and 1 cm, rather independently of the measuring distance and the surface material.

The sum of reflective coefficients can also be utilized as an indication of the light reflective properties of the surface being measured and, hence, as a characterization of its material. For example, Table 4 shows the average sum of the coefficients whose averages and standard deviations have been presented in Table 3.

The *Kolmogorov-Smirnov test* [6] has been applied in order to determine whether the sums computed for each material are significantly different to the sums corre-

Table 4. Average Sum of Reflective Coefficients of Different Materials Estimated at the Same Test Distance (10 cm)

Test Material	$\overline{C_0 + C_1}$	$\sigma_{C_0 + C_1}$
White thin cardboard (w_card)	9.270	0.433
Black thin cardboard (b_card)	7.307	0.234
Unfinished brown wood (u_wood)	10.502	0.328
Rugged gray wall (r_wall)	6.77	0.236
Shiny brown wood (s_wood)	4.420	0.095

sponding to every other material and, hence, it is possible to distinguish among the various materials. The K-S test applied to the samples belonging to every pair of tested materials returns a significance level that indicates whether the two samples are drawn from either the same or different probability distributions.

In particular, small significance levels indicate that the datasets belong to different distributions. In our case, the significance level obtained for every pair of materials is below 0.10 (significance levels below 0.15 imply different distributions, while those close to one indicate a same distribution). Hence, there is significant difference among the five materials that have been tested. This implies that the sum of reflective coefficients is a good characterization of the surface's material.

4 Conclusions

A new technique for computing the distance to a surface and characterizing its material exclusively through infrared sensors without a priori knowledge about the materials that can be encountered has been presented. Further work will determine the shape of the surfaces that are measured by analyzing the shape of the acquired angular scans, such as in [1]. We will also consider how redundancy provided by integrating adjacent sensors can help reduce the measurement errors present when a single sensor is utilized. Finally, more sophisticated models of light reflection [3] will be studied.

References

1. Aytaç, T., Barshan, B.: Differentiation and Localization of Target Primitives Using Infrared Sensors. IEEE/RSJ Int. Conf. on Intelligent Robots and Systems (2002) 105-110
2. Benet, G., Blanes, F., Simo, J.E., Perez, P.: Using Infrared Sensors for Distance Measurement in Mobile Robots. Robotics and Autonomous Systems 40 (2002) 255-266
3. Blinn, J.F.: Models of Light Reflection for Computer Synthesized Pictures, SIGGRAPH'77, Vol.11, No.2, July (1977) 192-198
4. Novotny, P.M., Ferrier, N.J.: Using Infrared Sensors and the Phong Illumination Model to Measure Distances. IEEE Int. Conf. on Robotics and Automation (1999) 1644-1649
5. Phong, B.T.: Illumination for Computer Generated Pictures. Communications of the ACM, 18(6), June (1975) 311-317
6. Press, W.H., Teukolsky, S.A., Vetterling, W.T., Flannery, B.P.: Numerical Recipes in C++: The Art of Scientific Computing, Cambridge University Press, 2nd Ed. (2002)

A Novel Search Strategy for Autonomous Search and Rescue Robots

Sanem Sarıel[1] and H. Levent Akın[2]

[1] Istanbul Technical University, Dept. of Computer Engineering, Istanbul, Turkey
sariel@cs.itu.edu.tr
[2] Boğaziçi University, Dept. of Computer Engineering, Istanbul, Turkey
akin@boun.edu.tr

Abstract. In this work, a novel search strategy for autonomous search and rescue robots, that is highly suitable for the environments when the aid of human rescuers or search dogs is completely impossible, is proposed. The work area for a robot running this planning strategy can be small voids or possibly dangerous environments. The main goal of the proposed planning strategy is to find victims under very tight time constraints. The exploration strategy is designed to improve the success of the main goal of the robot using specialized sensors when available. The secondary goals of the strategy are avoiding obstacles for preventing further collapses, avoiding cycles in the search, and handling errors. The conducted experiments show that the proposed strategies are complete and promising for the main goal of a SR robot. The number of steps to find the reachable victims is considerably smaller than that of the greedy mapping method.

1 Introduction

The disasters such as earthquakes make clear the necessity of having robust, dynamic, and intelligent planning systems and powerful human-machine interaction. In general, the scale of the disaster and the speed of changing situations are far beyond the capabilities of human-based mission planning [3]. Often the lack of qualified rescue workers is a major problem. In addition, there are many difficulties for human team members in disaster areas such as dangerous or unreachable places. Although rescue dogs could help reduce the human risk by searching smaller voids in the rubble than a human can, in many cases a video camera or any structural-assessment equipment is more useful [6]. In overcoming the difficulties mentioned above, robots can be very sutiable [2].

The research on this area has become very attractive during the last decade. RoboCup-Rescue League has been started in 2001 [7]. Many search and rescue (SR) robot architectures, the variety of which is very large [3], have been proposed. In these proposals, the focus is typically on mechanical design, and the planning strategies are not addressed in detail.

As the most important part of a SR robot, the planning layer should be capable of generating effective plans for finding victims in a short period of time

D. Nardi et al. (Eds.): RoboCup 2004, LNAI 3276, pp. 459–466, 2005.

while taking care of constraints. The uncertainties about the environment and the effects of actions, the environmental constraints, goal interactions, and time and resource constraints make the problem harder. Therefore, flexible search strategies should be designed to make the problem more manageable. Time is a very hard constraint while finding the humans suffering from the disaster. There is a tradeoff between effective planning without dead ends and the fast plans for finding the victims in the environment. Partial-planning or re-planning [5] strategies are known to be very effective. However the use of either of them as such is not suitable for real implementations of the SR robots. However, key ideas of both approaches could be used in the design of a planning layer design of a SR robot to take advantages of both. Architectures such as three layered or BDI (Belief-Desire-Intentions) may be used to implement in the planning layer design depending on the specific application [8].

Since the difficulties with the environment make the current SR robot proto-types to be designed as semi-autonomous or highly dependent on human robot in-teraction, autonomous robot designs are rarely encountered in the literature [1]. In this study, an autonomous planning strategy suitable for the environments when the aid of human rescuers or search dogs is completely impossible is proposed. The work area for a robot running this planning strategy can be small voids or possibly dangerous environments with gas leakages or under danger of possible explosions or further collapses. The layered model of InteRRap and the belief update proce-dures of BDI type architecture are combined, and exception handling strategies are attached to the proposed planning layer action selection mechanism to cope with the uncertainties on the environment. Reactive actions are executed directly ac-cording to the sensory information. This proposal of the hybrid planning strategy is believed to be useful in promoting further improvements in SR robot designs.

The rest of this paper is organized as follows: Section 2 introduces the pro-posed planning module design and the proposed strategies. In Section 3 the experimental results are given. Finally, Section 4 presents future work and con-cludes the paper.

2   Search and Rescue Planning Module Design

2.1   Requirements

The primary goal of the Planning and Behavior Module (PBM) of a SR robot is to find victims in the disaster area within a minimum possible time interval. The secondary goals are avoiding obstacles, avoiding risking resources or trig-gering a further collapse, searching the area effectively, avoiding cycles in the search, turning and moving to the directions of locations in which it is believed that humans to be rescued are located, turning and moving to the directions of locations in which it is reported by other robots or a dispatcher that humans to be rescued are located.

The PBM of a SR robot should be able to determine a plan based on the current state of the environment, the robot's location, the hierarchical structure of the desires (goals), and the constraints. It should also perform re-planning when necessary while executing a plan based on the intermediate changes on the environment or the internal state.

There may be static or dynamic obstacles, other robots, and the human workers of the rescue team in the environment. The environment structure and the map are unknown.

2.2   Design

The proposed PBM design consists of mainly three layers which interact with each other to achieve the same end as in the InteRRap architecture. The reactive layer interacts directly with the sensory modules, and produces an appropriate action. The planning layer constructs the plan for the robot to implement its task in an optimal way. The strategies for avoiding cycles, forming beliefs to direct the search space, and exception handling are implemented by this layer. The communication layer is the interaction layer of the module and implements the robot communications. It is responsible for informing other robots according to the planning layer outputs, and receiving the incoming information. The bandwidth requirements are very small.

PBM interacts with other modules of the robot to make decisions and to convert these decisions into actions. The actions based on the plans produced by the PBM are sent to the motor interface unit of the robot.

In the design, the following assumptions about the environment are made:

- The assigned rubble sub-area for the robot is a grid like environment.
- Robot's visibility space contains 8 current neighbor cells.
- The corner neighbor cells cannot be reached if the neighboring cells are obstacles
- The obstacles are modeled as gaps or untraversable huge obstacles
- The area surrounded by untraversable obstacles is not considered

The only assumption for the architecture of the robot is that it is battery powered and equipped with some special simple sensors, and it can operate several hours.

In the design of the PBM, effective algorithms for both exploration and exploitation are proposed. SR robot works in a dynamic and unknown environment. Initially, the only information available to the robot is the preloaded knowledgebase. In the main loop of the planning strategy, if there is some information about the possibilities of victims, and their possible locations, this information is used. Otherwise the area is explored effectively. In this manner, the strategy allows searching of the locations farthest from the visited locations to explore unknown locations of the victims. This loop continues whenever the state is recoverable and it is believed that there may be victims in the area. During search, the obstacle avoidance is implemented before collisions occur.

2.3   Planning Strategy for Victim Locations

In the strategy for moving to the locations believed to contain victims, the shortest path is considered. Therefore the robot initially turns to the direction

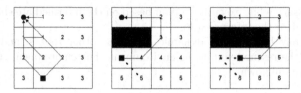

Fig. 1. The update strategy of cell distances to the target on the planned path

of the victim and tries to move forward continuously. Exceptions can occur when the robot encounters an obstacle. It can determine whether the object is dynamic or static by means of sensory inputs. Since the proposed approach is suitable for all SR robot architectures including ones lacking of such sensory equipments, the object type test is implemented blindly by waiting for a short time. If the object still stands in front, it is assumed to be a static object. If the prediction is wrong, in later steps, this information can be corrected because of the adaptive world knowledge update mechanism. For more advanced architectures having efficient sensors, this test process is not needed. During search, alternative directions should be tried. After alternative path selection, the shortest victim location choice is again executed. The *Locating Human Beings* Module may indicate another victim location closer than the target. Such information can also be sent from another robot. In this case, the target is changed, and the closest victim is chosen to rescue to satisfy the primary goal. When the target is changed, the information related to the previous target is sent to the dispatcher so that other robots can save the victim.

If there are no more targets, and it is believed that nobody is alive in the assigned sub-area a *finished* message, or if the situation cannot be recoverable, a *help* message is sent to the dispatcher. If the deadline ends before reaching the target, this information is sent to the dispatcher, and a new target is chosen.

The nearest victim path-planning strategy provides the robot to traverse a path, visiting victim places by using the shortest path. The predicted cell distances to the selected victim target are updated at every movement, and also when the world knowledge is updated as in Free Space Assumption planning theory [4]. An illustrative example of the update strategy of cell distances on the planned path can be seen in Fig. 1, where, the square represents the current location, and the circle the target. The lines with arrows are the alternative paths. The dotted line is the path taken by the robot. The numbers in the cells are the guessed distance values to the target. Initially the environment is completely unknown, assuming the sensor range is one-cell distance. After updating the world knowledge, the observed obstacle information is taken into account to predict the distance values close to real, and to provide an incremental search strategy.

The probability information for victim locations is stored as a vector in the long-term-memory. If the robot finds a victim in a location, it clears the probability information, and adds the location to the rescued location vector and sends a *saved* message to the dispatcher. Such a message can also arrive at the

NearestVictimPathPlan
    do while there is a victim probability
        ChooseClosestVictim;
        UpdatePathInformation;
        if a path to the target is not available
            Assign this target as unreachable;
            Send a declaration message to the
            dispatcher informing the victim location;
        else
            DefineDeadline;
            SelectClosestRotation;
            Move;
            if PlanFails
                HandleErrors;
    update worldKnowledge;

UpdatePathInformation
    from target expanding to all neighbor cells
        assign ∞ for the current distance;
        choose the smallest neighbor distance
        and update the own;
HandleErrors
    if there are obstacles on the path
        UpdatePathInformation;
        turn to the cell closest to the target
    if more than one cell with the same distance value
        choose the least visited;
    if the situation cannot be recoverable
        send a help and go to the idle mode;

Fig. 2. Moving to the nearest victim strategy

communication layer of the current robot. If the location in the message is the current target, the robot clears the probability information for this location and selects another target location in the next step. The obstacles information is also stored in the long-term memory. However since the environment is dynamic, the planning strategy takes into account of local changes while moving. In this case, the information related to the obstacles is updated. Therefore the wrong beliefs related to the dynamic obstacles labeled as static are corrected. The searching strategy algorithm is given in Fig. 2.

2.4   Search Strategy for Exploration

When the robot has nothing to do better, it tries to search the environment in an effective manner. This situation occurs when there is no motivation about a possible location believed to contain victims.

A selective pressure value is defined for unvisited locations, which is updated for each movement. The added value to the previous pressure value is a function of the distance between the cell and robot's location, and the number of unvisited neighbors of the cell. A pressure value of a cell is updated based on Eq. 1

$$pr_i^{k+1} = \begin{cases} 0 & \text{if } i \text{ is within the sensor range} \\ pr_i^k + dist(i(x,y,z), R(x,y,z)) + 10 * n_u/n_t & \text{otherwise} \end{cases} \quad (1)$$

where pr_i^k is the pressure value of the cell i at step k, n_u is the number of unvisited neighbors and n_t is the total number of neighbors in the sensor region, $dist()$ returns the Euclidian distance between two points in 3D, $i(x,y,z)$ is the cell location, and $R(x,y,z)$ is the robot location. The selective pressure values of cells within the sensor range are set to 0.

In the exploration strategy, a cell having the highest selective pressure value is selected as the new target. After selecting the target, a path is planned containing maximum number of unvisited cells. For the current cell, the three neighbors closest to the line between the current cell and the target cell forms a sub-region. An unvisited cell from this sub-region is selected. If there is not an unvisited sub-region neighbor, the other neighbors are examined and the unvisited neighbor

LeastVisitedLocationPlan
  Choose unvisited location with the
  highest pressure value;
  do while (target is not reached)
    if there is victim information
      break;
    UpdatePathInformation;
    ChooseTheSubRegionNeighbor;
    Move;
    UpdatePressureValues;
  if PlanFails

HandleExplorationErrors;
  update worldKnowledge;
HandleExplorationErrors
  if there are obstacles on the path
    UpdatePathInformation;
  if short term memory indicates that
  the situation cannot be recoverable
    send a help and go to the idle mode;
  if a path to the target is not available
    Assign the target as unreachable;

Fig. 3. Exploration Algorithm

closest to the target is chosen. If all the neighbors are visited before, the sub-region cell having the minimum visit value is chosen finally. The visited locations are removed from unvisited vector. As soon as any information about a victim location is determined, the nearest victim path-planning strategy is switched immediately. The proposed exploration algorithm is given in Fig. 3.

3   Experimental Results

A simulation environment in C++ using OpenGL Library was implemented to measure the performance of the proposed algorithms. The environment is constructed with random locations of obstacles and victims. The dynamic objects representing the refugees or human team members are allowed to walk or run in the environment. The number for sensor range indicating the maximum cell distance to sense any victim probability can be adjusted.

Since there is no similar algorithm, the designed algorithm was only compared with the greedy mapping method. Mine sweeping, vacuum cleaning or lawn moving planning approaches [1] could also be selected for comparison. In these tasks the cells should be visited at least once. However these algorithms are designed without time constraints whereas in SR case this is crucial.

The Greedy Mapping Method [4] is an adaptive method running on an unknown environment. While selecting the shortest unvisited cell in each step, the method traverses the shortest path from initial state through the goal state. The proposed nearest victim path-planning strategy is combined with this strategy to conduct tests.

The experiments were conducted with randomly generated 20x20 grids or mazes. The proposed strategy provides a complete search with an effective exploration capability. The exploration time (Coverage) or the number of steps to find victims (Goal) change based on the number of obstacles and the structure of the environment. The results of the proposed strategy can be seen in Table 1. Each row represents different obstacle densities.

As expected, the number of victims found decreases when the number of obstacles in the environment increases. The exploration time to visit all of the

Table 1. The performance of the proposed strategy for different densities

Real Obstacles		Detected Obstacles		Coverage		Goal		Number of Victims rescued	
μ	σ	μ	σ	μ	σ	μ	σ	μ	σ
56.04	7.10	56.16	7.15	531.72	40.31	119.6	50.96	4.92	0.28
122.40	8.81	129.2	13.06	548	103.78	171	61.97	4.9	0.31
199.80	10.12	237.90	24.62	341.70	96,05	102.90	56.60	2.80	1.48

Table 2. Comparison of the exploration strategy with greedy mapping method

Sensor Range	Greedy Mapping Exploration				Proposed Exploration			
	Coverage		Goal		Coverage		Goal	
	μ	σ	μ	σ	μ	σ	μ	σ
2	411.52	23.97	270.28	72.22	563.76	41.42	173.64	80.44
3	407.44	16.07	266.68	50.97	540.88	37.46	92.62	43.38
4	402.08	13.50	231.96	71.46	518.72	26.86	65.68	31.10

cells at least once decreases while obstacles density increases. The numbers of real and detected obstacles are different because of unreachable places formed by the combinations of the obstacles. Although it can be inferred that as the number of obstacles increases the number of steps to find the victims decreases, the number of victims rescued decreases drastically due to the unreachable places containing victims.

Table 2 presents the results for the exploration strategy for both the proposed and the greedy mapping exploration methods. The proposed nearest victim path-planning strategy is applied for both methods. The same randomly generated grids are used to compare the methods. The number of victims in the environment is 5 and the average number of cells containing obstacles is 29.54 with standard deviation 5.39. The results are average of 25 independent runs for each sensor range. The algorithms are compared for both the number of coverage steps (Coverage) and the number of steps to find all reachable victims (Goal).

It can be noted that the number of steps to find the victims in the environment decreases when the robot's sense region is increased. Therefore this result shows that the selection of the specialized sensors is very important for SR mission. As can be seen from the table, for the proposed exploration method the number of steps to reach the goal is smaller than that of greedy mapping. However the number of steps to cover the whole reachable places is greater. But for SR operation, to reach the goal in a minimum possible time is more important than covering the area with a smaller number of steps.

4   Conclusions

In this work, a novel autonomous planning strategy suitable for all types of SR robot architectures is proposed. The planning layer design can be used in the environments where the aid of human rescuers or search dogs is completely impossible. The conducted experiments show that the proposed strategies are complete and promising for the main goal of a SR robot. The number of steps to find the reachable victims is considerably smaller than that of the greedy mapping method. In the proposed exploration strategy, the environment is explored to find the victims as soon as possible by selecting the least visited cell as a new target. Although, the proposed strategy completes a single coverage of the environment with greater number of steps, this is acceptable due to the nature of the SR mission. The environment is searched till it is believed that nobody is still alive in the area. While visiting the least visited neighbors on the way to the target, the environmental changes can also be tracked.

As a future work, 3D real life experiments with real robots in a disaster area should be implemented and extending the ideas given in this paper to the multi-agent case can be considered.

References

1. Casper J. and Murphy R.R.: Human-Robot Interactions During the Robot-Assisted Urban Search and Rescue Response at the World Trade Center, *IEEE Transactions on Systems, Man an Cybernetics-Part B*, Vol. 33, No.3 (2003)
2. Davids, A.: Urban Search and Rescue robots: From Tragedy to Technology, *IEEE Intelligent Systems*, Vol. 17, No. 2, (2002) 81-83
3. Kitano, et. al.: RoboCup-Rescue: Search and Rescue in Large Scale Disasters as a Domain for Autonomous Agents Research, *IEEE Conf on Man, Systems, and Cybernetics* (1999)
4. Koenig S. et al.: Greedy Mapping of Terrain, *Proc. of the International Conference on Robotics and Automation,* (2001) 3594-3599
5. Murphy R. R.: *Introduction to AI Robotics,* England: The MIT Press, (2000)
6. Murphy R. R.: Marsupial and Shape-Shifting Robots for Urban Search and Rescue, *IEEE Intelligent Systems*, Vol. 15, No. 2, (2000) 14-19
7. Takahashi T. and Tadokoro S.: Working with Robots in Disasters, *IEEE Robotics & Automation Magazine,* Vol. 9, No. 3 (2002) 34-39
8. Weiss G.: *Multi Agent Systems: A Modern Approach to Distributed Artificial Intelligence,* England: The MIT Press, (1999)

World Modeling in Disaster Environments with Constructive Self-Organizing Maps for Autonomous Search and Rescue Robots

Çetin Meriçli, I. Osman Tufanoğulları, and H. Levent Akın

Boğaziçi University,
Department of Computer Engineering,
34342 Bebek, Istanbul, Turkey
{cetin.mericli, akin}@boun.edu.tr
osmantuf@garanti.com.tr

Abstract. This paper proposes a novel approach for a Constructive Self-Organizing Map (SOM) based world modeling for search and rescue operations in disaster environments. In our approach, nodes of the self organizing network consist of victim and waypoint classes where victim denotes a human being waiting to be rescued and waypoint denotes a free space that can be reached from the entrance of debris. The proposed approach performed better than traditional self-organizing maps in terms of both the accuracy of the output and the learning speed. In this paper the detailed explanation of the approach and some experimental results are given.

Keywords: Search & Rescue Robotics, Self-organizing Maps, Mobile Robotics, World Modeling.

1   Introduction

Search and rescue (SR) robotics is one of the promising areas of mobile robotics. The main aim of the SR robots is exploring the debris after a disaster (especially, after an earthquake) and locating the living victims in the collapsed buildings, if any. Since a map of the debris is usually not available, the robot has to make the map of the environment simultaneously with the exploration and victim detection process while marking the locations of victims to be saved on the generated map.

The robot should have the ability to determine its position and orientation in the debris by using sensory inputs. One of the most important methods for localization is using natural or artificial landmarks in the environment[1] [2] [3]. Since there are no known landmarks in an unknown environment, this method can not be used. Odometry sensors providing the relative displacement of the robot with respect to its initial position can also be used for localization. Because of friction, slippage and encoder errors in the locomotion parts, odometry data are fairly noisy. This noise increases the error in estimation cumulatively. In order to overcome this problem, position information must be corrected periodically. Using GPS is another alternative for position estimation but accuracy of commercially available GPS receivers are not so high. In this work, we

D. Nardi et al. (Eds.): RoboCup 2004, LNAI 3276, pp. 467–473, 2005.

assumed that the robot is equipped with a GPS receiver with high accuracy. Since the accuracy of the GPS is limited (outputs of GPS sensors are also noisy), the map making algorithm should be able to use noisy inputs.

Some approaches for world modeling and position estimation using self-organizing networks for mobile robots have been proposed in literature. Marques *et al* have proposed a system which uses sensory layer inputs for training a self-organizing network and after the training phase, finding the most similar neuron for each perception from sensors, and assuming the position of the robot is the position of the winner neuron in the network [4]. Nehmzow *et al* have used self organizing feature maps for position estimation in which they are feeding the network with history of motor actions instead of sensory data[5].

Topological map representation is appropriate for mapping the unknown environments because of its property of learning both distribution and topology of the data. In this work, we have used a constructive variant of Kohonen's Self Organizing Map for marking accessible free locations in the universe (which we call a Waypoint) and the locations of detected humans (which we call a Victim). We have used a constructive network architecture because it is assumed that we do not have any prior information about the disaster environment[6] so we should make a map of the environment in order to be able to mark detected humans and free spaces for reaching them on the map.

The rest of the paper is organized as follows: Brief background information about self-organizing networks is given in Section 2. Detailed description of proposed approach is given in Section 3. Section 4 covers experimental work and the last section is dedicated to conclusions.

2   Background

2.1   Self-Organizing Maps

Self-organizing maps are a special kind of neural networks that can learn to detect regularities and correlations in their input and adapt their future responses according to that input [7]. Competitive learning is used to learn to recognize groups of similar input vectors in such a way that neurons physically near each other in a layer respond to similar input vectors. For a given input, the most similar neuron (called The Winner) is selected based on the distance between input and the neurons (neurons compete with each other in order to be the winner). Then, the winner neuron and the neurons in a certain neighborhood of the winner neuron are updated with a rule called Kohonen Learning Rule:

$$w_{ij}(t) = w_{ij}(t-1) + h(w_i, w_g) \cdot \alpha \cdot (p_j(t) - w_{ij}(t-1)) \tag{1}$$

where, w_{ij} is the jth weight of ith neuron, p_j is the jth component of input, α is the learning rate, $h(w_i, w_g)$ is the vicinity function that depends on the distance between the updating neuron and the winner neuron, w_i is the updating neuron and w_g is the winner neuron. The vicinity function decreases as the distance of the updating neuron to the winner neuron increases, and becomes zero if the neuron is not in the vicinity of the winner neuron. Each time a neuron is updated, all the neurons in the vicinity of the neuron are also updated. Self-organizing maps learn both the distribution and the topology of the input vectors which they are trained on[8].

3   The Proposed Approach

In our approach, we have used a constructive self-organizing map for representing the free spaces and victims in the environment, accessibility of victims by using connectivity among free spaces and connectivity between victims and free spaces. It is assumed that, the sensorial layer of the robot supplies two kinds of signals: obstacle information and people detection information. Both of the signals are real numbers in the range of [0, 1] and represent the confidence about existence of either victim or free space ahead of the robot. Receiving a victim signal strengthens the belief about the existence of a victim node in the observed location and receiving an obstacle signal weakens the belief about the existence of a free space (a waypoint) in the observed location. Since usually the debris has more obstacles than free spaces (we can consider it as a maze-like environment), it is reasonable to keep track of free spaces and the connectivity of these spaces instead of marking obstacles on the map. For this purpose, we divide the nodes into two types: $Waypoints$ and $Victims$. Waypoints denote the free spaces that the robot can pass through and Victims denote detected humans. The nodes contains information about node type, the 3D position information, the number of hits of the node and the average confidence value of the node. Since only one network is used for map generation and the map contains two classes of nodes, class specific update rules and environment specific linking methods among and between classes have been developed.

In this work, we assumed that there are obstacle and victim detection modules that supply the probability of encountering an obstacle or a victim. Each time an observation is received, only the nodes belonging to the same class with the observation are updated (i.e. a waypoint node is neither updated nor selected as a winner when an observation of waypoint signal is received even if it is in the vicinity of the winner or it is the winner itself in terms of distance to the observation). If none of the nodes belonging to the same class are closer to the observation point than a certain threshold, a new node of the observation type is introduced to network. When a new node is constructed, neighborhood information of the nodes should be updated. In this phase, not only the distances between the nodes but also the physical accessibility from one node to another is considered. If a node is not accessible from another node, it is not added to the neighbor list of that node even if it is in the neighborhood range. Whenever a victim observation is made with a confidence greater than a certain threshold, the victim nodes are updated according to Kohonen Learning Rule but unlike the conventional rule, the amount of update is multiplied by the confidence of the signal. This prevents the nodes from diverging from the correct position when a signal with a low confidence is received for a long time. Each node, waypoint or a victim, can be of fixed type or variable type. A fixed node is not updated through learning even if it is in the vicinity of the winner neuron. A node is set as fixed if it has a number of wins greater than a certain number and its average confidence is greater than a certain threshold. The idea is that if there are more than a certain number of observations denoting that there is a victim in a location with a confidence over a certain threshold, that node should be fixed, and should not be updated anymore. The network starts with an initial waypoint (the first node of the system) representing the entrance point of the robot into the debris and is set to be fixed. A pseudo-code of the algorithm is given in Figure 1.

```
Initialize the network by defining a starting node
at (0,0,0) and fixed.

If an observation is received
    Find the winner node by comparing the distances
    between nodes and incoming observation

        If distance between winner and observation is
        less than a threshold,
            Update the nodes in te vicinity of the winner
            node belonging to the same type with observation
            type and set the node status to variable

            Update hits and average confidence fields of
            the winner

    Else
            Create a new node with type of observation and
            set the status to variable

            Update the neighborhood information for the new node
            and the nodes in its vicinity.

    End

        If the winner has a count of hits over a threshold and
        its average confidence is over threshold,
            set the node status of winner to fixed

End
```

Fig. 1. Pseudo-code implementation of the proposed approach

After the exploration is finished and training of the network is completed, the resultant network is a graph consisting of free spaces and victims and the links between nodes denoting the accessibility of nodes.

4   Experimental Work

4.1   Simulation Environment

For the experimental work, The Webots™ Mobile Robot Simulator version 3.2.22 is used as the testing environment. Webots™ uses a VRML97 compliant scene representation scheme and allows the user to develop C/C++ and Java robot controllers by providing API for accessing the simulator functions[9].

We have used a model of a Khepera robot equipped with a color camera as the prototype of SR robot and the controller for the robot is written in Java. For the experimentation, the robot is guided remotely and we mimic the signals from the people detection module by performing image processing. The light grey cylinders represent the victims in the environment. The proportion of the number of light grey pixels in the image obtained from the camera of the robot to the number of all pixels in the image is a number between [0, 1] indicating the confidence of the existence of a victim in the visual field of the robot. The people detection module of the robot is assumed to return the estimated coordinates of the detected people (if any) and the confidence about this estimation. In our experiments, the location of the robot is considered as the signal location and the results from image processing are considered as the confidence level of the signal. As the robot wanders around, the observations are passed to the network depending on the strength of the signals received. For the waypoint signal production, the information from infrared (IR) range sensors in front of the robot is used. The IR sensors of Khepera return a number in the range $[0, 1]$ where 0 means that there are no physical obstacles in the range of the sensor and 1 denotes a very close obstacle to the sensor. In our experiments, we have used the average of values obtained from four front IR sensors of the robot as the obstacle information which is again a number in the interval $[0, 1]$. Since an obstacle signal with a value of 1 means that the robot is confronted with a very close obstacle, we expressed the amount of free space ahead of the robot as $1 - \gamma$, where γ is the obstacle signal obtained by taking average of four frontal infrared sensors. It is assumed that the robot is equipped with a GPS receiver for its self localization. To mimic the noise in the GPS system and IR sensors, uniformly distributed random numbers are added to the exact coordinates of the robot and IR data.

4.2   Experiments

For comparison, two different networks are used in the experimentation phase: A conventional Kohonen's Self Organizing Map and our modified version of self organizing network. For the conventional SOM part, Matlab Neural Network Library is used whereas our modified version of the network is implemented in Java. The conventional SOM used is a network of 5 x 1 x 5 nodes since the displacements in the Y axis can not be obtained due to the limitations of the simulator. Our network starts with one initial node. The SOM network is trained using victim observations in 100 epochs. 1876 victim observations were obtained during the simulation by making the robot to follow the path given in Figure 2 where the circles represent the victims in debris. The number of the observations depends on the running time of the simulation.

Both the outputs of our implementation in one epoch and the output of SOM in 100 epochs can be seen in Figure 3. Here, light grey nodes denote waypoints, dark grey nodes denote victims, circles denote variable nodes and squares denote fixed nodes. It can be seen easily that the output of our approach has some major advantages over the classical self-organizing map approach. Our approach can represent both the victim locations, the free spaces and the connectivity of these free spaces for reaching victims and because of its custom update rules and spatially constrained vicinity definition, the victim locations are found with a higher accuracy than classical SOM. The training phase of the traditional SOM took approximately 10 minutes on a Pentium 4 based computer whereas, our approach does not need such a training phase since it generates the map online while the robot is wandering in the debris.

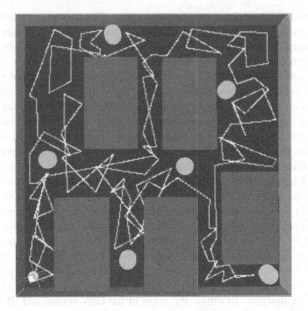

Fig. 2. Path of the robot in debris

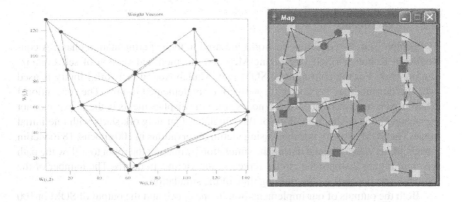

Fig. 3. Comparison of outputs of (a) traditional SOM and (b) our approach

5 Conclusions

SR robotics is gaining importance as the disasters causes large scale loss of human life and most of this loss is due to inefficiency in SR methods. By using SR robots in collapsed buildings, the chance of detecting the existence and locations of victims in debris can be increased.

In this work, an incremental self-organizing network based map generation methods for a search and rescue robot is presented. The main aim of this work was to come up with a world modeling algorithm that can represent the locations of human beings in a collapsed building and a list of free spaces that allows reaching those victims. The proposed approach is implemented on a simulator and compared with a traditional self-organizing map. Since debris is a maze-style environment, it is not easy to define neighborhood between neurons. In the definition of neighborhood functions, spatial constraints such as physical accessibility between nodes are used in addition to Euclidean distance between nodes. Using spatial accessibility prevents updating unrelated nodes even if they are in the vicinity of the winner neuron in terms of euclidean distance. There are two types of nodes in the network: Waypoints and Victims. Waypoints denote the free space that the robot is passed on in the exploration and Victim denotes a detected living human being waiting for to be rescued. Since we have two classes of signals to be learned and there are some spatial constraints in vicinity functions, a traditional self-organizing map is not sufficient for representing the world model. With its spatial constrained neighborhood definition and partial update method for updating different types of nodes, our proposed approach has brought a novel idea to the map making area in unknown environments and performed better than traditional SOM in both accuracy and learning speed.

Acknowledgements

This project is supported by Boğaziçi University Research Fund project 03A101D.

Special thanks to Prof. Ethem Alpaydın and Hatice Köse for valuable discussions and reviews.

References

1. A. Arsenio, "Active Laser Range Sensing for Natural Landmark Based Localization of Mobile Robots", M.Sc. Thesis, IST, Tech U Lisbon, 1997.
2. M.I. Ribeiro and J. G. M. Goncalves, "Natural Landmark Based Localization of Mobile Robots Using Laser Range Data", *Proceedings of the 1st Euromicro Workshop on Advanced Mobile Robots,* Kaiserslautern, Germany, 1996.
3. H. L. Akın *et al,* "Cerberus 2003 Team Report", Bogazici University, Istanbul, 2003.
4. R. Marques, E. Zalama, J. G. García-Bermejo, and J.R. Peran, "World Modeling and Position Estimation for a Mobile Robot Using Self-Organizing Networks" *4th IFAC International Symposium on Intelligent Components and Instruments for Control Applications,* SICICA 2000, Buenos Aires, Argentina, 2000.
5. U. Nehmzow, T. Smithers and J. Hallam, "Location Recognition in a Mobile Robot Using Self-Organising Feature Maps", *Information Processing in Autonomous Mobile Robots,* Springer-Verlag, 1991.
6. S. Zrehen and P. Gaussier, "Why Topological Maps Are Useful for Learning in an Autonomous Agent", *In Proceedings PerAc, IEEE Press,* Lausanne, September 1994.
7. T. Kohonen, "Self-Organization and Associative Memory.", *Series in Information Science, Vol. 8. Springer-Verlag,* Berlin, Heidelberg, New York, 1984.
8. R. O. Duda, P. E. Hart and D. G. Stork, *Pattern Classification,* Second Edition, Wiley & Sons, 2001
9. "Webots Mobile Robot Simulator", *http://www.cyberbotics.com,* 2003.

Approaching Urban Disaster Reality: The ResQ Firesimulator

Timo A. Nüssle, Alexander Kleiner, and Michael Brenner

Institut für Informatik, Universität Freiburg, 79110 Freiburg, Germany
{nuessle, kleiner, brenner}@informatik.uni-freiburg.de

Abstract. The RoboCupRescue Simulation project aims at simulating large-scale disasters in order to explore coordination strategies for real-life rescue missions. This can only be achieved if the simulation itself is as close to reality as possible. In this paper, we present a new fire simulator based on a realistic physical model of heat development and heat transport in urban fires. It allows to simulate three different ways of heat transport (radiation, convection, direct transport) and the influence of wind. The protective effects of spraying water on non-burning buildings is also simulated, thus allowing for more strategic and precautionary behavior of rescue agents. Our experiments showed the simulator to create realistic fire propagations both with and without influence of fire brigade agents.

1 Introduction

The RoboCupRescue Simulation League aims at simulating large scale disasters and exploring new ways for the autonomous coordination of rescue teams [2]. These goals are socially highly significant and feature challenges unknown to other RoboCup leagues, like long-term planning of rescue missions involving heterogenous agents. Moreover, the environment these agents act in is a large-scale simulation which is both highly dynamic and only partially observable by a single agent.

It is due to the latter features of the environment that real disaster situations seldomly can be predicted and, in turn, are often not adequately dealt with when they actually occur. Therefore, it must be one of the main goals of the RoboCupRescue Simulation League to develop realistic disaster simulators that allow agents to develop realistic mission plans. In this paper, we describe a new fire simulator that progresses towards this goal while not exceeding the run-time limitations of the RoboCupRescue simulation system. Previous approaches to firesimulation outside the RoboCupRescue domain are reviewed in an extended version of this paper [4].

The RoboCupRescue simulation system is a modular framework based on a Geographic Information System (GIS) describing a city map, and a kernel which acts as a communication hub and integrator of changes to the world model as proposed by the various agents and simulators connected to the kernel. (In the extended paper [4] we describe a direct communication interface for simulators that allows to share internal physical data and thus to model complex interactions, like fire causing the collapse of a house, without overloading the kernel communication channels).

Some of the new features of the introduced simulator are the calculation of heat development in burning houses as well as the simulation of three significant ways of

D. Nardi et al. (Eds.): RoboCup 2004, LNAI 3276, pp. 474–482, 2005.
© Springer-Verlag Berlin Heidelberg 2005

heat transportation *between* buildings. Especially, the influence of wind on the spread of fire is taken into account. Another step towards greater realism is achieved by the possibility to limit fire spread by "preemptive extinguishment", i.e. the spraying of water on non-burning buildings in order to temporarily protect them from catching fire. These new features do not only add to the realism of the simulation, but will also allow rescue agents to act more strategically and precautionary than before. Interacting with an adequate earthquake/collapse simulator, even the starting of fires can be simulated without need for artifical "ignition points" as used in the current simulation, hence supporting the automated generation of realistic disaster situations of varying difficulty.

The remainder of this paper is structured as follows. Section 2 introduces the physical theory underlying the simulation, whereas section 3 describes its implementation. Section 4 demonstrates some of the new features of the simulator and section 5 provides an outlook to further developments.

2   Fire Simulation

2.1   Physical Theory

Since fire produces and is ignited by heat, we have to familiarise our self's with heat and heat transportation. The following paragraph presents simplified relations, for more detail see the extendet version of this paper [4].

The temperature of an object is a measurement for the inherent heat energy depending on the objects heat capacity. The heat capacity describes the change of temperature in dependency from the energy change. Whenever two objects with different temperatures are joined, heat energy is transported from the warmer to the colder (figure 1a). The amount of transfered energy is proportional to the temperature difference and the exposition duration. This effect is called *direct transport*. Objects are emitting respectively receiving heat energy even when they are not directly connected. Objects are emitting *heat radiation* in dependency from there temperature, which is nothing else then light, typically in the infrared spectrum and hence energy. Other objects sharing a line of sight with the emitter are receiving a part of this energy, depending on the distance and the size

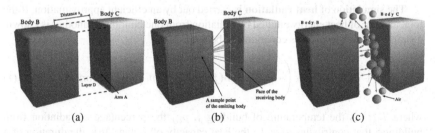

(a)   (b)   (c)

Fig. 1. Three ways of heat transportation: *direct heat transport* (a), two bodies (B and C) exchanging heat by area A, *radiation* (b), the receiver absorbs the energy from beams hitting it's surface and *convection* (c), heat is transported by air. (The more complex distance and layer computations shown in fig. (a) are described in the extended paper [4])

of the exposed areas (figure 1b). A third kind of heat transportation, called *convection*, can take place within gases. When a region of gas is warmed up its volume increases and therefore its mass density decreases. The result is an ascending of warmer volumes while colder descend(figure 1b).

3   Implementation

Due to practical reasons, the amount of computing power in the RoboCupRescue League is limited on both the server and the client side. Each simulation connecting to the kernel has to finish all calculations and network communication within a discrete time step of 500 milliseconds. Therefore efficient algorithms are a necessary requirement. Particularly, if simulators mutually depend on the results of their calculations, the worst case cost must never exceed the given time constraint. Computational complexity of a simulation is usually reduced by an appropriate discretization of the world. In the RoboCupRescue domain the discretization is already given by the level of detail of the provided GIS data. This data is distributed in entities, such as buildings, streets and civilians (of which currently only buildings are relevant for the fire simulation). Buildings are defined by a polygon describing their footprint, the number of floors, the area at ground level as well as the type of construction, i. e. steel frame, reinforced concrete or wood. As global properties, the wind direction and speed are provided [3]. Since this model does not suffice for the simulation of all physical effects, the simulator additionally implements a discrete model of the air temperature, that will be described in the subsequent section.

3.1   Discrete Model

The high complexity of urban fires, i.e. due to unpredictable air streams and an inhomogeneous distribution of fuel, can only be simulated with strong restrictions. Complex gas flux calculations are beyond question as well as air-flow pattern computation considering the influence of buildings. We restricted the model to two dimensions, like the rescue domain itself. The O_2 concentration level is assumed as constant and as sufficiently available for combustion.

The simulation of **heat radiation** is carried out by an efficient approximation. If the total amount of heat energy emitted by radiation from one building to another is known, a simple and fast to process equation can be used:

$$T_j(t+1) = T_j(t) + \left(-radiation(j) + \sum_{i \in B, i \neq j} p_{i,j} \cdot radiation(i) \right) \cdot \frac{\Delta t}{\Gamma_j} \quad (1)$$

where $T_j(t)$ is the temperature of building j, $p_{i,j}$ the percentage of radiation from building i that contributes to j, Γ_j the heat capacity of j, and Δt is the duration of a time step in the simulated world. Table 1 provides reasonable values of mass densities, whereas the specific heat capacity values for different construction types are currently set by the user in the configuration file. From these values the building specific heat capacity Γ_j is calculated.

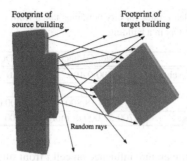

Fig. 2. Randomly emitted rays: The percentage of rays hitting the target building determines the amount of transfered energy by radiation

Equation 1 is based on the assumption that the temperature is constant during a simulation interval and that transportation takes place only between cycles. The value of each $p_{i,j}$ is not calculated completely but randomly sampled with a Monte Carlo method. Each building is broken down into its outer walls as they are provided by the GIS. For each wall a number of random rays originating from this wall are generated. The percentage of rays emitted from building i that hit building j is taken as a stochastic approximation for $p_{i,j}$ (see figure 2). Since the mapping of $p_{i,j}$ is assumed to be constant, these calculations are done offline during the simulation start-up. To enhance the performance, a few improvements have been implemented. The $p_{i,j}$ data for the loaded map is written to hard disc and linked with a hash code that is calculated from the building's unique longitude and latitude. Buildings with a distance exceeding a threshold, which can be set in the configuration file, are left out of the calculation. The expected error from this simplification is comparably low, since the energy density from a point source at distance r is proportional to $\frac{1}{r^2}$ and thus negligible. The radiation function $radiation(i)$ is calculated utilising the Stefan-Boltzmann-Law.

The simulation of **direct heat transport** and **convection** is limited to a single layer situated above the ground. Higher layers are ignored because we assume them to have a small effect regarding the emitting building. The layer is implemented by a two dimensional grid that discretizes the air's continuous heat distribution. The resolution of the grid is currently set to five meters, but may be set differently in the config file of the simulator. The standard ambient temperature is $20°$ Celsius which is the initial default value for all cells.

The update of a the temperature $s_i(t)$ of air cell i with respect to set of cells R within the air transmission range of cell i, and buildings B_i intersecting with cell i, is calculated by:

$$s_i(t+1) = s_i(t) + \left(\frac{\sum\limits_{j \in R, j \neq i} s_j(t) \cdot w_{i,j} + \sum\limits_{u \in B_i} T_u(t) \cdot a_{u,i}}{\sum\limits_{j \in R, j \neq i} w_{i,j} + \sum\limits_{u \in B_i} a_{u,i}} - s_i(t) \right) \cdot l_a \cdot \Delta t \quad (2)$$

Table 1. Typical energy release rates for city buildings taken from Chandler's investigation [1]. N denotes the number of floors in a building

Type of Fuel	Fuel Load (GJ/hectare)	Mass Density (Kg/hectare)
Dwellings, offices, schools	3,700-9,400	202,000-504,000
Apartments	$8,900 \cdot N$	$490,000 \cdot N$
Shops	9,400-18,800	500,000-1,010,000
Industrial & Storage	5,700-57,000 or more	300,000-3,000,000 or more

where $w_{i,j}$ weighs the temperature influence on cell i from surrounding cells according to their distance, $a_{u,i}$ weighs the influence of buildings B_i intersecting with air cell i, $T_u(t)$ is the temperature of building u and l_a is the heat exchange coefficient. In order to keep the original temperature values from time step t in memory, the implementation of formula 2 is carried out by employing two arrays that are swapped after each update.

Besides the air-to-air and building-to-air temperature exchange, also the air-to-building exchange has to be considered. Therefore an equation similar to equation 2 is introduced that accounts for the different heat capacities of buildings:

$$T_u(t+1) = T_u(t) + \left(\frac{\sum\limits_{i \in S_u} s_i(t) \cdot a_{u,i}}{\sum\limits_{i \in S_u} a_{u,i}} - T_u(t) \right) \cdot \frac{l_b \cdot \Delta t}{\Gamma_u} \qquad (3)$$

where $T_u(t)$ is the temperature of building u, Γ_u is the heat capacity of u, l_b is the heat exchange coefficient and S_u is the set of all air cells intersecting with u.

Furthermore every air cell loses heat to the atmosphere due to convection. The amount of heat loss for each cell $s_i(t)$ is approximated by:

$$s_i(t+1) = T_0 + (s_i(t) - T_0) \cdot c_{loss} \Delta t \qquad (4)$$

where T_0 denotes the ambient temperature and c_{loss} is a constant approximating a realistic average degree of heat loss.

The effect of global wind is simulated by shifting air cells accordingly to wind velocity and direction on the grid. However, since it is possible that the newly calculated position will not match the grid discretization, grid values, intersecting the shifted cell, have to be recalculated accordingly. The new value of a grid cell is calculated from the weighted average of all cells overlapping due to the shift.

Every building with a temperature above the ignition point and sufficient fuel is considered as burning. Then, during each cycle, a certain percentage of its initial fuel is transformed to energy and added to the building's energy value. Empirical data of fuel densities, as presented in table 1, is utilized for the calculation of the initial fuel values.

3.2   Extinguishing Fires

The action *extinguish building* in the RoboCupRescue domain is realized, for both extinguishing and preemptive extinguishing, by increasing an internal value for each building that represents the amount of water used on it by fire brigades. The fire simulator ensures that all necessary preconditions for this action are met, which are a sufficient amount of water in the fire brigade's tank, a position close enough to the fire and a maximum amount of water that may be emitted per round.

From the amount of water in a building a fraction, linearly proportional to the temperature of the building, is considered to be vaporizing and by this cooling the building during each cycle. The heat energy reduction is calculated by the product of the amount of vaporizing water and its vaporization constant.

Like in reality, preemptive cooling will protect buildings from catching fire temporarily but will not make them completely fire-proof as long as surrounding houses are burning. Thus, preemptive extinguishment offers new strategic possibilities for fire brigades but does not relieve them of the duty to stop fires completely.

4   Experiments

Due to the fact that real data of urban fires is hardly available and if so, is specific to a particular fuel distribution and wind, a close-to-reality evaluation seems to be impossible. Therefore we present a visualization of the new fire simulator's general behavior and compare it to that of the old one. In comparison to the old simulator, which tends to create a circular wall of fire, the new simulator spreads fire in a more realistic way (see figure 3). The dynamic fire propagation matches the complex behavior of real urban fires. As shown in figure 3, the fire spread of the new simulator depends on the density of buildings situated in the area. A high density of buildings leads to a rapid fire spread, whereas larger open spaces behave as fire barriers. The new feature allows fire brigades to predict the most likely fire spread by reasoning about the fire danger of certain districts. Prediction makes it possible to concentrate forces on jeopardized locations and to naturally exploit open spaces. For competitions, the fire barrier effect is not always wanted since an unextinguished fire should continue to grow in order to make a difference between successful and unsuccessful agents. This barrier can be overstepped by the activation of the wind feature included in the new simulator.

In real disaster situations, it might happen that the number of fire brigades is not sufficient for extinguishing a certain fire but may be high enough to control its spread. This is usually accomplished by preemptively watering non-burning buildings close to the fire border ("preemptive extinguishment"). In the new simulator, the amount of water used on a building will accumulate, then vaporize and thus cool the building. If the building is not yet burning this may even prevent it from catching fire. This new feature is visualized by the series of pictures in figure 4. As can be seen by the lower series, preemptive extinguishing of the diagonal row of buildings in the center prevents the ignition of all buildings behind.

In the RoboCupRescue domain fire brigades are allowed to use more than one nozzle during one *extinguish* command. By this it is possible to distribute water on more than one building at the same time (but note that the amount of water maximally allowed to be emitted during one cycle remains the same). Since extinguishing an ignited building requires virtually always more water than a single fire brigade can emit, this feature offered no tactical advantage so far. Together with the new feature of preemptively

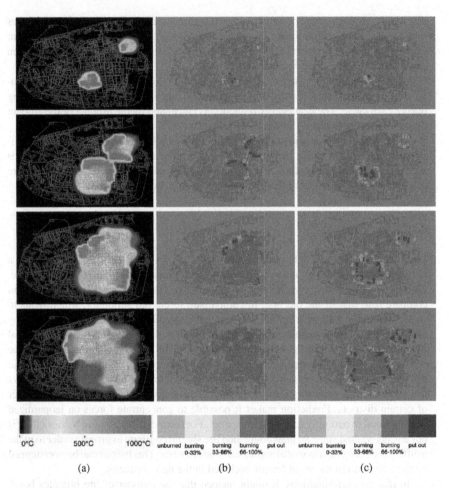

0°C   500°C   1000°C   unburned burning   burning   burning   put out   unburned burning   burning   burning   put out
                                       0-33%   33-66%   66-100%   0-33%   33-66%   66-100%

(a)   (b)   (c)

Fig. 3. From top to bottom: progress of fire spread of the new simulator, displayed in the internal model (a), in the RoboCupRescue world model (b) and compared to the progress of the old simulator (c)

extinguishing, however, it is possible for a single fire brigade to protect multiple buildings from ignition, since to protect a building requires less water than to extinguish it.

The simulator's runtime behavior has been evaluated on the three standard city maps used for the competition, which are *Kobe*, *Virtual City* and *Foligno*. On each map, we simulated, under the same settings, ten times a fire outbreak for a duration of 300 cycles. The simulations where carried out within a Java virtual machine (Blackdown Java HotSpot 1.4.1) on an AMD Athlon 700 MHz computer running a Linux operation system. Table 2 summarizes the average computing time for one cycle of the simulation on all of the three maps. Although this measurements do not include network communication

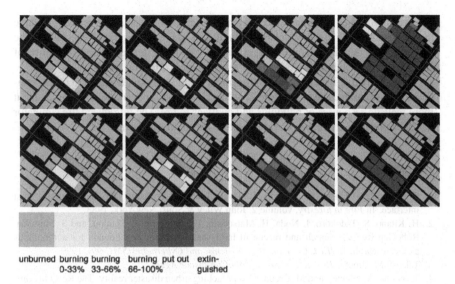

unburned burning burning   burning put out extin-
          0-33%   33-66%   66-100%   guished

Fig. 4. From left to right: the new feature *preemptive extinguishing*. Upper row: fire spread without prevention. Lower row: fire spread prevention due to watering the diagonal building row in the center (blue) in advance

Table 2. Runtime measurements of the new simulator. The first row provides the employed city map, whereas the second row provides it's complexity, denoted by the number of buildings and air cells involved in the simulation. The other rows provide average, standard deviation, maximum and minimum of calculation time within one cycle of the simulation

Map	Complexity	Average	Standard Deviation	Max	Min
kobe	733 buildings, 5896 cells	10.6ms	7.8ms	73ms	6ms
virtual city	1269 buildings, 6972 cells	12.8ms	8.1ms	85ms	8ms
foligno	1085 buildings, 17214 cells	24ms	9.6ms	85ms	17ms

time, it can clearly be seen that the new simulator complies with the domain's time constraint of $500ms$.

5 Outlook

The introduced fire simulator makes a clear step towards close-to-reality simulation of urban disasters. However, due to the high complexity of urban fire spread, this step is just the beginning. With increasing computing power we will be able to contribute more detail to the domain: Firstly, the simulation of fire could be carried out within entities smaller than houses, such as floors and rooms. This feature would make it easier for fire fighters to decide which part of the building they should extinguish in order to avoid fire trespassing to other buildings. Secondly, the air grid model could be realized in three dimensions, leading to a more realistic simulation of fire propagation, especially in the case of higher buildings. Thirdly, the simulator could be extended by the simulation of smoke trails, which have a physical and psychological effect on civilians.

Particularly the first and third improvement are likewise fundamental to other simulators in the domain. The collapse and civilian simulator, for example, might implement more realistic responses of buildings and civilians to fire. The introduced fire simulator has been prepared for being extended towards those improvements.

References

1. C. Chandler, P. Cheney, P. Thomas, L. Trabaud, and D. Williams. Fire at the urban-forest interface. In *Fire in forestry*, volume 2. John Wiley & Sons, New York, 1983.
2. H. Kitano, S. Tadokoro, I. Noda, H. Matsubara, T. Takahashi, A. Shinjou, and S. Shimada. RoboCup Rescue: Search and rescue in large-scale disasters as a domain for autonomous agents research. In *IEEE Conf. on Man, Systems, and Cybernetics(SMC-99)*, 1999.
3. Takeshi Morimota. *How to Develop a RoboCupRescue Agent*.
4. T. Nüssle, A. Kleiner, and M. Brenner. Approaching urban disaster reality: The ResQ firesimulator. Technical Report 200, Inst. for Computer Science, Univ. Freiburg, Germany, 2004.

Stochastic Map Merging in
Rescue Environments

Stefano Carpin and Andreas Birk

School of Engineering and Science,
International University of Bremen – Germany
{s.carpin, a.birk}@iu-bremen.de

Abstract. We address the problem of merging multiple noisy maps in
the rescue environment. The problem is tackled by performing a stochastic search in the space of possible map transformations, i.e. rotations
and translations. The proposed technique, which performs a time variant Gaussian random walk, turns out to be a generalization of other
search techniques like hill-climbing or simulated annealing. Numerical
examples of its performance while merging partial maps built by our
rescue robots are provided.

1 Introduction

One of the main tasks to be carried out by robots engaged in a rescue scenario
is to produce useful maps to be used by human operators. Among the characteristics of such environment is the lack of a well defined structure, because of
collapsed parts and debris. Robots are supposed to move in uneven surfaces and
to face significant skidding while operating. It follows that maps generated using
odometric information and cheap proximity range sensors turn out to be very
inaccurate. In the robotic systems we have developed this aspect is even more
emphasized by our choice to implement simple mapping algorithms for their real-
time execution on devices with possible low computational power [1],[2]. One of
the possible ways to overcome this problem is to use multiple robots to map
the same environment. The multi-robot approach has some well known advantages in itself, most notably robustness [3]. In the rescue framework, multi-robot
systems are even more appealing because of the possibility to perform a faster
exploration of the inspected area, thus increasing the chances to quickly locate
victims and hazards. As the goal is to gather as much information as possible, it
is evident that the maps produced by different robots will only partially overlap,
as they are likely to spread around in different regions and not to stick together
for the whole mission. It is then a practical issue of enormous importance to
merge together such partially overlapping maps before they are used by the human operators. To solve the map matching problem we have borrowed some
ideas from a recent randomized motion planning algorithm recently developed
by one of the authors which have turned out to work very efficiently [4],[5]. The
algorithm performs a Gaussian random walk, but its novel aspect is that it updates its distribution parameters so that it can take advantage from its recent

D. Nardi et al. (Eds.): RoboCup 2004, LNAI 3276, pp. 483–490, 2005.
© Springer-Verlag Berlin Heidelberg 2005

history. Section 2 formally defines the problem and describes the algorithmic machinery used to solve it, together with convergence results. Next, section 3 offers details about the implementation of the proposed technique and numerical results. A final discussion is presented in section 4.

2   Theoretical Foundations

We start with the formal definition of a map.

Definition 1. *Let N and M be two positive real numbers. An $N \times M$ map is a function*

$$m : [0, N] \times [0, M] \to \mathbb{R}.$$

We furthermore denote with $I_{N \times M}$ the set of $N \times M$ maps. Finally, for each map, a point from its domain is declared to be the reference point. *The reference point of map m will be indicated as $R(m)$.*

The function m is a model of the beliefs encoded in the map. For example, one could assume that a positive value of $m(x, y)$ is the belief that the point (x, y) in the map is free, while a negative value indicates the opposite. Moreover, the absolute value indicates the degree of belief. The important point is that we assume that if $m(x, y) = 0$ no information is available. From now on, for sake of simplicity, we will assume $N = M$, but the whole approach holds also for $N \neq M$.

Definition 2. *Let x, y and θ be three real numbers and $m_1 \in I_{N \times N}$. We define the $\{x, y, \theta\}$-transformation to be the functional which transforms the map m_1 into the map m_2 obtained by the translation of $R(m_1)$ to the point (x, y) followed by a rotation of θ degrees. We will indicate it as $T_{x,y,\theta}$, and we will write $m_2 = T_{x,y,\theta}(m_1)$ to indicate that m_2 is obtained from m_1 after the application of the given $\{x, y, \theta\}$-transformation.*

Definition 3. *A dissimilarity function ψ over $I_{N \times N}$ is a function*

$$\psi : I_{N,N} \times I_{N,N} \to \mathbb{R}^+ \cup \{0\}$$

such that

- $\forall m_1 \in I_{N,N}$ $\psi(m_1, m_1) = 0$
- *given two maps m_1 and m_2 and a transformation $T_{x,y,\theta}$, then $\psi(m_1, T_{x,y,\theta}(m_2))$ is continuous with respect to x, y and θ.*

The dissimilarity function measures how much two maps differ. In an ideal world, where robots are able to build two perfectly overlapping maps, their dissimilarity will be 0. When the maps cannot be superimposed the ψ function will return positive values.

Having set the scene, the map matching problem can be defined as follows.

Given $m_1 \in I_{N,N}$, $m_2 \in I_{N,N}$ and a dissimilarity function ψ over $I_{N \times N}$, determine the $\{x, y, \theta\}$-transformation $T_{(x,y,\theta)}$ which minimizes

$$\psi(m_1, T_{(x,y,\theta)}(m_2)).$$

The devised problem is clearly an *optimization* problem over $\mathbb{R}^3$. Traditional AI oriented techniques for addressing this problem include genetic algorithms, multipoint hill-climbing and simulated annealing (see for example [6]). We hereby illustrate how a recent technique developed for robot motion planning can be used to solve the same problem. In particular, we will also show that multipoint hill-climbing and simulated annealing can be seen as two special cases of this broader technique.

From now we assume that the values x, y and θ come from a subset of $S \subset \mathbb{R}^3$ which is the Cartesian product of three intervals. In symbols,

$$(x, y, \theta) \in S = [a_0, b_0] \times [a_1, b_1] \times [a_2, b_2].$$

Also, to simplify the notation we will often indicate with $s \in S$ the three parameters which identify a transformation, and we will then write T_s. Before moving into the stochastic part, we define a probability space [7] as the triplet (Ω, Γ, η) where Ω is the sample space, whose generic element is denoted ω. Γ is a $\sigma - algebra$ on Ω and η a probability measure on Γ.

Definition 4. *Let $\{f_1, f_2, \ldots\}$ be a sequence of mass distributions whose events space consists of just two events. The* random selector *induced by $\{f_1, f_2, \ldots\}$ over a domain D is a function*

$$RS_k(a, b) : D \times D \to D$$

which randomly selects one of its two arguments according to the mass distribution f_k.

Definition 5. *Let ψ be a dissimilarity function over $I_{N \times N}$, and RS_f be a random selector over S induced by the sequence of mass distributions $\{f_1, f_2, \ldots\}$. The* acceptance function *associated with ψ and RS_f is defined as follows*

$$A_k : S \times S \to S$$

$$A_k(s_1, s_2) = \begin{cases} s_2 & \text{if } \psi(m_1, T_{s_2}(m_2)) < \psi(m_1, T_{s_1}(m_2)) \\ RS_k(s_1, s_2) & \text{if } \psi(m_1, T_{s_2}(m_2)) > \psi(m_1, T_{s_1}(m_2)) \end{cases}$$

From now on the dependency of A on ψ and RS_f will be implicit, and we will not explicitly mention it. We now have the mathematical tools to define Gaussian random walk stochastic process, which will be used to search for the optimal transformation in S.

Definition 6. *Let t_{start} be a point in S, and let A be an acceptance function. We call* Gaussian random walk *the following discrete time stochastic process $\{T_k\}_{k=0,1,2,3,\ldots}$*

$$\begin{cases} T_0(\omega) = t_{start} \\ T_k(\omega) = A(T_{k-1}(\omega), T_{k-1}(\omega) + v_k(\omega)) & k = 1, 2, 3, \ldots \end{cases} \quad (1)$$

where $v_k(\omega)$ is a Gaussian vector with mean μ_k and covariance matrix Σ_k.

From now on the dependence on ω will be implicit and then we will omit to indicate it.

Assumption. We assume that there exist two positive real numbers ε_1 and ε_2 such that for each k the covariance matrix Σ_k satisfies the following inequalities:

$$\varepsilon_1 I \leq \Sigma_k \leq \varepsilon_2 I. \quad (2)$$

where the matrix inequality $A \leq B$ means that $B - A$ is positive semidefinite. The following theorem proves that the stochastic process defined in 1 will eventually discover the optimal transformation in S. The proof is omitted for lack of space.

Theorem 1. *Let $\hat{s} \in S$ be the element which minimizes $\psi(m_1, T_s(m_2))$, and let $\{T_0, T_1, \ldots, T_k\}$ the sequence of transformations generated by the Gaussian random walk defined in equation 1. Let T_b^k be the best transformation generated among the first k elements, i.e. the one yielding the smallest value of ψ. Then for each $\varepsilon > 0$*

$$\lim_{k \to +\infty} \Pr[|\psi(m_1, T_b^k(m_2)) - \psi(m_1, T_{\hat{s}})(m_2))| > \varepsilon] = 0 \quad (3)$$

Algorithm 1 depicts the procedure used for exploring the space of possible transformations accordingly to the stochastic process illustrated. As the optimal value

---

```
 1: k ← 0,   t_k ← t_start,   Σ_0 ← Σ_init,   μ_0 ← μ_init
 2: c_0 ← ψ(m_1, T_{t_start}(m_2))
 3: loop
 4:     Generate a new sample s ← x_k + v_k
 5:     c_s ← ψ(m_1, T_s(m_2))
 6:     if  c_s < c_k  OR  RD(t_k, s) = s  then
 7:         k ← k + 1,   t_k ← s,   c_k = c_s
 8:         Σ_k ← Update(t_k, t_{k-1}, t_{k-2}, . . . , t_{k-M})
 9:         μ_k ← Update(x_k, t_{k-1}, t_{k-2}, . . . , t_{k-M})
10:     else
11:         discard the sample s
```

Algorithm 1: Basic Gaussian Random Walk Exploration algorithm

---

of the dissimilarity is not known, practically the algorithm will be bounded to a certain number of iterations and it will return the transformation producing the lowest ψ value.

We wish to outline that this algorithm is a modification of the Adaptive Random Walk motion planner we have recently introduced [4]. The fundamental difference is that in motion planning one has to explore the space of configurations in order to reach a known target point, while in this case this information is not available.

3   Numerical Results

The results presented in this section are based on real-world data collected with the IUB rescue robots. A detailed description of the robots is found in [1]. We describe how we implemented the algorithm described in section 2 and we sketch the results we obtained. In our implementation a map is a grid of 200 by 200 elements, whose elements can assume integer values between -255 and 255. This is actually the output of the mapping system we described in [2]. According to such implementation, positive values indicate free space, while negative values indicate obstacles. As anticipated, the absolute value indicates the belief, while a 0 value indicates lack of knowledge. The function ψ used for driving the search over the space S is defined upon a map distance function borrowed from picture distance computation [8]. Given the maps m_1 and m_2, the function is defined as follows

$$\psi(m_1, m_2) = \sum_{c \in C} d(m_1, m_2, c) + d(m_2, m_1, c)$$

$$d(m_1, m_2, c) = \frac{\sum_{m_1[p_1]=c} \min\{md(p_1, p_2)|m_2[p_2] = c\}}{\#_c(a)}$$

where

- C denotes a set of values assumed by m_1 or m_2,
- $m_1[p]$ denotes the value c of map m_1 at position $p = (x, y)$,
- $md(p_1, p_2) = |x_1 - x_2| + |y_1 - y_2|$ is the Manhattan-distance between p_1 and p_2,
- $\#_c(m_1) = \#\{p_1|m_1[p_1] = c\}$ is the number of cells in m_1 with value c.

Before computing D, we preprocess the maps m_1 and m_2 setting all positive values to 255 and all negative values to -255. In our case then $C = \{-255, 255\}$, i.e., locations mapped as unknown are neglected. A less obvious part of the linear time implementation of the picture distance function is the computation of the numerator in the $d(m_1, m_2, c)$-equation. It is based on a so called distance-map $d\text{-}map_c$ for a value c. The distance-map is an array of the Manhattan-distances to the nearest point with value c in map m_2 for all positions $p_1 = (x_1, y_1)$:

$$d\text{-}map_c[x_1][y_1] = \min\{md(p_1, p_2)|m_2[p_2] = c\}$$

The distance-map $d\text{-}map_c$ for a value c is used as lookup-table for the computation of the sum over all cells in m_1 with value c. Figure 1 shows an example of a distance-map. Algorithm 2 gives the pseudocode for the three steps carried out to built it, while the underlying principle is illustrated in 2.

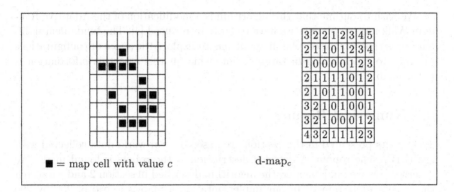

Fig. 1. A distance-map $d\text{-}map_c$

```
1:  for  y ← 0 to n − 1  do
2:    for  x ← 0 to n − 1  do
3:      if  M(x, y) = c  then
4:          d-map_c[x][y] ← 0
5:      else
6:          d-map_c[x][y] ← ∞
7:  for  y ← 0 to n − 1  do
8:    for  x ← 0 to n − 1  do
9:      h ← min(d-map_c[x − 1][y] + 1, d-map_c[x][y − 1] + 1)
10:     d-map_c[x][y] = min(d-map_c[x][y], h)
11: for  y ← n − 1 downto 0  do
12:   for  x ← n − 1 downto 0  do
13:     h ← min(d-map_c[x + 1][y] + 1, d − map_c[x][y + 1] + 1)
14:     d-map_c[x][y] = min(d-map_c[x][y], h)
```

Algorithm 2: The algorithm for computing $d\text{-}map_c$

It can be appreciated that to build the lookup map it is necessary just to scan the target map for three times. In this case it is possible to avoid the quadratic matching of each grid cell in m_1 against each grid cell in m_2.

While implementing the Gaussian random walk algorithm one has to choose how to update μ_k, Σ_k and the sequence of mass distributions $\{f_1, f_2, \ldots\}$ used to accept or refuse sampled transformation which lead to an increment in the dissimilarity function ψ. For the experiments later illustrated we update μ_k at each stage to be a unit vector in the direction of the gradient. Only two different Σ_k matrices are used. If the last accepted sample was accepted, $\Sigma_k = 0.1I$, where I is the 3×3 identity matrix. This choice pushes the algorithm to perform a gradient descent. If the last sample has not been accepted, $\Sigma_k = 10I$. This second choice gives the algorithm the possibility to perform big jumps when it has not been able to find a promising descent direction. The random decisor accepts a sampled transformation s with probability

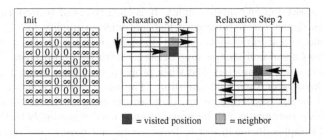

Fig. 2. The working principle for computing $d\text{-}map_c$

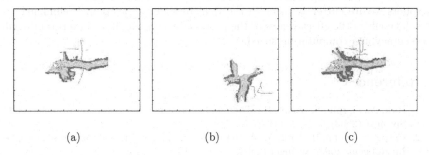

(a)   (b)   (c)

Fig. 3. Subfigures a and b illustrate the maps created by two robots while exploring two different parts of the same environment. To make the matching task more challenging the magnetic compass and the odometry system were differently calibrated. Subfigure c shows the best matching found after 200 iterations of the search algorithm

$$\frac{2(BD - \psi(m_1, T_s(m_2)))}{BD}$$

where BD is the best dissimilarity value generated up to current step. Figure 3 illustrates the result of the search procedure.

In particular, subfigure d shows the trend of the dissimilarity function for the sampled transformation generated by the algorithm. Many recurrent gradient descents stages can be observed, interleaved by exploring stages where wide spikes are generated as a consequence of samples generated far from the point currently being explored. In the devised example 200 iterations are enough to let the search algorithm find a transformation almost identical to the best one (which was determined by applying a brute force algorithm).

4   Conclusions

We addressed the problem of map fusion in rescue robotics. The specific problem is to find a good matching of partially overlapping maps subject to a significant amount of noise. This means finding a suitable rotation and translation which optimize an overlapping quality index. It is then required to perform a search in order to detect the parameters which optimizes a quality index. We introduced a theoretical framework, called Gaussian random walk. We outlined that it generalizes some well known approaches for iterative improvement. In fact, it incorporates a few parameters that when properly tuned can result in techniques like hill-climbing or simulated annealing. The novel aspect of the proposed algorithm is in the possibility to use time variant random distributions, i.e. the distributions' parameters can be updated and tuned accordingly to the already generated samples. It has in fact to be observed that most of the random based approaches use stationary distributions, or distributions whose time dynamics is not influenced by the partial results already obtained.

The proposed algorithm has been applied for fusing maps produced by the robots we are currently using in the Real Rescue competition. Preliminary results confirm the effectiveness of the proposed technique, both in terms of result accuracy and computation speed.

References

1. Birk, A., Carpin, S., Kenn, H.: The iub 2003 rescue robot team. In: Robocup 2003. Springer (2003)
2. Carpin, S., Kenn, H., Birk, A.: Autonomous mapping in the real robot rescue league. In: Robocup 2003. Springer (2003)
3. Parker, L.: Current state of the art in distributed autonomous mobile robots. In Parker, L., Bekey, G., J.Barhen, eds.: Distributed Autonomous Robotic Systems 4. Springer (2000) 3–12
4. Carpin, S., Pillonetto, G.: Motion planning using adaptive random walks. IEEE Transactions on Robotics and Automation (To appear)
5. Carpin, S., Pillonetto, G.: Learning sample distribution for randomized robot motion planning: role of history size. In: Proceedings of the 3rd International Conference on Artificial Intelligence and Applications, ACTA press (2003) 58–63
6. Russel, S., Norwig, P.: Artificial Intelligence - A modern approach. Prentice Hall International (1995)
7. Papoulis, A.: Probability, Random Variables, and Stochastic Processes. McGraw-Hill (1991)
8. Birk, A.: Learning geometric concepts with an evolutionary algorithm. In: Proc. of The Fifth Annual Conference on Evolutionary Programming, The MIT Press, Cambridge (1996)

Orpheus – Universal Reconnaissance Teleoperated Robot

Ludek Zalud

Department of Control and Instrumentation,
Faculty of Electrical Engineering and Communiacation,
Brno University of Technology, Bozetechova 2,
612 66 Brno, Czech Republic
zalud@feec.vutbr.cz
http://www.feec.vutbr.cz/UAMT/robotics/index.html

Abstract. Orpheus mobile robot is a teleoperated device primarily designed for remote exploration of hazardous places and rescue missions. The robot is able to operate both indoors and outdoors, is made to be durable and reliable. The robot is remotely operated with help of visual telepresence. The device is controlled through advanced user interface with joystick and head mounted display with inertial head movement sensor. The functionality and reliability of the system was tested on Robocup Rescue League 2003 world championship in Italy where our team placed on 1st place.

1 Introduction

The Orpheus robotic system have been developed in our department from the beginning of year 2003. The project is a natural continuation of "mainly research" U.T.A.R. project [1], [2], [3] and is intended as a practically usable tool for rescue

Fig. 1. Orpheus

teams, pyrotechnists and firemen. The Orpheus robotic system consists of two main parts: Orpheus mobile robot itself and operator's station.

D. Nardi et al. (Eds.): RoboCup 2004, LNAI 3276, pp. 491–498, 2005.

2 Mobile Robot Description

The robot itself (see Fig. 1) is formed by a box with 430x540x112mm dimensions and four wheels with 420 mms diameter.

The maximum dimensions of the robot are 550x830x410mm. The weight of the fully equipped robot with batteries is 32.5Kg. The mechanical construction of the robot is made by aluminium.

2.1 Locomotion Subsystem

Our department has developed a new Skid-steered Mobile Platform (SSMP) for the Orpheus mobile robot. The SSMP is intended to be both indoor and outdoor device, so its design was set up for this purpose. Another our important goal was to design the device easy-to-construct because of our limited machinery and equipment.

Finally we decided to make the platform like shown in Fig 2. The base frame of SSMP is a rectangular aluminium construction. Two banks of two drive wheels are each linked to an electrical motor via sprocket belt. The two drive assemblies for the left and right banks are identical but they operate independently to steer the vehicle. The motors can be driven in both directions, thus causing the vehicle to move forward, backward, right or left. Motors are equipped with incremental encoders and can be controlled either in velocity loop. .

Two 24V DC motors with integrated incremental encoders and three-stage planetary gearheads are used.

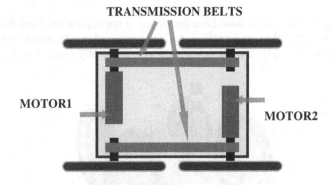

Fig. 2. Simplified Scheme of the Locomotor

2.2 Electronics

Most of the electronics is developed on our department. See Fig. 3 for a photo of the electronic subsystem of the Orpheus mobile robot.

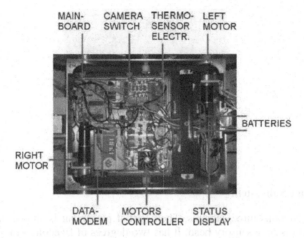

Fig. 3. Interior of the Orpheus

Microcontroller System

The Orpheus microprocessor system consists of 8 microcontrollers. Atmel AVR micro and Mega 8-bit RISC microcontrollers were used. The processors communicate by RS-232 serial interface using TTL levels.

Communication Processor - this processor serves to make an interface between ELPLRO datamodem and main processor. The main purpose of this processor is to transform the messages to and from the datamodem.

Main Processor - the main processor controls the whole system. It receives messages from the operator (through datamodem and communication processor). The architecture is master-slave, so the processor cyclically asks the other processors for the data.

Thermosensor Controller - the processor makes an interface between the thermosensor's RS-232 protocol and robot's internal RS-232 line. It is necessary because the thermosensor's protocol is too easy and has not any ID – responds to each message.

Servo Controller - this processor makes 6-channel standard modeller servo controller. The pulse-width may vary from 1.0 to 2.0 ms, is repeated each 20ms and may be set for each channel.

LCD Controller - four databit transfer mode for standard Hitachi LCD drivers is implemented in this microcontroller. The used LCD has four lines with 20 characters each, but the driver is universal and may be reconfigured for different LCDs.

Camera Switch and Analog Measurement Processor - Up to four cameras may be connected to the system. The camera switch board may switch among them. The processor also may be used for analog-to-digital conversion of various analog signals.

Motor Controllers - To control wheel velocities and direction a control and power switching board was developed. It consists of PID controller, H-bridge controller and full MOSFET H-bridge. There is one channel for each motor.

Fig. 4. Spatial Placement of Cameras

2.3 Sensory Subsystem

The robot contains three cameras. Their spatial placement is shown on Fig. 4. The main camera is on a sensory head. It has two degrees of freedom – may move left to right and up to down. The movements limits are similar to the ones of a human head.

The camera is a sensitive high resolution color camera with Sony chip. The other two cameras are black&white high-sensitive cameras with one degree of freedom. The front camera has IR light to work in complete darkness.

Fig. 5. The Sensory Head with Main Camera, Directional Microphone and Thermosensor

An infrared thermosensor Raytek Thermalert MID is used for object temperature measurement. The sensor provides three independent temperatures – the object temperature measured by IR, the sensory head temperature (the sensor measures the difference between temperatures in principle, so the derivation of this temperature is crucial to know if the measurement is precise), and the temperature of the electronics box, which we use to measure the temperature inside the robot. The thermosensor is placed beside the main camera and rotates with it (see Fig. 5). It causes the temperature of the object in the center of the camera picture is measured.

Standard walkie-talkie is used for one-directional audio transmission. One of two microphones displaced on the robot may be used during a mission.

The first one is integrated in the sensory head (see Fig. 5) and is directional. Since it moves with the main camera, the operator can hear the sounds from the direction he/she is looking to. The operator also may use the second – omnidirectional – microphone placed on the body of the robot.

We have made several experiments with two microphones and stereo audio perception on our older reconnaissance robotic system U.T.A.R. and we plan to use it on Orpheus as well, since it seems to be much more natural for the operator to have stereo perception similar to real-world than one omnidirectional microphone with perception axes parallel to the optical one.

It also seems to be very profitable to use electronically amplified microphones with user-variable gain.

2.4 Communication

Two independent devices are used for wireless communication with Orpheus: analog video transmitter for one-way video transmission and digital datamodem for bi-directional data communication between the robot and operator's station.

Fig. 6. Orpheus Operator's Station for Telepresence Control

3 Operator's Station

The operator's station for remote control of the mobile robot consists of several main parts: notebook, Imperx PCMCIA grabber, SAITEK Cyborg 3D Gold joystick, head mounted display: I-Glasses SVGA, INTERTRAX 2 headtracker – Intersense, Elpro Datamodem, video-receiver.

3.1 User Interface

The robot is controlled by operator with help of so called visual telepresence (see Fig. 6). The operator has a head mounted display with inertial head movement sensor. His/her movements are measured, transformed and transmitted to Orpheus. The user interface of Orpheus mobile robot system is programmed in C++ programming language under Microsoft Windows XP system.

The main advantage of the used user interface is that the digital data may be easily displayed over the video, so the operator does not need to switch among displays. The principle is that the added data are painted to small dark windows and these windows are blitted to the video image. The windows are semi-transparent, so the objects in video (or at least some of them) can be seen through the windows.

In the following text the small windows with additional data are called as displays.

Three main windows with different level of displayed data were designed. The full view, the quick view and the empty view.

Full View

All the accessible data are displayed on this display (see Fig. 7).

In the center part there is a Head Mounted Display Heading Display. This display shows the relative rotation difference between the camera and the body of the robot. This difference is derived from the operator's head movements.

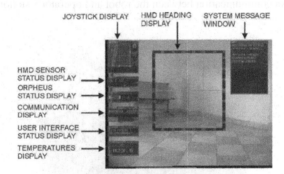

Fig. 7. Full view

Fig. 8. Quick view

The System Message Window represents the system messages like overall system status, list of devices currently connected to the system (joysticks, grabber, etc.). This is a tool to show events that happen once rather than continuous display (as against all of the other displays). The data are expressed as a text messages that roll on an "infinity paper roll".

From our experiments and testing it became obvious the full view is too complicated for operator in most standard situations and the operator may become overloaded by the amount of not-so-important data. For this reason a more simplistic quick view was developed (see Fig. 8). It contains only the most important data needed for the operator: user interface status display, HMD heading cross, and several indicators.

4   Conclusions and Perspectives

The system proved to work reliably during five-days period of the Robocup 2003 competition. Although the operator (author) had not preceeding experience of the system's remote control (the whole system was constructed in less than five months and they was no time for training) the control by visual telepresence was precise and reliable. It may be said the operator has a good view of the situation in the robot's environment. The control of camera movements by operator's head proved to be both intuitive and reliable. The design of the user interface including the virtual HUDs placed over the video image in head mounted display seems to be a good way to display important digital data to operator.

The Orpheus robotic system will be significantly improved in future. The new version marked as Orpheus-X1 is already in progress. The improvements include mechanical construction (see Fig. 9), electronics and user interface.

Fig. 9. Orpheus-X1with 2-DOF manipulator

Acknowledgement

This project was supported by the Ministry of Education of the Czech Republic under Project LN00B096.

References

1. Mlud L. (2001), Universal Autonomous and Telepresence Mobile Robot Navigation, In: 32nd International Symposium on Robotics - ISR 2001, pp 1010-1015, Seoul, Korea, April 19-21, 2001
2. Mlud L., Kopečný L., Neužil T. (2002), Laser Proximity Scanner Correlation Based Method for Cooperative Localization and Map Building, In: Proc. 7th International Workshop on Advanced Motion Control. Maribor, Slovenia: University of Maribor pp 480-486
3. Neužil, T.; Kopečný, L. & Mlud, L., Vision System of Universal Telepresence and Autonomous Robot, In: The 12th International Danube Adria Association for Automation & Manufacturing Symposium, pages 323-324, Jena, Germany, 24 - 27th October 2002, ISBN 3-901509-19-4
4. Wise, E., (1999). Applied Robotics, Prompt Publications, USA, ISBN: 0-7906-1184-8.
5. Everett, H.R., (1995). Sensors for Mobile Robots, Theory and Applications, AK Peters, Ltd., USA, ISBN: 1-56881-048-2.
6. Sheridan, T.B., (1992). Telerobotics, Automation, and Human Supervisory Control, MIT Press, Cambridge, USA.

Navigation Controllability of a Mobile Robot Population

Francisco A. Melo, M. Isabel Ribeiro, and Pedro Lima

Institute for Systems and Robotics,
Instituto Superior Técnico, Lisboa, Portugal
{fmelo, mir, pal}@isr.ist.utl.pt

Abstract. In this paper, the problem of determining if a population of mobile robots is able to travel from an initial configuration to a target configuration is addressed. This problem is related with the controllability of the automaton describing the system. To solve the problem, the concept of navigation automaton is introduced, allowing a simplification in the analysis of controllability. A set of illustrative examples is presented.

1   Introduction

Robotic navigation is a central topic of research in robotics, since the ability that a robot has to accomplish a given task may greatly depend on its capability to navigate in the environment. The related literature presents numerous works on the subject, and proposes different navigation strategies, such as Markov Models [1], dynamic behaviours [2] or Petri Nets [3].

In the last decades a great effort has been addressed to the subject of multi-robot systems. A common approach to the multi-robot navigation problem is the extension of known strategies for single robot navigation to the multi-robot case, for which there are several examples presented in the literature [4]. However, in the multi-robot navigation framework, new topics of investigation emerged, such as cooperation and formation control or flocking [5].

In this paper, the problem of multi-robot navigation is addressed. We analyze the problem of driving a robot population moving in a discrete environment from some initial configuration to a target configuration. We develop analysis strategies in order to determine under which situations the target configuration becomes non-achievable, in order to prevent those situations. In a previous work [6], this analysis has been conducted for a set of homogeneous robots([1]). This paper extends those results to a set of heterogeneous robots, i.e., where robots with different capabilities may intervene.

The robot population is modeled as a finite-state automaton (FSA) and the main contribution of this paper is the analysis of the blocking and controllabili-

---

[1] We consider a set of robots to be homogeneous when all robots are alike, i.e., they have the same capabilities.

D. Nardi et al. (Eds.): RoboCup 2004, LNAI 3276, pp. 499–507, 2005.
© Springer-Verlag Berlin Heidelberg 2005

ty properties of this automaton. Since it models the movement of the complete robot population in the environment, from a start configuration to given goal configuration, properties such as blocking and controllability have direct correspondence with the successful completion of this objective. For example, a blocking state corresponds to a distribution of the robots from which the desired goal configuration is not achievable (because one of the robots has reached a site from where it cannot leave, for example). Controllability means that such blocking states are avoidable: it is possible to disable some actions to prevent the robots from reaching blocking configurations.

The results presented in this paper relate the blocking and controllability properties of the automaton modeling the multi-robot system (which can be a large-dimension automaton, for complex systems) with the blocking and controllability properties of smaller automata, named as *navigation automata*, that model the navigation of each individual robot in the population.

The paper is organized as follows. In Section 2, some basic concepts are introduced and the problem under study is described. Section 3 approaches the problem of determining the blocking properties of the automaton describing the system. In Section 4, the results regarding controllability are presented for generic systems. Section 5 presents a set of illustrative examples. Finally, Section 6 concludes the paper and presents directions for future work. The proofs of all results in Sections 3 and 4 can be found in [7].

2   Navigation Automata and the Multi-robot system

In this section, some basic concepts regarding automata are introduced and the notation used throughout this paper is described.

Notation Regarding Automata [8]
An automaton Q is a six-tuple $Q = (X, E, f, \Gamma, x_0, X_m)$, where

- X is the state space;
- E is the set of possible events;
- $f : X \times E \longrightarrow X$ is the transition function;
- $\Gamma : X \longrightarrow 2^E$ is the active event function;
- x_0 is the initial state;
- X_m is the set of marked states.

The languages generated and marked by Q are denoted, respectively $\mathcal{L}(Q)$ and $\mathcal{L}_m(Q)$. An automaton is called *unmarked* if $X_m = \emptyset$.

An automaton Q is said to be *non-blocking* if $\overline{\mathcal{L}_m(Q)} = \mathcal{L}(Q)$ and *blocking* if $\overline{\mathcal{L}_m(Q)} \subsetneq \mathcal{L}(Q)$.

A set $X_C \subset X$ of states is said to be closed if $f(x, s) \in X_C$, for any $s \in \mathcal{L}(Q)$ and $x \in X_C$. A blocking automaton verifies $X_C \cap X_m = \emptyset$.

The Problem. Consider a system of N robots, moving in a discrete environment consisting of M distinct sites. This is referred as a N-R-M-S situation (N robots and M sites) or a N-R-M-S system. The set of sites is denoted by $\mathcal{S} = \{1, \ldots, M\}$.

Generally, when in site i, a robot will not be able to reach all other sites in a single movement. The function $\Omega_k : \mathcal{S} \to 2^{\mathcal{S}}$ establishes a correspondence between a site i and a set $\mathcal{S}_i \subset \mathcal{S}$ of sites reachable from i in one movement of robot k. If $j \in \Omega_k(i)$, then, for robot k, site j is *adjacent* to site i. Function Ω_k is called the *adjacency function* for robot k.

This paper addresses the problem of driving the robots from an initial configuration C_I to a final or target configuration C_F. The set of sites containing at least one robot in the final configuration is denoted by $\mathcal{S}_T$. The sites in $\mathcal{S}_T$ are called *target sites*. From the point of view of final configuration, no distinction is made among the robots, i.e., it is not important which robot is in each target site.

Navigation Automata. A robot k moves in the environment defined by the topological map according to its own adjacency function and is described by an unmarked automaton $G_k = (Y_k, E_k, f_k, \Gamma_k, y_{0k})$. Y_k is the set of all possible positions of robot k, verifying $Y_k = \mathcal{S}$. E_k is the set of all possible actions for robot k. Actions consist of commands leading to the next site for the robot to move to. Since all robots have the same event space, to avoid ambiguity, action i issued to robot k is denoted by $Go_k(i)$, where i is the next site for the robot k. Therefore, all events in the system correspond to movements of the robots. It is assumed that only one robot moves at a time. The active event function Γ_k when robot is in state i corresponds to the sites reachable from i in one movement. This means that $\Gamma_k = \Omega_k$.

Definition 1 (Navigation Automaton). *Given a robot k moving in a discrete environment consisting of M distinct sites, the navigation automata for this robot are the marked automata $G_k(Y_m) = (Y_k, E_k, f_k, \Gamma_k, y_{0k}, Y_m)$, where:*

- *Y_k, E_k, f_k and Γ_k are defined as above;*
- *Y_m is a set of target states, $Y_m \subset \mathcal{S}_T$.*

In the case of a homogeneous set of robots, in which $\Omega_1 = \ldots = \Omega_N$, all G_i are alike, except for the initial condition y_{0k}. In this situation, when the initial condition is clear from the context or not important, a navigation automaton will simply be denoted by $G(Y_m) = (Y, E_m, f_m, \Gamma_m, y_0, Y_m)$.

2.1   The Multi-robot System

If there are no constrains on the number of robots present at each site, each of the robots can be in M different positions, and there are M^N different possible configurations.

The system of all robots can be described by a FSA, $G = (X, E, f, \Gamma, x_0, X_m)$, where X is the set of all possible robot configurations, yielding, in the most general situation, $|X| = M^N$. Each state $x \in X$ is a N-tuple $(x_1, x_2, ..., x_N)$ and x_i is the site where robot i is.

Also, there is a set $X_F \subset X$ of states corresponding to the target configurations. Notice that, in automaton G, $X_m = X_F$.

As seen before, robot k has an available set of actions E_k, denoted by $Go_k(i)$. Therefore, the multi-robot system has a set of actions $E = \bigcup_k E_k$, all consisting of $Go_k(i)$ actions.

3   Blocking

Let G be the automaton modeling a N-R-M-S system.

If G is blocking, there is a closed set of states C, called *blocking set*, such that $C \cap X_m = \emptyset$. This, in turn, means that whenever the robots reach a configuration corresponding to a state $x \in C$ it is not possible to drive them to the desired configuration anymore.

Usually, blocking is checked by verifying $\overline{\mathcal{L}_m(G)} \subsetneq \mathcal{L}(G)$ exhaustively. In the present case, as the system can lead to relatively large automata for not so large M and N, a more effective way to check the blocking properties of G is desirable. This section addresses this problem.

3.1   Blocking and Navigation Automata

Let G be the automaton modeling a N-R-M-S system. If the system is homogeneous, Result 2 of [6] holds.

For a generic (non-homogeneous) system, if G is non-blocking, then so is each of the navigation automaton G_i, with all target states as marked states. However, the converse is not true. Theorem 1 follows.

Theorem 1. *In a generic N-R-M-S system, for its automaton G to be non-blocking, all the navigation automata $G_i(Y_m)$ must non-blocking, with $Y_m = \mathcal{S}_T$. Similarly, if G is blocking, there is at least one i and one target state $y_m \in Y_m$ such that $G_i(y_m)$ is blocking.*

Proof. See in [7].   □

3.2   Blocking Information Matrix

From section 3.1 one can conclude that, generally, it is not possible to determine the blocking properties of G simply by taking into account the blocking properties of each navigation automaton individually, since these properties depend on the relation between them. It is, however, possible to determine whether or not G is blocking, by *comparing* the blocking properties of each automaton $G_i(y_m)$.

In an N-R-M-S system, let $K(m)$ be the number of robots in target site m in the final configuration C_F. If $U \subset \mathcal{S}_T$ is a set of target sites, define $K(U) = \sum_{m \in U} K(m)$. Define as $B(m)$ the number of robots that block with respect to target site m. If $U \subset \mathcal{S}_T$ is a set of target sites, define $B(U)$ as the number of robots *simultaneously* blocking the sites in U. In general, $B(U) \neq \sum_{m \in U} B(m)$.

If the number of robots blocking simultaneously the sites in some set $U \subset \mathcal{S}_T$ is such that $N - B(U) < K(U)$, then G blocks, and therefore, $N - B(U) < K(U)$. This means that there are not enough "free" robots to go to the sites in M. This condition may be easily verified using the *blocking information matrix*.

Definition 2 (Blocking Information Matrix). *Given a generic N-R-M-S system, the* blocking information matrix *(BIM)* $\mathbf{B}_N$ *is a $N \times N$ matrix such that element (k, m) is 0 if $G_k(y_m)$ is blocking and 1 otherwise.*

Each of the N lines of matrix $\mathbf{B}_N$ corresponds to a different robot. On the other hand, if a site m has $K(m)$ robots in the target configuration, matrix $\mathbf{B}_N$ will have $K(m)$ columns corresponding to this site. Matrix $\mathbf{B}_N$ is easily computed from the analysis of the navigation automata $G_k(y_m)$, and the following result can be proved.

Theorem 2. *Given the Blocking Information Matrix $\mathbf{B}_N$ for a N-R-M-S system, the automaton G describing the overall system is blocking if and only if there is a permutation matrix $\mathbf{P}$ such that $\mathbf{PB}_N$ has only ones in the main diagonal.*

Proof. See in [7]. □

4   Supervisory Control

In this section, the problem of controllability of G is addressed. The controllability problem is related to the design of a supervisor S, such that, when applied to the original system, the resulting system marks some desired language $\mathcal{K}$.

Although the automaton G describing the system already marks the desired language, in a situation where the automaton is blocking, it is not desirable that the system reaches a blocking state, since this will prevent the final configuration to be reached. The presence of a supervisor S in the system under study will necessarily relate to this situation where blocking must be prevented. It is important to determine the existence of such a supervisor, i.e., it is important to determine if the system is controllable.

In the following analysis, the existence of unobservable events is disregarded even if they make sense from a modeling point of view, as described in [6].

4.1   System Controllability

Consider the automaton G describing a N-R-M-S system, which is assumed blocking, and suppose that there is a non-empty set of uncontrollable events $E_{uc} \subset E$. These events may correspond to accidental movements of the robots which cannot be avoided. If the system is homogeneous, Result 3 of [6] holds.

Consider, now, a heterogeneous system. As stated before, blocking in G is related to the number of robots "available" to fill each target site, when considering blocking sets. Controllability relates with the ability of a supervisor to disable strings of events driving a robot to a blocking set.

Let $E_{uk} \subset E_k$ be the set of uncontrollable events for robot k. It is possible to include controllability information in matrix $\mathbf{B}_N$ in order to conclude about the controllability of G. If $G_k(y_m)$ is blocking but controllable with respect to the language $\mathcal{K} = \mathcal{L}_m(G_k(y_m))$, then the element (k, m) of matrix $\mathbf{B}_N$ is set to -1. This motivates the following and most general form of Theorem 2.

Theorem 3. *Given the Blocking Information Matrix* $\mathbf{B}_N$ *for a generic N-R-M-S system, the automaton G describing the overall system is blocking if and only if there is a permutation matrix* $\mathbf{P}$ *such that* $\mathbf{P}\mathbf{B}_N$ *has only ones in the main diagonal.*

If G is blocking, but there is a permutation matrix $\mathbf{P}_1$ *such that* $\mathbf{P}_1\mathbf{B}_N$ *has only non-zero elements in the main diagonal, then G is controllable with respect to the language* $\mathcal{K} = \mathcal{L}_m(G)$.

Proof. See in [7]. □

Observe the relation between Theorem 3 and Result 3 of [6]. In fact, the latter can be derived from Theorem 3 when the robot set is homogeneous.

5   Examples

We present three examples of the application of Theorem 3 in a simple indoor rescue situation. Consider a non-homogeneous set of three robots in which:

The Crawler. *(Cr)* has tracker wheels and is capable of climbing and descending stairs. It is able to open doors only by pushing;

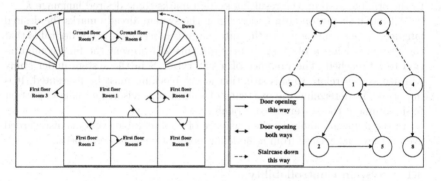

Fig. 1. Map of the environment

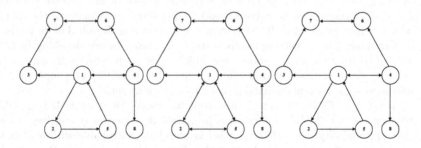

Fig. 2. Automata for the robots

The Puller. (Pl) is a wheeled mobile manipulator, able to open doors either by pushing or pulling. However, it is not able to climb stairs;

The Pusher. (Ps) is a wheeled robot, able to open doors only by pushing. It cannot climb stairs.

The rescue operation takes place in the indoor environment depicted in Figure 1 (e.g., a fire scenario). On the left is the physical map of the place, and on the right is the corresponding topological map. Each of the robots is described by a different automaton, as represented in Figure 2.

The robots will leave Room 1 to assist three different victims, somewhere in the building. The doors open as shown in Figure 1 which limits the robots access to the different rooms. When in Rooms 6 or 7, only the Crawler can go upstairs. Finally, when in Rooms 3 and 4, all the robots may fall downstairs, i.e., events $Go_k(6)$ and $Go_k(7)$ are uncontrollable for all k. The following examples illustrate the practical use of Theorem 3. We determine if there are configurations that prevent the success of a given rescue operation which, in terms of the framework proposed in this paper, correspond to blocking configurations. The situation where there are victims in sites a, b and c is referred to as the $a - b - c$ Rescue.

5.1   6 − 7 − 8 Rescue

In this situation, the BIM for the system is:

$$\mathbf{B}_3 = \begin{bmatrix} -1 & -1 & 0 \\ 1 & 1 & 0 \\ -1 & -1 & 1 \end{bmatrix}, \tag{1}$$

where the lines correspond to Pusher, Puller and Crawler and the columns correspond to target sites 6, 7 and 8, respectively.

Note that both Ps and Cr, once inside Room 8, are not able to leave. Then, $G_{Ps}(6)$, $G_{Ps}(7)$, $G_{Cr}(6)$ and $G_{Cr}(7)$ are blocking. However, by disabling the events $Go_{Ps}(8)$ and $Go_{Cr}(8)$, blocking can be prevented and $\mathbf{B}_3(1,1) = \mathbf{B}_3(1,2) = \mathbf{B}_3(3,1) = \mathbf{B}_3(3,2) = -1$.

If Ps or Pl get downstairs, they cannot go back upstairs. However, they cannot get to Room 8 without going through Room 4 and eventually falling to Room 6, which cannot be avoided, since $Go_k(6)$ is uncontrollable. Then, $G_{Ps}(8)$ and $G_{Ps}(8)$ are blocking and uncontrollable, and $\mathbf{B}_3(1,3) = \mathbf{B}_3(2,3) = 0$.

Finally, Pl can always reach Rooms 6 and 7, and Cr can always reach Room 8. $G_{Pl}(6)$, $G_{Pl}(7)$ and $G_{Cr}(8)$ are non-blocking, and $\mathbf{B}_3(2,1) = \mathbf{B}_3(2,2) = \mathbf{B}_3(3,3) = 1$.

From Theorem 3 the system is blocking but controllable. For example in the configuration where Crawler and Pusher are in room 8, it is impossible to reach the target configuration. However, this can be prevented, by disabling, for example, $Go_{Ps}(8)$, which is a controllable event.

5.2   2 − 8 − 8 Rescue

In this situation, the BIM for the system is:

$$\mathbf{B}_3 = \begin{bmatrix} -1\ 0\ 0 \\ -1\ 0\ 0 \\ -1\ 1\ 1 \end{bmatrix}, \tag{2}$$

with the columns corresponding to target sites 2, 8 and 8, respectively. It becomes evident that the system is blocking but, unlike the previous example, it is uncontrollable. Since two robots are required for site 8 and the only way to reach Room 8 is through Room 4, they will eventually move to Room 6 instead of moving to Room 8 (since $Go_k(6)$ is uncontrollable), once they get to Room 4. In this situation, it may be impossible to assist both victims in site 8. In fact, as long as there is more than one victim in site 8 ($K(8) > 1$), this problem will always exist. This happens because there are two robots which "helplessly" fall downstairs, blocking site 8 ($B(8) = 2$). Then, $N - B(8) = 3 - 2 = 1 < K(8)$, and the system is blocking. Since the only way to Room 8 is through Room 4, this situation cannot be prevented.

6   Conclusions and Future Work

The problem of analyzing the navigation of a set of mobile robots operating in a discrete environment was approached. Relevant results have been derived, that allow the use of small dimension automata (navigation automata) to infer about the blocking and controllability properties of the automaton that describes the complete system. In a situation where a specific configuration is aimed for a set of robots, the presented results allow to determine, using global information, if the global objective is achievable, and if blocking configurations are avoidable.

An important extension of the present work is the determination of the relation between the blocking properties of the navigation automata and the ergodicity of the Markov Chain which can be used to model the complete system, when a probabilistic uncertainty is associated to the events representing the movements of the robots. Other interesting issue is the use of this local information in an optimal decision process, when a decentralized system is considered.

Acknowledgements

Work partially supported by Programa Operacional Sociedade de Informação (POSI) in the frame of QCA III and by the FCT Project *Rescue—Cooperative Navigation for Rescue Robots* (SRI/32546/99-00). The first author acknowledges the PhD grant SFRH/BD/3074/2000.

References

1. Simmons, R., Koenig, S.: Probabilistic Robot Navigation in Partially Observable Environments. Proceedings of the International Joint Conference on Artificial Intelligence, Montreal, Canada (1995) 1080–1087
2. Steinhage, A.: Dynamical Systems for the Generation of Navigation Behaviour. PhD thesis, Institut für Neuroinformatik, Ruhr-Universität, Bochum Germany (1997)
3. Hale, R.D., Rokonuzzaman, M., Gosine, R.G.: Control of Mobile Robots in Unstructured Environments Using Discrete Event Modeling. SPIE International Symposium on Intelligent Systems and Advanced Manufacturing, Boston, USA (1999)
4. Balch, T., Hybinette, M.: Social Potentials for Scalable Multirobot Formations. IEEE International Conference on Robotics and Automation (ICRA-2000), San Francisco, USA (2000)
5. Balch, T., Arkin, R.C.: Behavior-based formation control for multi-robot teams. IEEE Transactions on Robotics and Automation **14** (1998) 926–939
6. Melo, F.A., Lima, P., Ribeiro, M.I.: Event-driven modelling and control of a mobile robot population. Proceedings of the 8th Conference on Intelligent Autonomous Systems, Amsterdam Netherlands (2004)
7. Melo, F.A., Ribeiro, M.I., Lima, P.: Blocking controllability of a mobile robot population. Technical Report RT-601-04, Institute for Systems and Robotics (2004)
8. Cassandras, C.G., Lafortune, S.: Introduction to Discrete Event Systems. The Kluwer International Series On Discrete Event Dynamic Systems. Kluwer Academic Publishers (1999)

Sharing Belief in Teams
of Heterogeneous Robots

Hans Utz[1], Freek Stulp[2], and Arndt Mühlenfeld[3]

[1] University of Ulm, James-Franck-Ring, D-89069 Ulm, Germany
[2] Technische Universität München, Boltzmannstr. 3, D-85747 München, Germany
[3] Technische Universität Graz, Inffeldg. 16b/II, A-8010 Graz, Austria
hans.utz@informatik.uni-ulm.de
stulp@in.tum.de
muehlenf@igi.tugraz.at

Abstract. This paper describes the joint approach of three research groups to enable a heterogeneous team of robots to exchange belief. The communication framework presented imposes little restrictions on the design and implementation of the individual autonomous mobile systems. The three groups have individually taken part in the RoboCup F2000 league since 1998. Although recent rule changes allow for more robots per team, the cost of acquiring and maintaining autonomous mobile robots keeps teams from making use of this opportunity. A solution is to build mixed teams with robots from different labs. As almost all robots in this league are custom built research platforms with unique sensors, actuators, and software architectures, forming a heterogeneous team presents an exciting challenge.

1 Introduction

Due to scientific as well as pragmatic reasons, there is a growing interest in the robotics field to join the efforts of different labs to form mixed teams of autonomous mobile robots. In RoboCup, the pragmatic reasons are compelling. The recent rule change in the F2000 league allows for more robots per team, and in the RoboCup Rescue league a group of heterogeneous robots with diverse capabilities is likely to perform better than one system that tries to encapsulate them all. However, the limited financial resources and the additional maintenance effort for further robots exceeds the capabilities of many research labs. Also, the threshold for new research groups to participate in RoboCup is lowered if they only need to contribute one or two robots to a mixed team, instead of having to build an entire team. Mixed teams are also motivated from a scientific perspective. They introduce the research challenge of cooperation within teams of extremely heterogeneous autonomous mobile systems.

As most robots in the F2000 league are custom built, or at least customised commercial research platforms with unique configurations of actuator and sensor configurations, mixed teams from different laboratories are extremely heteroge-

D. Nardi et al. (Eds.): RoboCup 2004, LNAI 3276, pp. 508–515, 2005.

neous. There are few commonly used high level libraries for sensor data processing and reactive actuator design in the community. Furthermore there is a multitude of methods and schools, each deliberately designing the control architecture of their robots fundamentally different to their competitors. This makes the unification of the software of the different robots of a potential mixed team almost impossible without substantial rewriting of at least one of the team's software. In our opinion it is also undesirable. Why should an autonomous mobile robot have to commit to any kind of sensor processing or control paradigm to be able to cooperate with another team mate, if both are programmed to interact in the same problem domain?

For cooperation between robots, the sharing of information about the environment is initially sufficient for successful cooperation. If all robots share both the same belief about their environment, as well as the same set of goals, similar conclusions should be drawn. This has proven to be a successful way of coordinating behaviour in RoboCup scenarios [1, 2]. Therefore, a central prerequisite for successful team cooperation is the unification of the beliefs about the world of the different agents. The limitations of the individual sensors usually provide each robot with quite limited information about the state of its environment. So it is unlikely that the beliefs derived solely from the robots' own sensors are automatically sufficiently similar to coordinate behaviour in a shared environment. Sharing of information might solve this problem.

This paper presents the early stage of an approach of robotic labs from three different research groups, of the universities of Ulm, Munich and Graz, to be able to play interchangeably with their different robot platforms in a mixed team. The main contributions of this paper are along two research directions. First, the design and implementation of a communication framework for sharing information within a team of extremely heterogeneous autonomous robots is presented. It is hardware and software independent, and can extend existing software architectures in a transparent way. Major code rewrites are not necessary to use this framework. Second, we specify an expressive shared belief state that our robots can communicate using this framework, to complement own beliefs, and to coordinate behaviour.

The remainder of this paper is organised as follows. Section 2 presents the three robotic teams. Section 3 shows the design of belief exchange between these teams. The implementation of the communication framework is presented in section 4. Related work is discussed in section 5, and we present future work and conclude with section 6.

2 An Overview of the Three Teams

In this section we will describe the differences between the three teams, emphasising those that are relevant to the sharing of a belief state. One of the most important differences lies in the sensors used. On the level of belief state representation, we have found that all teams use different methods of modelling uncertainty and inaccuracy. This is mostly caused by the different ways in which the sensors capture the world. Also, some teams use an egocentric frame of reference, and others allocentric. Furthermore, fusing information from other players or the coach into the belief states of individual robots is treated in different ways.

Ulm: The Ulm Sparrows [3] are custom built robots, with infrared based near range finders and a directed camera. The available actuators are a differential drive, a pneumatic kicking device and a pan unit to rotate the camera horizontally (180°). The robots act upon an egocentric belief state, and uses inaccuracy and uncertainty about observations. Fusion takes place for the ball.

Munich: The AGILO *RoboCuppers* [4, 2] are customised Pioneer I robots, with differential drive and a fixed forward facing colour CCD camera. They act upon an allocentric belief state, which models inaccuracy, and which they communicate and fuse locally. Opponent observations are fused with a central Multiple Hypothesis Tracker (MHT).

Graz: The Mostly Harmless [5] are custom-built robots with a modular design of hardware and software, motivated by the intention of using the robots in different domains. They have an omni-directional camera, and laser range finder. The belief state is allocentric, and fusion is done locally using a MHT.

3   Design of Belief Exchange

The scenario depicted in figure 1 shows how belief exchange can be useful. Player 1 has just shot at the goal, but the keeper has deflected the ball, which is now lying behind player 2. This player has a good chance at scoring, but cannot see the ball because it is outside of the field of view of its camera. Fortunately, two defenders observe the ball. One has a laser range finder, and can determine the location of the ball

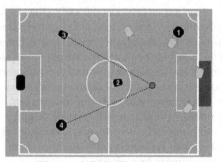

Fig. 1. A RoboCup Scenario

accurately. However, since colour information is lacking, it is not that certain about what the object is. Fortunately, the second defender has a colour camera, with which it recognises the ball, even if the localisation of it is not very accurate. Player 2 receives this information from the two defenders, and fuses it to derive there is a ball behind it. All players compute that player 2 is closest to the ball, and player 2, determined by locker-room agreement, will turn and try to score a goal.

This scenario shows that exchanging information allows individual robots with a limited field of view to acquire information about hidden parts of the state, and to generate a more consistent and certain belief about the world by fusing observations from different robots, possibly with different sensors.

To share beliefs, the teams must agree upon structures that encapsulate this information. In this section we will discuss the design of these structures. The three main elements are a time-stamp, the probabilistic dynamic pose of the robot itself, and a list of observed objects. The main design principle is that we want an expressive belief exchange structure, but not simply a superset of all the information that the three teams have used so far.

Time-stamped message-based communication. The beliefs of the robots are exchanged within the team by a message-based protocol. Messages that are corrupted by packet loss in the wireless network do not influence each other, and can safely be ignored. As a basic requirement for sharing information in a highly dynamic environment each message is accurately time-stamped. This allows for interpolation of the beliefs to minimise the estimation error between messages.

The own dynamic pose. Many sensor data processing algorithms assume that the probability of measuring a certain quantity is distributed according to a Gaussian normal distribution. We also use this concept, and represent the robot's pose by a three-dimensional vector and covariance matrix. The vector represents the mean of the Gaussian distribution, with $(0,0,0)$ being the centre of the field, facing the opponent goal. The covariance matrix holds the associated uncertainty as a three-dimensional Gaussian distribution around this mean. How these values can be computed is discussed more elaborately in [6]. This probabilistic representation allows robots to communicate not only where they belief they are, but also how certain they are about it. Apart from its pose, the robot also communicates its velocity as a tuple (v_x, v_y, ω). There are no uncertainty measures for the speed. We call the combination of pose, covariance matrix and speed a probabilistic dynamic pose.

Observed objects. Apart from their own pose, the shared belief state also contains a representation of all the observations of objects the robots have made. Each object is represented by the same dynamic pose and covariance matrix, which have been discussed in the previous section. Although this representation is somewhat redundant (goal-posts will never have a velocity), it is very general. The observed objects are projected in an egocentric frame of reference, in which the observer always has pose $(0,0,0)$.

Each observation must be mapped to one of the objects that can be expected to be encountered in the robot's environment. On a soccer field these objects are the ball, the teammates, the opponents, the corner-flags and the goal-posts. Any other objects (referee, assistants, flashers) are deemed obstacles. Since observations can usually not be mapped to one object with perfect certainty, each observation is accompanied by a list of all possible objects, and the estimated probability that the observation was this object. Observation-object assignments with a low probability are excluded from this list, as they are not likely to influence the decisions of a teammate, and would only use up bandwidth.

3.1   Design Issues

Allocentric vs. Egocentric. The most obvious way to fuse observations of different robots is by the use of a shared frame of reference. Therefore the robots use the allocentric frame of reference defined in the previous section, to communicate their poses. However, this approach requires knowing where you are with sufficient accuracy, something that cannot always be guaranteed. For many tasks (shooting a goal, passing to a teammate) it is sufficient to know the relative location of certain objects to the robot. For these reasons we have decided to use an egocentric frame of reference for the observations.

Inaccuracy and uncertainty. The differences between the sensor systems of the teams allow them to derive different information about the objects they observe. The laser range finder of the Graz team provides accurate information about object locations, but the uncertainty as to what kind of object it is is not as certain. The other colour-camera based teams are much more certain about the type of object (*orange* ball, *yellow* goal), but cannot locate the objects as precise. We have chosen to represent both aspects in the shared belief state, to allow all teams to express their different types of uncertainty. Appropriate fusion of this distributed multi-modal perception increases the accuracy and certainty of each robot's local belief state.

Object superclasses for observation assignments. To make the list of observation-object more compact, superclasses of the individual objects have been defined. Often it is the case that a robot recognises a robot as being an opponent, but not which number it has. Instead of sending that it can be Opponent 1,2,3 or 4, each with the same chance, the robot has the option of sending the superclass Opponent. The full tree of objects and superclasses can be seen in figure 2.

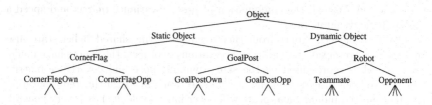

Fig. 2. Tree of Object Classes

4   Implementation of the Communication Framework

The team communication uses a message-based, type safe high-level communications protocol that is transfered by IP-multicast, as such a protocol keeps the communicated data easily accessible and prevents subtle programming errors that are hard to trace through different teams. As the communication in a team of autonomous mobile robots has to use some kind of wireless LAN, that is notoriously unstable especially in RoboCup tournaments, a connection-less message based protocol is mandatory. This way, network breakdowns and latencies do not block the sending robot. To save bandwidth, IP-multicast is used, since this way each message has only to be broadcasted once, instead of n times for n clients.

The implementation uses the notify multicast module (NMC) of the Middleware for Robots (MIRO) [7]. MIRO provides generalised CORBA based sensor and actuator interfaces for various robot platforms as well as higher level frameworks for robotic applications. Additionally to the method-call oriented interfaces, MIRO also uses the event driven, message-based communications paradigm utilising the CORBA Notification Service. This standardised specification of a publisher/subscriber protocol is part of various CORBA implementations [8]. Publishers (suppliers in their terminology) offer events. The so-called consumers subscribe to those events. They then receive the events supplied by the publisher through an event channel (EC). The data exchanged is specified in the CORBA interface definition language (IDL). Standardised mappings from IDL to most modern programming languages (C, C++, Java) exist.

CORBA uses a connection oriented (TCP/IP based) communication layer by default, the NMC module therefore plugs into the Notification Service architecture and exchanges events between the robots of a team transparently, using IP-multicast. For this purpose a service federation quite similar to the one described in [9] is used. An EC instance is run locally on each robot. A "NMC event consumer" subscribes for all events that are offered only locally but subscribed by other team mates and sends them to the multicast group. A "NMC event supplier" in turn listens to all events published via IP-multicast and pushes those into the local event channel, that are subscribed but not offered locally. To keep track of the offered and subscribed message types, NMC utilises two fields of the standard event message format: The domain name and the type name. By convention, the domain name contains the name of the robot producing the event. The type name describes its payload. As these fields are also part of the native offer/subscription management and filtering protocol of the notification service, robots can easily determine whether events they offer are currently subscribed in the team, and skip their production entirely if there are no subscribers.

Figure 3 illustrates a sample configuration of the notification channel setup. Two robots (A, B) produce two types of events $(1, 2)$, the resulting events are $\{A1, A2, B1, B2\}$. The events in the supplier and consumer boxes denote the offered and subscribed events. The events labelling the arrows denote the actual flow of events. Note that

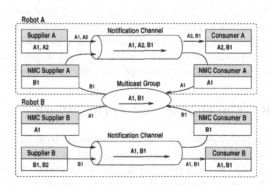

Fig. 3. A Federated Notification Channel Setup

suppliers and consumers can offer/subscribe for multiple events.

Communicating the IDL-specified belief state discussed in section 3 at 10Hz with all teammates uses, on average, less than 10% of the available bandwidth of a standard 802.11b WLAN (11 MBit/s). This should be available, even on heavily loaded networks, such as those in RoboCup tournaments.

We would like to emphasise that even though our main goal is to communicate belief states, this framework is not limited to this information only. In principle it is possible to communicate any type of information, for instance role assignments, utilities, coaching advice.

5   Related Work

Team wide information sharing is a well-known concept in RoboCup. CS Freiburg used a global world model fused on a central coach computer by the use of Kalman and particle filters [10]. It was then sent back to the team mates. The system was based on a very accurate LSR-based self localisation. High accuracy was achieved by using the walls that enclosed the field up to 2001 as landmarks. An early attempt of information sharing in the legged-robot league is described in [11]. However, this approach suffered from the severe network latencies of the communication device available on this platform.

The idea of cross team cooperation has some tradition within the RoboCup leagues. In the simulation league, the source code of many teams was published on the Internet allowing new participants to base their new team on previous participants of simulation league tournaments.

One of the most successful mixed teams in RoboCup has been the German-Team, which participates in the legged-league [12]. The GermanTeam is a cooperation of five universities participating with one team and one code repository. The exchange and integration of software is enabled by a standardised hardware platform, as well as a modular software design.

The most similar mixed team cooperation effort was done by the Azzurra Robot Team. They built a middle size league national team from various Italian universities. They also used a (proprietary) publisher/subscriber communication protocol, utilising UDP. However, their focus was on explicit role assignment and coordination strategies among the field players [13]. Unfortunately the Italian national team was dissolved after the RoboCup tournaments in 2000.

6   Conclusion and Future Work

Currently each team has its own fusion methods. The data exchanged is designed in such a way that each team should be able to use it for their fusion, as well as supply the data required by the other teams. Sharing the same belief state will allow us to compare the different methods better, as the belief state abstracts from the sensory differences of the teams. In the long run, a unified fusion can be expected, as all teams will try to get the best out of the shared information.

In this paper a CORBA based communication framework for sharing information within a team of extremely heterogeneous autonomous robots is presented.

Our three teams have extended their existing software architectures with this framework, enabling transparent communication between these teams. We have also introduced the shared belief state we communicate using this framework, and discussed the underlying design principles and motivations.

The design of the mixed team communication for sharing beliefs was started on the altruistic ground of scientific experimentation. Nevertheless it might turn into a necessity, as it seems that not all three teams can afford to attend this years Robot Soccer World Cup in Lisbon as an independent team.

Acknowledgements

The work described in this paper was partially funded by the German Research Foundation (DFG) in the SPP-1125 . Both MIRO and the shared belief state specification are freely available at: `smart.informatik.uni-ulm.de/Miro/` `sharedbelief.sourceforge.net/`

References

1. Tews, A., Wyeth, G.: Thinking as one: Coordination of multiple mobile robots by shared representations. In: Intl. Conf. on Robotics and Systems (IROS). (2000)
2. Beetz, M., Schmitt, T., Hanek, R., Buck, S., Stulp, F., Schröter, D., Radig, B.: The AGILO 2001 robot soccer team: Experience-based learning and probabilistic reasoning in autonomous robot control. to appear in Autonomous Robots, special issue on Analysis and Experiments in Distributed Multi-Robot Systems (2004)
3. Kraetzschmar, Mayer, Utz, et al: The Ulm Sparrows 2003. In: Proc. of RoboCup-2003 Symposium (to appear). LNAI, Springer-Verlag (2004)
4. Beetz, M., Gedikli, S., Hanek, R., Schmitt, T., Stulp, F.: AGILO RoboCuppers 2003: Computational principles and research directions (2004)
5. Steinbauer, G., Faschinger, M., Fraser, G., Mühlenfeld, A., Richter, S., Wöber, G., Wolf, J.: Mostly Harmless Team Description (2004)
6. Schmitt, T., Hanek, R., Beetz, M., Buck, S., Radig, B.: Cooperative probabilistic state estimation for vision-based autonomous mobile robots. IEEE Trans. on Robotics and Automation **18** (2002) 670–684
7. Utz, H., Sablatnög, S., Enderle, S., Kraetzschmar, G.K.: Miro – middleware for mobile robot applications. IEEE Trans. on Robotics and Automation **18** (2002) 493–497
8. Schmidt, D.C., Gokhale, A., Harrison, T., Parulkar, G.: A high-performance endsystem architecture for real-time CORBA. IEEE Comm. Magazine **14** (1997)
9. Harrison, T.H., Levine, D.L., Schmidt, D.C.: The design and performance of a real-time CORBA event service. In: Proc. of OOPSLA '97, Atlanta, ACM (1997)
10. Dietel, M., Gutmann, S., Nebel, B.: Cs freiburg: Global view by cooperative sensing. Volume 2377 of LNAI., Springer-Verlag (2002) 133–143
11. Roth, M., Vail, D., Veloso, M.: A world model for multi-robot teams with communication (2002)
12. Röfer, T.: An architecture for a national RoboCup team. Volume 2752 of LNAI., Springer-Verlag (2002) 417–425
13. Castelpietra, C., Iocchi, L., Nardi, D., Piaggio, M., Scalzo, A., Sgorbissa, A.: Communication and coordination among heterogeneous mid-size players: ART99. Volume 2019 of LNAI., Springer-Verlag (2001) 86–95

Formulation and Implementation of Relational Behaviours for Multi-robot Cooperative Systems

Bob van der Vecht[1,2] and Pedro Lima[2]

[1] Department of Artificial Intelligence, University of Groningen
Grote Kruisstraat 2/1, 9712 TS, Groningen, Netherlands
[2] Institute for Systems and Robotics, Instituto Superior Técnico,
Av. Rovisco Pais, 1 - 1049-001 Lisboa, Portugal
{bvecht, pal}@isr.ist.utl.pt

Abstract. This paper introduces a general formulation of relational behaviours for cooperative real robots and an example of its implementation using the pass between soccer robots of the Middle-Sized League of RoboCup. The formulation is based on the Joint Commitment Theory and the pass implementation is supported by past work on soccer robots navigation. Results of experiments with real robots under controlled situations (i.e., not during a game) are presented to illustrate the described concepts.

1 Introduction

Showing cooperation among robots from a team is probably one the top goals of RoboCup related research. Furthermore, it is desirable to formulate cooperative behaviours within a formal framework extendable to applications other than robotic soccer. Surprisingly, not many references to work on these two topics can be found in the RoboCup-related literature or in other publications referring cooperation among *real* robots.

An example of a tool that enables the development of applications based on the Joint Commitment Theory [1], including communications, has been thoroughly reported by Tambe, e.g., in [9]. Yokota et al. use explicit communication to achieve cooperation and synchronization in real robots. However, no explicit logical commitments are described [10]. Emergent cooperative behaviour among virtual agents is also described in [7], where a pass behaviour by implicit communication (observing the other robots behaviour) is implemented in a team of RoboCup Simulation League, and [6], where also a pass between RoboCup Simulation League agents is described such that conditions to pass are learned by a neural network. Again, there is no commitment between players, therefore the relational behaviour may be kept by one of the team mates even if the other has to withdraw its relational behaviour.

In this paper a general formulation of relational behaviours for cooperative real robots is introduced. Most of the paper describes an example of implementation of this formulation using the pass between soccer robots of the Middle-Sized

D. Nardi et al. (Eds.): RoboCup 2004, LNAI 3276, pp. 516–523, 2005.

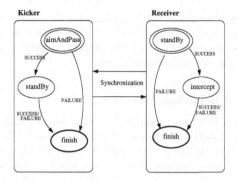

Fig. 1. Simple example of the pass commitment

League (MSL) of RoboCup. The formulation is based on the Joint Commitment Theory [1], and the pass implementation is supported by past work of the ISocRob team on soccer robots navigation [3].

In the pass relational behaviour, two participants set up a long term commitment, in which several individual behaviours are executed. One of the robots is referred to as the *kicker*; he starts having the ball and will try to kick the ball in the direction of the other robot, the *receiver*, who has to intercept the ball. In order to accomplish a pass successfully, two components of the commitment should be working well: the synchronization of both players' actions and the execution of their individual skills. The synchronization has to be achieved by communication. In Fig. 1, an example of the pass commitment has been illustrated by two state machines.

In the following sections, the theory behind the pass commitment as shown in Fig. 1 will be described. First the individual decision making and the behaviour synchronization will be explained and then the individual primitive behaviours. Results of experiments with real robots under controlled situations (i.e., not during a game) are also presented to illustrate the key primitive behaviours described.

2   Decision Making and Synchronization

In Fig. 1, several individual behaviours can be found within the commitment. At any time the participants have to select the correct primitive behaviour individually. The architecture that is used for behaviour coordination and decision making considers three types of behaviours: organizational, relational and individual [4], [5]. Organizational behaviours concern decisions involving the whole team, e.g., player role selection such as defender or attacker. The relational behaviours concern more than one player. Commitments among team mates are established here. Finally, at the individual level, primitive behaviours are selected, and motor commands are used to influence the environment in the desired way.

Behaviour selection is done in a module called the *logic machine*, explained in [8]. In this module, only the organizational and the individual levels were implemented. The work described here addresses the relational layer. The layers are processed sequentially. This means for the robot that he first chooses a role, next he selects a commitment, and the individual behaviour is selected after knowing the robot's role and commitment.

2.1   Relational Behaviour Selection

Joint Commitment Theory is used in this work to select relational behaviours [1]. Predefined logical conditions can establish a commitment between two agents. Once a robot is committed to a relational behaviour, he will pursue this task until one or more conditions become false, or until the goal has been accomplished. In the described project, the initiative for a relational behaviour is taken by one of the agents, who sets a *request* for a relational behaviour. A potential partner checks if the conditions to *accept* are valid. If so, the commitment is established. During the execution of the commitment the changing environment can lead to failure or success at any time. In that case the commitment will be ended.

In general, within a commitment three phases can be distinguished: *Setup*, *Loop* and *End*. During the setup and ending of a commitment, a robot is not executing a relational behaviour. The *logic machine* will not select any relational behaviour, so the commitment will be ignored during the primitive behaviour selection. Only in the *Loop* phase participants will select primitive behaviours concerning the commitment in order to achieve their joint goal.

2.2   Primitive Behaviour Selection

The selection of a primitive behaviour within the *Loop* phase of a commitment will be explained using the example of the pass commitment of Fig. 1. Three individual behaviours can be found there; **standBy** for both participants, **aimAnd-Pass** for the kicker and **intercept** for the receiver.

The pass commitment has been split up in several states from the beginning until the end, referred to as *commitment states*.

– *request* and *accept* in the *Setup* phase.
– *prepare* and *intercept* in the *Loop* phase.
– *done* and *failed* in the *End* phase.

In general, the states in the *Setup* and *End* phase will be the same for all commitments, only the *Loop* changes. Splitting the *Loop* phase in states allows synchronized execution of the pass. Here, each commitment state is linked to (a set of) primitive behaviours for both robots, see Table 1. When the commitment runs as planned, the pass states will be run through sequentially, from *request* until *done*. An error at any time can lead to the state *failed*. Note that pass states in the *Setup* and *End* phase do not lead to a primitive behaviour from the relational pass. Table 1 describes the pass commitment of Fig. 1. New commitments or other versions of commitments can be created under the same framework. One can change the individual behaviours that have been linked to the commitment states, or extend the *Loop* with more pass states and behaviours.

Table 1. Primitive behaviour selection by the *logic machine* in all pass states during the pass commitment

	Setup		Loop		End		Default
Commitment States:	*request*	*accept*	*prepare*	*intercept*	*done*	*failed*	*none*
Kicker	—	—	aimAndPass	standBy	—	—	—
Receiver	—	—	standBy	Intercept	—	—	—

2.3 Synchronization

To synchronize the behaviours, the participants will use explicit communication. Four variables, containing the identities of the participants and their commitment states, are kept in the agent's memory. Each of these four variables will be sent to the other participant in the relational behaviour when it is changed.

Following setup conditions, one agent sets a *pass request*, and a partner can enter the *accept* state. When the commitment has been established, the following rules will be looped:

- **Synchronize** commitment state with the state of the parter, if partner has moved on to a next state.
- **Switch** to new state by yourself, if predefined *switch conditions* allow that.
- Select a basic behaviour using Table 1.

This happens until the *done* or *failed* state is reached. Then the commitment will be finished and all variables will be cleared.

Synchronization is achieved at each moment an agent switches to a new commitment state, since the partner will always follow. By synchronizing commitment states each iteration, the commitment states of both agents can only be one step away from each other, and this difference will be corrected in the next loop. It is possible to let agents execute a sequence of primitive behaviours in one state. They will run asynchronously.

3 The Individual Behaviours

The primitive behaviours are running in a *control* module. Here the world situation is evaluated continuously, and motor commands will be send to the robot in order to influence the environment. The world model of the robot can be seen as a map with all identified objects in a xy-coordinate system. In the **standBy** behaviour, the robot will stay at the same position and will try to keep its front towards the ball position. This behaviour has been implemented earlier in the ISocRob team [5]. The other two behaviours, **aimAndPass** and **intercept**, had to be developed from scratch.

3.1   The AimAndPass Behaviour

When the pass commitment is started, the *kicker* has the ball. He wants to rotate with the ball to a certain direction and then shoot. This has been implemented in the **aimAndPass** behaviour.

In earlier research a controller has been developed to dribble with a ball to a goal posture [3]. A navigation algorithm described in [3] is used which takes the robot to a goal posture while avoiding obstacles, using a modified potential fields method that takes into account the non-holonomic nature of the robot. The motor commands given by this navigation algorithm are passed to a dribble filter which adapts them to the extra constraints of keeping the ball close. Parameters of the navigation algorithm can be changed to return a very strong angular acceleration towards the desired direction. Since the dribble filter stays with the dribble constraints, a controller is achieved which rotates with ball to the desired direction at the maximum turn angle.

3.2   The Intercept Behaviour

In order to intercept a moving ball, the **intercept** behaviour controller has been designed. The literature on visual servoing has solutions for similar problems [2], but only for a situation when the robot tracking the object is not moving within an environment cluttered with obstacles. Therefore, the solution described here is based on the previous mentioned navigation algorithm, [3], using a modified potential fields method. The intercept controller continuously estimates the interception point given the positions of ball and robot, and their predicted path. After a certain time t, position of ball and robot should be the same. The corresponding xy-coordinates indicate the interception point and they will be passed to the navigating algorithm. Interception point estimation is shown in Fig. 2.

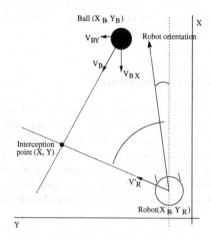

Fig. 2. Interception point calculation

Figure 2 shows a ball path estimation with position (x_b, y_b) and velocity (v_{bx}, v_{by}), and a robot with position (x_r, y_r). For the estimation, a constant velocity is assumed.

The robot path is estimated with the assumptions that the robot will move with a constant velocity in a straight line towards the interception point. This implies that the orientation of the robot will not change during the movement to the interception point and thus the initial orientation is ignored. An assumed average forward speed v_r' is used. Now, the robot path can be given as function of time, using the robot's current position (x_r, y_r), its absolute speed v_r' (given in the x-direction) and the orientation towards the interception point α, as shown in Figure 2. Since the interception point is still to be calculated, α is still unknown.

Variables α and t can be eliminated from respectively the x and y-component from both path estimations. The calculated t represents the time it takes for the ball and the robot to get to the interception point. By replacing t in the ball path estimation, xy-coordinates for the interception point are achieved.

Note that the assumption that the robot moves in a straight line to the interception point, may lead to an error in the estimation of the interception point, since the robots are non-holonomic. However, the estimation is iteratively applied and becomes more accurate at each new iteration step, until when the robot faces the correct heading, where the straight line motion assumption is fully correct.

4   Implementation and Results

4.1   Implementation of the Commitment

The theory for joint commitments has been implemented in Nomadic Super Scout II robots of the Robocup Middle-Sized League team ISocRob. The decision making and the synchronization have been implemented successfully. The framework for joint commitments requires the definition of the following aspects of a relational behaviour:

- List of commitment states
- Setup conditions of the commitment
- Switch conditions to switch between commitment states
- Related individual behaviours within each commitment state

The framework takes care of all communication and the synchronized execution of the relational behaviour.

For the implemented version of the pass commitment, mentioned in section 1, the following pass situation has been defined: a defender has the ball on its own half of the field and he will pass the ball to an attacker, who is on the other half of the field. The division in states and the related primitive behaviours are shown in Table 1. When the *kicker* has successfuly executed the aimAndPass behaviour, he switches to the *intercept* state. The *receiver* moves to the same state, and starts the intercept behaviour.

Note that in this case the *kicker* switches the pass state to *intercept*, and by doing so he tells the *receiver* to start the intercept behaviour. More reactive approaches of the pass commitment can define conditions that allow the *receiver* to switch to the *intercept* state, following his own observations.

The results of the execution of the pass behaviour are strongly related to the results of the individual behaviours. AimAndPass and intercept will be analysed in the following sections.

4.2   AimAndPass

The **aimAndPass** behaviour has been implemented following the methods described in previous sections. The performance of the behaviour is as expected. Figure 3(a) shows the result of a test with a real robot on a MSL soccer field. The robot start position is at the middle of the field, and his target is the middle of the goal which is located diagonally behind him. The robot turns with the ball and shoots in the desired direction. However, since the algorithm works with coordinates, errors can occur when the robot is badly localized. A small difference in the kicker orientation leads to a significant spatial error when the distance to the target grows. Those errors caused by localization can be avoided if the camera is used directly to determine the target direction.

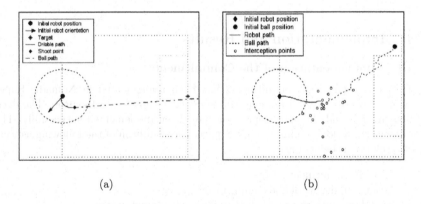

(a)   (b)

Fig. 3. Individual behaviour results: a) aimAndPass; b) intercept ball

4.3   Intercept

For the **intercept** behaviour, the assumed average robot speed, v_r', has been set to 0.5 m/s. An example of the interception of a moving ball by a real robot in a MSL soccer field is shown in Fig. 3(b). Clearly it can be seen that the robot takes the ball velocity into account, and moves directly to the interception point. In this example, the ball rolls with a speed of 1.2 m/s in a straight line. The ball path that is shown is the path as it is observed by the robot. Balls with a high velocity are likely to bump away after the interception.

5   Conclusions

In this paper the general formulation of relational behaviours among real robots, based on the Joint Commitment Theory, has been introduced through an illustrative example concerning a pass behaviour between RoboCup MSL robots. The formulation uses individual decision making and behaviour synchronization among intervening robots and has been tested successfully during laboratory games without an opponent. Results of the implementation of the key individual behaviours (**aimAndPass**, **intercept**) in real robots were presented.

Future work will concern the development of new relational behaviours under this framework, as well as the refining of the individual behaviours, particularly by using team mates visual recognition to eliminate the self-localization error and the required communication during the **aimAndPass** behaviour, and by providing the robots with force-controlled kicking ability, so as to enable ball interception by the receiver robot, by reducing the ball speed for pass kicks, as compared to goal kicks.

References

1. P. R. Cohen, H. J. Levesque, "Teamwork". *Nous*, Vol 35, (1991)
2. P. Corke, *Visual Control of Robots: High-performance Visual Servoing*, Research Studies Press LTD, (1996)
3. B. Damas, L. Custódio, P. Lima, "A Modified Potential Fields Method for Robot Navigation Applied to Dribbling in Robotic Soccer", *Proceedings of RoboCup 2002 Symposium*, (2003)
4. A. Drogoul, A. Collinot, "Applying an Agent-Oriented Methodology to the Design of Artificial Organizations: A Case Study in Robotic Soccer", *Autonomous Agents and Multi-Agent Systems*, Vol 1, (1998)
5. P. Lima, L. M. Custódio, et al., "ISocRob 2003: Team Description Paper", *Proceedings of the RoboCup 2003 Symposium*, (2003)
6. H. Matsubara, I. Noda and K. Hiraki, "Learning of Cooperative Actions in Multi-Agent Systems: a Case Study of Pass in Soccer.", *Adaptation, Coevolution and Learning in Multi-agent Systems: Papers from the AAAI-96 Spring Symposium*, (1996).
7. E. Pagello A. D'Angelo, F. Montsello, F. Garelli, C. Ferrari, "Cooperative Behaviors in Multi-Robot Systems Through Implicit Communication", *Robotics and Autonomous Systems*, Vol 29, (1999)
8. V. Pires, M. Arroz, L. Custódio, "Logic Based Hybrid Decision System for a Multi-robot Team", *Proceedings of 8th Conference on Intelligent Autonomous Systems*, (2004)
9. M. Tambe, "Towards Flexible Teamwork", *Journal of Artificial Intelligence Research*, Vol 7, (1997)
10. K.Yokota, K. Ozaki, N. Watanabe, A. Matsumoto, D. Koyama, T. Ishikawa, K. Kawabata, H. Kaetsu, and H. Asama, "Cooperative Team Play Based on Communication.", *Proceedings of RoboCup 1998 Symposium*, (1999)

Cooperative Planning and Plan Execution in Partially Observable Dynamic Domains

Gordon Fraser and Franz Wotawa

Graz University of Technology, Institute for Software Technology,
Inffeldgasse 16b/II, A-8010 Graz, Austria
{fraser, wotawa}@ist.tu-graz.ac.at

Abstract. In this paper we focus on plan execution in highly dynamic environments. Our plan execution procedure is part of a high-level planning system which controls the actions of our RoboCup team "Mostly Harmless". The used knowledge representation scheme is based on traditional STRIPS planning and qualitative reasoning principles. In contrast to other plan execution algorithms we introduce the concept of plan invariants which have to be fulfilled during the whole plan execution cycle. Plan invariants aid robots in detecting problems as early as possible. Moreover, we demonstrate how the approach can be used to achieve cooperative behavior.

1 Introduction

Artificial Intelligence (AI) planning techniques as decision making layer for autonomous agents acting in fast paced environments have traditionally been discriminated in favour of reactive approaches, such as behavior-based methods [1].

While undoubtedly being a very powerful technique applicable to countless problems, there are drawbacks with regard to computational complexity [2]. Decades of research on this matter have yielded numerous highly efficient planning algorithms. Unfortunately, the speed and advantages these algorithms offer is mostly dependent on certain properties of the domains they are applied to.

Considering such properties that distinguish the real world from toy domains used for planning research and experiments, the drawbacks of planning become even more imminent. Uncertainty, incomplete knowledge, the necessity to cooperate in multi-agents systems (MAS), limited resources, real-time requirements, etc. have all caused the creation of special planning extensions. Often, these extensions are incompatible among each other, usable only for the special cases they are forged for, and usually they aggravate the task of knowledge engineering and formulating tasks for human operators.

In this paper we present a framework that does not utilize such complex, special tailored planning algorithms or even reactive systems. Knowledge representation is done exclusively using first-order logic, uncertainty is intentionally neglected. Instead of optimizing plan creation, plan monitoring is focused upon. Plan invariants are presented as a means to efficiently monitor plan execution, and it is shown that efficient plan monitoring is sufficient for reactive behavior

D. Nardi et al. (Eds.): RoboCup 2004, LNAI 3276, pp. 524–531, 2005.

(assuming low-level tasks such as obstacle avoidance or path-planning are not performed by the planning system). It is further shown how they can be used to achieve cooperative behavior in a MAS consisting of independent planning agents, without extending the planning algorithm or the representation language.

Examples given throughout this paper use the RoboCup [3] Middle-Size league domain which features numerous qualities making AI planning a difficult task: it is a fast paced multi-agent environment hosting teams of four fully autonomous robots to compete in soccer games. All data are gathered by sensors the robots must bear themselves, and therefore much of the research in this league focuses on yet-to-solve problems concerned with vision and actuating. Processing power is limited as it is needed for other complex computation tasks (e.g. self localization, object tracking, etc.). Data are usually incomplete and uncertain. On the whole, the Middle-Size league offers a domain that comes very close to the real world.

2   Plan and Knowledge Representation

Following classical STRIPS [4] representation, a planning problem is defined as consisting of an initial state I, a goal description G and the domain theory A which describes the actions the agent can perform. STRIPS representation has seen plenty of useful extensions since it was first introduced. Applying some of these (e.g. quantification, disjunctions, etc.) allows the usage of first-order logic instead of plain propositional conjunctions as were used for original STRIPS planning (with the exception of disjunctions not being allowed for action-effect descriptions, as that would make actions non-deterministic).

2.1   Representation Language

State descriptions are given as logical sentences. Actions consist of a precondition and an effect, both again first-order logic sentences (without disjunctions for the effect). An agent also needs knowledge about the world and its assignment in it. Knowledge about the state of the world is drawn by observations that need to be mapped to predicates. In the presented framework, an agent's possible tasks are organized in a hierarchical way:

Strategy is shared by all agents of a multi-agent system, and consists of roles.
Role consists of a precondition and an invariant that are used to decide if a role is suitable. Furthermore a role contains a set of possible tasks an agent performing a role should pursuit.
Task is an extended planning problem. It consists of a precondition (initial state of planning problem), a goal description, and in addition an invariant that has to be true at all times during plan execution.

2.2   Example Situation

To illustrate the representation language, an example from the RoboCup domain is given. The situation in Figure 1 is described by using constants *Ball*, *OwnGoal*

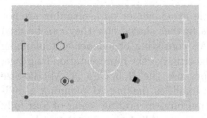

HasBall(*X*): Robot *X* is in ball possession
InReach(*X*, *Y*): Object *X* is in kicking range
of *Y*
Blocked(*X*): There is an obstacle between
the robot evaluating the statement and *X*
that prevents ball passing
AwaitingPass(*X*): *X* is awaiting a pass
At(*X*, *Y*): *X* is at position *Y*

Fig. 1. Example robotic soccer situation. Image is a screen-shot of Simsrv [5], a Middle-Size league simulator. Connected agents are running the implementation of the described planning framework

and *OpponentGoal* for own and opponent goal respectively, *Attacker* and *Helper* for the two octagonal robots, where the one close to the ball is assumed to be *Attacker*.

Attacker and *Helper* share the same strategy consisting of two roles, an offensive and an assisting one, having the preconditions *HasBall*(*Self*) and ¬*HasBall*(*Self*) respectively, where *Self* stands for the agent evaluating the statement. As *HasBall*(*Self*) is true for *Attacker*, we assume it chooses the offensive role while *Helper* chooses the assisting role. *Attacker* chooses the goal *At*(*Ball*, *OpponentGoal*), and its plan has an invariant ¬*Blocked*(*Goal*). *Helper* chooses the goal *InReach*(*Attacker*) ∧ ¬*Blocked*(*Attacker*).

2.3   Plan Invariants

The actual planning problem that has to be solved by a planning algorithm is created by choosing a task. The initial state I is determined by the agent's observations that have to fulfill the task's precondition in order for the task to be selected. The goal description G is contained in the task, and a plan p is calculated by a planning algorithm using the domain theory A. In contrast to classical planning we extend the classical planning problem definition by plan invariants.

An extended plan for a planning problem (I, G, A) is a tuple (p, inv), where p is a solution plan for (I, G, A) and inv is a logical sentence which states the plan invariant. The invariant inv is a manually defined logical sentence that has to hold in the initial and all subsequent states. Previously, invariants have been used for planning in order to cut down the size of the search space and thus speeding up planning ([6]) However, such invariants are pure domain constraints and unlike the plan invariants used in this context.

Section 4 shows examples of how advantages can be gained by using plan invariants. Plan invariants are unlike plan preconditions in that the truth value of the invariant is not changed by applying operators. The precondition, on the other hand, is more like a trigger that describes the state the world has to be in, in order for a plan to start, and this state of the world needs to be altered in order to reach a goal.

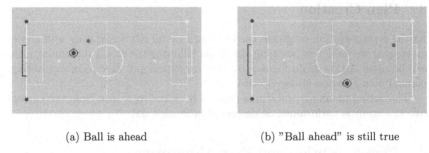

(a) Ball is ahead   (b) "Ball ahead" is still true

Fig. 2. Qualitative Reasoning, uncertainty does not need to be handled

2.4   Uncertainty and Nescience

Data collected by robots acting in real world environments are very error prone, assuming realistic, affordable equipment. It has been argued [7] that uncertainty and stochastic methods are necessary to process such uncertain data. However, we claim that certainty is not a requirement for decision making. Hence, choosing a sufficiently abstract vocabulary for the representation language the aspect of uncertainty is of secondary importance. The remaining problem is to find a sufficient level of abstraction, and moreover how to be sure that it is sufficient.

As an example, consider the RoboCup environment. Using numerical comparisons for distance measures as predicates is inseparably bound to uncertainty. In order to make a decision, however, it is sufficient to review the relative disposition of objects. Is the ball behind or in front of the robot (Figure 2)? Is a teammate to the left or to the right? At this level of abstraction it is irrelevant if a distance measure is incorrect. This idea is adhered to by the research field of qualitative reasoning [8]. Of course, a resulting binary decision might be wrong, especially in border cases caused by impreciseness. Even so, this risk is justifiable. If a decision is indeed wrong, plan execution will fail, and quickly detecting failure is exactly what the presented framework focuses on.

Finally, as a simple counter-measure to incomplete knowledge a new predicate is defined in order to make nescience explicit. Often a robot is not able to observe all of its environment, and so data necessary to calculate the truth value of some predicates might be lacking. For example, if a RoboCup robot does not actually see the ball it cannot tell anything about dispositions of the ball with regard to other objects. A predicate $Unknown(Ball)$ changes its truth value with to true. Such a predicate is a very useful addition to a plan's invariant. For example, a plan for offensive soccer play might only be feasible as long as the ball is observed, or a plan to assist a teammate might only be feasible as long as the teammate is seen.

On the whole this makes the representation language very straight-forward yet powerful and flexible.

3 Plan Creation

As the representation language used is based on STRIPS and only well-known extensions are used, any planning algorithm following the classical planning ideas can be used. As the main focus is on plan execution and execution monitoring, manual plan creation is also possible.

The actual planning problem is created by selecting a task from the current role the agent is performing, as described in Section 2.3

4 Plan Execution and Plan Monitoring

Assumptions made at plan creation time (e.g., atomic actions that always succeed) cannot hold in the real world, where the plan is executed. Actions fail, other agents do interact with the world, and plans are likely to fail. However, failure is not the only reason for a plan to be invalid. Possible reasons that invalidate a plan are:

Inexecutable Actions: It might not be possible to execute actions at any point of the plan execution. This is detected when trying to execute the action, as its precondition is not fulfilled.

However, an action might be inexecutable also at later points in the plan. For example, a soccer robot might be busy with communication or some movement action, while at a later point in its plan there is a kicking-action that cannot be executed because the robot has lost the ball. If the current action does not check ball possession with its precondition, the plan is executed up to the kick action where the failure is detected. Adding a predicate *HasBall* that is true if the agent is in ball possession to the plan invariant, immediately invalidates the plan as soon as the agent looses the ball, and enables it to quickly react and re-plan.

Failed Actions: Actions might fail to achieve the intended effect. This can easily be detected by validating the action's effect description with the world state after the execution is finished.

Unreachable Goal: Even if all actions of a plan are executable, i.e. their preconditions are fulfilled (unless they depend on other action's effects), a goal might not be reachable. This is caused by exogenous events which are not considered at the time of plan creation.

For example, in a RoboCup soccer game a referee might call charging. Clearly it would not be feasible to include the requirement of charging not being called in all actions' preconditions, besides for other plans the same actions could still be executed.

As another example, there are other agents that interact with the same environment. Any of these other agents might at any time invalidate effects of previously executed actions, thus making the goal unreachable.

Unfeasible Goal: Even though all actions are executable and the goal reachable, a plan might not be necessarily feasible. In a dynamic domain the objectives of an agent might quickly change. This becomes especially apparent when considering multi-agent interaction, as described in Section 5.

Clearly, plan execution has to be monitored to ensure plan validity and quick reaction to unexpected changes. During execution of an action, its precondition is checked, and only if it is fulfilled the action is executed. If an action's effect is fulfilled before executing the action, the action is skipped and plan execution is continued with the next action. The plan invariant has to be monitored at all times of plan execution. If the invariant is invalidated, an action fails or is not executable, the agent cannot continue to execute the current plan. Plan reparation or re-planning needs to be performed.

Figure 3 shows the robots from the example given in Figure 1 after executing their plans for a while. At some point during execution, *Attacker* detects that it cannot fulfill its plan as the current plan invariant is no longer valid (*Blocked(Goal)* becomes true). It has to consider a new task.

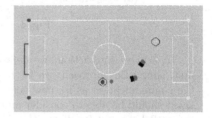

Fig. 3. Invalidated plan invariant

5   Cooperative Planning

This section shows how to achieve cooperation with simple means with a focus on monitoring, it is not meant as a fully fledged Distributed AI framework.

The first step towards cooperative behavior is determined by the concept of strategies and roles as explained earlier. A strategy implicitly defines a team's overall goals. Roles define the agents' tasks in a way to enable cooperative behavior. For example, in the RoboCup domain, subsidiary roles could be attacker and assistant player. Preconditions and invariants assigned to these roles are defined so that assuming all agents share the identical view of the world no two agents choose the same role. However, it is not very likely that robots acting in a real world environment share the same view. Therefore each role has a utility function. When an agent picks a role it broadcasts this decision together with the utility value. If another agent chooses the same role but has a higher utility, the other agent has to choose a new role. This approach for role coordination has already successfully been used for robotic soccer teams in the RoboCup Middle-Size league as shown in [9, 10].

Complementary roles let the agents choose such goals that might depend on each other. For example, the RoboCup assistant robot might try to stay close enough to the attacking robot in order to be able to receive a pass at any time. A plan invariant for this plan could be the requirement that the attacker is in ball possession (*HasBall(X)*). If the attacker looses the ball, the invariant of the assistant's plan becomes invalid and the assistant has to choose a new task.

This consideration reveals the need for such predicates that describe what other cooperating agents are doing, and then include these predicates to plan invariants to monitor cooperating agents. In a RoboCup scenario, such predicates could be *HasBall(X)* to indicate that *X* has the ball or *AwaitingPass(Y)* to indicate that *Y* is awaiting a pass.

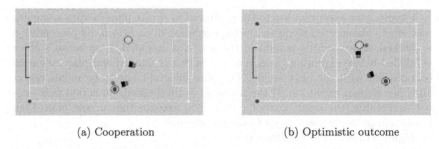

(a) Cooperation   (b) Optimistic outcome

Fig. 4. Continued example

To achieve interaction, we define communication as regular actions. As a first, simplistic approach, we let the queried agent consider the fulfillment of the communication as its new goal. Of course this approach is very crude, an agent should be able to consider multiple goals, perform communication actions in parallel etc. but for illustration it is sufficient.

As an example, the action $RequestAwaitPass(X)$ requests agent X to await a pass, hence the effect of this action is $HasBall(X)$. The agent performing the request waits until agent X is done with its plan to fulfill the request. This might cause two problems: firstly, the agent with the ball might be threatened by an opponent (thus invalidating the plan), and protecting the ball from the opponent by moving away might be the preferable solution. To cover this case, the plan invariant needs to check if waiting is safe. The second problem is that the other agent might fail to achieve the requested result, e.g. as communication might not be reliable. Again, a plan invariant could be used to verify that the other agent is actually trying to help.

Figure 4 concludes the example started in Figure 1, illustrating execution monitoring and communication as described in the preceding sections. *Attacker* and *Helper* are two robots of the same team, where *Attacker* has chosen an offensive role, and *Helper* a supportive one. In Figure 3, *Attacker*'s current plan invariant was invalidated. As a result, a new plan was calculated: RequestAwait-Pass(Helper), Pass(Helper), Goto(ScorePosition), AwaitPass(Helper), Kic(Goal).

Executing the request-action tells *Helper* to choose the goal:

$InReach(Attacker) \land \neg Blocked(Attacker)$.

Figure 4(a) shows that *Helper* executes a plan to fulfill the goal requested by *Attacker*. At the end of the plan execution, the effect of *Attacker*'s request-action becomes true, and the *Pass* action can be performed. As soon as *Helper* has the ball, it chooses a new goal:

$\neg Blocked(Attacker) \land \neg Blocked(Goal, Attacker) \land$

$Inreach(Goal, Attacker) \land Inreach(Attacker)$, while *Attacker* continues executing its old plan. Finally, assuming quite dumb opponents for the sake of the example, *Attacker* and *Helper* are ready to pass and score a goal. (Figure 4(b))

6   Conclusion and Further Research

We have presented an integrated framework for plan creation, execution and monitoring based on traditional STRIPS planning and qualitative reasoning principles. The advantages offered by such a framework have been outlined. Plan invariants have been introduced as a means to detect execution problems as early as possible. The application of communication sketched in Section 5 underlines the usefulness of plan invariants, while further research is necessary to improve cooperation through communication.

Since first-order logic is used as representation language and no special-tailored extensions are used, the agent programming (i.e. defining goals, roles, etc.) is very straight-forward and intuitive. Applying the framework to other domains is a mere matter of defining new predicates and constants to be used for action, goal and role descriptions. Currently, we are investigating the application of our soccer robots in an office environment to perform service tasks.

References

1. Brooks, R.A.: Intelligence Without Reason. In Myopoulos, J., Reiter, R., eds.: Proceedings of the 12th International Joint Conference on Artificial Intelligence (IJCAI-91), Sydney, Australia, Morgan Kaufmann publishers Inc.: San Mateo, CA, USA (1991) 569–595
2. Bylander, T.: The Computational Complexity of Propositional STRIPS Planning. Artificial Intelligence **69** (1994) 165–204
3. Kitano, H., Asada, M., Kuniyoshi, Y., Noda, I., Osawa, E.: RoboCup: The Robot World Cup Initiative. In Johnson, W.L., Hayes-Roth, B., eds.: Proceedings of the First International Conference on Autonomous Agents (Agents'97), New York, ACM Press (1997) 340–347
4. R.E.Fikes, Nilsson, N.J.: STRIPS: A New Approach to the Application of Theorem Proving to Problem Solving. Artificial Intelligence **2** (1972) 189–208
5. Kleiner, A., Buchheim, T.: A Plugin-Based Architecture For Simulation In The F2000 League (2003)
6. Kautz, H.A., Selman, B.: The Role of Domain-Specific Knowledge in the Planning as Satisfiability Framework. In: Artificial Intelligence Planning Systems. (1998) 181–189
7. Collins, G., Pryor, L.: Planning under uncertainty: Some key issues. In: Proceedings of the 14th International Joint Conference on Artificial Intelligence (IJCAI). (1995) 1567–1573
8. Weld, D., de Kleer, J., eds.: Readings in Qualitative Reasoning about Physical Systems. Morgan Kaufmann (1989)
9. Castelpietra, C., Iocchi, L., Nardi, D., Piaggio, M., Scalzo, A., Sgorbissa, A.: Communication and Coordination among Heterogeneous Mid-size Players: ART99. In: RoboCup 2001: Robot Soccer World Cup V. Volume 2019 of LNCS., Springer (2002) 86–95
10. Weigel, T., Auerbach, W., Dietl, M., Dümler, B., Gutmann, J.S., Marko, K., Müller, K., Nebel, B., Szerbakowski, B., Thiel, M.: CS Freiburg: Doing the Right Thing in a Group. In: RoboCup 2001: Robot Soccer World Cup V. Volume 2019 of LNCS., Springer (2002)

Exploring Auction Mechanisms for Role Assignment in Teams of Autonomous Robots

Vanessa Frias-Martinez[1], Elizabeth Sklar[1], and Simon Parsons[2]

[1] Department of Computer Science,
Columbia University,
1214 Amsterdam Avenue, New York, NY 10027, USA
{vf2001, sklar}@cs.columbia.edu
[2] Department of Computer and Information Science,
Brooklyn College, City University of New York,
2900 Bedford Avenue, Brooklyn, NY 11210, USA
parsons@sci.brooklyn.cuny.edu

Abstract. We are exploring the use of auction mechanisms to assign roles within a team of agents operating in a dynamic environment. Depending on the degree of collaboration between the agents and the specific auction policies employed, we can obtain varying combinations of role assignments that can affect both the speed and the quality of task execution. In order to examine this extremely large set of combinations, we have developed a theoretical framework and an environment in which to experiment and evaluate the various options in policies and levels of collaboration. This paper describes our framework and experimental environment. We present results from examining a set of representative policies within our test domain — a high-level simulation of the RoboCup four-legged league soccer environment.

1 Introduction

Multi agent research has recently made significant progress in constructing teams of agents that act autonomously in the pursuit of common goals [12, 15]. In a multi agent team, each agent can function independently or can communicate and collaborate with its teammates. When collaborating, the notion of *role assignment* is used as a means of distributing tasks amongst team members by associating certain tasks with particular roles. The assignment of roles can be determined *a priori* or can change dynamically during the course of team operation.

Collaboration enables a team of agents to work together to address problems of greater complexity than those addressed by agents operating independently. In general, using multiple robots is often suggested to have several advantages over using a single robot [4, 7]. For example, [11] describes how a group of robots can perform a set of tasks better than a single robot. Furthermore, a team of robots can localize themselves better when they share information about their environment [7]. But collaboration in a team of robots may also add undesirable delays through the communication of information between the agents.

D. Nardi et al. (Eds.): RoboCup 2004, LNAI 3276, pp. 532–539, 2005.
© Springer-Verlag Berlin Heidelberg 2005

We are exploring — within dynamic, multi-robot environments — the use of auction mechanisms to assign roles to agents dynamically and the effect of different approaches to collaboration within the team. In order to evaluate this set, we have developed a theoretical framework and a simulation environment. The theoretical framework helps us to identify the space of possibilities, and the simulation environment helps us to evaluate the various degrees of collaboration.

This paper begins by highlighting some background material on auctions and the use of auction mechanisms in multi agent systems. Then we describe our theoretical framework. Next we detail our experimental environment — a high-level simulation of the RoboCup Four-Legged Soccer League. We then present results of simulation experiments evaluating both collaborative and non-collaborative models of information sharing as well as various auction policies. Finally, we close with a brief discussion and directions for future work.

2   Auctions

Following Friedman [9], we can consider an *auction* to be a mechanism that regulates how commodities are exchanged by agents operating in a multi agent environment. An *auction mechanism* defines how the exchange takes place. It does this by laying down rules about what the traders can do — what *messages* they can exchange in an interaction — and rules for how the allocation of commodities is made given the actions of the traders. Auctions have been used in different environments for resource allocation, such as electronic institutions [6], distributed planning of routes [13] or assignment of roles to a set of robots to complete a common task [10].

3   Theoretical Framework

In our auction, there are two types of agents: the *auctioneer* and the trader — a *player* in the RoboCup soccer game. The player makes an *offer* and the auctioneer's job is to coordinate the offers from all the players and perform role assignment. There are five main components to our model.

First, we define $\mathcal{R}$ to be the set of possible roles: $\mathcal{R} = \{PA, OS, DS\}$, where PA is a primary attacker, OS is an offensive supporter, and DS is a defensive supporter. Note that the goalie is not considered a role to be assigned in this manner, since it cannot change during the course of the game.

Next, we define $\mathcal{P}$ to be a set of player attributes: $\mathcal{P} = \{d_{ball}, d_{goals}, d_{mates}, d_{opps}\}$ where d_{ball} contains the distance from the player (who is making the offer) to the ball; d_{goals} contains the distance from the player to each goal; d_{mates} contains the distance from the player to each of its teammates; and d_{opps} contains the distance from the player to each player on the opposing team.

Third, we define F to be a set of functions which define the method for sharing perception information between agents. This information could be shared with teammates, the auctioneer, or both. Fourth, we define $\mathcal{M}$ to be a *matching function*, the method used by the auctioneer for clearing the auction, i.e., matching the offers with roles. In other words, the matching function captures the coordination strategy. Finally, we define an *auction*, $\mathcal{A}$, to be: $\mathcal{A} = \langle P, R, M, f \rangle$ where $P \subseteq \mathcal{P}$ and $P \neq \emptyset$; $R \subseteq \mathcal{R}$ and $R \neq \emptyset$; $M \subseteq \mathcal{M}$ and $M \neq \emptyset$; and $f \in F$.

Our work is systematically exploring the space of all possible auctions $\mathcal{P} \times \mathcal{R} \times \mathcal{M} \times F$. $\mathcal{B}$ denotes the set of possible types of offers in a particular auction, $A \in \mathcal{A}$: $\mathcal{B} = \{\mathbf{r}, \mathbf{w}\}$ where: $\mathbf{r} \subseteq R$ is a set of roles for which the player bids; $\mathbf{w}$ is a set of real-valued weights, one weight corresponding to each of the roles in $\mathbf{r}$ (a weight of 0 means that the player is not interested in making an offer for the corresponding role); and $f(p)$, $p \subseteq P$, is the mechanism by which perceptual data is used to determine $\mathbf{r}$ and $\mathbf{w}$.

To date, we have defined two different types of auctions within this framework — a *simple* auction [5] and a *combinatorial* auction [3]. We can define a simple auction $b_t \in \mathcal{B}$ as: $b_t = \{r, w\}$, where the role r and w are singletons(unique offer). And a *combinatorial auction*, is defined as: $b_t = \{(r_0, r_1, r_2), (w_0, w_1, w_2)\}$ where r_i and w_j are singletons. Using different combinations of weights allows the agent to bid for different combinations of roles, and this makes the auction combinatorial [1].

4   SimRob: Our Simulated Approach to a RoboCup Game

We are using RePast[14] to implement our environment. RePast allows us to build a simulation as a state machine in which all the changes to the state machine occur through a schedule. In order to model a RoboCup soccer game in RePast, we need to define the agents, the environment and the state machine that RePast will execute at each scheduled *tick*, i.e., simulated time step.

4.1   Agent Parameters

The RoboCup Four-Legged League environment has four Sony AIBO robots per team and a bright orange ball. Each one of the robotic agents is associated with an array containing the values that define their perception and localization:

$$(x, y, \phi, d_{ball}, d_{goals}, d_{opps}, d_{mates}, b_{ball}, b_{goals}, b_{opps}, b_{mates}) \tag{1}$$

where (x, y) are the 2D coordinates of the robot on the field[1]; ϕ is the orientation of the robot[2]; d_{ball} is the distance from the robot to the ball, d_{goals} is the distance from the robot to each goal, d_{opps} is an array containing the distance from the robot to each opponent, and d_{mates} is an array containing the distance from the robot to each teammate. The boolean values in the second half of equation (1) indicate if the ball has been detected by the player (b_{ball}), if each goal has been detected by the player (b_{goals}), if each opponent has been detected nearby (b_{opps}) and if each teammate has been detected nearby (b_{mates}).

---

[1] The field itself is broken down into the same discretized grid that we use for localization on the AIBOs.

[2] The 360° of orientation are divided into eight 45° sections, numbered 0 through 7.

4.2   Simulation Skeleton

We use RePast in order to simulate the development of a game with the agents. At the beginning of the simulation, we define four agents (per team) and a ball in the field. Each of the agents is defined as explained above, by means of an array as in equation (1). The simulation run in RePast can be divided into the following steps:

(1a) Generation of the agent parameters In this first step, we obtain the parameters of each of the agents in the field. The localization of the robot is expressed with the coordinates (x, y) in a 2D field. We also obtain the distances to the ball d_{ball}, to the goal d_{goals} and to the opponents d_{opps}.

(1b) Amount of information shared by the agents The information shared by the agents is: $mingoal$, a boolean variable that is true when the agent is the one closest to the goal. This variable can be defined when the agents share the variable d_{goals} among them. $maxopp$ is a boolean variable that is true when the agent is farthest away from the opponents in the field. This variable can be defined when the agents share d_{opps}. And $maxball$ is a boolean variable that is true when the agent is farthest away from the ball. This value can be defined when the variable d_{ball} is shared among the agents.

(2a) Defining a bidding policy for the agents For each simulation tick of the game play, the agent's bid will be the role associated by the policy being tested to the set of perceptions gathered by the agent at that simulation tick.

(2b) Defining an auction policy for the auctioneer The auction is responsible for distributing the roles between the agents on the field. The auctioneer will go through the different roles in the bid until one of the roles in the array is assigned to the agent, meaning that the bid is won.

(3) Game Play Once the agent-roles are defined, we have to actually simulate the joint task to be developed by the agents. As stated before, our aim is that of simulating a soccer game. The game model is very simple. Each role has a state graph that will output a certain behavior depending on the perceptions gathered by the agent:

- PA BEHAVIOR: If the agent sees the goal and the ball, then it kicks the ball, otherwise it turns to look for the ball without losing track of the goal.
- OS BEHAVIOR: If the ball is seen, the agent kicks it.
- DS BEHAVIOR: If the ball is seen, the agent follows it in order to prevent an agent from the opposing team scoring.

Finally, if a goal is scored, the robots are sent back to their initial positions and the ball randomly changes location. Then, the three step (parameter generation, auction execution and game play) simulation is run again.

5   Experiments

This section describes our experimental work to date. We have started to explore the range of possible auctions and their effect on the coordination of a team, as measured

Table 1. Example non-collaborative simple(S) and combinatorial (C) auctions

Ball seen	Opponent seen	Mate seen	Role(S)	Role(C)
0	0	0	DS	[OS,.7,DS,.2,PA,.1]
0	0	1	DS	[OS,.7,DS,.2,PA,.1]
0	1	0	DS	[OS,.7,DS,.2,PA,.1]
0	1	1	DS	[OS,.7,DS,.2,PA,.1]
1	0	0	DS	[OS,.7,DS,.2,PA,.1]
1	0	1	DS	[OS,.7,DS,.2,PA,.1]
1	1	0	OS	[OS,.7,DS,.2,PA,.1]
1	1	1	OS	[OS,.7,DS,.2,PA,.1]

by their performance in simulated games. We have experimented with four very simple types of coordination and describe policies that we have used for experimentation, chosen somewhat *ad hoc*. In current work, we are learning policies [8].

5.1 Non-collaborative Simple Auction

This approach defines a team of agents that don't share any perception data. Hence, each one relies on the information that it gathers independently of the others. The offers made by the agents follow the policy in Table 1 column Role(S). This shows that we have defined the agent to offer to be OS when both ball and opponent are seen. In any other case, our agent will offer to be DS. We have chosen a simple matching policy that just associates a fixed role to each of the possible sets of perceptions.

5.2 Non-collaborative Combinatorial Auction

In this case there is still no sharing of perception, but the bid now contains a vector defining the agent's role preferences For our experiments, we have defined two different bidding policies. The *offensive* policy, defined in Table 1, column Role(C), represents a team with an attacking approach, always looking for the goal and aiming to score. The other policy is more defensive. The offensive policy assigns the array of roles [DS,.7,OS,.2,PA,.1] to each of the agents. The matching is the same as before.

5.3 Collaborative Simple Auction

In this case, the agents share all the perception data. Hence, when defining the bids, we can also share the three variables related to the minimum and maximum distances to the ball, opponents and goal. The table defining the bidding policy is huge. In Table 2, column Role(S), we show a few lines to give the sense of it, but it is deliberately similar to the policy for the non-collaborative auction to give a reasonable comparison. When no elements are seen by any of the agents, the agent bids for the role DS. When everything is seen and the distances are minimum, the agents bid to be OS. The matching policy is also the same as for the non-collaborative examples.

Table 2. Collaborative simple(S) and combinatorial(C) auctions

Ball seen	Opp seen	Mate seen	MinGoal	MaxOpp	MaxBall	Role(S)	Role(C)
0	0	0	0	0	0	[DS]	[OS,.7,DS,.2,PA,.1]
0	0	0	0	0	1	[DS]	[OS,.7,DS,.2,PA,.1]
...	.	.	.	.	.	.	.
1	1	1	1	1	1	[OS]	[OS,.7,DS,.2,PA,.1]

Table 3. Results

Number of goals scored after 2000 ticks.

	unique		not unique	
	offensive bid	defensive bid	offensive bid	defensive bid
noncollab simple	16	–	16	–
noncollab comb	33	43	30	47
collab simple	40	–	67	–
collab comb	49	37	78	67

5.4   Collaborative Combinatorial Auction

Here the bidding Table 2, column Role(C), is similar to the previous one, but contains a vector of bids and weights instead of only one role, and this vector is like that for the non-collaborative combinatorial auction. Again we ran experiments with an attacking bidding policy and a defensive bidding policy, and the matching table is the one used in the previous examples.

5.5   Results

Teams using each of the types of coordination described above (including separate offensive and defensive techniques in the combinatorial auction) were run in simulation against the same, simple, opponent in order to evaluate the effectiveness of the collaboration policy. The opposing team moved randomly around the field, but was not intended as serious opposition, rather it was intended as a baseline against which all mechanisms could be judged equally. For each coordination mechanism, we ran two sets of experiments. In one, the "unique" experiments, we made the auctioneer assign unique roles to agents. In the "not unique" experiments, the auctioneer was allowed to assign duplicate roles. The average number of goals scored for each of the different kinds of collaboration are given in Table 3, and plots of the goals scored over time for a sample game are given in Figures 1.

In the non unique approach, collaborative teams score almost double the number of goals of the non-collaborative teams. In the unique role approach, differences in the score of the games between the collaborative and non-collaborative approaches for both simple and combinational auctions are not so marked. This is due to the fact that our

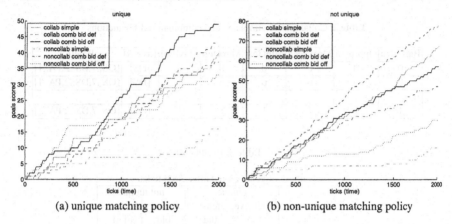

| (a) unique matching policy | (b) non-unique matching policy |

Fig. 1. Goals scored over the course of a game

matching policy is very demanding and since we do not allow repeated roles, the auctioneer often ends up distributing roles randomly. In order to prove this last assertion, we defined a parameter called *success ratio* in the simulation. The success ratio is associated with the acceptance of the bids made by an agent. The higher the ratio, the more times its bid has been accepted. In the not uniqueness experiments, we obtained very low ratios, meaning that the agents almost never won a bid, and so, the roles were distributed randomly.

6   Conclusions and Future Work

This paper has described our preliminary work in exploring the use of auction mechanisms to coordinate players on a RoboCup team. While this work is only just beginning, we believe that the results demonstrate the potential of the approach to capture a wide range of types of coordination, and to be able to demonstrate their effectiveness through simulation. In addition, this approach makes it simple to explore more complex, and potentially more flexible, kinds of role allocation than have been previously used in the legged-league, for example [2, 16].

Our longterm work is to build on this foundation and explore a wide range of possible auctions through simulation and on real (physical) robots. We are currently using learning techniques to automatically explore the space of auctions. We further intend to implement the most effective bidding and matching policies developed on our real Legged-League team.

Acknowlegements

This work was made possible by funding from NSF #REC-02-19347 and NSF #IIS 0329037.

References

1. C. Boutilier and H. H. Hoos. Bidding languages for combinatorial auctions. In *Proceedings of the 17th International Joint Conference on Artificial Intelligence*, pages 1211–1217, San Francisco, CA, 2001. Morgan Kaufmann.

2. D. Cohen, Y. Hua, and P. Vernaza. The University of Pennsylvania Robocup 2003 Legged Soccer Team. In *Proceedings of the RoboCup Symposium*, 2003.

3. S. de Vries and R. Vohra. Combinatorial auctions: A survey. *INFORMS Journal of Computing*, (to appear).

4. G. Dudek, M. Jenkin, E. Emilios, and D.Wilkes. A taxonomy for multi-agent robotics. *Autonomous Robots*, 3(4), 1996.

5. R. Engelbrecht-Wiggans. Auctions and bidding models: A survey. *Management Science*, 26:119–142, 1980.

6. M. Esteva and J. Padget. Auctions without auctioneers: distributed auction protocols. In *Agent-mediated Electronic Commerce II, LNAI 1788*, pages 20–28. Springer-Verlag, 2000.

7. D. Fox, W. Burgard, H. Kruppa, and S. Thrun. Collaborative multi-robot localization. In *Proceedings of the 23rd German Conference on Artificial Intelligence*. Springer-Verlag, 1999.

8. V. Frias-Martinez and E. Sklar. A team-based co-evolutionary approach to multi agent learning. In *Proceedings of the 2004 AAMAS Workshop on Learning and Evolution in Agent Based Systems*, 2004.

9. D. Friedman. The double auction institution: A survey. In D. Friedman and J. Rust, editors, *The Double Auction Market: Institutions, Theories and Evidence*, Santa Fe Institute Studies in the Sciences of Complexity, chapter 1, pages 3–25. Perseus Publishing, Cambridge, MA, 1993.

10. B. Gerkey and M. Mataric. Sold!: Auction methods for multirobot coordination. *IEEE Transactions on Robotics and Automation*, 2000.

11. D. Guzzoni, A. Cheyer, L. Juli, and K. Konolige. Many robots make short work. *AI Magazine*, 18(1):55–64, 1997.

12. G. A. Kaminka, D. V. Pynadath, and M. Tambe. Monitoring deployed agent teams. In Jörg P. Müller, Elisabeth Andre, Sandip Sen, and Claude Frasson, editors, *Proceedings of the Fifth International Conference on Autonomous Agents*, pages 308–315. ACM Press, 2001.

13. T. L. Lenox, T. R. Payne, S. Hahn, M. Lewis, and K. Sycara. Agent-based aiding for individual and team planning tasks. In *Proceedings of IEA 2000/HFES 2000 Congress*, 2000.

14. Repast. http://repast.sourceforge.net.

15. M. Tambe. Towards flexible teamwork. *Journal of Artificial Intelligence Research*, 7:83–124, 1997.

16. M. Veloso and S. Lenser. CMPAck-02:CMU's Legged Robot Soccer Team. In *Proceedings of the RoboCup Symposium*, 2002.

A Descriptive Language for Flexible and Robust Object Recognition

Nathan Lovell and Vladimir Estivill-Castro

School of CIT, Griffith University, Nathan 4111 QLD, Australia

Abstract. Object recognition systems contain a large amount of highly specific knowledge tailored to the objects in the domain of interest. Not only does the system require information for each object in the recognition process, it may require entirely different vision processing techniques. Generic programming for vision processing tasks is hard since systems on-board a mobile robots have strong performance requirements. Such issues as keeping up with incoming frames from a camera limit the layers of abstraction that can be applied. This results in software that is customized to the domain at hand, that is difficult to port to other applications and that is not particularly robust to changes in the visual environment.

In this paper we describe a high level object definition language that removes the domain specific knowledge from the implementation of the object recognition system. The language has features of object-orientation and logic, being more declarative and less imperative. We present an implementation of the language efficient enough to be used on a Sony AIBO in the Robocup Four-Legged league competition and several illustrations of its use to rapidly adjust to new environments through quickly crafted object definitions.

1   Introduction

Most object recognition systems use hard-coded, domain specific knowledge. For example, all leagues in Robocup rely on the ball being orange and spherical. If it were changed to be a non-uniform colour or a non-uniform shape then most object recognition systems would have to be largely re-coded[1]. We saw how devastating this was in last years challenge in the four-legged league which required the robots to locate a black and white ball where only eight of the twenty-four teams managed to even identify the ball and no team passed the challenge[2].

This paper presents a higher level descriptive language of the objects we are likely to recognise in a particular domain and thus leave the underlying vision

---

[1] This is certainly true for the team Griffith 2003 code. There are many other examples of systems that are programmed in this way because they are based on the vision algorithms developed by Carnegie Mellon University[1] for the year 2000 competition.

[2] http://www.openr.org/robocup/challenge2003/Challenge2003_result.html

D. Nardi et al. (Eds.): RoboCup 2004, LNAI 3276, pp. 540–547, 2005.

system programming unchanged. We claim that the approach allows rapid expansion to newer or changing vision domains. The case study for this work is the Sony AIBO platform in the Robocup Four Legged league. By modern standards in vision processing systems, these inexpensive robots are not computationally rich. We will present both the high-level descriptive language called XOD (XML Object Description) and the techniques used to translate it to C++ code. We will also present several illustrations where we have used our system to recongise vastly changed objects on the soccer field.

2   Related Work

There have been a considerable number of studies on different methods for representing objects generically in such a way that they can be located within target images. Much of this work focuses on trying to learn object representations from sample images and, as such, the generic representation usually takes the form of a statistical model [2] or some subspace representation of the features of the object [3]. These methods rely on supervised learning techniques. While these systems do function reasonably well in variant lighting conditions and poses of objects, they are still not easily applied to the task of mobile robotics. They work well in situations where the object, though possibly unknown, is in a controlled environment within the image (such as when the background of the image is known and the object features, therefore, can be easily extracted) but are less tolerant to object occlusion and background uncertainty.

Another closely related problem, which our work also addresses, is that of how to combine and apply existing vision processing techniques to the object recognition task. For example, some object A may require that edges be extracted and the texture of the internal pixels analysed for identification while another object B is more easily identified by colour segmentation and connected region analysis. Draper [4] proposes that the task of object recognition is a goal driven task and the user of the system should not need to specify which combination of vision processing techniques are applicable on a per-object basis. Instead Draper proposes that the vision processing tasks themselves can be treated as primitives and their correct combination and application learned by the system for each object, again by supervised learning techniques. Our work addresses this problem by the static transformation of user-created object descriptions into a vision processing pipeline for each object which will use only applicable techniques according to the description.

The difference between the problems discussed above and the problem addressed by this paper is that in the field of mobile robotics one usually has the added difficulty of determining *if* any of some known set of objects is present in an image and, if so, *where*. Our work simplifies the task of describing the objects - our system does not learn object descriptions but rather they are given to us by the user. The end of this is similar to other systems in that the domain specific knowledge is still removed from the vision processing module and the details of the vision processing itself are not needed to be understood by the user. The advantage of this approach, however, is that our system allows for quick and robust adaptation to varying visual domains with no prior sample images. Supervised machine learning techniques require large databases of classified samples and substantial amounts of processing time, neither of which is necessarily available in the field of mobile robotics.

The idea of codifying the domain specific knowledge in a goal driven and code independent manner is not new. The German team in Robocup 2003 [5] presented an XML based specification language for agent behaviors. Our paper presents a similar system applied to the task of visual object recognition.

3   The XOD

XOD stands for "XML Object Description" and it is the first version of a language used to describe the objects we expect our robot to see in a code-independent way. The language works with several types of primitives - points, lines, blobs and objects - as well as collections of these primitives. When we formalise the language as a logic where these primitives are the terms of the logic. The relationships between the primitives of our language become the predicates of the logic. We also take advantage of the overlaps between object-orientation and knowledge representation in AI to allow our primitives to have properties defined on them. We now define what we mean by these primitives.

A *point* is used to represent a pixel. It is not the responsibility of the vision system to convert items from pixel into world coordinates so a point's location (x, y) in an image is one of its properties. We can use points to specify properties of other objects such as line intersections or centers of objects.

A *line* is simply a collection of connected points, for example a connected border between two different colours. There are two stages of processing on lines - edge detection and vectorisation. It is sometimes useful to represent lines in raster form and sometimes useful to represent them in vector form so our language allows descriptions for both representations. Our language operates on the properties of lines and the relationships between lines and blobs to find objects. Note that a property of a line could also be an object (as in the line's first point).

A *blob* represents a bounding rectangle on a connected set of similarly coloured pixels as well as some other properties of these pixels. Our language also operates on the properties and relationships between blobs to find objects. Both the definition and implementation of blobs in our system are a little different to that of other vision systems[1]. Blobs are often defined as a connected set of similarly coloured pixels and implemented with tree-based union-find operations. We have found that simply maintaining bounding rectangles with some other useful properties allows for faster performance with no penalty on blob identification.

Thus an *object* is a named entity in our language. It can be a blob, a line or a point or a set thereof. Objects are made externally available for post-processing by other modules on the robot. We may, for example, have an

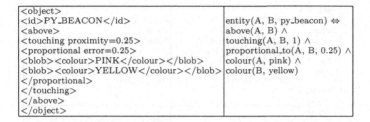

| <object>
<id>PY_BEACON</id>
<above>
<touching proximity=0.25>
<proportional error=0.25>
<blob><colour>PINK</colour></blob>
<blob><colour>YELLOW</colour></blob>
</proportional>
</touching>
</above>
</object> | entity(A, B, py_beacon) ⇔
above(A, B) ∧
touching(A, B, 1) ∧
proportional_to(A, B, 0.25) ∧
colour(A, pink) ∧
colour(B, yellow) |

Fig. 1. A simple XOD representation and its logical equivalence for a pink/yellow beacon in RoboCup Four-Legged League

object called *ball* or we may also have an object called *field_edge* which is actually a set of vectorised lines representing the edge of the field.

3.1 Declarative Elements

The task of XOD is to represent objects of interest in a way that allows C++ code to be automatically generated for the task of locating the objects within an image if they are present. We use an XML based language because of its transportability and readability by humans and machines (but the language can be represented as logic clauses for even more readability, see illustrations and figures). The main construct in this language is the tag <object> which encapsulates the definition of an object. All objects have a property of a name indicated by the <id> tag. Many more features or properties are possible a la Object-Orientation or Frames.

To illustrate the language assume now that the object the vision system will attempt to recognise is to be identified via the use of colour (a common case in Robocup), and not by lines or edges. Then the object will be encapsulated (circumscribed/bounded) by a blob (if it is a single colour) or encapsulated by two or more overlapping blobs of different colours. Our language allows us to search for interesting blobs by both their properties (for example the <colour> tag) and the relationships between blobs (<above>, <in_front>, <touching>, etc) . For example, the XOD in Figure 1 could be used to locate a pink blob touching the top of and proportional to a yellow blob, useful if we were looking for the pink on yellow beacon in RoboCup Four-Legged League.

XOD also allows limitations on the blobs that we use based on the properties of the blob. For example if we were looking for a large orange blob that has at least 50% of the pixels within the bounding rectangle of the correct colour then we might write something like the XOD in Figure 2. Figure 2 also illustrates how the language will permit us to check relationships (or apply predicates) with other, previously defined objects such as the field edge. The truth value of these predicates are determined differently depending on the type of object being used. For example, the truth value of *below(A, FIELD_EDGE)* will need to be calculated differently depending whether *FIELD_EDGE* is an edge set or

`<object>`
`<id>FIELD_EDGE</id>`
`<edge>`
`<source>FIELD_BLOB</source>`
`<colour>WHITE,GREEN</colour>`
`<colour>YELLOW,GREEN</colour>`
`<colour>BLUE,GREEN</colour>`
`<vectorise>line</vectorise>`
`</edge>`
`</object>`
	entity(A, field_edge) $\Leftrightarrow$
	$a \in A \land$
	bounded_by(a, field_blob) $\land$
	[between_colours(a,white,green) $\lor$
	between_colours(a,yellow,green) $\lor$
	between_colours(a,blue,green)] $\land$
	straight_line(a)

`<object>`
`<id>CLOSE_BALL<id>`
`<below>`
`<blob>`
`<colour>ORANGE</colour>`
`<area op=gt>5000</area>`
`<pixels op=gt>2500</pixels>`
`</blob>`
`<object>FIELD_EDGE</object>`
`</below>`
`</object>`
	entity(B, close_ball) $\Leftrightarrow$
	below(B, field_edge) $\land$
	
	colour(B, orange) $\land$
	greater_than(area(B), 5000) $\land$
	greater_than(pixels(B), 2500)

Fig. 2. A simple XOD representation and its logical equivalence for a large, primarily orange blob below the field edge line

`<object>`
`<id>FAR_BALL</id>`
`<blob>`
`<colour>ORANGE</colour>`
`<pixels op=lt>500</pixels>`
`</blob>`
`<select>`
`<area op=gt></area>`
`</select>`
`</object>`
	entity(A, far_ball) $\Leftrightarrow$
	
	colour(A, orange) $\land$
	less_than(pixels(A), 500) $\land$
	
	$\forall x$ greater_than(area(A), area(x)) $\land$
	colour(x, orange) $\land$ less_than(pixels(x), 500)

Fig. 3. As an approximation to finding a far orange ball, here is an XOD definition finding the largest orange blob less than 500 pixels with its equivalent logic

a blob. In Figure 2 it is implemented as a set of edges to illustrate that objects do not necessarily need to be formed from blobs.

3.2   Imperative Elements

Our language goes beyond declarative statements to simplify using quantifiers. This is illustrated when more than one blob in an image matches the criteria in an object definition. In this situation we have the option to specify how to select the correct one or to save the list for post-processing. We do this via the `<select>` tag. We may choose either to keep the entire set or to select one object according to some criteria. Figure 3 illustrates the example where we want to choose the largest orange blob with less than 500 pixels.

We can choose to select by any of the attributes of the blob or by comparing to some other named entity. For example, it is possible to select the closest red blob to the object identified already to be the yellow goal. In the four legged league this could be useful in identifying the red goal keeper. In the absence of

Fig. 4. The conceptual schematic pseudo-code of figure 1. The actual implementation has more efficient control flow

any select statement the default action is dependent on the type of object. If the object is a blob then the largest blob by area is selected however if the object is a point or an edge then the set is kept and the object remains a collection.

There are other imperative statements in the language used to modify or create objects rather than define or select them[3].

3.3   Implementation

Each XOD is converted to C++ code that performs the following three-step meta-procedure:

1. Build subsets of the universal blob set according to the properties specified in the <blob> section of the definition.
2. Apply predicates forming sets of composite blobs which represent intermediary and candidate objects.
3. Apply the operators according to the <select> statements.

Figure 4 illustrates the meta-code generated from the XOD example in Figure 1 (searching for a pink on yellow beacon). Firstly subsets containing respectively all the pink and yellow blobs are collated from the universal blobset. If there were other property restrictions defined on these blobs then they would also be checked at this stage. The second step is to compare every blob in list 1 against every blob in list 2 to see if the predicates defined by the relationship tags evaluate to true. If they do, then a new blob is created that bounds both of these blobs and it is added to the universal blobset as well as a new set containing those blobs generated in this stage. Properties are generated from the two source blobs and it is given a colour unique to this XOD. Finally the largest blob (by area) is selected from the set of composite blobs and named PY_BEACON because the XOD in Figure 4 has no <select> statement so the default action applies (select largest by area).

For efficiency, our implementation does not actually manipulate sets of blobs but rather indices to blobs in an array. All blobs (including composite blobs created by our procedure) are stored in a single, unchanging (with the exception of adding new blobs) array. We also perform several other important optimisations.

---

[3] <create>, <copy>, <set>

4   XOD Illustrations

4.1   Black and White Ball

One of the Robocup four-legged league challenges last year required teams to locate a black and white ball and then score a goal with it. The challenge was not performed successfully by any team in the competition. The XOD language allows us to specify a definition for the ball that requires only very minor alterations in the ball definition and the system will identify balls of non-uniform colour easily; in this case, a black and white ball. The XOD used to identify the black and white ball in Figure 5 differed only by three lines to the XOD used to locate an orange one.

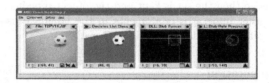

Fig. 5. The XOD used to locate the orange ball differs only by three lines to the XOD used to locate the the black and white ball

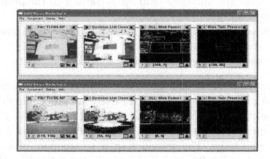

Fig. 6. AIBO Vision Workshop 2 running sideways beacon detection code. The XOD to find sideways beacons (flags) differs only in one word to the XOD used to locate normal beacons

4.2   Flag Instead of Beacon

One proposal to move Robocup more towards real soccer is to replace the navigation beacons with corner flags. The pictures in Figure 6 illustrate that our XOD system is capable of easily adapting to this change. Turning the beacons on the side requires only that the <above> relationship in the beacon definition be replaced with a <left_of> relationship.

5 Conclusion

XOD is an XML based language used to represent object descriptions in a code-independent way. This frees the vision system from requiring domain dependent knowledge to locate useful objects within an image. We have presented both the XOD language as well as an efficient implementation quick enough to be used on the Sony AIBO platform during a Robocup competition.

XOD enables us to quickly adapt to changing circumstances in the vision domain. For example, if the ball were changed to be a non-uniform colour (such as black and white) or the beacons were changed to be flags then the XOD definition of the objects can quickly be adapted to the new situation. The XOD also enables us to extend our vision processing application to domains other than the one it was originally designed for. We illustrated this by detecting an air-hockey puck. It is possible to use XOD to describe any object which can be easily identified using either colours or lines or by its relationship to other objects. This allows us to use XOD to identify many objects in widely different vision applications.

References

1. Bruce, Balch, Veloso: Fast and inexpensive color segmentation for interactive robots. In: International Conference on Intelligent Robots and Systems, IEEE Computer Society Press (2000)
2. Hornegger, J., Niemann, H.: Statistical learning, localization, and identification of objects. In: International conference on computer vision, IEEE Computer Society Press (1995) 914–919
3. Nayar, S., Murase, H., Nene, S.: Parametric appearance representation. In Nayar, S., Poggio, T., Nayar, P., eds.: Early Visual Learning. Oxford University Press (1996)
4. Draper, B.: Learning control strategies for object recognition. In Ikeuchi, Veloso, eds.: Symbolic Visual Learning. Oxford University Press (1996)
5. Lotzsch, M., Bach, J., Burkhard, H., Jungel, M.: Designing agent behaviour with the extensible agent behaviour specification language xabsl. In: 7th International Workshop on RoboCup 2003 (Robot World Cup Soccer Games and Conferences), Lecture Notes in Artificial Intelligence, Padova, Italy, Springer (2004)
6. Draper, B., Ahlrichs, U., Paulus, D.: Adapting object recognition across domains: A demonstration. Lecture Notes in Computer Science **2095** (2001) 256

Modular Learning System and Scheduling for Behavior Acquisition in Multi-agent Environment

Yasutake Takahashi[1], Kazuhiro Edazawa[2], and Minoru Asada[1]

[1] Emergent Robotics Area, Dept. of Adaptive Machine Systems,
Graduate School of Engineering, Osaka University,
Yamadaoka 2-1, Suita, Osaka 565-0871, Japan
{yasutake, asada}@ams.eng.osaka-u.ac.jp
[2] eda@er.ams.eng.osaka-u.ac.jp

Abstract. The existing reinforcement learning approaches have been suffering from the policy alternation of others in multiagent dynamic environments such as RoboCup competitions since other agent behaviors may cause sudden changes of state transition probabilities of which constancy is necessary for the learning to converge. A modular learning approach would be able to solve this problem if a learning agent can assign each module to one situation in which the module can regard the state transition probabilities as constant. This paper presents a method of modular learning in a multiagent environment, by which the learning agent can adapt its behaviors to the situations as results of the other agent's behaviors. Scheduling for learning is introduced to avoid the complexity in autonomous situation assignment.

1 Introduction

There have been an increasing number of work to robot behavior acquisition based on reinforcement learning methods [1, 2]. The conventional approaches need an assumption that the environment is almost stationary or changing slowly so that the learning agent can regard the state transition probabilities as constant during its learning. Therefore, it seems difficult to apply the reinforcement learning method to a multiagent system because a policy alteration of other agents may occur, which dynamically changes the state transition probabilities from the viewpoint of the learning agent. RoboCup provides such a typical situation, that is, a highly dynamic, hostile environment, in which an agent has to obtain purposive behaviors.

There are a number of studies on reinforcement learning systems in a multiagent environment. Asada et al. [3] proposed a method which estimates the state vectors representing the relationship between the learner's behavior and those of other agents in the environment using a technique of system identification, then reinforcement learning based on the estimated state vectors is applied to obtain a cooperative behavior. However, this method requires a global learning

D. Nardi et al. (Eds.): RoboCup 2004, LNAI 3276, pp. 548–555, 2005.

schedule in which only one agent is specified as a learner and the rest of agents have a fixed policies. Therefore, the method cannot handle the alternation of the opponents policies. This problem happens because one learning module can maintain only one policy. A modular learning approach would provide one solution to this problem. If we can assign multiple learning modules to different situations in each of which module can regard the state transition probabilities as constant, then the system could show a reasonable performance.

Jacobs and Jordan [4] proposed the mixture of experts, in which a set of the expert modules learn and the gating system weights the output of the each expert module for the final system output. This idea is very general and has wide applications. Singh [5, 6] has proposed compositional Q-learning in which an agent learns multiple sequential decision tasks with a number of learning modules. Each module learns its own elemental task while the system has a gating module which learns to select one of the elemental task modules. However, there are no such measure to identify the situation that the agent can switch modules corresponding to the change of the situation. Tani and Nolfi [7, 8] extended the idea to mixture of recurrent neural network and introduced it to predict sensory flow pattern under a navigation task. Their scheme, however, doesn't have any control learning structure, which makes it difficult to acquire a purposive behavior by itself. Doya et al. [9] have proposed MOdular Selection and Identification for Control (MOSAIC), which is a modular reinforcement learning architecture for non-linear, non-stationary control tasks. Their idea was applied to relatively simple tasks/dynamic environment, however, it is uncertain that it is possible to assign modules automatically in the multi-agent system that has highly dynamic ones.

We adopt the basic idea of the mixture of experts into an architecture of behavior acquisition in the multi-agent environment. In this paper, we propose a method by which multiple modules are assigned to different situations and learn purposive behaviors for the specified situations which are expected as the result of other agent's behavior under different policies. Takahashi et al. [10] have shown preliminary experimental results under same domain, however, the learning modules were assigned by the human designer. In this paper, scheduling for learning is introduced to avoid the complexity in autonomous situation assignment.

2   A Basic Idea and an Assumption

The basic idea is that the learning agent could assign one behavior learning module to each situation which is caused by the other agents and the learning module would acquire a purposive behavior under the situation if the agent can distinguish a number of situations in which the state transition probabilities are constant. We introduce a modular learning approach to realize this idea. A module consists of learning component that models the world and an execution-time planning component. The whole system performs these procedures simultaneously.

- find a model which represents the best estimation among the modules,
- update the model, and
- calculate action values to accomplish a given task based on dynamic programming (DP).

As a experimental task, we prepare a case of ball passing behavior without interception by the opponent player (Figs. 3,5). In the environment there are a learning agent (passer), a ball, an opponent, and two teammates (receivers). The problem here is to find the model which can most accurately describe the opponent's behavior from the viewpoint of the learning agent and to execute the policy which is calculated under the estimated model. It may take a time to distinguish the situation, therefore, we put an assumption : The opponent continues the one of its policies during one trial and changes after the trial.

3   A Multi-module Learning System

Fig. 1 shows a basic architecture of the proposed system, that is, a multi-module reinforcement learning system. Each module has a forward model (predictor) which represents the state transition model, and a behavior learner (policy planner) which estimates the state-action value function based on the forward model in a reinforcement learning manner. This idea of combination of a forward model and a reinforcement learning system is similar to the H-DYNA architecture [11] or MOSAIC [9]. The system selects one module which has the best estimation of a state transition sequence by activating a gate signal corresponding to a module while deactivating the gate signals of other modules, and the selected module sends action commands based on its policy.

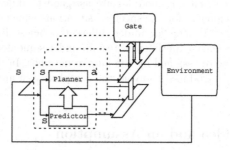

Fig. 1. A multi-module learning system

3.1   Predictor

Each learning module has its own state transition model. This model estimates the state transition probability $\hat{\mathcal{P}}^a_{ss'}$ for the triplet of state s, action a, and next state s':

$$\hat{\mathcal{P}}^a_{ss'} = Pr\{s_{t+1} = s' | s_t = s, a_t = a\} \tag{1}$$

Each module has a reward model $\hat{\mathcal{R}}^a_{ss'}$:

$$\hat{\mathcal{R}}^a_{ss'} = E\{r_{t+1}|s_t = s, a_t = a, s_{t+1} = s'\} \tag{2}$$

We simply store all experiences (sequences of state-action-next state and reward) to estimate these models.

3.2   Planner

Now we have the estimated state transition probabilities $\hat{\mathcal{P}}^a_{ss'}$ and the expected rewards $\hat{\mathcal{R}}^a_{ss'}$, then, an approximated state-action value function $Q(s,a)$ for a state action pair s and a is given by

$$Q(s,a) = \sum_{s'} \hat{\mathcal{P}}^a_{ss'} \left[\hat{\mathcal{R}}^a_{ss'} + \gamma \max_{a'} Q(s',a') \right] , \tag{3}$$

where $\hat{\mathcal{P}}^a_{ss'}$ and $\hat{\mathcal{R}}^a_{ss'}$ are the state-transition probabilities and expected rewards, respectively, and γ is discount rate.

3.3   Module Selection

The gating signal of the module becomes larger if the module does better state transition prediction during a certain period, else it becomes smaller. We assume that the module which does the best state transition prediction has the best policy against the current situation because the planner of the module is based on the model which describes the situation best. In our proposed architecture, the gating signal is used for gating the action outputs from modules. We calculate the gating signals g_i of the module i as follows:

$$g_i = \prod_{t=-T+1}^{0} e^{\lambda p_i^t}$$

where p_i is an occurrence probability of the state transition from the previous $(t-1)$ state to the current (t) one according to the model i, and λ is a scaling factor.

3.4   New Module Assignment

If all modules show worse prediction of state transition, that means all gating signals g_i of the modules become small, the system add one learning module and feed data of sensory-motor sequence to this modules for a while.

4   Task and Assumption

The task of the learning agent is to pass the ball to one of the teammates while it avoids interception by the opponent. The game is like a three on one; there are

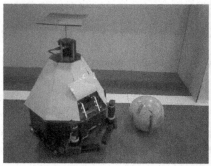 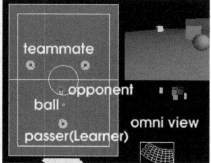

Fig. 2. A real robot   **Fig. 3.** A simulation environment

one opponent and other three players. The player nearest to the ball becomes to a passer and passes the ball to one of the teammates while the opponent tries to intercept it.

Fig. 2 shows a mobile robot we have designed and built. Fig. 3 shows the simulator of our robots and the environment. The robot has an omni-directional camera system. A simple color image processing is applied to detect the ball area and opponent ones in the image in real-time (every 33ms). The left of Fig. 3 shows a situation in which the agent can encounter and the bottom right shows the simulated image of the camera with the omni-directional mirror mounted on the robot. The robot consists of an omni-directional vehicle of which motion (any translation and rotation on the plane) can be controlled.

The state space is constructed in terms of the centroid of the ball on the image, the angle between the ball and the opponent, and the angles between the ball and the teammates (see Fig. 4 (a) and (b)). We quantized the ball position space 11 by 11 as shown in Fig. 4 (a) and the each angle into 8. As a result, the number of state becomes $11^2 \times 8 \times 8 \times 8 = 61952$. The action space is constructed in terms of desired three velocity values (x_d, y_d, w_d) to be sent to the motor controller (Fig. 4 (b)). Each value is quantized into three, then the number of action is $3^3 = 27$. The robot has a pinball like kick device, and it automatically kicks the ball whenever the ball comes to the region to be kicked. It tries to estimate the mapping from sensory information to appropriate motor commands by the proposed method.

The initial positions of the ball, the passer, the opponent, and teammates are shown in Figs. 5. The opponent has two kinds of behaviors; it defend the left side, or right side. The passer agent has to estimate which direction the opponent will defend and go to the position in order to kick the ball to the direction the opponent does not defend. From a viewpoint of the multi-module learning system, the passer agent will estimate which situation of the module is going on, select the most appropriate module to behave. The passer agent acquires a positive reward when it approach to the ball and kicks it to one of the teammate dodging the opponent.

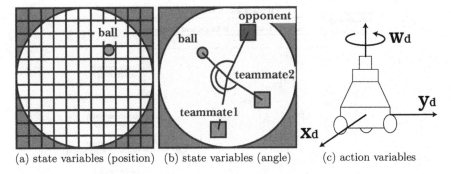

(a) state variables (position)   (b) state variables (angle)   (c) action variables

Fig. 4. A state-action space

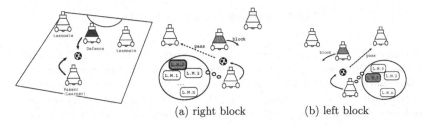

(a) right block   (b) left block

Fig. 5. Task : 3 on 1   **Fig. 6.** Module switching

4.1 Learning Scheduling

We prepare a learning schedule composed of three stage to show its validity. The opponent fixes its defending policy as right side block at the first stage. After 250 trials, the opponent changes the policy to block the left side at the second stage and continues this for another 250 trials. Then, the opponent changes the defending policy randomly after one trial.

4.2 Simulation Result

We have applied the method to a learning agent and compared it with one module learning system. We have also compared the performances between the methods with and without the learning scheduling. Fig. 7 shows the success rates of those during the learning. The success indicates that the learning agent successfully kick the ball without interception by the opponent. The success rate indicates the rate of the number of successes in 50 trials. The multi-module system with scheduling shows better performance than the one-module system. The "mono. module" in the figure indicates "monolithic module" system and it tries to acquire a behavior for both policies of the opponent with one learning module. The monolithic module with scheduling means that we applied learning scheduling mentioned in **4.1** even though the system has only one learning module. The

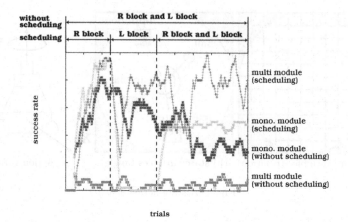

Fig. 7. Success rate during the learning

performance of this system is similar with multi-module system until the end of first stage (250 trials), however, it goes down at the second stage because the obtained policy is biased against the experiences at the fist stage and cannot follow the policy change of the opponent. Since the opponent takes one of the policies at random at the third stage, the learning agent obtains about 50% of success rate. "without scheduling" means that we do not applied learning scheduling and the opponent changes its policy at random from the start. Somehow the performance of the monolithic module system without learning scheduling is getting worse after the 200 trials. The multi-module system without learning schedule shows the worst performance in our experiments. This result indicates that it is very difficult to recognize the situation at the early stage of the learning because the modules has too few experiences to evaluate their fitness, then the system tends to select the module without any consistency. As a result, the system cannot acquires any valid policies at all.

5   Conclusion and Future Work

In this paper, we proposed a method by which multiple modules are assigned to different situations which are caused by the alternation of the other agent policy and learn purposive behaviors for the specified situations as consequences of the other agent's behaviors. We have shown reffectiveness of the proposed method with a simple soccer situation and the importance of the learning scheduling.

References

1. M. Asada, S. Noda, S. Tawaratumida, and K. Hosoda. Purposive behavior acquisition for a real robot by vision-based reinforcement learning. *Machine Learning*, 23:279–303, 1996.

2. Jonalthan H. Connell and Sridhar Mahadevan. *ROBOT LEARNING*. Kluwer Academic Publishers, 1993.
3. M. Asada, E. Uchibe, and K. Hosoda. Cooperative behavior acquisition for mobile robots in dynamically changing real worlds via vision-based reinforcement learning and development. *Artificial Intelligence*, 110:275–292, 1999.
4. R. Jacobs, M. Jordan, Nowlan S, and G. Hinton. Adaptive mixture of local experts. *Neural Computation*, 3:79–87, 1991.
5. Satinder Pal Singh. Transfer of learning by composing solutions of elemental sequential tasks. *Machine Learning*, 8:323–339, 1992.
6. Satinder P. Singh. The effeicient learnig of multiple task sequences. In *Neural Information Processing Systems 4*, pages 251–258, 1992.
7. Jun Tani and Stefano Nolfi. Self-organization of modules and their hierarchy in robot learning problems: A dynamical systems approach. Technical report, Technical Report: SCSL-TR-97-008, 1997.
8. J. Tani and S. Nolfi. Self-organization of modules and their hierarchy in robot learning problems: A dynamical systems approach. Technical report, Sony CSL Technical Report, SCSL-TR-97-008, 1997.
9. Kenji Doya, Kazuyuki Samejima, Ken ichi Katagiri, and Mitsuo Kawato. Multiple model-based reinforcement learning. Technical report, Kawato Dynamic Brain Project Technical Report, KDB-TR-08, Japan Science and Technology Corporation, June 2000.
10. Yasutake Takahashi, Kazuhiro Edazawa, and Minoru Asada. Multi-module learning system for behavior acquisition in multi-agent environment. In *Proceedings of 2002 IEEE/RSJ International Conference on Intelligent Robots and Systems*, pages CD-ROM 927–931, October 2002.
11. Satinder P. Singh. Reinforcement learning with a hierarchy of abstract models. In *National Conference on Artificial Intelligence*, pages 202–207, 1992.

Realtime Object Recognition Using Decision Tree Learning*

Dirk Wilking[1] and Thomas Röfer[2]

[1] Chair for Computer Science XI, Embedded Software Group, RWTH Aachen
wilking@informatik.rwth-aachen.de
[2] Center for Computing Technology (TZI), Universität Bremen
roefer@tzi.de

Abstract. An object recognition process in general is designed as a domain specific, highly specialized task. As the complexity of such a process tends to be rather inestimable, machine learning is used to achieve better results in recognition. The goal of the process presented in this paper is the computation of the pose of a visible robot, i.e. the distance, angle, and orientation. The recognition process itself, the division into subtasks, as well as the results of the process are presented. The algorithms involved have been implemented and tested on a Sony Aibo.

1 Introduction

Computer vision is a multistage task. On the lowest level, the digitized image is searched for certain, domain-dependent features. The next step combines these features to more meaningful objects. Thereafter, a classification of the features computed so far must take place, but the concrete implementation concerning this classification is not as predefined as the low-level segmentation techniques.

Fig. 1. Colored markers used on Sony Aibo robots

---

* The Deutsche Forschungsgemeinschaft supports this work through the priority program "Cooperating teams of mobile robots in dynamic environments".

D. Nardi et al. (Eds.): RoboCup 2004, LNAI 3276, pp. 556–563, 2005.

In order to find an algorithmic base for the recognition of robots, the problem of pose determination is solved regarding an image of a single robot. The most obvious feature of a robot in the Sony Four-Legged Robot League (SFRL) are the colored markers placed on different parts of the body (cf. Fig. 1).

Apart from the unique color which can be used easily to find a robot in an image, the geometric shapes of the different parts provide much more information about the position of the robot. The shapes themselves can be approximated using simple line segments and the angles between them.

2   The Recognition Process

The framework GT2003 that is used for this work is provided by the German-Team ([2]). The process is embedded into a recognition module that receives the robot's current image as input and delivers the poses of the robots perceived on the field. As shown in Fig. 2, the preprocessing stage consists of the segmentation and surface generation. The pose recognition deals with the higher level functions of attribute generation, classification, and analysis of the symbols generated by the classification.

The recognition begins with iterating through the surfaces that have been discovered by the preprocessing stage. For every surface, a number of segments approximating its shape and a symbol is generated (e. g. head, side, front, back, leg, or nothing) as shown in Fig. 3. The symbols are inserted into a special 180° symbol memory which is shown in Fig. 3d) at a position that depends on the rotation of the head of the robot, e. g., a robot looking to the right will insert all

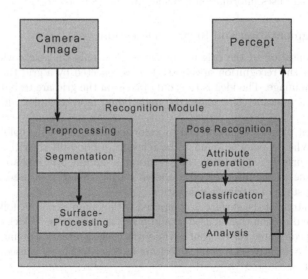

Fig. 2. Overview of the recognition module

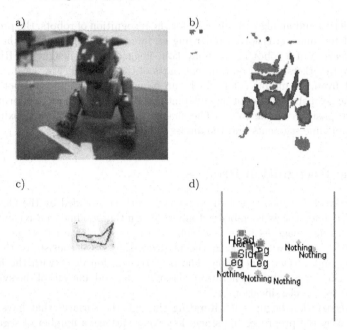

Fig. 3. From original image to symbol representation. a) Original image. b) Result of segmentation. c) Boundary of a singe segmented surface. d) The 180° symbol memory.

symbols on the right side of the memory. The analysis step that computes the real pose only uses information provided by the 180° memory.

2.1   Segmentation and Surface Detection

The segmentation of the original image is done only on pixels which are important for the recognition of robots. This is reflected by a grid that is used to process the image. The idea is that only pixels on the grid are tested and if they belong to a relevant color class, the underlying, more complex algorithms are started.

Relevant pixels are determined by color segmentation using color tables (see e.g. [1]). When a pixel is found to be relevant, a flood-fill algorithm is started. This way, surfaces—along with their position, bounding box, and area—are computed. Some of the information gathered during this step is used in the later processing stage of attribute generation.

The contour of the surface is computed with a simple contour following algorithm. The direction that is used to add pixels is changed in a clockwise fashion.

Having computed the contour of the surface, the generation of line segments is started. The iterative-end point algorithm as described in [6] is used to compute the segments.

2.2   Attribute Generation

As an important part of machine learning, the choice of the attributes to be learned must be considered. Different kinds of attributes have been implemented. Simple attributes, as proposed in [3] and [4], comprise, e. g., color class, area, perimeter, and aspect ratio. Regarding the representation of the surface as line segments, the number of corners, the convexity and the number of different classes of angles between two line segments are provided as attributes. In addition, the surface is compared to a circle and a rectangle with the same area. The discretization of continuous values is achieved with a simple algorithm that searches for an optimal split value in a brute force manner.

A different approach to compute shape based attributes is presented in [5]. The underlying idea, described as conjunctions of local properties, consists of the merging of neighboring features as one single attribute. This is adopted as sequences of adjacent angles.

2.3   Classification

As classification algorithm, the decision tree learning algorithm (cf. [7]) is chosen. This algorithm creates a tree consisting of attributes as described above and symbols which form the leafs of the tree. The tree is built by calculating the attribute with the highest entropy which depends on the number of occurrences of different attribute values.

The problem of over-fitting is solved using χ^2-pruning. The basic idea behind χ^2-pruning is to determine whether a chosen attribute really detects an underlying pattern or not. For the probability that the attribute has no underlying pattern a value of 1% is assumed.

2.4   Analysis

The analysis starts with inserting symbols into the 180° memory (cf. Fig. 4). Using this memory, groups of symbols representing a robot each are combined using the following heuristics:

- Find a side, front, or back symbol in the 180° memory. This symbol is the first symbol.
- Add all head symbols that are near to the first symbol. Test the distance, size, and relative position of the head.
- Add all leg symbols and perform the same tests on them.
- Add other body related symbols and perform the same tests on them.

The surface area of a group is used to determine the distance to the robot. The direction to the robot is computed by the group's position in the 180° memory. The relative position of the head within the group and the existence of front or back symbols indicate the rough direction of the robot. A more precise value is calculated using the aspect ratio of the initial symbol.

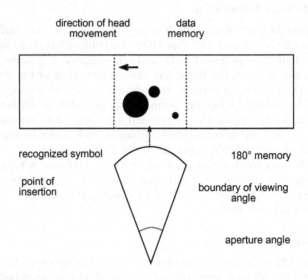

Fig. 4. Structure of the 180° memory used to process symbols

3   Results

The following results have been achieved with an opponent robot as recognition target under the normal lighting conditions of the SFRL.

Fig. 5. Performance results for different subtasks

Classification

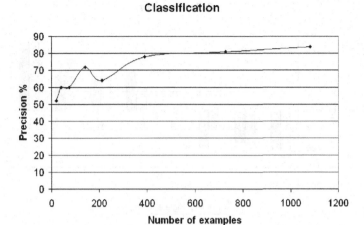

Fig. 6. Classification precision of the recognition process

The first question concerns the overall speed of the process. With an average execution time of 27 milliseconds, the usage on the Aibo robots of the SFRL is possible. Fig. 5 shows that the slowest subtask is the generation of attributes. The reason for that is mainly the usage of the iterative-end point algorithm which consumes much of the execution time using many floating point operations. In contrast, the decision tree classification as well as the analysis of symbol groups

Range error

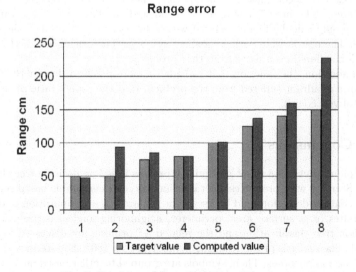

Fig. 7. Precision of the calculation of the range

Rotation error

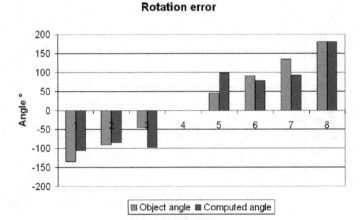

Fig. 8. Precision of the calculation of the rotation

are quite fast. Although only started when required and driven by a flood fill algorithm, the segmentation constantly uses about a fifth of the execution time. The decision tree classification has a precision of 84 % using 1080 examples to classify 5 classes (cf. Fig. 6). The importance of the number of examples decreases fast. The difference in precision between 730 and 1080 examples is 2 %. Even with the low number of 25 examples a classification precision of over 50 % is reached.

To determine the error of the range and rotation calculation, eight different positions have been measured about ten times each. The averaged results are presented in Figures 7 and 8. The second and the last case of the range test are the only calculations associated with a high error compared to the original value. The reason were wrong classified symbols. The calculation of the robot's rotation suffered even more from that problem.

In addition, the function that provided a more precise value for the rotation within a quadrant suffered from the problem that the aspect ratio of the initial symbol was coarse.

4   Conclusions

This paper presents a robot recognition process based on decision tree classification. Starting with preprocessing tasks such as color class table based segmentation, contour detection, and line segment generation, a computation of different attributes (e. g. surface area, perimeter, neighboring angles) is presented. The decision tree classification and the related χ^2 pruning are discussed in greater detail. The analysis of the resulting symbols uses a 180° short-term memory for combination purposes. Then, symbols are grouped together based on a heuristics that makes extensive use of the structure of the target objects. Finally, symbol groups are processed to get the distance and direction to the robot. The rotation is computed again based on the structure of symbols and the aspect ratio of the main surface.

However, due to the complexity and length of the process, some parts could be streamlined. Especially the heuristics used during the analysis step can be improved using a skeleton template based, probabilistic matching procedure. This procedure could deal both with the problem of occlusion and missing symbols. In addition, improvements concerning the speed of the attribute generation can be achieved. Considering the process architecture, the dependence of the recognition tasks among each other must be examined. Especially the classification task could benefit from a slowly, rather parallel[1] growing recognition process that provides expected values for the classification.

References

1. J. Bruce, T. Balch, and M. Veloso. Fast and inexpensive color image segmentation for interactive robots. In Proceedings of the 2000 IEEE/RSJ International Conference on Intelligent Robots and Systems (IROS 00), volume 3, pages 2061–2066, 2000.
2. Burkhard, H.-D., Düffert, U., Jüngel, M., Lötzsch, M., Koschmieder, N., Laue, T., Röfer, T., Spiess, K., Sztybryc, A., Brunn, R., Risler, M., v. Stryk, O. (2001). GermanTeam 2001. Technical report (41 pages, only available online).
3. Shepherd, B. A.. An apprasial of a Decision Tree approach to Image Classification. In Proc. of the Eighth International Joint Conference on Artificial Intelligence, pages 473–475, 1983.
4. Gerhard-Helge Pinz. Bildverstehen. Springer, 1994.
5. Cho, K., Dunn, S. M.. Learning Shape Classes. In IEEE Transactions on Pattern Analysis and Machine Intelligence, 1994.
6. Duda, R. O., Hart, P. E.. Pattern Classification and Scene Analysis. Wiley, New York, 1972.
7. Quinlan, J. R.. C 4.5: Programs for Machine Learning. Morgan Kaufman Publishers, San Mateo, 1993.

---

[1] In contrast, the symbols are currently generated without any context knowledge.

Optimizing Precision of Self-Localization in the Simulated Robotics Soccer

Vadim Kyrylov, David Brokenshire, and Eddie Hou

Simon Fraser University – Surrey,
Surrey, British Columbia V3T 2W1 Canada
{vkyrylov, brokenshire, ehou}@sfu.ca
http://www.surrey.sfu.ca/

Abstract. We show that previously published visual data processing methods for the simulated robotic soccer so far have not been utilizing all available information, because they were mainly based on heuristic considerations. Researchers have approached to estimating the agent location and orientation as two separate tasks, which caused systematic errors in the angular measurements. Further attempts to get rid of them (e.g. by completely neglecting the angular data) only aggravated the problem and resulted in the losses in the accuracy. We utilize all the potential of the visual sensor by jointly estimating the agent view direction angle and Cartesian coordinates using the extended Kalman filtering technique. Our experiments showed that the achievable average error limit for this particular application is about 25-33 per cent lower than that of the best algorithms published by far.

1 Introduction

We started this work in 2001, developing the SFUNLEASHED team for the simulated robotic soccer competitions. While trying to improve the BASIC UVA team [2, 5] used as a prototype, we realized that better, or maybe best possible, utilization of the information provided by the visual sensor was important. Player self-localization is one of related tasks.

The visual sensor of the simulated soccer agent returns measurements of the polar coordinates of objects, i.e. the range and the direction, with regular time intervals. The range to the landmarks used for self-localization (borderlines and flags) is measured relative to current agent location in the field, and the direction is measured relative to the agent neck orientation. The precise data are distorted by rounding errors. With current Soccer Server settings, quantization of the angular measurements results in the random error uniformly distributed within [-0.5, 0.5] degrees. The magnitude of the landmark range measurement error is [-0.5%, 0.5%] of the distance [3].

By far, processing visual sensor information with the purpose of self-localization has been addressed in different ways in the simulated robotic soccer community. The related works could be split into two groups, elaborating either on the deterministic or stochastic approaches.

D. Nardi et al. (Eds.): RoboCup 2004, LNAI 3276, pp. 564–573, 2005.

1.1 Deterministic Methods

Indeed, for given agent position, the visual sensor in the simulated soccer always supplies deterministic data. This feature has recently given birth to a method based on two-dimensional interval arithmetic, implemented in the LUCKY LÜBECK simulated soccer team [4]. However, because such methods are risk-aversive, the interval estimate tends to have an exaggerated size of the area to which the observed object is believed to belong. It is also unknown if this method provides the best possible accuracy, since there is just no explicitly stated optimality criteria.

One more recent work done at the Humboldt University of Berlin, uses the optimization of the mismatch between the readings of several landmarks using a gradient descent function [1]. This method offers a mathematically elegant solution, which has resulted in significantly higher precision than some simple heuristic algorithms. However, the weak point is that the angular measurements are being completely ignored in this implementation, which means that all the potential of the simulated visual sensor has not been completely utilized.

1.2 Stochastic Methods

Subtle notions of using stochastic estimation methods are spread over all the RoboCup literature. However, very few, if any, comprehensive overviews have been published. One early exception is the detailed description of the visual sensor data processing algorithm in [6]. For self-localization, the agent neck direction angle is estimated first. Then a weighted sum of the individual estimates of the agent position relative to different flags is used. The weights are inversely proportional to the distances from the agent to these landmarks.

A rather thorough investigation of several methods can be found in [2]. In this work, the so-called particle filter has been selected for visual sensor data processing over two simple heuristic methods and the classic Kalman filter. This method was implemented in the BASIC UVA soccer team [5], which we are using as a prototype. Compared to the nearest-flag method for agent self-localization, the particle filter reportedly yielded almost three times better accuracy. The downside was the two order of magnitude increase in the execution time required to obtain this gain. With this reservation, it looks like using the particle filter for performing the agent self-localization task is overkill, and the grounds on which less resource-demanding Kalman filter had been rejected are not all convincing.

1.3 Previously Mistreated Factor: The Agent Neck Direction Error

All deterministic and stochastic methods published by far result in that the agent neck direction is estimated separately. This adds a systematic component δ to all angular measurements of the landmarks used for determining the Cartesian coordinates (Fig.1).

In Fig.1 precise directions are shown as thin lines, while the thicker lines show the measurements distorted by the systematic error. With the self-localization algorithms proposed in [2,4, 5,6], the estimated agent location would be somewhere close to the center of the shaded triangle, which even does not cover the true agent location **A**. Adding more flags would hardly improve the situation, because each angular measurement is equally distorted by δ. We believe that this was exactly the reason of why some authors were reluctant about using angular data in self-localization and were relying on the range measurements only in the simulated soccer.

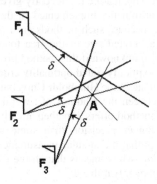

Fig. 1. The ambiguity area with the systematic error δ added to the angular measurements

This allows us to conclude that not all the potential gain has been obtained from the application of stochastic methods to visual sensor data processing in the simulated soccer. It is still unclear, where the limit of the perfection lies and how much could be in principle obtained by improving the data processing algorithms.

The **purpose of this paper** is two-fold. Firstly, we want to synthesize near-optimal algorithms for estimating object locations by a soccer agent completely utilizing all the potential information in the multiple readings received in one cycle. Secondly, we want to find the limits of perfection for visual sensor data processing methods in the simulated robotic soccer.

We are pursuing these goals by a comprehensive stochastic analysis of the visual sensor of the simulated soccer player. Here we consider the static case only and deliberately do not use filtering data over time, leaving the multi-cycle case for the future. Because we are using the extended Kalman filtering technique, it can be generalized for the multi-cycle case rather easily.

2 Using the Extended Kalman Filtering Algorithm

The complete utilization of the information supplied by the visual sensor could be achieved by jointly estimating the neck direction angle with the agent Cartesian coordinates using the extended Kalman filter [7]. It is recursively yielding an optimal estimate of the state vector **x** based on a set of measurements. The optimality criterion is the maximal likelihood of $\hat{\mathbf{x}}_k$ [8].

Assuming that the filter has been already applied $(k-1)$ times, on the k-th step its input parameters are, the measurement vector $\hat{\mathbf{z}}_k$ and its error covariance matrix $\mathbf{R}_k$. Other inputs also are the optimal state vector estimate obtained on the previous step $\hat{\mathbf{x}}_{k-1}$ and its covariance matrix $\mathbf{P}_{k-1}$. The filter returns an optimal estimate $\hat{\mathbf{x}}_k$ of the state vector and its covariance matrix $\mathbf{P}_k$ on current k-th step. In our context, k is the index of the source providing raw data at the same instant of time. It is assumed that the state vector is changing with k as

$$\mathbf{x}_k = \mathbf{f}(\mathbf{x}_{k-1}, \mathbf{w}_{k-1}) , \qquad (1)$$

where $\mathbf{f}$ is the state transition function, and $\mathbf{w}$ is the white Gaussian random sequence having zero mean and covariance matrix $\mathbf{Q}_k$.

The measurement model has the following form:

$$\hat{\mathbf{z}}_k = \mathbf{h}(\mathbf{x}_k, \mathbf{v}_k) , \qquad (2)$$

where $\mathbf{h}$ is the observation transformation function and $\mathbf{v}$ is a white Gaussian measurement noise having zero mean and covariance matrix $\mathbf{R}_k$.

The Kalman filtering algorithm is executed in three steps. First, the *a priori* estimate $\tilde{\mathbf{x}}_k$ of the state variable vector and its covariance matrix $\tilde{\mathbf{P}}_k$ are computed using (1):

$$\tilde{\mathbf{x}}_k = \mathbf{f}(\hat{\mathbf{x}}_{k-1}, 0) , \qquad (3)$$

$$\tilde{\mathbf{P}}_k = \mathbf{A}_k \mathbf{P}_{k-1} \mathbf{A}_k^T + \mathbf{W}_k \mathbf{Q}_{k-1} \mathbf{W}_k^T , \qquad (4)$$

where $\mathbf{A}_k$, and $\mathbf{W}_k$ are the Jacobian matrices [9]:

$$A_{[i,j]} = \frac{\partial f_{[i]}}{\partial x_{[j]}}(\hat{x}_{k-1}, 0) , \quad W_{[i,j]} = \frac{\partial f_{[i]}}{\partial w_{[j]}}(\hat{x}_{k-1}, 0) .$$

Second, the Kalman matrix 'weight' is calculated:

$$\mathbf{K}_k = \tilde{\mathbf{P}}_k \mathbf{H}_k^T (\mathbf{H}_k \tilde{\mathbf{P}}_k \mathbf{H}_k^T + \mathbf{V}_k \mathbf{R}_k \mathbf{V}_k^T)^{-1} , \qquad (5)$$

where $\mathbf{H}_k$, and $\mathbf{V}_k$ are the Jacobian matrices:

$$H_{[i,j]} = \frac{\partial h_{[i]}}{\partial x_{[j]}}(\tilde{x}_k, 0) , \quad V_{[i,j]} = \frac{\partial h_{[i]}}{\partial w_{[j]}}(\tilde{x}_k, 0) .$$

Third, the optimal state variable vector estimate and its covariance matrix are computed:

$$\hat{\mathbf{x}}_k = \tilde{\mathbf{x}}_k + \mathbf{K}_k (\hat{\mathbf{z}}_k - \mathbf{h}(\tilde{\mathbf{x}}_k, 0)) , \qquad (6)$$

$$\mathbf{P}_k = (\mathbf{I} - \mathbf{K}_k \mathbf{H}_k)\tilde{\mathbf{P}}_k . \qquad (7)$$

Noteworthy that this estimate is only optimal when the error statistics is Gaussian. In Section 3 below we analyze how non-Gaussian statistics in the simulated soccer affects the accuracy.

Now we customize the general Kalman filtering algorithm to the particular case of estimating the simulated soccer agent state vector. For the purpose of the self-localization, the **state vector variable** contains two agent Cartesian coordinates, x, y and the neck direction angle, β:

$$\mathbf{x} = \begin{bmatrix} x & y & \beta \end{bmatrix}^T .$$ (8)

Now consider the **model of the state vector dynamics** (1). Because we are dealing with a snapshot, this allows us to reduce (3) to the trivial linear model

$$\mathbf{x}_k = \mathbf{x}_{k-1} .$$ (3a)

Moreover, the Jacobian matrices $\mathbf{A}_k$, and $\mathbf{W}_k$ in (4) are both zero matrices and we get:

$$\tilde{\mathbf{P}}_k = \mathbf{P}_{k-1} ,$$ (4a)

The **measurement model** (2) is substantially nonlinear. The observation vector for k-th landmark is the pair of polar coordinates, with the direction (azimuth) d_k measured relative to the agent neck direction and the range r_k measured from the agent to the landmark:

$$\mathbf{z}_k = \begin{bmatrix} r_k & d_k \end{bmatrix}^T .$$ (9)

From the geometry of the relationship between $\mathbf{z}$ and $\mathbf{x}$ we get:

$$\mathbf{z}_k = \mathbf{h}_k(\mathbf{x}, \mathbf{v}_k) ,$$ (2a)

where $\quad r_k = h_k^{(1)}(\mathbf{x}, \mathbf{v}_k) = \sqrt{(x - x_{Fk})^2 + (y - y_{Fk})^2} + v_{rk} ,$

$$d_k = h_k^{(2)}(\mathbf{x}, \mathbf{v}_k) = \tan^{-1}\left(\frac{y - y_{Fk}}{x - x_{Fk}} \right) - \beta + v_{dk} ,$$

x_{Fk}, y_{Fk} are the Cartesian coordinates of the k-th reference point $\mathbf{F}_k$, and $\mathbf{v}_k = \begin{bmatrix} v_{rk} & v_{dk} \end{bmatrix}^T$ is the random measurement error vector.

The low index k in (2a) underscores that, in general, for different landmarks the measurement model could be different. For all visible flags the covariance matrix is,

$$\mathbf{R}_k = \begin{bmatrix} r_k^2 \sigma_r^2 & 0 \\ 0 & \sigma_d^2 \end{bmatrix} ,$$ (10)

where $\sigma_d^2 = 1/(12 \cdot (180^\circ / \pi)^2)$ and $\sigma_r^2 = (0.01)^2 / 12$ for static objects [3].

For the field borderlines, however, the measurement model differs in that the range measurements are unavailable and only the direction can be measured. To keep the same model for both types of landmark, in the covariance matrix (10) for borderlines we assume that $\sigma_r^2 = +\infty$.

For calculating the **matrix weight** (5) we need two Jacobian matrices, $\mathbf{H}_k$, and $\mathbf{V}_k$. They immediately follow from (2a) and (10).

$$\mathbf{H}_k = \begin{bmatrix} \dfrac{x}{\sqrt{(x-x_{Fk})^2+(y-y_{Fk})^2}} & \dfrac{y}{\sqrt{(x-x_{Fk})^2+(y-y_{Fk})^2}} & 0 \\[3mm] -\dfrac{y-y_{Fk}}{(x-x_{Fk})^2+(y-y_{Fk})^2} & \dfrac{x-x_{Fk}}{(x-x_{Fk})^2+(y-y_{Fk})^2} & -1 \end{bmatrix}, \tag{11}$$

$$\mathbf{V}_k = \mathbf{V} = \begin{bmatrix} 1 & 0 \\ 0 & 1 \end{bmatrix}. \tag{12}$$

Therefore, the sought matrix weight can be calculated, as follows:

$$\mathbf{K}_k = \mathbf{P}_{k-1}\mathbf{H}_k^T(\mathbf{H}_k\mathbf{P}_{k-1}\mathbf{H}_k^T + \mathbf{R}_k)^{-1}. \tag{5a}$$

This weight is used to compute the optimal state variable vector estimate

$$\hat{\mathbf{x}}_k = \hat{\mathbf{x}}_{k-1} + \mathbf{K}_k(\hat{\mathbf{z}}_k - \mathbf{h}_k(\hat{\mathbf{x}}_{k-1},0)), \tag{6a}$$

and its covariance matrix

$$\mathbf{P}_k = (\mathbf{I} - \mathbf{K}_k\mathbf{H}_k)\mathbf{P}_{k-1}. \tag{7a}$$

Expressions (3a-7a), coupled with (10-12) yield the sought optimal algorithm for agent localization in the simulated soccer. For N visible landmarks it must be executed for $k=1,2,...,N$.

The **initial state** $(\hat{\mathbf{x}}_0,\mathbf{P}_0)$ for this algorithm is extrapolated from the previous $(k-1)$th cycle, which increases the self-localization error 2..3 times. With more than three flags visible on k-th cycle, this allows ignoring measurements made in the previous cycle and just set

$$\mathbf{P}_0 = \begin{bmatrix} +\infty & 0 & 0 \\ 0 & +\infty & 0 \\ 0 & 0 & +\infty \end{bmatrix}. \tag{13}$$

This results in that (5a) assigns very large weight to the innovation in (6a) and we do not care about $\hat{\mathbf{x}}_0$ at all. However, less than three flags are visible 9 per cent time for 90-degree visual sector and 1 per cent time for 180-degrees. Because ignoring history data in these cases would result in the increased errors, the diagonal elements in $\mathbf{P}_0$ are set to finite positive values.

3 Simulation Results

To determine the precision of the proposed method, we have run five sets of simulations. Each of them contained total 2000 experiments by placing the agent in the random points which were uniformly spread over the field. The neck direction was uniformly distributed in the $[0°, 360°)$ interval. The agent viewing sector was either 90 or 180 degrees wide, with average number of visible flags 5.7 and 11.4, respectively. The results (Table 1) are accurate with less than ±0.9% error, 19 times of 20.

Experiment #1 estimates **the achievable accuracy of the proposed algorithm** and provides a benchmark for comparing it with the alternative self-localization methods. Experiment #2 estimated **the loss in precision which resulted from violating the assumption that the visual sensor measurement errors are Gaussian** as required by Kalman filter theory We compared the self-localization errors in two cases: (a) default Soccer Server rounding errors (±0.5° for the angle and ±0.5% for the range), and (b) Gaussian errors having zero means and same standard deviations (i.e. 0.289° and 0.289%, respectively).

Table 1. Experiments with the self-localization algorithms (90-dergee agent viewing sector)

Experi ment	Description	Objective	Localization error, meters
1	Kalman Filter as proposed in Section 2	Estimating absolute accuracy	0.091 (100%)
2	Kalman Filter with Gaussian measurement error statistics	Estimating loss of precision from non-Gaussian statistics	0.090 (98.9%)
3	Kalman Filter with ignored angular measurements	Estimating the gain from using angular data	0.13 (143%)
4	All-flags, with systematic angular error present	Comparison with the widely used algorithm	0.11 (121%)
5	Nearest flag algorithm	Comparison with the simplest method	0.30 (330%)

We found that in the Gaussian case, for which the Kalman filter is strictly optimal, the mean error is 98.9% of the error obtained in the default case. This gives the idea of the price paid for the sub optimality. Since the difference is very close to the accuracy limit of our experiments, it is clear that the non-Gaussian statistics only insignificantly affects the estimator quality. Presumably this unreasonable concern was the main reason for using particle filter in [2].

Experiment #3 evaluated **the loss in precision when angular measurements are ignored,** as suggested by some of our predecessors. The test was run by setting in (10) the angular variance σ_d^2 1000 times greater that it originally was. This forced the Kalman filter to ignore angular measurements and resulted in a 43% increase of the error.

Experiment #4 **evaluates a very common self-localization algorithm.** It is using one border line for estimating the neck direction angle and then is utilizing all visible flags to estimate the agent location. No care is taken of the systematic angular error, like shown in Fig.1. Although this algorithm has been developed by us, it is combining the ideas, which are widely spread in the simulated soccer community. We speculate that it could be somewhat more accurate than each of its predecessors, because it is taking correlation of the Cartesian coordinates measurements into account. Still compared to the Kalman filter, it results in the self-localization error which is greater by 21%. We have also run similar experiment with a 180 degree visual sector, which resulted in a 45% increase of the error. Assuming that 90- and 180-degree sectors are being used with equal probability, the average localization error would be greater by 33%.

Experiment #5 gives the reference point for the achievable accuracy of the proposed method as **compared to the simplest, the nearest-flag self-localization algorithm**. The Kalman filter offers a three-fold accuracy gain in this case. This is consistent with the reported gain from using the particle filter [2].

It is safe to say that Table 1 contains the minimal achievable average self-localization errors in the simulated soccer and could be used as the reference point.

In a separate set of the experiments we assessed **the benefits of two-agent ball tracking**. Two agents were simultaneously viewing the ball from different locations. Each agent knew its location and orientation precisely, and could relay his estimate of the ball coordinates to the second agent without increasing the estimation error. The latter agent was using two sets of ball (x,y) coordinate measurements to estimate the ball location more precisely. Noteworthy that, unlike landmarks, ball Cartesian coordinate measurement errors may be highly correlated.

We compared two algorithms, heuristic and the Kalman filter (Table 2). The first algorithm was calculating the ball Cartesian coordinates based on the range and direction measurements by the agents. Then two x,y pairs of the ball coordinates were being merged into the final estimate using the near-optimal method based on the maximal-likelihood criterion [10] and its two modifications. One modification ignored the correlation between the Cartesian coordinates, and the second ignored the angular data. The Kalman filter was using the joint estimate of the ball location based on the observations of its polar coordinates by the two agents.

Agents were placed 2000 times 10 meters apart in random locations at the distance from the ball uniformly distributed in the 5 to 15 meter interval. The results are accurate with less than ±1.1% error, 19 times of 20. They show that, while estimating the ball location in a 90-degree viewing sector, the assistance by a teammate can potentially reduce the mean linear error about two times. Ignoring correlation would reduce this gain to just 1.3 times. Ignoring angular measurements would be counterproductive.

Table 2. One- and two-agent ball tracking average linear error (in meters) using different estimation algorithms (90-degree sector)

One agent	Two agents			
	Ignore angles	Ignore-correlation	Near-optimal	Kalman Filter
1.17 (193%)	4.10 (676%)	0. 78 (129%)	0.606 (100%)	0.601 (99.2%)

In other experiments reported elsewhere [10] we have also found that the joint estimation of the Cartesian coordinates and the agent orientation would offer more significant gains in the robotics application, where the systematic error is greater than in the simulated soccer (0.5°). In particular, we have found that the average localization error can be decreased 4..8 times, if the angular bias is uniformly distributed in the interval as big as ±4.0°. This implies that **the significance of this work is not**

limited to the simulated soccer. Rather, with only minor changes the proposed algorithm can be reused in other robotics applications, where joint estimation of the robot location and its orientation could result in significant gains.

4 Conclusion

We have derived algorithms for determining absolute Cartesian coordinates of objects using imprecise readings of local polar coordinates, supplied by a visual sensor in a system like robotic soccer. The proposed solution is based on the Kalman filtering technique.

The innovation is in the rigorous treatment of this problem from the positions of the stochastic estimation theory. In particular, by jointly estimating the agent location and orientation, we have taken into account the correlation of the raw measurements which is emerging after converting them into Cartesian coordinate system. By far, this correlation has been neglected in the simulated soccer applications, which resulted in some losses. Although in self-localization this loss is negligible, in the related task of ball tracking using agent communication negligence of correlation increases the linear error by 30 per cent.

Using the joint estimation of the agent location and orientation allows walking around the problem of the systematic error present in the angular measurements. This problem has been persisting in the self-localization algorithms published so far and limited their accuracy. In particular, we have shown that ignoring the angular data as the way to getting rid of the systematic error would result in about 43 per cent increase in the localization error.

We have demonstrated that the non-Gaussian statistics of the raw measurement errors, which presumably was a concern for some researchers who have been reluctant about using Kalman filter in the simulated soccer on these grounds, is indeed not an obstacle. Our experiments has shown that replacing the non-Gaussian errors with the equivalent 'true' normal noise would not result in statistically significant differences in the self-localization accuracy. Compared to the best published algorithms, the new method can reduce the average error by 25-33 per cent.

Because we have shown that the assumption made about Gaussian statistics does not reduce the precision, we can guarantee that, the solution found cannot be tangibly improved in terms of the mean error of the location estimate. We thus indeed have utilized all the potential of the visual sensor, as it applies to single simulation cycle.

The computational effort required for implementing this algorithm is proportional to the number of the landmarks used for self-localization. Compared with the nearest flag method, the increase in the computation time is roughly six-fold. We believe that it is a fair cost for gaining a three times higher accuracy than the nearest-flag algorithm.

We also have found that the proposed algorithm could potentially offer even more gains in cases when the systematic angular errors are greater that those present in the simulated soccer. Therefore we hope that after some modifications, our algorithm could be reused in some other robotics applications.

The results, however, are limited to data processing in a single simulation cycle. Our future work will be targeted at similar comprehensive study of the coordinate data processing over time using all the power of methods offered by the Kalman filtering technique.

References

1. Bach, J., Gollin, M.: Self-Localization revisited. In: Birk, A., Coradeschi, S., Tadokoro, S. (eds.): RoboCup 2001, Lecture Notes in Artificial Intelligence, Vol. 2377. Springer-Verlag Berlin Heidelberg New York (2002) 251-256
2. de Boer, R., Kok, J.: The Incremental Development of a Synthetic Multi-Agent System: The UvA Trilearn 2001 Robotic Soccer Simulation Team. M.S. Thesis. Faculty of Science, University of Amsterdam (2002)
3. Foroughi. E., Heintz, F., Kapetanakis, K., Kotiadis, K., Kummeneje, J., Noda, I., Obst, O, Riley, P., and Steffens, T.: RoboCup Soccer Server User Manual. Versions 7.06 and later (2001) http://sourceforge.net/projects/sserver
4. Haker, M., Meyer, A., Polani, D., and Martinez, T.: A Method for Incorporation of New Evidence to Improve World State Estimation. In: Birk, A., Coradeschi, S., Tadokoro, S. (eds.): RoboCup 2001, Lecture Notes in Artificial Intelligence, Vol. 2377. Springer-Verlag Berlin Heidelberg New York (2002) 362-367
5. Kok, J., de Boer, R., Groen, F., Vlassis, N.: UvA Trilearn 2001 - Soccer Simulation Team. Faculty of Science, University of Amsterdam (2001) http://carol.wins.uva.nl/~jellekok/robocup/2001/index_en.html
6. Müller-Gugenberger P., Wendler, J.: AT Humboldt 98 -- Design, Implentierung und Evaluierung eines Multiagentensystems für den RoboCup-98 mittels einer BDI-Architektur. Diploma thesis. Huboldt University of Berlin (1998) http://www.informatik. hu-berlin.de/~wendler/paper/gugenberger Wendler:diplom:1998.ps.gz
7. Catlin, D. (1989) Estimation, Control, and the Discrete Kalman Filter (Applied Mathematical Sciences, Vol 71) Springer Verlag, ISBN: 038796777X
8. MathWorld: Maximal Likelihood, http://mathworld.wolfram.com/Maximum Likelihood.html
9. MathWorld: Jacobian, http://mathworld.wolfram.com/Jacobian.html
10. Kyrylov, V., Brokenshire, D., Hou, E. Optimizing Precision of Self-Localization in Simulated Soccer Agents. Technical report SIAT-TR-VK-01-2003.

Path Optimisation Considering Dynamic Constraints

Marko Lepetič, Gregor Klančar, Igor —krjanc,
Drago Matko, and Bo•tjan Potočnik

Faculty of Electrical Engineering, University of Ljubljana
Trža•ka 25, SI-1000 Ljubljana, Slovenia
marko.lepetic@fe.uni-lj.si.si

Abstract. Path planning technique is proposed in the paper. It was developed for robots with differential drive, but with minor modification could be used for all types of nonholonomic robots. The path was planned in the way to minimize the time of reaching end point in desired direction and with desired velocity, starting from the initial state described by the start point, initial direction and initial velocity. The constraint was acceleration limit in tangential and radial direction caused by the limited grip of the tires. The path is presented as the spline curve and was optimised by placing the control points trough which the curve should take place.

1 Introduction

Mobile, autonomous robots are about to become an important element of the "factory of the future" [12]. Their flexibility and their ability to react in different situations [9] open up totally new applications, leaving no limit to the imagination. To drive the mobile robot from its initial point to the target point, the robot must follow previously planned path. Well-planned path together with robot capabilities assure desired efficiency of the robot. The path could be optimised considering different aspects such as minimum time, minimum fuel, minimum length and others [4, 7, 10]. When the path is planned in details, the robot capabilities are exactly known and that makes an advantage when coordinating several mobile robots [3].

This paper deals with time optimal path planning considering acceleration limits. The proposed technique is presented on the robot soccer system (Fig. 1), which became very popular recently. It is an excellent test bed for various research interests such as path planning [4, 7, 10], obstacle avoidance [4], multi-agent cooperation [3, 11], autonomous vehicles, game strategy [2, 8], robotic vision [6], artificial intelligence and control. The robot soccer has also proven to be excellent approach in engineering education, because it is attractive and through the game the students get immediate feedback about the quality of their algorithms.

Mirosot is one of the games, for which the rules are provided by FIRA (Federation of International Robot-soccer Association). The robot size is limited with the cube of 7.5 cm side length. The navigation of the robots is provided with the vision system. The obtained positions of the robots and the ball are used for calculating the commands that are then sent to each robot radio transmitter. There are

D. Nardi et al. (Eds.): RoboCup 2004, LNAI 3276, pp. 574–585, 2005.
© Springer-Verlag Berlin Heidelberg 2005

two leagues of Mirosot. Small league is a game of 3 against 3 robots on the playground of 1.5 m x 1.3 m, while 5 robots of each team play middle league on the playground sized 2.2 m x 1.8 m.

Fig. 1. The robot soccer system

The problem for which the solution is presented in this paper is the following: We want to find the path for the robot that would give the robot minimum time to move from the start point (SP) to the end point (EP) where the robot kicks the ball. Besides SP and EP, also the orientation and velocity in both points should be considered. The robot should stay inside its acceleration limits all the time. It could be said the paper presents an anti-skid path design.

The paper is organized as follows: Section 2 presents the mathematical model of the robot and its limitations. A quick overview of curve synthesis and analysis is given in Section 3. Section 4 describes the proposed technique. Case study is presented in Section 5 and application aspects are discussed in Section 6. Section 7 gives the conclusions.

2 Robot Model and Limitattions

The robot is cubic shape with the side of 7.5 cm. It is driven with the differential drive, which is located at the geometric centre. This kind of drive allows zero turn-radius. The front and/or the back of the robot slide on the ground. For more detailed description see Fig. 2. The commands that the computer sends to the robot are reference for linear and angular velocity. The microprocessor on the robot calculates the reference angular velocities of the left and right wheel. The motors that drive the wheels contain encoders so the microprocessor also knows actual velocities. The PID controller in the microprocessor then calculates the needed voltage for both motors. The PID controller together with powerful motors causes sliding of the wheels if the desired velocity makes step change. This knowledge is important when modelling the robot.

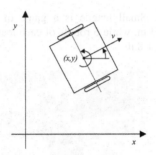

Fig. 2. The robot

The movement of the robot can be modelled with the following equations:

$$\dot{x} = v_{real} \cos(\varphi)$$
$$\dot{y} = v_{real} \sin(\varphi) \qquad (1)$$
$$\dot{\varphi} = \omega_{real}$$

where x, y and φ stand for position and orientation respectively, vreal is real linear velocity and wreal is real angular velocity. If the wheels are not sliding, both velocities are very close to the reference velocities that have been sent to the robot. With these assumptions the real velocities from eq. (1) can be substituted with the ones, which has been sent as commands. We get:

$$\dot{x} = v \cos(\varphi)$$
$$\dot{y} = v \sin(\varphi) \qquad (2)$$
$$\dot{\varphi} = \omega$$

Only this simplified model will be used and all other dynamics will be neglected. It must not be forgotten, that this model is good only when the wheels don't slide or in other words, when the robot is not forced with too large acceleration. The overall acceleration can be decomposed to tangential acceleration and radial acceleration. The tangential acceleration is the derivative of velocity with the respect to time and is caused with desire to increase or decrease speed.

$$a_{tang} = \frac{dv}{dt} \qquad (3)$$

The radial acceleration is caused by turning at certain speed and is the product of linear and angular velocity

$$a_{rad} = v \cdot \omega \qquad (4)$$

Since tangential and radial acceleration are orthogonal, the overall acceleration is the Pythagoras sum as follows:

$$a \quad \sqrt{a_{tang}^2 \quad a_{rad}^2} \qquad\qquad (5)$$

The overall acceleration is limited with the friction force. The limit of tangential acceleration differs from the limit of radial acceleration. That happens, because the gravity centre of the robot is on certain height above ground level. When accelerating in linear direction, the robot leans on the rear slider, which takes over a part of the robot weight. That means that the wheels of the robot press on the ground with the force that is smaller than gravity force. We know that the friction force is product of the force orthogonal to the ground and the friction index. Comparing tangential acceleration to the radial, the orthogonal force is smaller, which causes lower acceleration limit.

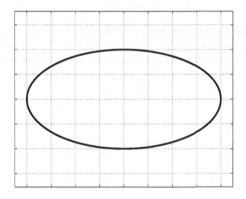

Fig. 3. Acceleration limits

The acceleration limits have been measured in our case. To measure radial acceleration limit, the angular velocity was set to a certain value and then the linear velocity was slowly increased. The slipping moment was determined visually. The maximal radial acceleration was then calculated from eq. (4). Tangential acceleration limit measurement was little more complicated. In this case slipping cannot be determined visually, so the vision system was used. Several experiments were made. During each experiment the robot was forced with the constant acceleration. The acceleration at each next experiment was slightly increased comparing to the previous experiment. Real acceleration of the robot was measured as second derivation of robot's position, which was obtained using the vision system. Measured maximum tangential acceleration was 2 m/s^2 and maximum radial acceleration 4 m/s^2, so the overall acceleration should be somewhere inside the ellipse as it is shown in Fig. 3.

3 Curve Design and Analysis

There are many possible ways to describe the path. Spline curves are just one of them. The corresponding theory has been presented in number of books and papers [1,5] so in this paper a quick overview will be given. The two dimensional curve is got by combining two splines, x(u) and y(u)., where u is the parameter along the curve. Each spline consists of one or more segments – polynomials. The point of tangency of two neighbour segments is called knot. The spline could be interpolated through desired points in (u,x) or (u,y) domain, where also the derivative conditions can be fulfilled. When the knots are set, the spline parameters can be obtained by solving a linear equation system. If the p-th order spline consists of m segments, than the number of parameters to determine is

$$m(p \quad 1) \tag{6}$$

Number of linear equations is

$$n \quad (m \quad 1)p \tag{7}$$

where n is number of explicitly defined points and derivative conditions at these points, $(m-1)$ is number of knots and p is number of continuous derivatives at the knots. The number of searched parameters should be equal to the number of linear equations what leads to:

$$m \quad n \quad p \tag{8}$$

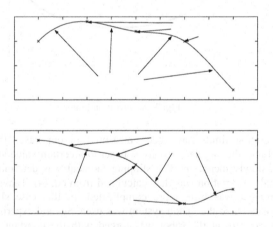

Fig. 4. The splines

This equation presents the general spline condition, and if the constructor is not careful, some segments can be over- and other can be under-defined. To avoid this problem the knots were set to fit in the proposed interpolation points. These points are called control points (CP).

Fig. 4 shows the sample of set conditions to design the splines x(u) and y(u). Splines from Fig. 4 are joint to the curve y(x) that is shown in Fig. 5. There are 7 conditions (n=7) to define each of splines and each of splines consists of 4 segments (m=4). According to eq. (8) this leads to the cubic spline. New inserted CP raises n and m for 1 and eq.(8) remains fulfilled.

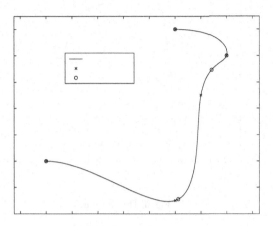

Fig. 5. The spline curve

The orientation at the start and the end point (SP and EP) are given as angles, but should be transformed to the derivative conditions. The following can be written

$$\text{SP} \quad arctg \frac{y\,(u_{\min})}{x\,(u_{\min})} \quad , \quad \text{EP} \quad arctg \frac{y\,(u_{\max})}{x\,(u_{\max})} \tag{9}$$

where x'(umin), y'(umin), x'(umax) and y'(umax) are derivatives of splines x(u) and y(u) with the respect to parameter u at the start and the end point, and must be obtained knowing only the start and end direction. This leaves some free space, so the following was proposed:

$$\sqrt{x\,(u_{SP})^2 \quad y\,(u_{SP})^2} \quad \frac{dist\ SP, first\,CP}{u_{firstCP} \quad u_{SP}} \tag{10}$$

$$\sqrt{x\,(u_{EP})^2 \quad y\,(u_{EP})^2} \quad \frac{dist\ last\,CP, EP}{u_{EP} \quad u_{lastCP}}$$

Time optimal path planning requires robots to drive with high speed. For driving with high speed smooth path is necessary. The path smoothness is presented by the curvature . When dealing with spline curves in two dimensions k is given as follows:

$$(u) \quad \frac{x\,(u)y\,(u) \quad y\,(u)x\,(u)}{x\,(u)^2 \quad y\,(u)^2 \quad ^{3/2}} \tag{11}$$

The geometrical meaning of the curvature is inverted value of circle radius in particular point (1/R). The curvature for the curve shown in Fig. 5 is presented in Fig. 6.

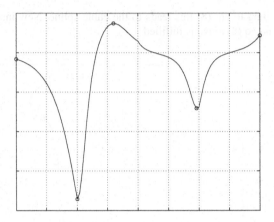

Fig. 6. The curvature

4 Finding the Optimal Path

In competition systems, such as robot soccer, the time needed by robots to get to desired points is most critical. So the problem to be solved is a minimum time problem where the time is calculated by integration of time differentials along the path

$$t = \int_{init.pos.}^{target} \frac{ds}{v} \tag{12}$$

Considering

$$ds = \sqrt{x(u)^2 + y(u)^2} \, du \tag{13}$$

Eq. (12) can be written as

$$t = \int_a^b \frac{\sqrt{x(u)^2 + y(u)^2}}{v(u)} du \tag{14}$$

To assure the real robot to follow the prescribed path, it must not slide, i.e. his accelerations must be within limits given in Fig. 3. It is well known that the time optimal systems operate on their limits, so the acceleration must be on the ellipse given in Fig. 3. The problem is solved by constraint numerical optimisation with control points as free parameters to be optimised. The optimisation procedure is as follows:

1. Choose initial control points and calculate the initial path. An example of this is shown in Fig. 4.
2. For given path the highest allowable overall velocity profile is calculated as follows:

Its curvature is calculated according to Eq. (11) as shown in Fig. 6.

The local extreme (local maximum of absolute value) of the curvature are determined and named turning points (TP). In these points the robot has to move with maximum allowable speed due to radial acceleration limit. Its tangential acceleration must be 0.

Before and after a TP, the robot can move faster, because the curve radius get bigger than in TP. Before and after the TP the robot can tangentially decelerate and accelerate respectively as max. allowed by (de) acceleration constraint. In this way the maximum velocity profile is determined for each TP and have the shape of "U" (or "V") as shown in Fig. 7. At some point the velocity profile becomes horizontal. The velocity there is so high, that the radial acceleration is out of limits. The part of the curve after that point is useless. This happens because the curvature starts increasing (the influence of the neighbour TP). But that neighbour TP requires lower speed in that area so the described problem doesn't really have meaning.

Similarly the maximum velocity profile (due to tangential acceleration/ deceleration) is determined for initial (SP) and final (FP) (if required) velocity respectively.

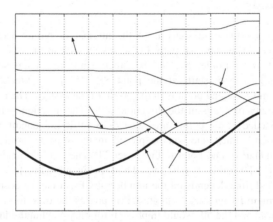

Fig. 7. Highest allowable velocity profile excluding start end end point

The highest allowable overall velocity profile is determined as the minimum of all velocity profiles, as indicated in Figs. 7 and 8 (bold curves)

The initial and final (if required) velocities must be on the highest allowable overall velocity profile (as it is in Fig. 8). If not, the given path cannot be driven without violating acceleration constraints. (The case in Fig. 7).

For given highest allowable velocity profile the cost function is calculated according Eq.(14).

3. Optimize the problem with control points as optimizing parameters.

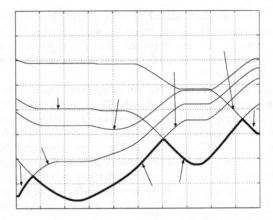

Fig. 8. Highest allowable velocity profile including start end end point limitations

5   Case Study

The objective of this case study is to the number of points needed to find good approximation of time optimal path. Let us take a look to the case for which we can say it is not very simple, but on the other hand we cannot say it is the most complicated. The robot starts at the point SP(-0.5, 1) in direction 225° with the velocity of 1 m/s. The end point is in the origin of the system. The robot should pass it with the velocity of 1 m/s in the direction 180°. The question is how many control points are needed. Two points are needed to fulfil the conditions of initial and terminal velocity. Each one can be placed in the way to ensure some minimum distance from start or end point to the closer TP. The test was made with the various number of CPs. The initial number was 2 and was increased up to 7 CPs. Fig. 9 shows how the needed time depends on the number of CPs. It can be seen that the use of 4 CPs are optimum in our case. The 4th CP improves the time for a tenth of a second (more than 6 %) and the 5th would improve it for only one hundredth of a second.

The resulted paths are shown in Fig. 10. The doted line presents 2 CP path, 3 CP path is shown with dashed line and 4 CP path with continuous line. 5, 6 or 7 CP paths are practically the same and are presented with the thick line. It can be seen where 2 and 3 CP paths spend too much time because of not well-defined path. 5 (or more) CP path is slightly different from the 4 CP one and the difference lies in the area where a large improvement cannot be done.

In some cases there would be more than 4 CPs needed to find path close to optimal. But the problem of using only 4 CPs is not critical. In case of not using enough CPs the result is not so close to optimal (time needed would increase). If the number of playing robots is taken into account, we can say that the robot with such complicated path would also need more time to reach the goal. The goal is usually to kick the ball and that is job just for one robot. The supervisory algorithm who controls the roles of the robots would choose the robot with minimum time needed to do that and would probably not choose the robot with complicated path.

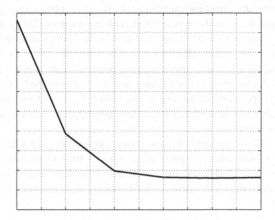

Fig. 9. Time needed according to the number of Control points

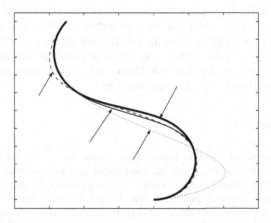

Fig. 10. Optimal paths constructed from the different number of Control points

6 The Application Aspect

The proposed technique uses optimisation to find optimal solution. As it is well known, the optimisation is very time-consuming. The particular problem becomes burning when the realisation is taken into account. The robot's control algorithm acts in the following way. First the path is planned, then the control action is calculated from planned path using the inverted model of the robot. This is repeated each time instant. The time allocated to the path planning is therefore shorter than sample time. In the dynamically changing environment, like it is robot soccer game, short sample time is required. Actually it is defined with the camera. Using the NTSC standard camera the sample time is 33 ms, and this is far shorter time than time needed for optimisation. The idea that solves this problem is called multi-parametric programming. For a grid of initial relative position of the robot regarding to the ball, the paths (CPs) are obtained in advance and are stored to the look-up table. Inputs are relative robot position, initial angle, initial and final velocity and outputs were the CPs. The table was determined for certain quantization. For the intermediate points, linear interpolation was used.

The use of look-up table also increases cooperating capabilities. Robots can very quickly determine which of them needs shorter time to perform an action. Shorter time if often closely related to the effectiveness. Such precision path planning offers a lot of support to the multi agent decision-making algorithm that is in charge for robot cooperation.

7   Conclussions

The path finding algorithm for nonholonomic mobile robots was proposed. The case study concerned slippery conditions in robot soccer environment. The path is presented as a spline curve and was got with the control points positioning. The control points were placed using the optimisation function where the criterion was needed time. The optimisation is very time-consuming process and cannot be done online, so the look-up table was built. Due to well defined future moving of all robots, the cooperation between players also improved.

References

1. The MathWorks Inc., Spline Toolbox User's Guide, Version 2, 1999.
2. Asada, M., Uchibe, E., Hosoda, K., Cooperative behavior acquisition for mobile robots in dynamically changing real worlds via vision-based reinforcement learning and development. Artificial Intelligence, Vol. 110, pp. 275-292, 1999.
3. Candea, C., Hu, H., Iocchi, L., Nardi, D., Piaggio, M., Coordination in multi-agent RoboCup teams. Robotics and Autonomous Systems, Vol.36, pp. 67–86, 2001.
4. Desaulniers, G., On Shortest paths for a car-like robot maneuvering around obstacles. Robotics and Autonomous Systems, Vol. 17, pp. 139–148, 1996.
5. Ginnis, A. I., Kaklis, P. D., Planar C2 cubic spline interpolation under geometric boundary conditions, Computer Aided Geometric Design, Vol. 19, Iss. 5, pp. 345-363, 2002.
6. Klančar, G., Orqueda, O., Matko, D., Karba, R., Robust and efficient vision system for mobile robots control-application to soccer robots. Electrotechnical Review, Journal for Electrical Engineering and Computer Science, Vol 68, No 5, pp 306-312, 2001.
7. Martin, C. F., Sun, S., Egerstedt, M., Optimal Control, Statistics and Path Planning. Mathematical and Computer Modelling, Vol. 33, pp. 237-253, 2001.

8. Matko, D., Klančar, G. Lepetič, M., A Tool For the Analysis of Robot Soccer Game, Proceedings of the 2002 FIRA Robot World Congress, Vol. 1, pp. 743-748, Seoul, May 26-29 2002.

9. Podsedkowski, L., Nowakowski, J., Idzikowski, M., Vizvary, I., A new solution for path planning in partially known or unknown environment for nonholonomic mobile robots. Robotics and Autonomous Systems, Vol. 34, pp. 145–152, 2001.

10. –krjanc, I., Klančar, G., Lepetič, M., Modeling and Simulation of Prediction Kick in Robo-Football. Proceedings of the 2002 FIRA Robot World Congress, Vol. 1, pp. 616-619, Seoul, May 26-29 2002.

11. –vestka, P., Overmars, M. H., Coordinated path planning for multiple robots. Robotics and Autonomous Systems, Vol. 23, pp. 125-152, 1998.

12. Ting, Y., Lei, W. I., Jar, H. C., A Path Planning Algorithm for Industrial Robots. Computers & Industrial Engineering, Vol. 42, pp. 299-308, 2002

Analysis by Synthesis, a Novel Method in Mobile Robot Self-Localization

Alireza Fadaei Tehrani[1, 2], Raúl Rojas[4], Hamid Reza Moballegh[1, 3],
Iraj Hosseini[1, 3], and Pooyan Amini[1, 3]

[1] Robotic Center of Isfahan University of Technology
http://robocup.iut.ac.ir
[2] Mechanical Engineering Department, Isfahan University of Technology (IUT)
mcjaft@cc.iut.ac.ir
[3] Electrical Engineering Department, Isfahan University of Technology (IUT)
iraj@sarv-net.com
{moballegh, pooyan_amini}@yahoo.com
[4] Freie Universität Berlin, Takustraße 9, 14195 Berlin, Germany
rojas@inf.fu-berlin.de

Abstract. Fast and accurate self-localization is one of the most important problems in autonomous mobile robots. In this paper, an analysis by synthesis method is presented for optimizing the self-localization procedure. In the synthesis phase of this method, the robot's observation of the field is predicted using the results of odometry. It is done by calculating the position of the landmarks on the captured image. In the analysis phase, the local search algorithms find the exact position of the landmarks on the image from which the best matching coordinates of the robot are determined using a likelihood function. The final coordinates of the robot are then obtained from the odometry sensor, using an integrated delay compensation and correction technique. Experimental results show that precise and delay-free results are achieved with a very low computational cost.

1 Introduction

Self-localization is one of the most challenging problems in the field of autonomous mobile robots. Accuracy of the obtained localization data directly affects the quality of the mission to be done by the robots. In most cases, self-localization must be done in a dynamic, uncertain environment containing a great amount of noise. Such an environment leads to use reliable and precise sensors for data acquisition.

The common sensors used in this field are omni-directional and frontal cameras [3] [13], odometry [4] [7], ultrasonic sensors [6] [12], laser range finders (LRF) [10] [19], and infrared (IR) field surface detectors. Among these, the vision and odometry-based self-localization techniques are currently implemented in most mobile robotic projects [3] [4].

Vision-based self-localization suffers from several disadvantages such as high computational cost, low sample rate, unacceptable delay and inadequate accuracy. According to this, vision techniques are currently used together with other

D. Nardi et al. (Eds.): RoboCup 2004, LNAI 3276, pp. 586–593, 2005.

self-localization techniques such as odometry [3]. Several techniques have been developed to fuse odometry and vision outputs together in order to gain more precise and real time results, such as Kalman filter, Markov and Monte Carlo Localization (MCL) [17] [18] [19], and Complementary filtering [20]. In these techniques, vision and odometry algorithms run independently and another module is responsible for fusing their results and calculating the final position and orientation.

In this paper we have introduced a new approach to perform data fusion. In this approach, one of the sensors (usually the simpler sensor) is used to predict another one. This is called the synthesis phase. The position of the landmarks which appear on the captured image is predicted based on the results of the odometry sensor. Having this information, a fast local search can be performed in order to determine the exact position of the landmarks. This is similar to tracking methods except that in tracking, the search is done around the last position of the landmarks on the image. The search area in tracking techniques needs to become wider as the velocity of the robot increases while in our method the search area around the predicted position remains constant. The results of the search algorithm are then converted to the robot coordinates by a probabilistic map matching technique. An integrated delay compensation and correction method is then implemented to obtain the final coordinates of the robot.

The approach has been tested in the RoboCup scenario in which robot positioning is usually a requirement for successful coordination and overall team behavior. The team PERSIA obtained the 3rd place in the middle size league of RoboCup 2003 Italy.

2 Description of the System

Fig. 1 shows the block diagram of the designed self-localization system. In this method, localization data is provided from two sources: omni-directional camera and the odometry system.

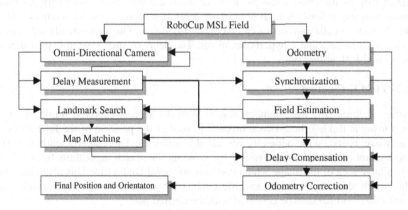

Fig. 1. Block diagram of the self-localization algorithm

The procedure first synchronizes the output of odometry with the camera and then predicts the position of the landmarks on the image based on the synchronized odometric results. Having a preliminary knowledge about the position of the landmarks, the procedure then refers to the image and performs an exact local search to calculate the accurate position of each landmark in the image. The calculated parameters form a local feature map which is matched to the global field map using a likelihood function that is maximized at the most probable coordinates of the robot. Finally an integrated delay compensation and correction technique is used to synchronize the two sources and also eliminate the delay of the final output. Further parts of this section describe the function of each module in detail.

2.1 Odometry

An approach to decrease the cumulative errors of the odometry sensors is presented in [7] by separating the driving and odometry wheels which increases precision of the sensor to a great extent. In the present method, odometry provides the preliminary estimation of the robot coordinates which is then improved by the further parts of the system using vision-based techniques. It is also used to eliminate the delay caused by the vision module.

2.2 Omni-Directional Camera

In many systems, vision is used as a proper complement to cooperate with odometry. However, the vision-based localization is robust enough to be implemented individually. The reason for fusing odometry and vision is high amount of delay, low precision and heavy process of the vision-based techniques.

Omni-directional cameras are preferred to frontal cameras because they can provide the robot with 360° observation range [13] [16].

Another shortcoming of the vision-based techniques is their delay, which is generated partly by the camera and its interface hardware and driver and partly by image processing algorithms. A high amount of delay can easily cause the control system that uses the localization data as its feedback to diverge. Our vision hardware is composed of a camera and a hyperbolic mirror. The delay of such system is measured to be around 100 milliseconds.

2.3 Field Estimation

After synchronizing the odometry results with the current captured frame, the next step towards linking these two sources is to predict the position of the landmarks on the image, based on the odometry generated coordinates. This can efficiently decrease the amount of needed computations, by avoiding the search algorithms from searching a large area of the image which contains no useful landmarks. This also simplifies the search algorithm for each specific landmark because the search area rarely contains other landmarks to be confused with the desired one.

The function that maps the surface of the field to the captured image is a radial transform. Under this transform, the viewing angle of the objects around the robot remains unchanged on the image, while their distance from the center of the image is not linearly related to its real value. Also, every line which is perpendicular to the field surface appears as a radial line in the captured image. According to this phenomenon, the field can be predicted as a set of distances and angles which describe each landmark as it appears in the polar coordinate system placed on the center of the image. The following objects are defined as landmarks; goal vertical bars, intersections between goal inner corners and the field surface, and poles.

The field prediction function is not only used to predict the position of the landmarks in the image, but also it forms the global feature map of the field in the map matching procedure.

2.4 Landmark Detection Algorithms

It is very complicated to search the image for a landmark while having no knowledge of its position because almost every landmark in the MSL field consists of blue and yellow parts and can not be simply distinguished from others according to its color.

Another problem which occurs during the landmark detection is the possibility of having a landmark occluded by another landmark or robots. For example the poles may be occluded by the goal sides from some observation points inside the field.

These problems can be solved by performing a local search around the result of field prediction procedure for each landmark. According to this predicted position, a proper search region can be defined for the detection algorithm that is not related to the velocity of the robot. For most of the landmarks, the search region contains no other landmarks. The region sometimes contains part of the neighbor landmark, so the detection algorithm in the worst case should only distinguish between two neighbor landmarks. This will result in further simplification of the algorithms which indeed, efficiently decreases their computational cost.

2.5 Map Matching

After finding the exact position of each landmark on the image, another module is needed to calculate the position and orientation of the robot. Two methods for performing this task are discussed in this section.

In the first method, the robot coordinates are calculated by triangulation as in [8]. One of the drawbacks of the triangulation method is the high sensitivity of the algorithm to incorrect input parameters. The reason for this sensitivity is that the triangulation algorithms use no means of redundancy in their calculations. Obviously, the robot can localize itself by knowing the position of at least 3 landmarks on the image. The triangulation methods typically use 3-5 parameters to calculate the robot coordinates while more than 15 parameters can be easily derived from the image.

The second method is the probabilistic map matching, which is designed to take advantage of redundancies in the input parameters to reduce the sensitivity of the algorithm to incorrect inputs which may be generated due to noise and occlusion. In this method the best match between the landmark detection results and the global field map is found using maximum likelihood estimation.

In order to compare the feature parameters in the global field map with the detected parameters, a maximum likelihood similarity measure is defined. The function must be designed so that it is not affected seriously if one or more parameters are presented incorrectly. Such a function can be composed as (1) in which S and S' are vectors of the measured and estimated features (angles and distances) and $N_i(.)$ is the ith normal function defined in (2) for comparing corresponding elements of S_i and S'_i. α_i presents the acceptable error range in the compared parameters. Using trial and error, the proper α_i for distances is found to be 0.02 and for angles is found to be 0.1.

$$L(S,S) \quad \sum N_i(S_i \quad S_i) \tag{1}$$

$$N_i(x) \quad e^{\quad x^2} \tag{2}$$

Fig. 2 shows the result of applying the likelihood function for comparing a set of features calculated from a typical point with the global field map.

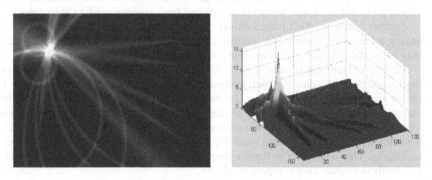

Fig. 2. The likelihood function is used to find the position of the robot

Applying the maximum likelihood measure to all possible coordinates of the robot will need a great amount of computations. In order to decrease the computational cost of the algorithm, the matching can be done in a neighborhood of the odometric coordinates of the robot.

2.6 Global Search

The final coordinates of the robot in the introduced algorithm are the vision-improved version of the odometric coordinates. Since odometry is only capable of tracking the displacement and rotation of the robot, the system must be initialized once with the correct coordinates in order to be able to localize the robot. In this case the local landmark detection algorithms are temporarily substituted with a global detection algorithm for a limited number of landmarks. It is known from the features of omni-directional cameras that both goals can always be viewed in a region which is restricted between two circles. A good approach is to detect the vertical goal bars on a circle inside this area, using an edge detector that finds the edges of the blue and yellow regions.

3   Experimental Results

In order to verify the improvement of the technique introduced in this article, we have compared several features of the analysis by synthesis technique with the previous version of the self-localization algorithm used by the robots. This algorithm is similar to many of the current algorithms in its performance, accuracy and computational cost. In this algorithm, the landmark detection is done by region growing and the coordinates of the robot are calculated using the triangulation method. Both algorithms have run on a 1.2 GHz Pentium III notebook. In the following sections two test procedures to compare accuracy and computational cost of the algorithms are introduced.

3.1   Accuracy

The accuracy of the self-localization systems can be compared based on two parameters; the average of error and its standard deviation. The standard deviation of the localization error is important because in many systems, a medium constant error is more acceptable than a lower value but rapidly varying one. The average values and standard deviation of position and orientation errors are given in Table 1.

Table 1. The average and standard deviation of the localization errors

	Position Err. Avg. (cm)	Position Err. SD (cm)	Orientation Err. Avg. (deg)	Orientation Err. SD (deg)
Old Method	24	12	5.1	2.4
New Method	3.2	0.7	1.5	0.3

3.2   Computational Cost

Computational cost of self-localization algorithms is the most important factor in implementation of the mobile robots because these algorithms are computationally much more expensive than the other algorithms running on the CPU of the robot. The analysis by synthesis method has reduced this usage to a great amount. In order to compare the computational cost of the algorithms the performance indicator of Windows 2000 is used.

Another feature of the algorithms which should be compared is the constancy of their amount of computations during the game. This feature can be measured by determining the CPU usage of both algorithms when localizing the robot in different positions inside the field. A comparison of CPU usage between the algorithms is made in Table 2. Three typical points inside the field are selected for this comparison, which are center of the field, inside the goals and near the poles.

Table 2. Comparison of CPU usage between new and old methods

	Center of the Field	Inside the Goals	Near the Poles
Old Method	53%	82%	64%
New Method	39%	36%	42%

4 Conclusion

Joint vision-odometry self-localization of moving robots has attracted the attention of many researchers. In this work, the performance of the mobile robot self-localization is improved by proposing a new method called analysis by synthesis. In this method, the vision sensor of the system is first predicted (synthesized) using the odometric coordinates of the robot. The image is then analyzed based on the preliminary prediction to achieve more accuracy in the robot coordinates. An integrated correction and delay compensation module then corrects the coordinates which are obtained from odometry sensor and obtains the final coordinates of the robot. This method has not only reduced the computational cost of the self-localization algorithm but also increased the precision and reliability of the outputs. Experimental results clearly verify the improvements.

References

1. P. MacKenzie and G. Dudek, Precise Positioning Using Model-Based Maps, *Proceedings IEEE International Conference on Robotics and Automation,* San Diego, CA, May 1994.
2. C. Marques, P. Lima, A Localization Method for a Soccer Robot Using a Vision-Based Omni-Directional Sensor, *The 4rth International Workshop on RoboCup,* 2000.
3. F. Marando, M. Piaggio and A. Scalzo, Real Time Self Localization Using a Single Frontal Camera, *The 9th International Symposium on Intelligent Robotic Systems (SIRS 2001),* LAAS-CNRS, Toulouse, France, July 2001.
4. A. Motomura, T. Matsuoka and T. Hasegawa, Self-Localization Method Using Two Landmarks and Dead Reckoning for Autonomous Mobile Soccer Robots, *The 7th International RoboCup Symposium,* Padua, Italy, July 2003.
5. F. V. Hundelshausen and R. Rojas, Tracking Regions, *The 7th International RoboCup Symposium,* Padua, Italy, July 2003.
6. C-C. Tasi, A Localization System of a Mobile Robot by Fusing Dead-Reckoning and Ultrasonic Measurements, *IEEE Transaction on Instrumentation and Measurement,* Vol.47, No.5, pp. 1399-1404, 1998.
7. A. Fadaei Tehrani, A. M. Doosthosseini, H. R. Moballegh, P. Amini, M. M. Daneshpanah, A New Odometry System to Reduce Asymmetric Errors for Omnidirectional Mobile Robots, *The 7th International RoboCup Symposium,* Padua, Italy, July 2003.
8. J. Ji, G. Indiveri, P. Ploeger, A. Bredenfeld, An Omni-Vision based Self-Localization Method for Soccer Robot, *Proceedings of the IEEE IV2003, Intelligent Vehicles Symposium,* Columbus, Ohio, USA, June 2003.
9. H. Utz, A. Neubeck, G. Mayer and G. Kraetzschmar, Improving Vision Based Self-Localization, *The 6th International RoboCup Symposium,* Fukuoka, Japan, June 2002.
10. L. ™Mlud, L. Kopečný, T. Neužil, Laser Proximity Scanner Correlation Based Method for Cooperative Localization and Map Building, *The 7th International Workshop on Advanced Motion Control (AMC'02),* Maribor, Slovenia, July 2002.
11. C. F. Olson, Probabilistic Self-Localization for Mobile Robots, *IEEE Transactions on Robotics and Automation,* Vol. 16, No. 1, pp. 55-66, February 2000.
12. A. Ohya, A. Kosaka and A. Kak, Vision-Based Navigation by a Mobile Robot with Obstacle Avoidance Using Single-Camera Vision and Ultrasonic Sensing, *IEEE Transaction on Robotics and Automation,* Vol. 14, No. 6, pp. 969-978, December 1998.

13. C. F. Marques, P. U. Lima, Vision-Based Self-Localization for Soccer Robots, *Proceedings of the 2000 IEEE/RSJ International Conference on Intelligent Robots and Systems (IROS 2000)*, Vol. 2, pp. 1193-1198, 2000.
14. M. Schmitt, M. Rous, A. Matsikis, K.-F. Kraiss, Vision-based Self-Localization of a Mobile Robot Using a Virtual Environment, *Proceedings of the 1999 IEEE International Conference on Robotics & Automation (ICRA 1999)*, Detroit, Michigan, pp. 2911-2916, May 1999.
15. F. V. Hundelshausen, M. Schreiber, F. Wiesel, A. Liers and R. Rojas, MATRIX: A force field pattern matching method for mobile robots, Technical Report B-08-03, *Free University of Berlin*, June 2003.
16. A. Bonarini, P. Aliverti and M. Lucioni, An Omnidirectional Vision Sensor for Fast Tracking for Mobile Robots, *IEEE Transaction on Instrumentation and Measurement*, Vol. 49, No. 3, pp. 509-512, June 2000.
17. J. Wolf and A. Pinz, Particle Filter for Self Localization using Panoramic Vision, *The 27th Workshop of the Austrian Association for Pattern Recognition (OEAGM 2003)*, June 2003
18. J.-S. Gutmann, D. Fox, An Experimental Comparison of Localization Methods Continued, *Proceedings of the 2002 IEEE/RSJ International Conference on Intelligent Robots and Systems*, pp. 454-459, EFPL, Lausanne, Switzerland, October 2002.
19. J.-S. Gutmann, T. Weigel, B. Nebel, A Fast, Accurate, and Robust Method for Self-Localization in Polygonal Environments Using Laser-Range-Finders, *Advanced Robotics Journal*, Vol. 14, No. 8, pp.651-668, 2001
20. S. You and U. Neumann, Fusion of Vision and Gyro Tracking for Robust Augmented Reality Registration, *Proceedings IEEE International Conference on Virtual Reality*, pp. 71-78, Yokohama, Japan, March 2001.

Robots from Nowhere

Hatice Köse and H. Levent Akın

Boğaziçi University,
Department of Computer Engineering,
34342 Bebek, Istanbul, Turkey
{kose, akin}@boun.edu.tr

Abstract. In this study, a new method called Reverse Monte Carlo Localization (R-MCL) for global localization of autonomous mobile agents in the robotic soccer domain is proposed to overcome the uncertainty in the sensors, environment and the motion model. This is a hybrid method based on both Markov Localization(ML) and Monte Carlo Localization(MCL) where the ML module finds the region where the robot should be and MCL predicts the geometrical location with high precision by selecting samples in this region. The method is very robust and fast and requires less computational power and memory compared to similar approaches and is accurate enough for high level decision making which is vital for robot soccer.

Keywords: Global localization, ML, MCL, robot soccer.

1   Introduction

The localization problem is estimation of the position of a robot relative to the environment, using its actions and sensor readings. Unfortunately these sensors and the environment are uncertain, so the results are typically erroneous and inaccurate. Consequently, localization still remains as a nontrivial and challenging problem and from the simplest geometric calculations which do not consider uncertainty at all, to statistical solutions which cope with uncertainty by applying sophisticated models, many solutions have been proposed for this problem [1], [2], [3]. Although some of these approaches produce remarkable results, due to the nature of the typical environments they are not satisfactory because fast solutions with less memory and computational resources are demanded. This is especially true for a real-time application in a dynamical soccer field using robots with onboard computational resources. Generally, solutions producing precise results suffer from slowness, and high memory usage. Whereas a fast solution in practice typically produces only coarse results. Even when they produce precise local results, some approaches like Kalman filters, fail to find the global position.

This work is a part of the Cerberus Team Robot soccer project [4], and aims to localize the legged robots in the soccer field globally, while solving problems mentioned above. There are a several limitations and assumptions related to the rules of the Robocup [5]. In this work, three approaches to solve this problem were developed in parallel. The first of these new approaches is a new geometrical localization algorithm, which is based on just a single landmark observation at a time. This approach is later extended

D. Nardi et al. (Eds.): RoboCup 2004, LNAI 3276, pp. 594–601, 2005.

to a ML based method. In addition, a novel hybrid approach called Reverse Monte Carlo Localization(R-MCL) combining the ML and MCL methods is designed and implemented.

The organization of the paper is as follows: In the second section, a survey of localization methods is presented. In the third section detailed information about the proposed approach can be found. In the fourth section, the results of the application of proposed approach are present. In the fifth section, conclusions and suggestions for future work are given.

2   Localization Methods

The simplest localization method depending on the range and bearing data is triangulation, which uses geometry to compute a single point that is closest to the current location. But in real world applications a robot can never know where it is exactly because of the uncertainty in its sensors, and the environment. Consequently, several different approaches which estimate the position of robot probabilistically were introduced to integrate this uncertainty into the solutions.

Kalman filter (Kalman-Bucy filter) is a well-known approach for this problem. This filter integrates uncertainty into computations by making the assumption of Gaussian distributions to represent all densities including positions, odometric and sensory measurements. Since only one pose hypothesis can be represented, the method is unable to make global localization, and can not recover from total localization failures [6], [7], [3].

Many works consider Markov localization (ML) [1], [8]. ML is similar to the Kalman filter approach, but it does not make a Gaussian distribution assumption and

Table 1. Comparison of Localization Methods

Method	Capability of global localization	Accuracy	Speed	Memory usage	Robustness to noise	Fast recovery from kidnapping
EKF	no	H	H	L	L	L
ML	yes	L**	M	H**	H	H
MCL	yes	M**	M	H**	M	M
SRL1***	yes	M**	M	M**	M	L
SRL2***	yes	M**	M	M**	L	H
A-MCL	yes	M**	M	M**	XH	H
M-MCL	yes	M**	M	M**	H	H
ML-EKF	yes	M**	M	H**	H	H
Fuzzy*	yes	L**	M	H**	H	H
Geometrical	yes	H**	H	L	L	L
R-MCL	yes	M**	H	H**	H	H

* Fuzzy method is the method implemented in [12].

** These are grid based and sample based methods. So accuracy and memory usage changes with the cell size, and the number of samples used. But they still remain in acceptable ranges.

*** SRL1 and SRL2 differ in their wish to accept additional samples on each noisy observation. This leads fast recovery from kidnapping but increase noise and decrease accuracy.

allows any kind of distribution to be used. Although this feature makes this approach flexible, it adds a computational overhead.

Monte Carlo Localization (MCL) is a version of Markov localization that relies on sample-based representation and the sampling/importance re-sampling algorithm for belief propagation [2], [9]. Beliefs are represented by a set of K weighed samples (particles) which are of type $((x, y, \theta), p)$, where p's are positive numerical weighting factors such that sum of all p is 1. Odometric and sensory updates are similar to ML. Most of the MCL based works suffer from the kidnapping problem, since this approach collapses when the current estimate does not fit observations. There are several extensions to MCL that solve this problem by adding random samples at each iteration. Some of these methods are Sensor Resetting Localization (SRL), Mixture MCL (Mix-MCL), and Adaptive MCL (A-MCL). In SRL, when the likelihood of the current observation is below a threshold, a small fraction of uniformly distributed random samples is added [10]. Mix-MCL additionally weights these samples with current probability density. This method has been developed for extremely accurate sensor information [3]. Adaptive MCL only adds samples when the difference between short-term estimate (slow changing noise level in the environment and the sensors) and the long-term estimate (rapid changes in the likelihood due to a position failure) is above a threshold. The key idea is to use a combination of two smoothed estimates (long term and short term) of the observation likelihoods [3].

ML-EKF method is a hybrid method aiming to make use of the advantages of both methods, taking into consideration the fact that ML is more robust and EKF is more accurate. So this method finds the location of the agent coarsely by grid based ML and then inside this area uses EKF to find a more accurate solution [3].

Although there have been only a few fuzzy logic based approaches, they appear to be promising [11], [12]. In these approaches, the uncertainty in sensor readings (distance and heading to beacons) is represented by fuzzy sets. The above mentioned localization approaches and their capabilities are summarized in Table 1.

Some of the comparisons of the known algorithms in this part are based on [3]. There is an ongoing work for testing of these methods with same data set for comparison.

3   The Proposed Approach

In robot soccer, teams of robots, that are capable of seeing and moving, play matches against each other, and the team with the highest goal score win the match. In order to do this, the player robots must detect their location, the goals, the ball, the members of their team and the opponent team members(optional for high level planning), and place the ball in the opponent team's goal to score a goal. A robot is typically expected to find its own location using the six distinguishable unique landmarks in the field, and then use this information to find the location of the ball and goal. Consequently, localization is a vital problem for robot soccer.

Since, as discussed in section 2, no single method satisfies all the needs in terms of single robot localization, three methods including a hybrid algorithm which tend to integrate the advantageous parts of single methods and overcome the deficiencies were developed in this work.

3.1   Geometrical Localization

The geometrical localization method assumes the input data is measured exactly (does not contain noise), and therefore does not need any error modeling. Our previous algorithm in [13] required at least two landmarks to be seen at any time to calculate the position accurately. Although it worked also for the one landmark case, it could not give satisfactory results. This new method is designed to work with one landmark information which is much more realistic within the new field sizes. So even if the robot sees more than one landmark, they are treated separately and one-landmark information is used at each step. The ratio of the distance between the predicted location and the observed location of the landmark is used to predict the new x and y coordinates of the robot.

The bearing is also found by using the new predicted x and y coordinates. Whenever a new visual data comes, the new position is calculated based on the measurement and old position. A point between the newly measured position and the old position is taken as the new position. This new position is placed between the two positions proportional to the belief of the robot on them. The assumption here is that: The more you believe in a position the closer you are to that position. This is used to reduce the effect of inaccurate measurements on the new position. When the odometric data arrives, the position is blurred among the moved distance and heading. The bearing is added to the original heading and it is normalized to give the new heading of the robot. The odometric data consists of the distance moved forward, left and the bearing of turn. This method assumes that the measurements are exact, or noise is below a threshold.

3.2   Markov Localization

The assumption made about error in the previous subsection is not correct in general. So we need a method to handle the uncertainty in the visual and odometric data. Since ML is a grid-based algorithm, it gives a coarse but robust result. Unfortunately it requires complex computations since at every step all grid cells should be taken into account. Although the accuracy of the result might not be sufficient for implementing high level planning which is required for robot soccer. It can converge faster than the sample based algorithms. So a ML based method is designed and implemented. This method works in a similar manner as the geometrical localization algorithm (Figure 1):

```
For all grid cells
        If visual data available
                Apply visual update
        Else if odometric data availabla
                Apply odometric update
        Calculate probability of each cell according to ML
        // use current beliefs and probability, and newly calculated belief and probability
        //of each cell and calculate the weighted average
return the best cells (with maximum probability)
```

Fig. 1. The ML based proposed algorithm

Unfortunately, as the uncertainty increases, the maximum probability decreases and the number of grid cells with the maximum probability increases, so the accuracy of the final position decreases. One solution is to use the averages of the centers of the cells, as the final position. But this parameter will not cover all space uniformly. So it would be useful to use a sample based method within the ML method, in order to increase coverage speed and accuracy. If a method is used to update only several cells but not all at each update, then the computational complexity would decrease drastically, too.

3.3   R-MCL

As indicated in the previous method, ML is robust and converges fast, but coarse and computationally complex. On the other hand, sample based MCL is not as computationally complex as ML, and gives accurate results. However, it can not converge to a position as fast as ML, especially in the case of an external impact on the position of the robot (such as kidnapping). In addition, the number of samples to be used is generally kept very high to cover all space and converge to the right position. There are several extensions for adaptive sample size usage, but these still do not solve the slow coverage problem. So it might be useful to converge to several cells by ML or another grid based method, then inside these bulk of grids, produce a limited number of samples to find the final position. The average of these samples would give the final position and the standard deviation might give the uncertainty of the final position as in the MCL based methods. The algorithm in Figure 2 simply works as shown in Figure 3:

In the original MCL, the number of samples is increased to decrease bias in the result. In R-MCL since we converge by selecting cells with maximum probability, so the bias is already decreased, therefore we do not need this to decrease bias.

After testing this version, some improvements were done on the current version of R-MCL and ML. In the modified ML, not only the distance but also the bearing information is used to find the best grids, so the number of chosen grids decrease and confidence increases. As a result of these modifications, the accuracy of the results improved considerably. Later when the samples are drawn, also the best samples are selected using distance and the bearing from these very good cells, and their average is returned as the current pose. Notice that, samples are taken into consideration only when the position reaches to a certainty level, in other words the number of chosen cells are

Apply visual/odometric update to all cells
Choose grid cells with probability > threshold as in ML //only choosing the cells with maximum probability might not be
//adequate due to test field's conditions
Apply resampling
       Calculate the number of cells to be produced according to your level of uncertainty
       Produce random samples of this amount from the chosen cells
       Pick up samples with probability> threshold from this sample set
Find the average of the positions of chosen samples and return as final position
Find standard deviation of the positions of samples and return as the uncertainty of the final position

Fig. 2. The R-MCL based proposed algorithm

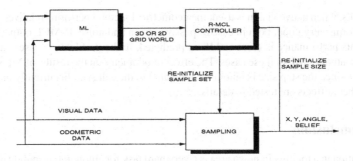

Fig. 3. The R-MCL working schema

below a limit(e.g. 50), and there is at least one very good cell which is below or equal to the minimum error in both distance and bearing limitations. Also if there are no samples which satisfy the minimum bearing and distance error condition then the results of ML are used instead. Also the bearing of the new pose is found by the ML module inside the R-MCL because it is more accurate and robust.

4   Tests and Results

A simulator was developed which is used to produce realistic data for testing the implementations. It produces both odometric and visual data according to the user's choice, besides it enables the addition of random noise to the produced data optionally, to make it more realistic. The new algorithms are tested on a set of tests based on fixed paths in the field. In the first group of tests there is active vision and in the second, the robot could take information about one landmark so it is a very hard and challenging case. Also noise with a magnitude of 20 cm (chosen very high-twice the cell size- to observe the effects easily) is added in some tests, and likewise odometry is included in a group of tests. The tests of R-MCL are repeated 50 times each, because the samples are drawn randomly and this affects the accuracy of the results. The average and standard deviation of the errors that are produced during the tests are presented in the Table 2. As seen from the

Table 2. Experimental results

test no	active vision	noise	odometry	geometrical	ml	r-mcl
1	yes	no	no	18.13±35.60	4.27±1.97	3.68±1.84
2	yes	yes	no	20.12±33.56	16.63±24.10	17.22±24.08
3	yes	no	yes	18.13±35.61	4.27±1.97	3.72±1.55
4	yes	yes	yes	23.48±31.70	11.30±2.73	9.13±3.46
5	no	no	no	168.60±1.68	5.70±0.68	4.06±1.75
6	no	yes	no	169.29±2.65	35.28±27.94	34.41±25.12
7	no	no	yes	168.76±1.75	5.70±0.68	3.37±2.49
8	no	yes	yes	167.94±3.01	5.69±0.68	3.90±1.77

test results, when active vision is used, the geometrical method performs relatively well. ML performs very good, both in case of passive vision and noise. R-MCL outperforms all, and its performance is close to ML as predicted, in the problematic cases, since it uses the output of ML in these cases. The effect of odometry on the results of ML seems ignorable since the step size is ignorable compared to the cell size. In currently ongoing works, these effects are tested in details.

5   Conclusions

Localization in a totally unknown area is a very hard task for autonomous mobile robots. This work aims to propose a fast, reliable, computationally and resource efficient solution to global localization problem. The solution should be successful in environments like the Robocup Games and the challenges which require very high accuracy and speed. For this reason in this paper several new localization algorithms are developed. The first method is a new geometrical localization method which is based on one observation at a time. This property makes this method more robust and realistic. Extending this idea, a ML based method is also developed. Next, to make use of the robustness of ML but to make the results more precise, a hybrid method called R-MCL method is implemented. This method recovers from kidnapping in a very fast manner and it is very robust to even high levels of noise and inadequate environmental information. It is higly accurate and its accuracy can be improved by improving resampling and choosing the best cells and samples. It performs quite well when compared with the outperforming methods such as ML-EKF, and A-MCL, and there is an ongoing research for testing it with these methods on the same data set from ERS 210 quadrupled robots.

Acknowledgements

This project is supported by the Boğaziçi University Foundation and by Boğaziçi University Research Fund project 03A101D.

References

1. W. Burgard, D. Fox, ,D. Hennig, and T. Schmidt, "Estimating the Absolute Position of a Mobile Robot Using Position Probability Grids," *Institut für Informatik III., Universtiat Bonn, Proc. of the Fourteenth National Conference on Artificial Intelligence (AAAI-96)*, 1996.
2. S. Thrun, D. Fox, W. Burgard, and F. Dellaert, "Robust Monte Carlo Localization for Mobile Robots," *Elsevier, Artificial Intelligence,* 128, pp. 99-141. 2001.
3. J.S. Gutmann, and D. Fox, "An Experimental Comparison of Localization Methods Continued," *In Proc. of the 2002 IEEE/RSJ Int. Conf. on Intelligent Robots and Systems (IROS'02)*, Page: 454-459, Lausanne, Switzerland October 2002.
4. Cerberus, *http://robot.cmpe.boun.edu.tr/aibo/home.php3,* 2003.
5. Robocup, , *http://www.robocup2003.org/,* 2003.
6. A.W. Stroupe, T. Balch, "Collaborative Probabilistic Constraint Based Landmark Localization," *Proceedings of the 2002 IEEE/RSJ Int. Conf. on Intelligent Robots and Systems IROS'02)*, Page: 447-452, Lausanne, Switzerland, October 2002.

7. A.W. Stroupe, K. Sikorski, and T. Balch, "Constraint-Based Landmark Localization," *Proceedings of 2002 RoboCup Symposium,* 2002.
8. D. Fox, W. Burgard, and S. Thrun, "Markov Localization for Mobile Robots in Dynamic Environments" *Journal of Artificial Intelligence Research 11,*pp. 391-427., 1999.
9. D. Schulz, and W. Burgard, "Probabilistic State Estimation of Dynamic Objects with a Moving Mobile Robot," *Elsevier, Robotics and Autonomous Systems 34,*pp. 107-115. 2001.
10. S. Lenser, and M. Veloso, "Sensor Resetting Localization for Poorly Modelled Mobile Robots" *Proceedings of ICRA 2000, IEEE,* 2000.
11. P. Buschka, A. Saffiotti, and Z. Wasik, "Fuzzy Landmark-Based Localization for a Legged Robot" *Proc. of the IEEE/RSJ Intl. Conf. on Intelligent Robots and Systems (IROS),* Takamatsu, Japan, pp. 1205-1210, 2000.
12. H. Kose, S. Bayhan, and H. L.Akin, "A Fuzzy Approach to Global Localization in the Robot Soccer Domain" *IJCI Proceedings of International XII Turkish Symposium on Artificial Intelligence and Neural Networks (TAINN 2003),* ISSN 1304-2386 Vol:1, No:1 pp.1-7, July 2003.
13. H.L. Akin, A. Topalov, and O. Kaynak, "Cerberus 2001 Team Description," *Robocup 2001: Robot Soccer World Cup V, Birk, A., Coradeschi, S., and Tadokoro, S. (Eds.),* LNAI 2377, pp.689-692, Springer Verlag. 2001.

Design and Implementation of Live Commentary System in Soccer Simulation Environment

Mohammad Nejad Sedaghat, Nina Gholami, Sina Iravanian,
and Mohammad Reza Kangavari

Intelligent Systems Lab, Computer Engineering Department,
Iran University of Science and Technology, Tehran, Iran
{mnsedaghat, sina_iravanian}@yahoo.com
n_gholami@mail.iust.ac.ir, kangavari@iust.ac.ir
http://caspian.iust.ac.ir

Abstract. Soccer simulation commentary system is a suitable test bed for exploring *real time systems*. The *rapidly changing* simulation environment requires that the system generates real time comments based on the information received from the *Soccer Server*. In this article, a three-layer architecture of Caspian Soccer Commentary system is presented, and each component of the system is briefly described. The emphasis of this paper is on design and implementation of the *Analyzer* and the *Content Selector* subsystems. The Analyzer takes advantage of the *State Machine* to keep track of the game situations. The *Scheduling* and *Interruption* mechanism is proposed to improve the efficiency of the Content Selector subsystem. The presented Commentary System together with the other Caspian presentation and analysis tools won the first place in RoboCup 2003 Game Presentation and Match Analysis competitions.

1 Introduction

The development of a live commentary system for Soccer Simulation requires dealing with *time pressure* issues. That is, once the commentator recognized the game situation, he has to report it in a small time interval. This is because of the fast rate of situation change in such environments. Using simulated soccer games, makes it possible to take advantage of rich simulator's log file, instead of dealing with challenges of image processing in real soccer matches. [1][2][3]

In order to have an influence on the audience, the artificial commentator should speak through the language used by a human commentator using his common jargon. In addition the more natural voice it has the more acceptances it will receive from the audience. To achieve this, it has been decided to use *prerecorded human report* statements. It is clear that using natural human voice has a great impact on the quality of communication with the audience, but the excitement of the game cannot be experienced without the existence of the *special sound effects* like chants, applause and referee whistle. Therefore it is important to generate appropriate special sound effects according to the game trend. More details on this process in Caspian commentator is presented in "Special Sound Effects Manager" section.

D. Nardi et al. (Eds.): RoboCup 2004, LNAI 3276, pp. 602–610, 2005.

In this article Caspian commentary system architecture is presented, including its subsystems and their functionality which are listed below:

1. Analyzer
2. Statistical Analyzer
3. Content Selector
4. Special Sound Effects Manager
5. Sound Manager

The emphasis of this paper is on the design and implementation of the *Analyzer* and the *Content Selector* subsystems. The Analyzer takes advantage of the *State Machine* to keep track of the game situations. The *Scheduling* and *Interruption* mechanisms are proposed to improve the efficiency of the Content Selector subsystem.

2 Related Work

So far, three soccer commentary systems have been developed:

1. Rocco from DFKI [4]
2. Byrne from Sony CSL [5]
3. MIKE from ETL [6][7]

The functionality of these three systems is that, after receiving data from the Soccer Server in each cycle, generate comments to describe the game situation. [8]

```
"kasuga" 9 kick off,
"andhill" 5, well done,
we are life from an exciting game, team "andhill" in red versus
"kasuga" in yellow, he finds "andhill" 9,
yellow 6 intercepts the pass from "andhill" 9, forward from red 7,
yellow 4 intercepts,
still number 4,
number 9 is arriving,
ball played forward by "kasuga" 11,
failed, good luck for "andhill",
the keeper kicks off the goal,
number 2 does well there,
```

Fig. 1. An instance of the Rocco's textual commentary

Generally, the transformation process from the Soccer Server data to an appropriate report statement is done through the following steps:

1. Game analysis
2. Topic control and content selection
3. Natural language generation

Although MIKE and Rocco produce disembodied speech, Byrne uses a face as an additional means of communication. Rocco uses a *template-based generator* instead of fully fledged natural language generation components. That is, the language is generated by selecting templates consisting of strings and variables that will be instantiated with natural reference to object delivered by nominal-phrase generator. MIKE (Multi-agent Interaction Knowledgably Explained) is an automatic real-time commentary system capable of producing output in English, Japanese, and French. Figure 1 illustrates an instance of the text commentary, generated by Rocco.

All these three systems generated natural-language utterances using a speech synthesizer. As the generated verbal comments have a noticeable difference with the human natural voice, these systems could not effectively catch the attention of the audience.

Our vision is to develop a live soccer commentary system, so that one can hardly recognize an artificial commentator is reporting the game. To achieve this, it has been decided to use *prerecorded human report* statements instead of generating text and then converting it to speech. Note that a soccer game consists of many similar situations that can be grouped together. For example many situations in a game can be described as "It is a definite chance!" Therefore it is possible to have some prerecorded report statements for each group of situations. Not only it doesn't limit the commentator functionality, but it also has an *effective influence* on the audience.

3 System Architecture

A three-layer architecture has been used for the Caspian Live Commentary system. The Analyzer and the Statistical Analyzer, form the bottom layer of our architecture. Above this layer, there is Content Selector and Special Sound Effects Manager. Sound Manager comprises the third layer of the proposed architecture as shown in figure 2.

The Analyzer receives information from the Soccer Server and determines the game status. Some examples of the games status determined by the Analyzer are "One and One", and "Scoring Chance". The Statistical Analyzer subsystem performs statistical analysis on data received from the Analyzer. The Content Selector subsystem takes the game states from the Analyzer and selects an appropriate statement to report the current situation of the game. Then, it sends a request to the Sound Manager to play the selected statement. The Special Sound Effects Manager works in parallel with the Content Selector and decides on the suitable environmental sounds for the current situation, and sends a request to the Sound Manager. Finally the Sound Manager organizes the submitted requests and plays the sounds in a consistent way.

4 Analyzer

The Caspian Commentary system is designed to report both live and replayed games. In order to report a live game, the commentary system connects to the Soccer Server and receives the same information that the monitor program gets for updating its visualization. The system uses the rcg log file of the Soccer Server to report on a replayed game. The rcg log file is a binary file generated by the Soccer Server during the time that the game is running and contains the data related to each cycle of the game. As a result, two different sources of data input, has been considered for the Analyzer:

1. Soccer Server: to report on a live match.
2. Log File: to report on a replayed match.

No matter which of these two input streams are used, the received information consists of:

1. players' locations and orientations
2. ball position and velocity
3. play modes such as goal, throw-in, free kick, and so on

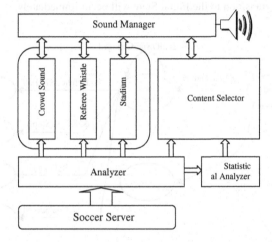

Fig. 2. System architecture and its interrelated components

Analyzer determines the game state using a *State Machine*, based on the data received from the input stream. There are two main points in designing the State Machine. One is to integrate a set of game states that can best cover *different possible situations* in a soccer game. The other one is to develop reliable *state transition functions*. For example, assume that the current game state is "One and one" this state can be followed by any of the following states:

1. Goal: goal is scored.
2. Out: the attacker kicked out the ball.
3. Defender: the opponent defenders got the ball.
4. Keeper: the opponent goal keeper caught the ball.

A portion of the state machine which detects "One and One" and the successive game situations is illustrated in figure 3.

4.1 Set of Game States

Game states in the state machine are divided into three categories:

1. **Primitive States:** The states that have only the Initial State as their predecessor. For example, the "One-And-One" is an instance of a primitive state, as shown in figure 3.
2. **Non Primitive States:** The states that can be reached by visiting at least one primitive or non-primitive state in the machine. In other words, a non-primitive state is dependent on the previous game state. For example, the transition to "Goal-After-OneAndOne" can happen only if the previous game state was "One-And-One".
3. **Final States:** The states that have no successor in the state machine. For example, "Throw-In" is a final state. Every time the machine reaches a final state a transition to the Initial State will occur immediately.

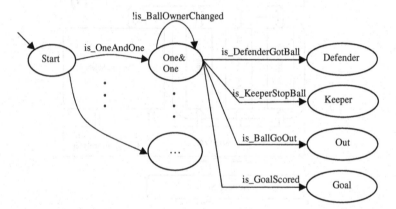

Fig. 3. A portion of the state machine which detects "One and One" and the successive game situations

4.2 State Transition Functions

A State transition function determines whether a transition to a specific new state can take place. This is done by verifying a set of preconditions. In other word, if the preconditions are satisfied then the transition to the new state will take place. For example consider the "One-And-one" state illustrated in figure 3. "is_OneAndOne" is a state transition function which determines whether a transition to the "One-And-One" state can take place.

There are two challenges that should be addressed in the design of each state transition function.

1. Which preconditions should be considered?
2. How to verify whether the precondition is satisfied?

Many transitions between different states of the state machine are carried out when the team in possession, looses the ball. So, it is very important to implement the function which detects the team in possession, quite well.

5 Statistical Analyzer

The Statistical Analyzer retrieves the statistical information based on the current game state determined by the Analyzer. Here are some instances of the statistical information: successful pass rate, number of shots, ball possession, number of offsides, etc. The results show that, the audience is really interested in the presented statistical information, especially those that cannot be easily retrieved by them. Also it can be used as a reliable metric to judge about the efficiency of the players' skills. For example, an increase in the successful pass rate shows that the agent's pass skill has been improved.

Some of the statistical information like number of offsides can be retrieved by keeping track of play mode changes (announced by the referee). On the other hand, there are some items like successful pass rate that should be extracted by analyzing the game.

Table 1. The classification of the commentator's report statements

> **Play Mode Comments:** comments on play modes, such as: offside, free kick, free kick fault, throw in, corner, and goal.
>
> **Play Situations:** situations like: good scoring chance, poor scoring chance, out after good scoring chance, good defending, poor defending, one and one, out after one and one, keeper caught after one and one, defender got the ball after one and one, good catch by the goal keeper, poor catch by the goal keeper, etc.
>
> **Players' Actions:** good tackle, good pass, good dribble, etc.
>
> **Misc.:** greeting comments (e.g. Hello every body....), full time (e.g. Full time here and the score is two-nil), after scoring (e.g. The dead lock broken there and the score is one-nil), dispossession (e.g. He lost the ball), long possession (e.g. The ball wasn't lost for long)

6 Content Selector

The Content Selector receives the current game situation and statistical information as an input, and decides on the statement to be reported. This module selects the appropriate utterance from a set of prerecorded report statements. The classification of the commentator's statements is given in table 1. Only those statements that satisfy the following criteria are picked up.

1. **Concise and Meaningful:** Since the commentary system has to keep up with a rapidly changing environment, it is important to use concise statements to describe the current situation. In fact, the current situation may change in every

simulation cycle and using long statements may lead to inconsistence commentary.

2. **Various and Exciting:** A commentator, who always expresses a specific situation by identical statements, is boring to the audience. For example it is not pleasing to announce "It is a corner now!" on every corner kick situation. For this purpose, various statements are considered in the set of prerecorded statements to report each situation. In addition, each statement is designed to be exciting so that the audience will experience the fun and excitement of the game.

3. **Impartial:** In fact, the commentator should not report biased statements. Consequently, a set of impartial prerecorded statements have been picked to achieve this goal.

Having an integrated set of prerecorded statements, the commentary system should decide which one is appropriate for the current situation. The selection procedure is a combination of the *Scheduling* and *Interruption* mechanisms which are described below.

6.1 Scheduling Mechanism

This mechanism is designed to set a suitable time interval between two successive report statements. This means that, the commentator may refuse to report a new state in order to meet time restrictions. But, there are some exceptions for important events, such as scoring, that should be considered in the design of this mechanism.

6.2 Interruption Mechanism

As it is mentioned in scheduling mechanism, there are some game states that are really important (e.g. scoring the goal). Therefore it is worth to interrupt the current reporting statement and announce the critical event. In other word, it is required to introduce the Interruption mechanism. Although the interruption mechanism is necessary for the commentary system, but having several interruptions during the game, makes the audience feel confused! For this reason, the interruption rate during the game should be in an acceptable range. Therefore, the interruption mechanism is considered only for critical events like scoring the goal.

Applying the described algorithm in the Caspian Commentary system results in a consistent report of the game, but it still has some shortcomings that will be described in Conclusion and Future Work section.

7 Special Sound Effects Manager

Having implemented the commentator, we found out although the commentator was doing well at reporting the game, it couldn't bring excitement to the audience. To address these problems, a new module named Special Sound Effects Manager was introduced which itself is made up of three sub modules.

This module receives the current game state as an input and picks up the appropriate environmental sounds including cheering of spectators, referee whistle and stadium announcer. Then it submits the sound requests to the Sound Manager.

This module plays a key role in conveying fun and excitement to the people who are watching the game.

7.1 Crowd Sound Effect

This is the most effective sound effect among the other ones. In the current implementation spectators are the soccer fans. They wisely keep track of the flow of the game, and make critical situations stand out by the sound effects associated to them. There are three sound effects implemented into this module, namely chant, applause, and scream.

7.2 Referee Whistle

According to the FIFA rules, there are several kinds of whistle blows for different events during a game. For example kick off, half time and corner kicks; each has its own style of blowing. The implemented referee whistle module, fully complies with the official FIFA rules.

7.3 Stadium Announcer

It announces the beginning and the end of a match. It also makes an announcement each time a goal is scored.

8 Conclusion and Future Work

The presented commentary system along with the other Caspian presentation tools, won the first place in RoboCup 2003 Soccer Simulation League, Game Presentation and Match Analysis Competitions, in Padua, Italy.

Caspian Commentator is designed to be an effective means of communication with the audience, by reporting the game facts at the right time and in a realistic way. It has been observed that the Caspian Commentator has a great impact on conveying the excitement to the people who are watching the game. More specifically, successful implementation of the State Machine in the Analyzer Module, leads to correct recognition and tracking of the game states. In addition, utilizing effective scheduling and interruption mechanism prevents the commentary system to overwhelm audience with his comments. But it has still some shortcomings and needs to be improved. One is that the audience is interested in receiving the meta-information while being informed about the general flow of the game. Some instances of the meta-information are history of the teams, how many times they play in front of each other, and what the results of previous matches were.

Furthermore, the audience is concerned about receiving technical information such as formation, player skills, and the commonly used strategies in a specific team. To meet this requirement, the Analyzer of the commentary system should be improved, so that it can retrieve the required information. Considering that "Team Modeling" is one of the major challenges in the Soccer Simulation Coach Competitions, it is possible to utilize the research studies in this domain, to improve the Commentary system's performance.

References

1. Kitano, H., Asada, M., Kuniyoshi, Y., Noda, I., Osawa, E., Matsubara, H.: RoboCup: A Challenge Problem For AI. AI Magazine(1) (1997) 73-85.
2. Noda, I., Matsubara, H.: Soccer Server and Researchers on Multi-Agent Systems. In Proceedings of IROS-96 Workshop on RoboCup. (1996) pages 1-7.
3. Cheng M., et. al. Soccer Server Manual version 7.07. http://sserver.sf.net August 2002
4. Voelz, D., André, E., Herzog, G., Rist, T.: Rocco: A RoboCup Soccer Commentator System. In M. Asada and H. Kitano (eds.), RoboCup-98: Robot Soccer World Cup II. Springer (1999) page 50-60
5. Binsted, K.: Character Design for Soccer Commentary. Paper presented at Second RoboCup Workshop, Paris, France. (1998)
6. Tanaka-Ishii, K., Hasida, K., Noda, I.: Reactive Content Selection in the Generation of Real Time Soccer Commentary. Paper presented at COLING-98, Montreal, Canada. (1998)
7. Tanaka-Ishii, K., Noda, I., Frank, I., Nakashima, H., Hasida, K., Matsubara H.: MIKE: An Automatic Commentary System for Soccer. Paper presented at the 1998 international Conference on Multi-agent Systems, Paris, France. (1998)
8. Andre, E., Binsted, K., Tanaka-Ishii, K., Luke, S., Rist, T.: Three RoboCup Simulation League Commentator Systems. AI Magazine (Spring 2000) Page 57-66.

Towards a League-Independent
Qualitative Soccer Theory for RoboCup[*]

Frank Dylla[1], Alexander Ferrein[2], Gerhard Lakemeyer[2], Jan Murray[3],
Oliver Obst[3], Thomas Röfer[1], Frieder Stolzenburg[4],
Ubbo Visser[1], and Thomas Wagner[1]

[1] Center for Computing Technologies (TZI), Universität Bremen, D-28359 Bremen
{dylla, roefer, visser, twagner}@tzi.de
[2] Computer Science Department, RWTH Aachen, D-52056 Aachen
{gerhard, ferrein}@cs.rwth-aachen.de
[3] Universität Koblenz-Landau, AI Research Group, D-56070 Koblenz
{murray, fruit}@uni-koblenz.de
[4] Hochschule Harz, Automation and Computer Sciences Department,
D-38855 Wernigerode
fstolzenburg@hs-harz.de

Abstract. The paper discusses a top-down approach to model soccer knowledge, as it can be found in soccer theory books. The goal is to model soccer strategies and tactics in a way that they are usable for multiple RoboCup soccer leagues, i.e. for different hardware platforms. We investigate if and how soccer theory can be formalized such that specification and execution is possible. The advantage is clear: theory abstracts from hardware and from specific situations in leagues. We introduce basic primitives compliant with the terminology known in soccer theory, discuss an example on an abstract level and formalize it. We then consider aspects of different RoboCup leagues in a case study and examine how examples can be instantiated in three different leagues.

1 Motivation

Thinking about the goal of the RoboCup community "to beat the human soccer champion by the year 2050" we start thinking about the human way of playing soccer. Talking to real experts in that field revealed that strategy and tactics play a major part in the game. But a computer scientist is more intrigued by available methods and restrictions that do exist for various reasons (e.g. expressivity of languages used). The following question arises: *Can we apply soccer theory to the RoboCup domain in a way that the majority of the leagues would benefit?* The motivation of this paper is therefore to take an adequate soccer theory book and examine its formalization.

Success in modern soccer games largely depends on the physical and tactical abilities of single players and on the overall strategy that coordinates team behavior whose

---

[*] This research has been carried out within the special research program DFG-SPP 1125 *Cooperative Teams of Mobile Robots in Dynamic Environments* and the Transregional Collaborative Research Center SFB/TR 8 on *Spatial Cognition*. Both research programs are funded by the German Research Council (DFG).

D. Nardi et al. (Eds.): RoboCup 2004, LNAI 3276, pp. 611–618, 2005.

goal is to sustain the strength of the individual players and to restrict the abilities of the opponents. Additionally, the use of an appropriate tactic is the foundation for co-ordinated team behavior. A big advantage of this approach is that the outcome can be applied to more than one RoboCup league. We go further and argue that it is possible to have a team of robots from different institutions that are able to play soccer together.

The paper is organized as follows: We motivated our approach in Sect. 1, introduce basic primitives compliant with the terminology known in soccer theory. We discuss an example on an abstract level formalizing it with   as specification language in Sect. 2. We then consider aspects of different RoboCup leagues in a case study and examine how examples can be instantiated in three different leagues in Sect. 3. We discuss our approach in Sect. 4.

2   World Modeling for the Soccer Domain

This section contains a description of how modern soccer knowledge is organized. Nowadays there are many textbooks on soccer theory. Here, we focus on Lucchesi's book [8], because it concentrates on the presentation of tactics (and not on training lessons). We derive basic primitives from [8] and formally specify some soccer tactics.

2.1   The Organization of Soccer Knowledge

According to [8], we interpret a soccer strategy as a tuple $str = \langle RD, CBP \rangle$. With RD as a set of *role descriptions* that describe the overall required abilities of each player position in relation to CBP, the set of *complex behavior patterns* is associated with the strategy. Given the strategy str, the associated role description $rd \in RD$ can be described by the defense tactics task, the offense tactics task, the tactical abilities, and the physical skills. Although soccer strategies in current literature [8, 10] are not as highly structured as strategies for American football, they provide sufficient structure to build up a top-level ontology with respect to specialization and aggregation. According to [8], the offensive phase can be structured into four sub-phases: *gaining ball possession*, *building up play*, *final touch* and *shooting*. In general, there are two ways to build up the play: either we introduce the phase in a counter-attack manner, fast and direct with a long pass or deliberately by a diagonal pass or a deep pass followed by a back pass. In the sequel, we will concentrate on the building-up phase.

2.2   Basic Primitives

Following the lines of [8], we distinguish between *role* (back, midfield, forward) and *side* (left, center, right) in soccer. This distinction is more or less independent from the pattern of play (e.g. 3-4-1-2 or 4-2-3-1). The combination of role and side (e.g. center forward) can be interpreted as *type* of a (human or robotic) soccer player or as *position* (region or point) on the soccer field. Therefore, we basically have nine different positions, as illustrated in Fig. 1(a).

The notions player type and position can be seen as instances or specializations of the notion of an abstract *address*, usually associated with its (actual) coordinates or a region on the soccer field. Also the ball (strictly speaking, its position) is an address, i.e. the parameter or goal of a test or operation of a soccer player (agent). A movable *object* in the context of soccer may be a player or the ball. An object is in a current *state*, which includes besides other data the current speed or view direction.

Although not explicitly mentioned, a *model of behavior* is assigned to every object, e.g. average or maximum speed or as a special case a deceleration rate for the ball. Additionally every player needs to hold data about other agents' states. We abstract this by the term *world model*. All this is summarized in the class diagram in Fig. 1(b).

In [8–p. ii] only few symbols are introduced that are used throughout the many diagrams in that book: players (in many cases only the team-mates, not the opponents are shown), the ball, passing, movement of the player receiving the ball, and dribbling. Conceptually, all symbols correspond to *actions*, which we abbreviate as *pass*, *goto*, and *dribble*. Since all actions are drawn as arrows starting at some player, naturally two arguments can be assumed: *player* and *address*. $goto(player[LF], region[CF])$ e.g. means that the left forward player moves in front of the opponent goal.

Although in most cases this is not explicitly mentioned in [8], actions require that certain prerequisites are satisfied, when they are performed. Since our approach aims at a very abstract and universal (league-independent) formalization of soccer, we restrict ourselves to only two tests: possession of ball and reachability. Each of them can be seen as *predicate* with several arguments: *hasBall* has the argument *player* (the ball owner); *reachable* has two arguments, namely an object and an address.

A pass e.g. presupposes reachability, i.e. it should be guaranteed that the ball reaches the team-mate. Clearly, the implementation of the reachability test is heavily dependent of the respective soccer league and its (physical) laws. Therefore, at this point, we only give a very general and abstract definition: Object o can reach an address a iff o can move to a and after that the ball is not in possession of the opponent team. This also covers the case of going to a position where the ball will be intercepted. We will go into further details in Sect. 2.5.

2.3   Towards a Formal Specification of Soccer Tactics

For specifying soccer moves we use the logic-based programming language   [6]. is a language for reasoning about actions and change and is based on the situation calculus [12]. Properties of the world are described by fluents, functions and relations with a situation term as their last arguments. The way actions change fluents is specified in terms of so-called *successor state axioms*, which also provide a solution to the frame problem. Together with action precondition axioms, axioms for the initial situation, a few foundational axioms and a domain closure and unique names assumption these form the basic action theories [12].   uses basic action theories to define the meaning of primitive actions. In addition it provides familiar control structures like sequence, if-then-else, or procedures to specify complex action patterns. Recent extensions dealing with concurrency, continuous change and time [2, 5] make the language suitable for the soccer domain.

While   has been and is used to implement soccer agents [3], we use it here merely as a specification language, because it comes equipped with a formal semantics. As we will see, the language allows a fairly natural representation of typical play situations. The primitive actions we consider here are *goto(player,region)*, *pass(player,*

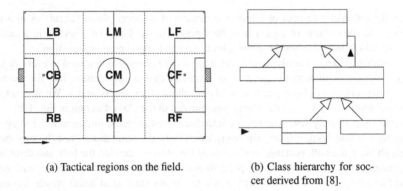

(a) Tactical regions on the field.   (b) Class hierarchy for soc-
                                            cer derived from [8].

Fig. 1. Tactical regions and address hierarchy derived from [8]. The field is divided into three rows (corresponding to player roles) back ( ), midfield ( ), and forward ( ) and three lanes (sides): left ( ), center ( ), right ( ). An address may be one of the nine regions or player types

region), and *dribble(player, region).* Further we need the action *intercept* which is a complex action built from the primitive ones. The arguments of the actions are *player* and *region* denoting that the particular player should go to, pass, or dribble the ball to the given position. For describing the properties of the world on the soccer field we need the fluents *reachable* and *hasBall(player)* among others.

2.4   Example

Fig. 2(a) depicts a possible move for a counter-attack. There, player movements are represented by arrows ($\rightarrow$ or $\curvearrowright$), passes are indicated by dashed arrows ($-->$), and squiggly arrows ($\rightsquigarrow$) stand for dribbling. Before we are able to formalize the whole manœuvre, we have to think about what *passing* means exactly. As in several action calculi, we introduce *constraints* associated with this action. A pass from player p to p' requires that beforehand p is in ball possession and the ball can be passed to p', i.e. the logical conjunction $hasBall(p) \wedge reachable(ball, p')$. Afterwards p' is in ball possession, i.e. $hasBall(p')$. In [8–p. 27], three different types of passes are mentioned that can be formalized by additional constraints: long pass with $p.role = B \wedge p'.role = F$, diagonal pass with $p.side \neq p'.side$, and deep pass with $p.role < p'.role$ where we assume that the roles (which can also be understood as rows in Fig. 1(a)) are ordered.

In Fig. 2(a), player 8 just captured the ball from the opponent team, dribbles toward the goal while the forwards (player 9 and player 11) revolve the opponent defense in order to get a scoring opportunity from both corners of the penalty area while player 10 starts a red herring by running to the center. The white circles represent the opponents.[1]

The counter-attack can be specified with   as shown in Fig. 2(b). The program is from the view of player 8, that is, all actions and tests are performed by this player. Player 8 gains the ball with an intercept action. He dribbles toward the center (denoted

---

[1] In the original figure (diagram 21 in [8]) there are no opponent players as well as no dedicated regions; we inserted them here for illustration purposes.

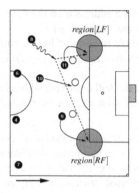

```
proc counterattack_21
  intercept;
  startDribble(region[CF]);
  waitFor(reachable(p₁₁, region[RF])∨
          reachable(p₉, region[RF])∨
          ∃x.Opponent(x) ∧ Tackles(x));
  endDribble;
  if reachable(p₁₁, region[LF])
    then pass(region[LF]);
    else if reachable(p₉, region[RF])
        then pass(region[RF]);
endproc
```

(a) Extended diagram 21 from [8].   (b) The specification in   .

Fig. 2. Counter-attack example

by *region*[*CF*])) until either player 11 or player 9 is able to receive the pass or an opponent forces player 8 to do another action (which is not specified in this example). In the specification above we use the action pair *startDribble* and *endDribble* instead of a single *dribble* action accounting for temporal aspects of that action. Splitting the dribble action into initiation and termination is a form of implicit concurrency, since other actions can be performed while dribbling. We omit further technical details and refer to [5].

The next step in the presented sequence is a *waitFor* construct. Its meaning is that no further actions are initiated until one of the conditions becomes true, i.e. player 11 or 9 are able to receive a pass in their respective region or an opponent tackles player 8, i.e., an opponent can intercept the ball (go-reachability). It is perhaps worth mentioning that during the blocking of the *waitFor* the dribbling of player 8 continues and sensor inputs are processed to update the relation *reachable*, which is discussed in more detail in Sect. 2.5. See [5] for details of how sensor updates can be formalized in   .

Finally, in the conditional we have to test which condition became true to choose the appropriate pass. Note that we do not choose an action in the case of neither player 9 nor player 11 can receive the pass as this would be the matter of another soccer move procedure. The counter-attack programs for the other players can be specified similarly.

2.5   Reachability

For our theory, reachability is central. As our theory aims at being a general one for different soccer leagues, we do not have a specific reachability relation. Building a specific reachability relation is dependent on the league and even within a league, it depends on abilities of single robots or agents. However, the different reachability relations share some properties independent of the league. In general, we can distinguish three different reachability relations:

1. a player p not being in ball possession will reach an address a on the field before any other player: $reachable_{go}(p, a)$ with prerequisite $\neg hasBall(p)$
2. a player p being in ball possession is able to dribble towards address a with high probability of still being in ball possession afterwards: $reachable_{dribble}(p, a)$ with prerequisite $hasBall(p)$
3. a player p being in ball possession is able to pass the ball b towards address a with high probability of a team-mate being in ball possession afterwards: $reachable_{pass}$ (b, a) with prerequisite $hasBall(p)$

We are aware of the fact that we need as precise world knowledge as possible, e.g. current positions and speed, for determining the future ball possession like above. Additionally we need assumptions on future behaviors, e.g. the ball path after being kicked. While for team-mates we know the agent's internal structure we may conclude possible future actions with high probability. About opponent agents a lot less is known and therefore predictions are more uncertain. The uncertainty of world data is quite different over the leagues. In the simulation league world data is quite reliable while in the four-legged league, e.g. position estimations are not very accurate.

Many different implementations of reachability can be thought of for the different leagues. The use of Voronoi diagrams and their dual, the Delaunay triangulation (see e.g. [1]) has proven useful in the past, especially in the simulation league. Here only direct neighboring players, team-mates and opponents, are connected. Note that a direct approach with Voronoi diagrams is only one possibility for implementing reachability. It will only be applicable for robotic soccer, if all agents more or less have the same physical abilities in each region on the soccer field.

3   The RoboCup as Case Study

So far, we have only presented a very abstract way of describing team-play and cooperative moves in soccer. We investigated the reachability relation, that forms a central part of the theory, and discussed some of the underlying models and assumptions, as well as the simplifications we made, but nothing has been said about the concrete problems that arise when one tries to actually carry out the specified moves. Thus, in this section we will discuss possible ways of realizing the abstract specifications in the mid-size, simulation, and legged league.

Mid-Size League. The design of robots in the mid-size league underlies only few restrictions like the maximum size of robots. As the robots are fully autonomous, one central problem is the perception of the environment and dealing with actuators like ball kicking devices. Therefore, many problems in this league rather deal with low-level problems, e.g. vision or ball handling, than with high-level aspects (team-play). Concerning the primitive actions as in the example in Sect. 2.4, goto, pass, and dribbling facilities are needed.

On the other hand, it was shown in [3] that   with extensions like decision-theoretic planning or probabilistic projection can be applied in the mid-size league in the *RoboCup* (2003 and 2004) and is really competitive and, thus, it should be possible to adapt the moves described in [8].

Simulation League. From a technical point of view, the simulation league is suited best for carrying out the tactics presented in [8]. First of all, this is the only league where teams of 11 players play against each other. So the number of players that is needed for making the presented moves is given. In addition, the skills of the players are developed well enough in this league, too, such that team-play can easily be realized.

As dribbling is an expensive and potentially unsafe behavior in the simulation league, and passing is preferred, we focus on describing possibilities of implementing *pass reachability* (see Sect. 2.5). The reachability of a pass partner is usually determined by checking a cone from the player with the ball towards a potential pass recipient. If this cone is free of opponent players, the recipient is reachable with a pass. A possible implementation is presented in [14].

Sony Four-Legged Robot League. As in the mid-size league, the four players are too few for carrying out tactical diagrams such as the ones in [8], and, moreover, with the current field size there is no need for passing the ball.

One method used in this league by the GermanTeam [13] to describe robot behavior is the *Extensible Agent Behavior Specification Language* (*XABSL*) [7]. It describes behavior in the form of a hierarchy of state machines, so-called *options*, using XML. In each option, the current *state* defines which sub-option or which *basic behavior* (some pre-coded routine such as *pass*) is active. So at each point in time, a path from the root option through several levels of other options to a basic behavior is active. This path changes whenever the current state of an option is changed to another one based on a decision tree. In principle, the behavior of player 8 in diagram 21 in [8] (counter-attack) can be modeled in XABSL. As the XML description would be out of proportion for this paper, we refer the reader to the long version [4].

4   Conclusions

As mentioned in Sect. 2.5, the concept of reachability is important. *Reachable* can be based on some qualitative information such as distance (e.g., near, intermediate, far away) or orientation (e.g., front-left, right, back-right). Examples of how to use qualitative spatio-temporal knowledge and reasoning for the RoboCup can be found in [9, 15, 16].

The intention of our investigation was to apply soccer theory as stated in soccer expert books to the RoboCup soccer domain. Our motivation was the significance of strategies and tactics in real soccer games. The main goal was to figure out whether we would be able to find an abstract level of formalization that enables us to bring benefit to multiple soccer leagues in RoboCup. We have chosen two examples for counter-attacks and used   as specification language. Please note that   is only one example for a specification language.

The biggest lesson learned is that we *are* able to formalize soccer theory on an abstract level. This might not be surprising, however, some of the concepts real soccer experts use are quite fuzzy and therefore difficult to define and implement. A prominent example is the concept of *reachability*, which is used in our examples. It turned out that the definition plays a crucial part in the implementation.

Our case study revealed that there is a lot of work to be done. There are still many problems concerning the low-level skills such as receiving the ball in the mid-size league. The simulation league has an excellent platform for this kind of experiments. The move has been implemented prototypically for the simulation league teams of both, RoboLog Koblenz[2] and Allemaniacs Aachen as a proof of concept. We will carry out systematic experiments in the future. The behavior of the German Team in the Sony legged league is modeled in XABSL as mentioned in the previous section. It turned out that the abstract behavior could be modeled and therefore implemented.

References

1. F. Aurenhammer and R. Klein. Voronoi diagrams. In J.-R. Sack and J. Urrutia, editors, *Handbook of Computational Geometry*, chapter 5, pages 201–290. North-Holland, 2000.
2. G. De Giacomo, Y. Lésperance, and H. J. Levesque. ConGolog, A concurrent programming language based on situation calculus. *Artificial Intelligence*, 121(1–2):109–169, 2000.
3. F. Dylla, A. Ferrein, and G. Lakemeyer. Specifying multirobot coordination in ICPGolog – from simulation towards real robots. In *Proc. of the Workshop on Issues in Designing Physical Agents for Dynamic Real-Time Environments: World modeling, planning, learning, and communicating (IJCAI 03)*, 2003.
4. F. Dylla, A. Ferrein, G. Lakemeyer, J. Murray, O. Obst, T. Röfer, F. Stolzenburg, U. Visser, and T. Wagner. Towards a league-independent qualitative soccer theory for RoboCup. Fachberichte Informatik 6/2004, Universität Koblenz–Landau, 2004.
5. H. Grosskreutz and G. Lakemeyer. On-line execution of cc-Golog plans. In *Proc. of IJCAI-01*, 2001.
6. H. J. Levesque, R. Reiter, Y. Lesperance, F. Lin, and R. B. Scherl. GOLOG: A logic programming language for dynamic domains. *Journal of Logic Programming*, 31(1-3), 1997.
7. M. Lötzsch, J. Bach, H.-D. Burkhard, and M. Jüngel. Designing agent behavior with the extensible agent behavior specification language XABSL. In Polani et al. [11].
8. M. Lucchesi. *Coaching the 3-4-1-2 and 4-2-3-1*. Reedswain Publishing, 2001.
9. A. Miene, U. Visser, and O. Herzog. Recognition and prediction of motion situations based on a qualitative motion description. In Polani et al. [11].
10. B. Peitersen and J. Bangsbo. *Soccer Systems & Strategies*. Human Kinetics, 2000.
11. D. Polani, B. Browning, A. Bonarini, and K. Yoshida, editors. *RoboCup 2003: Robot Soccer World Cup VII*, LNAI Series, volume 3020. Springer, Berlin, Heidelberg, New York, 2004.
12. R. Reiter. *Knowledge in Action*. MIT Press, 2001.
13. T. Röfer, I. Dahm, U. Düffert, J. Hoffmann, M. Jüngel, M. Kallnik, M. Lötzsch, M. Risler, M. Stelzer, and J. Ziegler. Germanteam 2003. In Polani et al. [11].
14. B. Riedel. Developing similarity measures for comparing game situations in RoboCup. Diplomarbeit, Knowledge Based Systems Group, RWTH Aachen, Aachen, Germany, 2002.
15. F. Stolzenburg, O. Obst, and J. Murray. Qualitative velocity and ball interception. In M. Jarke, J. Köhler, and G. Lakemeyer, editors, *KI-2002: Advances in Artificial Intelligence – Proceedings of the 25th Annual German Conference on Artificial Intelligence*, LNAI 2479, pages 283–298, Aachen, 2002. Springer, Berlin, Heidelberg, New York.
16. T. Wagner, O. Herzog, and U. Visser. Egocentric qualitative spatial knowledge representation for physical robots. In C. Schlenoff and M. Uschold, editors, *AAAI Spring Symposium*, Stanford, CA, 2004. AAAI Press. To appear.

---

[2] Thanks to Heni Ben Amor for the implementation.

Motion Detection and Tracking for an AIBO Robot Using Camera Motion Compensation and Kalman Filtering[1]

Javier Ruiz-del-Solar and Paul A. Vallejos

Department of Electrical Engineering, Universidad de Chile
{jruizd, pavallej}@ing.uchile.cl

Abstract. Motion detection and tracking while moving is a desired ability for any soccer player. For instance, this ability allows the determination of the ball trajectory when the player is moving himself or when he is moving his head, for making or planning a soccer-play. If a robot soccer player should have a similar functionality, then it requires an algorithm for real-time movement analysis and tracking that performs well when the camera is moving. The aim of this paper is to propose such an algorithm for an AIBO robot. The proposed algorithm uses motion compensation for having a stabilized background, where the movement is detected, and Kalman Filtering for a robust tracking of the moving objects. The algorithm can be adapted for almost any kind of mobile robot. Results of the motion detection and tracking algorithm, working in real-world video sequences, are shown.

1 Introduction

Movement analysis is a fundamental ability for any kind of robot. It is especially important for determining and understanding the dynamics of the robot's surrounding environment. In the case of robot soccer players, movement analysis is employed for determining the trajectory of relevant objects (ball, team mates, etc.).

However, most of the existing movement analysis methods require the use of a fixed camera (no movement of the camera while analyzing the movement of objects). As an example, the popular background subtraction movement detection algorithm employs a fixed background for determining the foreground pixels by subtracting the current frame with the background model. The requirement of a fixed camera restricts the real-time analysis that a soccer player can carry out. For instance, a human soccer player very often requires the determination of the ball trajectory when he is moving himself, or when he is moving his head, for making or planning a soccer-play. If a robot soccer player should have a similar functionality, then it requires an algorithm for real-time movement analysis that can perform well when the camera is moving. The aim of this paper is to propose such an algorithm for an AIBO robot. This algorithm can be adapted for almost any kind of mobile robot.

The rationale behind our algorithm is to compensate in software the camera movement using the information about the robot body and robot head movements.

---

[1] This project was partially funded by the FONDECYT (Chile) project 1030500.

D. Nardi et al. (Eds.): RoboCup 2004, LNAI 3276, pp. 619–627, 2005.

This information is used to correctly align the current frame and the background. In this way a stabilized background is obtained, although the camera is always moving. Afterward, different traditional movement analysis algorithms can be applied over the stabilized background. Another feature of our algorithm is the use of a Kalman Filter for the robust tracking of the moving objects. This allows to have reliable detections and to deal with common situations such as double detections or no detection in some frames because of lighting conditions.

2 Related Work

A large literature exists concerning movement analysis in video streams using fixed cameras. As an example, every year is held the PETS event, in which several state-of-the-art tracking and surveillance systems are presented and tested (see for example [4] and [5]). Different approaches have been proposed for moving object segmentation; including frame difference, double frame difference, and background suppression or subtraction. In the absence of any a priori knowledge about target and environment, the most widely adopted approach is background subtraction [3]. *Motion History* is another simple and fast motion detection algorithm. According to [8], the *Motion History* and *Background Subtraction* algorithms have complementary properties, and when possible it is useful their join use.

Image alignment using gradient descending is one of the most used alignment algorithms. It can be divided into two formulations: the additive approach, which consists on start from an initial estimation of the parameters, and iteratively find appropriates parameters increments until the estimated parameters converge [7]; and the compositional approach, which estimates the parameters using an incremental warp. This last approach iteratively solves the estimation problem using an incremental warp of the images to be aligned with respect to a template. This allows pre-computing the Jacobean more efficiently [1]. But the key for obtaining an efficient algorithm is switching the role of the image and the template. This leads to the formulation of the inverse compositional algorithm [2], where the most computationally expensive operations are pre-calculated, allowing a faster convergence. In [6] it was proposed the robust inverse compositional algorithm as an extension to the inverse compositional algorithm, allowing the existence of outliers into the alignment with almost the same efficiency.

Regarding moving objects tracking, Kalman Filtering, Extended Kalman Filtering and Particle Filtering (also known as Condensation and Monte Carlo algorithms) are some of the most common used algorithms. Due to its simplicity, the Kalman filter is still been used in most of the general-purpose applications.

The here-proposed motion detection and tracking system is based in the described algorithms: background difference and motion history for motion detection, robust inverse compositional algorithm for the image and background alignment, and Kalman filtering for the tracking of moving objects.

3 Proposed Motion Detection System

3.1 System Overview

In figure 1 is shown a block diagram of the proposed system. The system is composed by four main subsystems: *Image Alignment, Motion Detection, Detection Estimation, and Background Update*. In the *Image Alignment* module, the last updated background image (B_{k-1}) and the last frame image (I_{k-1}) are aligned with respect to the current frame image (I_k). The camera motion angles ( ) are employed in this alignment operation. Both aligned images, B_k^* and I_{k-1}^*, respectively, are then compared with I_k in the *Motion Detection* module for determining the current moving pixels. As a result of these comparisons the *Motion History* and *Background Subtraction* algorithms generate preliminary detections (a set of moving pixels), D_{H_k} and D_{B_k}, respectively. These detections are joined in the *Rejection Filter* module, and a single set of candidate blobs (in this case moving objects), built using adjacent moving pixels, Det_k, is obtained. The motion detections are analyzed in the *Detection Estimation* module using a Kalman Filter, and the final detections Det_k^* are obtained. Finally, the background is updated using B_k^*, I_k and Det_k^* (which defines the new foreground pixels) by the *Background Update* module.

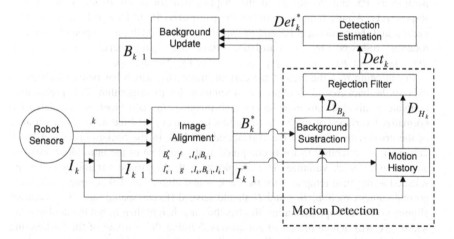

Fig. 1. Block diagram of the proposed system. Parameters are described in the main text

3.2 Image Alignment

The alignment of the last updated background image (B_{k-1}) and the last frame image (I_{k-1}) is implemented using the robust inverse compositional algorithm [6]. The alignment operation is implemented as a sequence of incremental warps (see section 2). The initial estimation of the warp is calculated based on the camera motion angles (stored in the $_k$ vector; they correspond to the tilt, pan and roll camera rotation angles). The initial estimated warp is a composition of a rotation followed by a displacement. The angle of rotation is estimated as the variation of the roll angle of the camera, while the displacement Dx/Dy in the X/Y axis corresponds to the pan/tilt angle:

$$Dx_{pan2 \, pan1} = \frac{R_{pan2} \sin BA_{tilt2} - pan1 \sin BA_{tilt1}}{180}$$

$$Dy = BA_{tilt2} \cos_{pan2} - BA_{tilt1} \cos_{pan1} \quad 180$$

(1)

where $_R$ represent the rotation in radians, Dx and Dy represent the displacement in pixels in their respective axis, BA is the angle of the body, and the angles $_{tilt1}$, $_{pan1}$, $_{roll1}$, $_{tilt2}$, $_{pan2}$, $_{roll2}$ are the tilt, pan and roll angles of the robot's head in the last and in the current image, respectively, measured in radians.

Then the warp is defined by a set of six parameters as:

$$Wx = 1 \; P1 \; x \; P3 \; y \; P5$$
$$Wy = P2 \; x \; 1 \; P4 \; y \; P6$$

(2)

where (Wx,Wy) define the new pixel coordinates which initial coordinates were (x,y). A pure displacement warp has the parameters P1 to P4 equals to zero, and the parameters P5 and P6 equals to the displacement in pixels in the x and y axis respectively. A pure rotation warp has the parameters P1 to P4 equals to the rotation matrix, and the parameters P5 and P6 equals to zero. Finally, a compound warp of a rotation followed by a translation have the parameters: $P1 \; \cos_R \; 1$, $P2 \; \sin_R$, $P3 \; \sin_R$, $P4 \; \cos_R \; 1$, $P5 \; Dx$, $P6 \; Dy$.

For aligning B_{k-1}, the area of the current image (I_k), which has being estimated to overlap the background, is chosen as a template for the algorithm. This preliminary template is divided into nine blocks (sub-images). In each block is calculated the normalized variance of its pixels (intra-block variance), and the normalized variance of the error with respect to the correspondent block in the background (inter-block variance). A variability factor is computed as the quotient of the intra-block variance and the inter-block variance. The six blocks with the largest variability factor are selected as the final templates for the background alignment. Taking into account the normal camera motion, B_{k-1} and I_k should have different spatial sizes for a correct alignment. In our implementation B_{k-1} has the same height than I_k, but the double of its width. We will denote the set of parameters defining the warping of the background $\mathbf{P_B}$. The algorithm for obtaining $\mathbf{P_B}$ is detailed described in [6].

For aligning I_{k-1} the calculated warp of B_{k-1} is employed as a first approximation. However, given that I_{k-1} and I_k have the same size, the calculated warp has to be actualized with a composition with a prior displacement to achieve the same spatial configuration of the background (I_{k-1} should be translated into background spatial coordinates), and then with a composition with a post displacement to reach the spatial configuration of the current image (the warped image should be taken back to its original coordinates). Thus, defining $\mathbf{P_1}$ as the set of parameters needed to produce a displacement equal to the last position of I_{k-1} inside the background, and defining $\mathbf{P_2}$ as the set of parameters needed to produce a displacement equal to the inverse of the estimated final position of I_k inside the background, the warp needed to align I_{k-1} is (the set of parameters defining this warping are $\mathbf{P_I}$):

$$W(\mathbf{x}, \mathbf{P_I}) \quad W(\mathbf{x}, \mathbf{P_2}) \circ (W(\mathbf{x}, \mathbf{P_B}) \circ W(\mathbf{x}, \mathbf{P_1})) \tag{3}$$

For simplicity on the notation, $\mathbf{x}$ denotes both spatial image coordinates. The function $W(\mathbf{x}, \mathbf{P^*})$ corresponds to a warping operation over $\mathbf{x}$ using the set of parameters $\mathbf{P^*}$.

3.3 Motion Detection

The *motion detection* module is composed by three algorithms, *Motion History* and *Background Subtraction* for movement detection, and *Rejection Filter* for filtering wrong detections and forming the movement blobs.

3.3.1 Motion History
The difference image DM_k is defined as:

$$DM_k(\mathbf{x}) \quad \begin{cases} mIncrement & \text{if } |I_k(\mathbf{x}) \quad I^*_{k-1}(\mathbf{x})| \quad T_m \\ 0 & \text{otherwise} \end{cases} \tag{4}$$

where *mIncrement* corresponds to a factor of increment in the motion and T_m corresponds to a motion threshold. DM_k contains the initial set of points that are candidate to belong to the MVOs (Moving Visual Object). In order to consolidate the blobs to be detected, a 3x3 morphological closing [10] is applied to DM_k. Isolated detected moving pixels are discarded applying a 3x3 morphological opening [10]. The motion history image MH_k, calculated from DM_k, is then updated as:

$$MH_k \quad MH_{k-1} * DecayFactor \quad DM_k \tag{5}$$

Finally, all pixels of MH_k whose luminance is larger than a motion detection threshold (T_h) are considered as pixels in motion. These pixels generate the detection image D_{H_K}. 3x3 morphological closing and opening are applied to D_{H_K}.

3.3.2 Background Subtraction
Foreground pixels are selected at each time k by computing the distance between the current image I_k and the current aligned background B_k^*, obtaining D_{B_K} as:

$$D_{B_k}(\mathbf{x}) \quad \begin{cases} 1 \; if \; |I_k(\mathbf{x}) \quad B_k^*(\mathbf{x})| \quad T_p \\ 0 \; otherwise \end{cases} \tag{6}$$

In order to consolidate the blobs to be detected, a 3x3 morphological closing is applied followed by a 3x3 morphological opening.

3.3.3 Rejection Filter

By means of 8-connectivity movement blobs, composed by connected candidate moving pixels, are built using D_{H_K} and D_{B_K}. For each blob b is defined a movement density MD_b as:

$$MD_b = \frac{\sum\limits_{x \, b} |I_k \, x \quad I_{k\,1} \, x|}{Area(b)} \tag{7}$$

MD_b measures the average change in the last frame for the blob b. Ghosts (groups of pixel that are not moving, detected like movement because they were part of a MVO in the past) should have a low MD_b, while the MVOs should have a large MD_b. Then, blobs with a small area (area $\leq T_b$), with a large area (area $\geq T_s$) and blobs with a small movement density ($MD_b \leq T_d$) are considered miss detections and discarded.

3.4 Detection Estimation

Targets (the MVOs) are tracked by keeping a list with the state of each of them. The state of a given target u includes: the position of the center of mass (x_u, y_u), the speed (Vx_u, Vy_u), the area (a_u) and the growing speed (Va_u). For each received movement blob is calculated the area and the center of mass. These variables are used as sensor measurements, and integrated across the different motion detections (the ones coming from D_{H_K} and D_{B_K}) and over time using a first order Kalman Filter [10]. This process includes 6 stages: *prediction*, *measure-target matching*, *update*, *detection of new targets*, *deprecated targets elimination* and *targets merge*. After those 6 stages the target state list, whose values are estimated by the Kalman Filter, corresponds to the final motion detections (Det_k^*).

Prediction. Using a first order cinematic model it predicts the state vector for each target, based on the last estimated state, and projects the error covariance ahead.

Measure-Target matching. In order to update the targets is imperative identifying which (blob) measure affects each target. For each measure-target combination is calculated a confidence value as the probability function given by the Kalman filter for the target evaluated on the measure. For each target, all measures with a confidence value over a threshold T_{tl} are associated with the target. If a measure does not have any associated target, then it is consider as a new target candidate and passed to the *Detection of new targets* stage. For each measure associated with a target, speeds (spatial speed: Vx and Vy, and growing speed Va) are to be estimated. This estimation is performed using the difference between the target state before the prediction and the measured state, divided by the elapsed time since last prediction.

Update. It computes the Kalman gain, updates the state vector for each target using their associated measures, and updates the error covariance. The targets without associated measures are not updated.

Detection of new targets. All measures without an associated target are considered as new target candidates. Their spatial speed is calculated as the distance from the image border, in the opposite direction of the image center, divided by the elapsed time since the last frame, and their growing speed is set to zero.

Deprecated targets elimination. Targets without associated detections in the last 2 frames are considered as disappeared MVO and eliminated from the target list.

Targets merge. For each target-target combination two confidence value are calculated as the probability function given by the Kalman filter for one target evaluated on the other target state. If any of this confidence values is over a threshold T_j, then the two targets are considered equivalents, and the target with the largest covariance (measured as the Euclidian norm of the covariance matrix) is eliminated from the target list.

3.5 Background Update

The background model is computed as the weighted average of a sequence of previous frames and the previously computed background:

$$
B_k(\mathbf{x}) \begin{cases} B_k^*(\mathbf{x}) & if(\mathbf{x}) \quad DET_k^* \\[2em] I_k(\mathbf{x}) \quad (1 \quad)B_k^*(\mathbf{x}) & otherwise \end{cases} \tag{8}
$$

4   Experiments

For the experiments, an AIBO robot using the motion software of the UChile1 AIBO soccer team [11], configured for allowing just head movements was used. The algorithm runs in the robot in real time. For analysis purposes, two video sequences were employed. In both, the robot moves its head in an ellipsoidal way, keeping the roll angle of the camera approximately aligned with the horizon. While the robot is moving its head, a ball is moving, once in the same direction of the camera movement, and once in the opposite direction. In figure 2 are shown the different stages of the algorithm while processing the frame 28 of the first video sequence.

The here-proposed motion detection algorithm can be enhanced using additional object information, such as color when detecting moving balls. Thus, in the *Rejection Filter* module was implemented a ball color filter applied to the blobs. This color filter uses the average U-V values (YUV color space) from each blob for filtering. If the Euclidean distance between the U-V average value of a blob and the ball U-V value (model) is larger than a threshold Tc, then the blob is discard. This filter decreases significantly the number of false positives errors. It should be stressed that this filter can be applied only after blobs have been already detected.

The first/second video sequence was 33/37 frames long. 11/14 frames contain a moving ball, but the first appearance of the ball cannot be detected because there is no way to know if the ball is moving or if it is stopped. Thus the relevant information are only 9/12 frames with moving ball, 6/9 of them were successfully detected, which correspond to a successful detections rate of 67%/75%. In table 1 are shown some statistics of the analysis of these video sequences, using and not using the color filter.

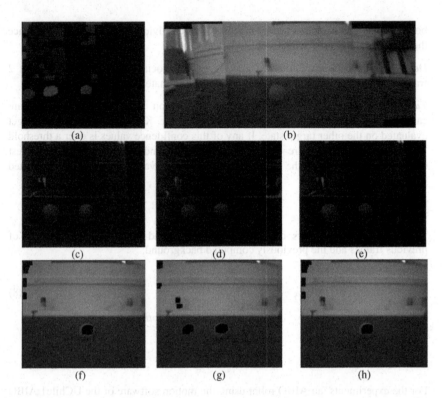

Fig. 2. Process stages at frame 28 in video sequence 1. (a) Motion history representation M_{Hk}. (b) Background model B_k^*. (c) Motion history error image. (d) Background subtraction error. (e) Current Image I_k. (f) Motion history detection D_{Hk} in black overlapped to current image. (g) Background subtraction detection D_{Bk} in black overlapped to current image. (h) Final detections DET_k^*, generated by the Kalman Filter

Table 1. Analysis of detections in the video sequence 1 and 2

Sequence number	1		2	
Number of frames	33		37	
Frames with a moving ball present	11		14	
Ball color filter	Off	On	Off	On
Frames with successful moving ball detection	6 (55%)	6 (55%)	9 (64%)	6 (43%)
Frames with a moving ball present but not detected	5 (45%)	5 (45%)	5 (36%)	8 (57%)
Detections corresponding to moving balls	7	6	9	6
Detections corresponding to ghosts	9	0	7	0
Detections corresponding to other moving objects	10	0	8	0
Fake detections (excluding ghosts)	296	2	265	5
Total number of detections	322	8	289	11
False detections average by frame, excluding ghosts	8,97	0,24	7.16	0,14

5 Conclusions

Results of the motion detection and tracking of objects in real-world video sequences using the proposed approach were shown. The system operates in real-time and the relevant moving objects, the ball in this case, are detected and tracked.

In a future work we will extend our system by using also body displacement. This extension would consider additional image displacements and rotations based on the robot joint angles. We will also share the ball tracking information between different robots by implementation a cooperative tracking algorithm. Another feature of the system to be improved is the high amount of false detections. We are working on a heuristic for reducing this kind of detections, beyond the use of a simple color filter.

References

1. S. Baker and I. Matthews, Equivalence and Efficiency of Image Alignment Algorithms, *Proc. of the 2001 IEEE Conference on Computer Vision and Pattern Recognition*, 2001.
2. S. Baker and I. Matthews, *Lucas-Kanade 20 years on: A unifying framework: Part 1*, Technical Report CMU-RI-TR-02-16, Carnegie Mellon University, Robotics Institute, 2002.
3. R. Cucchiara, C. Grana and A. Prati, Detecting Moving Objects and their Shadows - An evaluation with the PETS2002 Dataset, *Proc. of the 3rd IEEE Int. Workshop on PETS*, 18-25, Copenhagen, June 2002.
4. J. Ferryman (Ed.), *Proc. of the 3rd IEEE Int. Workshop on PETS*, Copenhagen, Denmark, June 2002.
5. J. Ferryman (Ed.), *Proc. of the Joint IEEE Int. Workshop on VS-PETS*, Nice, France, October 2003.
6. T. Ishikawa, I. Matthews and S. Baker, *Efficient Image Alignment with Outlier Rejection*, Technical Report CMU-RI-TR-02-27, Carnegie Mellon University, Robotics Institute, October 2002.
7. B. Lucas and T. Kanade, An Iterative image registration technique with an application to stereo vision, *Proc. of the Int. Joint Conf. on Artificial Intelligence*, 1981.
8. J. Piater and J. Crowley, Multi-Modal Tracking of Interacting Targets using Gaussian Approximations, *Proc. of the 2nd IEEE Int. Workshop on PETS*, Hawaii, USA, Dec. 2001.
9. S. Gong, S. McKenna, A. Psarrou, *Dynamic Vision From Images to Face Recognition*, 2000.
10. J. Russ, The Image Processing Handbook, Second Edition, IEEE Press, 1995.
11. J. Ruiz-del-Solar, P. Vallejos, J. Zagal, R. Lastra, G. Castro, C. Gortaris, I. Sarmiento, UChile1 2004 Team Description Paper, *Proc. of the 2004 RoboCup Symposium*.

The Use of Gyroscope Feedback in the Control of the Walking Gaits for a Small Humanoid Robot

Jacky Baltes, Sara McGrath, and John Anderson

Department of Computer Science,
University of Manitoba,
Winnipeg, Canada
jacky@cs.umanitoba.ca
http://www.cs.umanitoba.ca/~jacky

Abstract. This paper describes methods used in stabilizing the walking gait of TAO-PIE-PIE, a small humanoid robot given rate feedback from two RC gyroscopes. TAO-PIE-PIE is a fully autonomous small humanoid robot (30cm tall). Although TAO-PIE-PIE uses a minimal set of actuators and sensors, it has proven itself in international competitions, winning honors at the RoboCup and FIRA HuroSot competitions in 2002 and 2003. The feedback control law is based solely on the rate information from two RC gyroscopes. This alleviates drift problems introduced by integrating the RC gyroscope feedback in the more common position control approaches.

1 Introduction

Recent years have seen increased interest in humanoid robots, with many small humanoid robots emerging from research labs, hobbyists, and universities mainly in Japan, but also other countries ([4], [5], [3]).

This paper describes our first attempts at using feedback control to balance the walking gait of TAO-PIE-PIE. TAO-PIE-PIE was intended as a research vehicle to investigate methods for deriving control methods for stable walking patterns for humanoid robots. Stable walking, especially over uneven terrain, is a difficult problem. One problem is that current actuator technology (RC Servos, DC motors) generate less torque in comparison to their weight than human muscle. Another problem is that feedback from gyroscopes and actuators is very noisy. The necessary smoothing of the input signals makes it hard to use them in actively controlling the walking motion.

Cost was an important design criteria in TAO-PIE-PIE's development. Previous experience has shown us that the use of commonly available cheap components not only helps to keep the cost of a project down, but it also has led to the development of novel, versatile, and robust approaches to problems in robotics.

Another design goal was to reduce the number of degrees of freedom (DOF) of the robot. This reduces the cost of the humanoid robot as well as increases its robustness. Each DOF adds extra complexity in the mechanical design and the design of the control electronics. Furthermore, reducing the number of DOFs allows us to exploit the dimensions of the humanoid walking problem. The minimum set of DOFs that allow a humanoid robot to walk is also of interest, since it leads to energy efficient designs.

D. Nardi et al. (Eds.): RoboCup 2004, LNAI 3276, pp. 628–635, 2005.

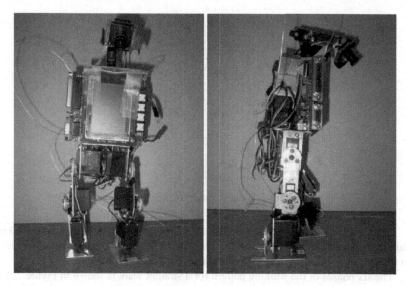

Fig. 1. Front and side view of TAO-PIE-PIE

TAO-PIE-PIE is the third generation of humanoid robots developed in our lab. Figure 1 shows the mechanical construction of TAO-PIE-PIE.

The actuators and sensors consist of widely available RC servos and RC gyroscopes for remote controlled cars and helicopters.

The Eyebot controller ([2]) was chosen as embedded processor, since it is relatively inexpensive, yet powerful enough to provide vision information. A small CMOS camera provides visual feedback for the robot.

The mechanical design was done in conjunction with Nadir Ould Kheddal's robotics group at Temasek Politechnic, Singapore. TAO-PIE-PIE is constructed out of 0.5mm aluminum, with RC servos used as structural components in the design.

Furthermore, TAO-PIE-PIE is intended to compete at international humanoid robotic competitions such as RoboCup and FIRA HuroSot ([1]. Among other things, this means that TAO-PIE-PIE must be able to actively balance, walk, run an obstacle course, dance, and kick a ball.

The remainder of this paper is structured as follows. The methodology used to develop and details of the implementation of the walking gaits are given in section 2, while section 3 presents an evaluation of this approach.

2   The Walking Gait

One of the fundamental problems in humanoid robots is the development of stable walking patterns. A walking pattern is dynamically stable if the center of pressure (COP) is within the supporting area. A statically stable walking pattern also has the center of mass (COM) within the supporting area.

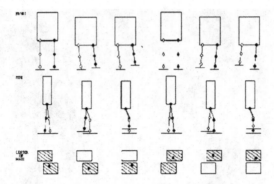

Fig. 2. Walking Pattern of TAO-PIE-PIE

We employ a divide and conquer approach and partition the statically stable walking gaitinto six phases: three for the right leg and three for the left. The phases were selected in such a way that the robot is statically stable at the end of eachphase.

The six phases of the walking pattern for a straight walk is shown in Figure 2. The bottom row of images in Figure 2 shows the approximate position of the COM in each phase. We describe these phases moving from left to right in the figure.

TAO-PIE-PIE starts in phase 1 — "Two Leg Stand" — where the right leg is in front and the left leg is behind. Both legs are on the ground and the COM is between the two legs.

From phase 1, TAO-PIE-PIE moves to phase 2 — "One Leg Stand" —. In this phase, the ankle servo generates a torque which moves the COM to the inside edge of the right leg. This also results in the back (left) leg to lift off the ground.

During the transition from phase 2 to phase 3 — "Ready for Landing" — is in static balance. TAO-PIE-PIE moves the free left leg forward and positions it so that it is ready for landing. The COM moves to the front of the supporting leg. This stabilizes the transition to phase 4.

During the transition from phase 3 to phase 4 — "Two Leg Stand Inverse" — the robot is in dynamic balance. The supporting leg extends its knee joint to shift the COM over the front edge of the supporting leg. The ankle servo of the supporting leg generates a torque to move the COM over the right side. The left leg will touch the ground in front of the right leg.

Phases 5 and 6 are the mirror images of phases 2 and 3 respectively. After phase 6, the motion continues with a transition to phase 1.

2.1   Sensor Feedback

The only feedback about the motion of TAO-PIE-PIE is provided by two gyroscopes that provide information about the angular velocity in the lateral plane and saggital plane respectively.

The raw sensor data of the gyroscopes is very noisy. We therefore compute a running average over five samples to smooth out the noise. Figure 3 shows the gyroscope readings for the lateral and saggital plane over approximately twenty steps.

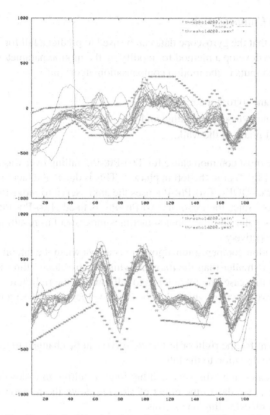

Fig. 3. Gyroscope Readings in the lateral and saggital plane over 10 Steps. Linear Approximation of the Safe Zone

Since TAO-PIE-PIE did not fall over during this extended walking trial, these gyroscope readings were used to determine a "safe zone" for the velocity feedback of the gyroscopes.

We then created a linear approximation of the "safe zone envelope" and generated minimum and maximum thresholds for the gyroscope readings. The approximation is shown using red and blue lines in Fig. 3.

2.2   Sensor Feedback in Detecting a Fall

Initially, we ran a series of experiments to verify the accuracy of the approximated "safe zone" by making TAO-PIE-PIE beep whenever the measured angular velocity was above or below the threshold in the saggital and lateral plane respectively. The goal was to show that TAO-PIE-PIE would beep just before falling over. These experiments proved very successful. TAO-PIE-PIE detected a fall with 95% accuracy with few ($< 5\%$) false positives.

2.3   Motion Compensation

After verifying that the gyroscope data can be used to predict a fall for TAO-PIE-PIE, the next step was to develop a method for modifying the motion parameters to avoid a fall. There are three inputs to the motion compensation algorithm:

1. Saggital plane gyroscope reading;
2. Lateral plane gyroscope reading; and
3. The current phase of the walk.

Initially, the most common cause for TAO-PIE-PIE falling over was a fall to the right in phase 2 (see Fig. 2) or to the left in phase 5. This is due to the fact that because of the limited number of DOFs, TAO-PIE-PIE uses the ankle servo to move the COM over the right or left foot. Since the torso of TAO-PIE-PIE is fixed, TAO-PIE-PIE is precariously balanced at this point and the robot sometimes moves to far, resulting in a fall to the right or left respectively.

The first motion compensation algorithm is active when the lateral plane gyroscope reading is larger/smaller than the maximum/minimum velocity threshold in phase 2/5 respectively. In this case, the robot tends to fall towards the right/left.

There are two ways in which the rotational velocity in the saggital plane can be controlled:

1. The set point for the right or left ankle servo can be changed to induce a torque in the opposite direction to the fall;
2. The robot can extend the knee and hip joint, resulting in a slowed down rotation. This effect is similar to the effect of slowing down the rotation of a chair while seated in it by extending one's arms.

We focus on modifying the angular velocity through the first method, since during a straight walk, the left-right velocity is mainly generated through the ankle servos. The second method is disadvantageous in that it also modifies the forward-backward balance of the robot. The set points for the servos are based on linear interpolations between a set of control points.

If the angular velocity is too large, then the motion compensator modifies the set point of the servo by moving it 10% closer to the start point of the pattern. Similarly, if the angular velocity is not large enough, then the set point is slightly extended.

The same approach is used when controlling falls in the saggital plane. In this case, however, there is no single servo that is responsible for the angular velocity. Instead, both set points for the knee and hip joint are modified by 90% to prevent a fall.

The feedback from the gyroscopes is also used to detect abnormal behavior. For example, if the robot's foot is caught on the carpet, instead of moving the leg forward, the robot will fall onto the leg too early. If this abnormal feedback is detected the robot attempts to stabilize itself by constraining all movement within the phases, in essence putting both feet on the ground as quickly as possible and straightening up its upper body. The constrainment will continue until both gyroscopes show appropriate angular velocities.

3   Evaluation

We evaluated the motion compensation algorithm by subjectively looking at the static walking pattern. The standard walking pattern of TAO-PIE-PIE is quite stable even without motion compensation. The robot did not fall during any of these experiments. However, the walking gait with motion compensation was more balanced resulting in a straight line walk. Without motion compensation, TAO-PIE-PIE would veer to the right significantly. The walking speed of the robot remains unchanged.

We also evaluated the motion compensation by subjectively by comparing the gyroscope feedback with and without motion compensation. The results of this comparison are shown in Fig. 4.

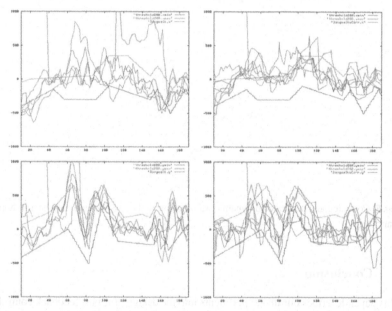

Fig. 4. Comparison of original walking gait (left column) and walking gait with motion compensation (right column) in the saggital (top row) and lateral (bottom row) plane

As can be seen from the plots, the motion compensation does constrain the walking gait so that the gyroscope feedback is more in the desired envelope. Most of the time, the walking gait remains in the desired velocity envelope.

Work is currently underway on developing a dynamic (shuffle-like) walk with correspondingly dynamic turns. When feedback correction was applied to walk, the walk covered more distance than without. Subjectively as well, the gyroscope feedback was not only better maintained within the desired envelope, but also formed a much more regular path, as shown in Fig. 5.

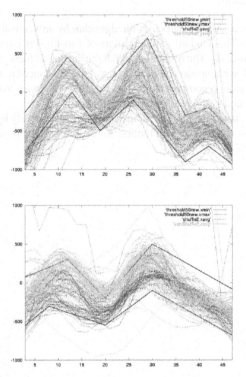

Fig. 5. Corrected and uncorrected dynamic walking gait in the saggital (top) and lateral (bottom) plane.

4   Conclusion

This paper describes our first experiments into the design of robust feedback control for walking of small humanoid robots. There is much work left to be done.

The current motion compensation algorithm is simple, but works surprisingly well in practice. We plan on investigating more complex methods for motion compensation and balancing in the future. For example, the motion compensation should not be a constant factor, but should be proportional to the current velocity.

We intend to extend this evaluation into more uneven terrains. The hope is that by using feedback, TAO-PIE-PIE is able to compensate for uneven terrain and adapt its walking gait. As well, these corrections could be applied to kicking and turning movements in order to correct for all movements the robot makes.

As noted in section 2, safe zones need to be developed from a previously stable walk. Thus speeding up a walk requires developing a stable walk in order to re-calibrate the safe envelopes. Investigation should proceed into the safe envelopes and their correlation to a change in speed or the desired walk pattern (fast, slow, left turn, right turn, etc). This could lead to implementing a gait or gait corrections without previously developing the stable walk.

TAO-PIE-PIE has shown itself to be a powerful and flexible platform for research into humanoid robotics. It has proven itself during international competitions winning a second place in theRoboCup anda technical merit award in the FIRA 2002 competitions.We have learned important lessons in the design of humanoid robots from TAO-PIE-PIE, which we will use in the design of the next generation humanoid robot HIRO. HIRO will use four additional DOFs (two in thehip and one for each leg). HIRO will also have more sensors,especially a set of force sensors in the feet. It also features a faster embedded processor (Intel Stayton), which allows us toimplementbetter on-board computer vision algorithms. One of the maingoals of the HIRO platform will be to investigate methods for augmenting the balancing of the robotusing visual feedback.

References

1. Jacky Baltes and Thomas Bräunl. *HuroSot Laws of the Game.* University of Manitoba, Winnipeg, Canada, May 2004. http://www.fira.net/hurosot.
2. Thomas Bräunl. Thomas bräunl's homepage. WWW, November 2002. http://robotics.ele.uwa.edu.
3. Jung-Hoon Kim, Ill-Woo Park, and Jun-Ho Oh. Design of a humanoid biped robot lower body. In *Proceedings of the 3rd International Workshop on Human-friendly Welfare Robotic Systems.* KAIST, January 20 - 22 2002.
4. Fuminori Yamasaki, Tatsuya Matsui, Takahiro Miyashita, , and Hiroaki Kitano. Pino the humanoid: A basic architecture. In Peter Stone, Tucker Balch, and Gerhard Kraetszchmar, editors, *RoboCup-2000: Robot Soccer World Cup IV*, pages 269–278. Springer Verlag, Berlin, 2001.
5. Changjiu Zhou. Linguistic and numeral heterogenous data integration with reinforcement learning for humanoid robots. In *Proceedings of the 1st IEEE Conference on Humanoid Robots*, 2000.

The UT Austin Villa 2003 Champion Simulator Coach: A Machine Learning Approach

Gregory Kuhlmann, Peter Stone, and Justin Lallinger

Department of Computer Sciences,
The University of Texas at Austin
Austin, Texas 78712-1188
{kuhlmann,pstone,hosty}@cs.utexas.edu
http://www.cs.utexas.edu/~{kuhlmann,pstone,hosty}

Abstract. The UT Austin Villa 2003 simulated online soccer coach was a first time entry in the RoboCup Coach Competition. In developing the coach, the main research focus was placed on treating advice-giving as a machine learning problem. Competing against a field of mostly hand-coded coaches, the UT Austin Villa coach earned first place in the competition. In this paper, we present the multi-faceted learning strategy that our coach used and examine which aspects contributed most to the coach's success.

1 Introduction

The Coach Competition is a fairly recent addition to the RoboCup Simulated Soccer League. The competition aims to encourage research in multiagent modeling and advice-giving. The challenge is to create a coaching agent that significantly improves a team's performance by providing strategic advice.

In the RoboCup simulator [1], an online coach agent has three main advantages over a standard player. First, a coach is given a noise-free omniscient view of the field at all times. Second, the coach is not required to execute actions in every simulator cycle and can, therefore, allocate more resources to high-level considerations. Third, in competition, the coach has access to logfiles of past games played by the opponent, giving it access to important strategic insights.

On the other hand, the coaching problem is quite difficult due to two main constraints. First, to avoid reducing the domain to a centralized control task, a coach agent is limited in how often it can communicate with its team members. In addition, the coach must give advice to players that have been developed independently, often by other researchers. For this to be possible, coaches communicate with coachable players via a standardized coach language called CLANG [1].

Our UT Austin Villa coach was a first time entry in the coach competition. Similarly to some previous approaches to coaching ([5],[6],[7]), we treat advice-giving as a machine learning problem. Competing against a field of mostly hand-coded coaches, the UT Austin Villa coach earned first place in the competition.

D. Nardi et al. (Eds.): RoboCup 2004, LNAI 3276, pp. 636–644, 2005.

In this paper, we present the multi-faceted learning strategy that our coach used and examine which aspects contributed most to the coach's success.

2   Coach Framework

The basic operation of our UT Austin Villa[1] coach is as follows. Prior to a match, the coach examines the provided logfiles of games played by the *fixed opponent*. We call this team the fixed opponent, because in competitions it is determined by the league organizers and is the common opponent for all coach entries. The logfiles contain data from the fixed opponent's previous games. In particular, the coach does *not* see log files of the coachable team itself playing against the fixed opponent. All coaches advise the same coachable team consisting of players that understand and react to CLANG messages.

The coach collects data about players on the fixed opponent team as well as the players on the team the fixed opponent is playing against. For each player, the coach collects aggregate data such as the player's average location, as well as data about high-level events, such as passes and dribbles. After every change in possession, the coach's game analysis module attempts to classify the prior possession as a sequence of high-level events. The details of the identification procedure are described in our team description [2].

The data collected during logfile analysis are fed into a group of learning algorithms that generate player models for both teams. The models are then used to produce three different kinds of advice: formational, offensive, and defensive. The learned advice is combined with a few hand-coded rules and sent to the coachable team at the beginning of the match.

In past years' competitions, the team to be coached consisted of players developed at a single institution. Even in the absence of a coach, the players constituted a coherent team. In order to magnify the impact of a coach, in the 2003 competition, coachable teams were assembled from players developed at three different institutions: UT Austin Villa (our own), Wyverns from Carnegie Mellon, and WrightEagle from USTC in China. Furthermore, the coachable players were designed with only limited default strategy.

As a result, it was necessary to provide the players with advice about general game play. After brief experimentation with the coachable players, we identified the basic skills that they were missing and added hand-coded rules to help them overcome these weaknesses.

While the coach is best able to reason about players in terms of their roles, CLANG requires players to be specified by their uniform numbers. For this reason, the coach maintains a mapping between roles and uniform numbers for each player on both teams. Learned rules and hand-coded advice are created with role variables in the place of uniform numbers. When the rules are sent, the coach uses the current role map to insert the uniform numbers corresponding to each role variable. If during the course of the game players change roles, the

---

[1] http:/www.cs.utexas.edu/~AustinVilla

affected rules are sent again with the updated player numbers. The details of how role mapping was used in the 2003 competition are described in our team description [2].

3   Learning

The core of the UT Austin Villa coach is its ability to learn player models from logfiles of past games. Similarly to Riley et al. [7], we break this problem down into learning three basic types of strategies: offensive (how the player should try to score), defensive (how they should act near their own goal), and formational (where the players should position themselves by default). This paper describes an independent formulation and implementation of these three basic strategies which differs in many of the particulars from previous work.

We assume that the set of available logfiles of the fixed opponent includes some games in which the opponent wins and some games in which it loses; in competition, we were given two of each. In the logfiles in which the fixed opponent performs well, we model the fixed opponent's offense and attempt to learn defensive advice to counter it. For the games in which the fixed opponent loses, we model the winning team and learn formational and offensive action selection advice.

For both offensive and defensive advice, the product of our learning algorithm is a classifier that is able predict the next high-level event to occur, given the current state of the game. To encode the simulator's state, we used a large set of features including the positions of all 22 players, the position of the ball, and the distances between them.

We used the J48 decision tree algorithm, implemented in the Weka machine learning software package [8], to train a series of decision trees, one for each modeled player. Because the structure of a decision tree is easily understandable, it is fairly straightforward to convert a tree into CLANG advice. The details of the example creation and advice generation procedures for the offensive and defensive advice are described in the following two sections. We then present the methods behind our formational advice learning.

3.1   Offensive Advice

When learning offensive advice, the coach attempts to model the behavior of the player with the ball. During the learning process, the coach builds a classifier for each player that tries to predict what that player will do with the ball in any given situation. For player i, we define the possible classes to be:

- $Pass(k)$: Pass to teammate with uniform number $k \in \{1..11\} - \{i\}$.
- $Shot$: Take a shot on goal.

During logfile analysis, when a shot or pass is identified, the state of the environment at the last kickable time is stored in the database along with the true class label and the player number: i. A classifier is then built for each player using only the examples corresponding to its own player number.

Once we have trained a decision tree for player i, we can convert it into advice to be given to our own corresponding player. To understand the advice generation process, consider the example decision tree for player 5 shown in Figure 1. Each leaf node of the decision tree is an action. The path from the root to the leaf defines a conjunction of conditions under which that action should be executed. Therefore we can construct a rule, {condition}→{action}, for each leaf node in the decision tree. For example, the rule for the leftmost leaf of the example decision tree is:

```
(BallX < 10) ∧ (BallY < 10)->Pass(6)
```
Or in CLANG:

```
(define
  (definerule OffRule1 direc
    ((and (bpos (rec (pt -52.5 -34) (pt 10 34)))
          (bpos (rec (pt -52.5 -34) (pt 52.5 10))))
     (do our {5} (pass {6}))))
  )
)
```

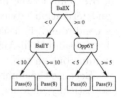

Fig. 1. Example decision tree learned for offensive advice

3.2 Defensive Advice

To generate defensive advice, we model the behavior of the opponent and attempt to foil its predicted strategy. Here, we aim to predict how a given player will acquire the ball. The set of classes is $Pass(k)$ where k is the uniform number of the player *by* whom the pass was made. Because we are interested in predicting a pass before it is made, we don't just record the state at the last kick time as we did in the offensive case. Instead, we record the 10 cycles (1 second) prior to the last kickable time and label each instance with the true class label and the player number of the pass receiver. An example tree learned for player 5 is shown in Figure 2.

We use a heuristic model to convert the learned predictions regarding opponent behaviors to defensive actions that can *prevent* that action. To prevent a pass, it is a good idea to position a defender along a passing lane closer to the intended receiver than to the passer. We found that positioning the defender at about 70% of the pass length away from the ball was a reasonable choice. Assuming that our player 7 is guarding opponent 5, then the CLANG rule corresponding to the leftmost branch of the decision tree in Figure 2 is:

```
(define
  (definerule DefRule1 direc
    ((and (bpos (rec (pt -52.5 -34) (pt 0 34)))
          (bpos (rec (pt -52.5 -34) (pt 52.5 10))))
     (do our {7} (pos (((pt opp 6) * (pt .7 .7)) +
                       (pt opp 5) * (pt .3 .3))))))
  )
)
```

Fig. 2. Example decision tree learned for defensive advice

3.3   Learning Formations

Our approach to learning a team formation is similar to our approach to learning offensive advice. The coach observes a team that can beat the opponent and then attempts to mimic that team's behavior. We model the formation as a home position (X, Y) and ball attraction vector (BX, BY) for each player. In CLANG, a formation is a positioning rule of the following form for each player, P:

```
(do our {$P} (pos ((pt $X $Y) + ((pt ball) * (pt $BX $BY)))))
```

The X and Y values are calculated as the average x and y coordinates of the observed player during the course of the game. Values for BX and BY were handpicked for each position and were found through brief experimentation. In some cases, we found that the ball attraction would cause the forwards to play too far towards the opponent goal, so to compensate, we manually moved the home positions back a bit.

4   Experimental Results

In this section we present the results of several experiments involving our learned coach, both in competition and in more controlled settings.

4.1   The Competition

The UT Austin Villa coach came in first place out of 12 entries in the 2003 RoboCup Coach competition. The competition consisted of three rounds. In each round, the coached team played three ten-minute games against a fixed opponent. Coaches were evaluated based on goal difference: the number of goals scored by the coachable team minus the number of goals scored by the opponent. The fixed opponents were all teams that competed in the main simulator competition: Boldhearts in round 1, Sirim in round 2, and EKA-PWr[2] in round 3.

The score differences and rankings for the top four finishing teams are shown in Table 1.[3] Our coach was ranked 7th after the first round. After making improvements to the hand-coded advice (but still retaining the learned offensive and formation advice as described above), we moved into first place after the second round. Four coaches (our own UT Austin Villa along with FC Portugal[4], Iraniansand Helli-Amistres[5]) progressed to the final round with UT Austin Villa coming out on top.

Because the number of games in the coach competition is too small to provide statistically significant results, we reran the final round for 50 games on our own. The advice sent by the other coaches was extracted from the logfiles of

---

[2] http://autonom.ict.pwr.wroc.pl/RoboCup/english/english.html
[3] Complete results are available from www.uni-koblenz.de/~fruit/orga/rc03/
[4] http://www.ieeta.pt/robocup/
[5] http://www.allamehelli.net/pages/robo.html

Table 1. Total scores and rankings for the top four finishing teams in the 2003 RoboCup coach competition. The score consists of the number of goals scored by the coached team followed by the number scored by the fixed opponent

Coach	1st Round (Boldhearts)		2nd Round (Sirim)		3rd Round (EKA-PWr)	
UT Austin Villa	0:19	7th	0:2	1st	8:2	1st
FC Portugal	1:21	8th	0:8	4th	7:3	2nd
Iranians	0:14	4th	0:5	3rd	3:2	3rd
Helli-Amistres	1:12	2nd	0:3	2nd	7:7	4th

Table 2. Summary of 50 runs of the final round of the 2003 RoboCup Coach Competition. Average goal differences are shown along with their standard deviations and overall ranking

Coach	Score	StdDev	Rank
UT Austin Villa	2.38	2.61	1st
FC Portual	2.24	1.53	1st
Iranians	-0.4	1.74	4th
Helli-Amistres	0.85	1.81	3rd

the competition and duplicated verbatim. We used the same team of coachable players as the one used in the competition so as to reproduce the exact conditions. The results of the comparision are summarized in Table 2.

In our tests, our coach had the highest average, but based on a two-tailed student's t-test the results are not statistically better than those of the second place team, FC Portugal. Even after 50 games, the results were not significant ($p > 0.9$). However, the scores for FC Portugal and our UT Austin Villa coach were significantly[6] better than the next best team: Helli-Amistres. Therefore, under controlled conditions our coach tied with FC Portugal for first place.

4.2   Additional Experiments

After the competition, we conducted additional controlled experiments to isolate the key components of our learned coach agent. For all tests, we used the same fixed opponents as in the competition (BoldHearts, Sirim, and EKA-PWr), and the same scoring metric. All reported scores have been averaged over 25 games.

Our first experiments were aimed at isolating the impact of each variety of advice given by our coach. We tested several different configurations of advice using a coachable team consisting of only our (UT Austin Villa) players.

The results of these first experiments are presented in Table 3. In the table, the column labled "w/ HC" indicates whether or not the hand-coded advice was

---

[6] For all of the results presented in this paper, significance was determined by using a two-tailed student's t-test with $p < 0.05$.

Table 3. Average goal differences with varying levels of advice

Opponent	w/ HC	None	Formation	Offensive	Defensive	Full
BoldHearts	N	-8.8	-3.3	-2.9	-2.9	-2.7
	Y	-6.8	-0.5	-1.4	-5.7	-6.5
Sirim	N	-4.1	2.6	1.2	0.9	1.7
	Y	-5.4	-1.6	-0.3	0.8	-0.4
EKA-PWr	N	-0.6	2.8	2.9	3.4	2.7
	Y	1.0	3.62	2	2.12	2.43

included (**Y**es/**N**o). The column labeled "Formation" contains the results for learned formation advice only. The "Offensive" and "Defensive" columns show the results of adding offensive and defensive advice, respectively. "Full" includes all three types of learned advice. These advice configurations were compared with the default behavior of the coachable players without any advice, labeled "None".

From the table, it is clear that both with and without hand-coded rules, across all opponents, the learned advice did significantly better than no advice at all ($p < 0.05$).

Another observation that is true across the board is that the learned formation advice had the most significant impact of all advice types. On the other hand, with the exception of EKA-PWr, it appears that the offensive and defensive advice conflicted with the hand-coded advice. During the competition, we noticed that this was occuring after the first round. As a result, we decided to turn off the learned defensive advice for the remaining rounds. In retrospect, this was a very prudent decision.

Except in the case of Boldhearts (with defensive advice removed), we would have probably achieved higher scores in the competition, had we not added the hand-coded rules. This is a surprise considering the preliminary tests we performed with the coachable players, which suggested that hand-coded advice was necessary.

5   Related Work

Some previous work has been done on learning to give advice to RoboCup simulated soccer players. Similarly to our own work, Riley et al. [7] approached advice-giving as an action-prediction problem. Both offensive and defensive models were generated using the C4.5 [3] decision tree learning algorithm. Their work also stressed the importance of learned formation advice. While our decomposition of the problem is similar to theirs, our model representations and advice-generation procedures are quite different. For example, whereas our approach learns the player's average position and then considers the positioning of the players with respect to the ball when giving advice, theirs ignores the ball and instead focuses on correlations between players. In addition, the semantics of our learned defensive rules, which aim to learn not what the player with the ball will do, but how it will get the ball in the first place differs from what was done previously.

In other work, Riley and Veloso [6] used Bayesian modeling to predict opponent movement during set plays. The model was used to generate adaptive plans to counter the opponent's plays. In addition, Riley and Veloso [5] have tried to model high-level adversarial behavior by classifying opponent actions as belonging to one of a set of predefined behavioral classes. Their system was able to classify fixed duration *windows* of behavior using a set of sequence-invariant action features.

ISAAC [4] is a game analysis system created as tool for simulated soccer team designers. Similar to a coach, this system analyzes logfiles of a game in order to suggest advice for how a team's play can be improved. However, this advice is meant to be understood by the team's developers instead of the agents themselves.

6   Conclusion and Future Work

We have presented our multi-facted learning approach to giving advice in RoboCup simulated soccer. Using this approach, our UT Austin Villa coach won first place in the 2003 RoboCup Coach Competition. Through controlled experiments, we found that our coach was significantly better than the third and fourth place finishing teams and at least as good as the second place finisher. In addition, we have identified the learned formation rules as the most effective type of advice.

In our research we are continuing to enhance and carefully test the learned defensive and offensive advice, and we plan to test the degree to which each type of learned advice is opponent-specific. Meanwhile, we plan to continue working on finding ways to learn better, more adaptive formations. In addition, we intend to explore various methods for generating set play advice. Finally, we will be adapting our learning strategy to include online learning.

Acknowledgments

We thank Patrick Riley for helpful comments and suggestions. This research is supported in part by NSF CAREER award IIS-0237699.

References

1. Mao Chen, Ehsan Foroughi, Fredrik Heintz, Spiros Kapetanakis, Kostas Kostiadis, Johan Kummeneje, Itsuki Noda, Oliver Obst, Patrick Riley, Timo Steffens, Yi Wang, and Xiang Yin.   Users manual: RoboCup soccer server manual for soccer server version 7.07 and later, 2003.   Available at http://sourceforge.net/projects/sserver/.
2. Gregory Kuhlmann, Peter Stone, and Justin Lallinger. The champion UT Austin Villa 2003 simulator online coach team. In Daniel Polani, Brett Browning, Andrea Bonarini, and Kazuo Yoshida, editors, *RoboCup-2003: Robot Soccer World Cup VII*. Springer Verlag, Berlin, 2004. To appear.

3. J. Ross Quinlan. *C4.5: Programs for Machine Learning*. Morgan Kaufmann, San Mateo, CA, 1993.
4. Taylor Raines, Milind Tambe, and Stacy Marsella. Automated assistants to aid humans in understanding team behaviors. In M. Veloso, E. Pagello, and H. Kitano, editors, *RoboCup-99: Robot Soccer World Cup III*, pages 85–102. Springer Verlag, Berlin, 2000.
5. Patrick Riley and Manuela Veloso. On behavior classification in adversarial environments. In *Proceedings of the Seventeenth National Conference on Artificial Intelligence (AAAI-2000)*, 2000.
6. Patrick Riley and Manuela Veloso. Recognizing probabilistic opponent movement models. In A. Birk, S. Coradeschi, and S. Tadokoro, editors, *RoboCup-2001: The Fifth RoboCup Competitions and Conferences*. Springer Verlag, Berlin, 2002.
7. Patrick Riley, Manuela Veloso, and Gal Kaminka. An empirical study of coaching. In H. Asama, T. Arai, T. Fukuda, and T. Hasegawa, editors, *Distributed Autonomous Robotic Systems 5*, pages 215–224. Springer-Verlag, 2002.
8. Ian H. Witten and Eibe Frank. *Data Mining: Practical Machine Learning Tools and Techniques with Java Implementations*. Morgan Kaufmann, October 1999. http://www.cs.waikato.ac.nz/ml/weka/.

ITAS and the Reverse RoboCup Challenge

Tarek Hassan and Babak Esfandiari

The Department of Systems and Computer Engineering,
Carleton University, Ottawa, Ontario, Canada, K1S 5B6
thassan@connect.carleton.ca,
babak@sce.carleton.ca

Abstract. ITAS is a tool that allows a human to play soccer in the RoboCup Soccer Simulator environment. This is essentially the reverse challenge to that of RoboCup. Instead of bringing the machine to the real world, ITAS strives to seamlessly interface man to the machine world. This presents a fundamental human-computer interaction design problem. This paper shows how the reverse RoboCup challenge can benefit the RoboCup community and what value it brings to robotics and AI research in large. An overview of the features of ITAS and its development using the Usability Engineering Lifecycle are then given, followed by a comparison with a related system, OZ-RP. ITAS is an open source project. The most recent releases are available at http://itas.sourceforge.net.

1 Introduction

ITAS, In The Agent's Shoes, is a RoboCup Simulation League soccer player that is controlled in real time by a human user. It provides the user with a representation of the sensory stimuli and the action commands that the RoboCup Soccer Simulator provides an agent. Originally developed as a tool with the purpose of logging a human player's interaction with the simulator as data for machine learning, it was quickly realized that developing a competitively performing human controlled player presented quite a challenge.

Since a human is as alien to the simulated environment as a robot is to the real world, this presents a fundamental human-computer interaction design problem. Development of ITAS has lead to the recognition of the Reverse RoboCup challenge as an important undertaking unto itself and meeting this challenge became the project's main objective. The following is the statement of the Reverse RoboCup Challenge and an invitation to others to take on this challenge to develop the best human interface to the RoboCup Soccer Simulator (RCSS) environment.

- *To develop a team of human controlled simulated soccer players*
  that can win against the RoboCup Simulation League champion team.Ž

2 Why the Reverse RoboCup Challenge?

2.1 Human Element in RoboCup Competition

Perhaps the most important benefit that the Reverse RoboCup Challenge brings to the RoboCup community is the introduction of a human element into the competition.

D. Nardi et al. (Eds.): RoboCup 2004, LNAI 3276, pp. 645–652, 2005.

Even though RoboCup's mission statement is to defeat the human world soccer champions, so far the evolution of RoboCup players has been mostly based on competition amongst robots/agents. How can the progress of RoboCup towards its original goal be validated without the introduction of a human challenger in some form?

Until meticulous, full scale humanoid robotics are a reality, this problem cannot be addressed in the physical domain. However, human intelligence, strategy, decision-making and team work can be introduced to the competition by allowing real soccer players to play in the simulated soccer environment of the RoboCup Simulation League. As such, ITAS strives to be a tool that reflects the soccer play experience with respect to the kinds of decisions that are made in real time and the key environment and state information required to make those decisions.

2.2 Machine Learning

Generating data for machine learning was the original purpose of ITAS. To have a human player perform soccer tasks in RoboCup's simulated environment can be of great value. Logs of the player's actions will be based on the simulator parameters, facilitating interpretation and processing of the data for machine learning. However, the value of the data collected will depend on the quality of the human player's performance, which in turn depends on the quality of the interface to the simulated player.

2.3 Telerobotics

Interfaces developed for the reverse challenge can be adapted into interfaces for telerobotics control systems. The most obvious field being the RoboCup robotics leagues, as a start. ITAS's interface approach is particularly well suited for dealing with dynamic or unfamiliar environments since it is not dependant on relating incoming data to a predefined global model and placing the user in it, but rather presents the state of the environment as it is perceived in real time. The open source nature of the project makes it very accessible to robotics teams for such implementations and we encourage them to explore that option.

2.4 Other Areas

The reverse challenge makes for a great medium for research in collaborative and competitive interaction between artificial agents and human users in simulated or virtual environments. This can be of interest to researchers from many disciplines ranging from cognitive science and human factors to computer science and artificial intelligence.

Another exciting application for the reverse challenge is exploring the potential of the hybrid agent. This is an agent that combines the efficiency and accuracy of an artificial agent at performing tasks in the simulated environment, such as passing, dribbling, shooting and evasion, with the real-time strategic input of an actual soccer player. Such an agent may provide valuable competition to RoboCup champions with regards to facing a more human-like competitor.

Other areas of interest include user input devices. The complexity of the task at hand demands intuitive user control over the actions of an agent in a fast paced environment, making it a good platform for testing new and different hardware. The OZ-RP group (another reverse RoboCup challenger) have implemented their interface on several platforms utilising a variety of control devices.

3 ITAS

3.1 Development Approach

ITAS development follows the processes defined by the Usability Engineering Lifecycle (UEL). The UEL consists of three phases, namely Requirement Analysis, Design/Verification/Development, and Installation, each composed of iterative sub phases. Perhaps the most valuable aspect of UEL to this project has been the focus on user profile and contextual task analysis that are part of the Requirement Analysis phase, as they have shaped the usability objectives of ITAS.

The contextual task analysis for this project considered two task domains which were important. The first is real life soccer. This is the native environment of the ideal user for ITAS, so understanding it goes a long way in understanding the user's expectations and actions in the simulated environment. The physical characteristics and constraints of the real world are of little interest here, but rather the situational characteristics, as they provide the context in which the user will perceive the environment, set objectives and make decisions. The second domain is that of the simulated environment, ranging from the parameters, data objects and commands RCSS handles to the constraints and requirements of the ITAS interface. Understanding and then relating the two task domains to one another formed the base of ITAS's interface design.

Soccer objectives change dynamically depending on game situations. We found the most significant state variable affecting a player's task priorities to be ball possession (both relative to the individual and team-wide ball possession state). Team possession affects the whole team's strategy whereas individual possession affects the particular player's decisions. Based on observation of and discussions with our test group of experienced soccer players and the aforementioned states, a soccer task organization model was developed.

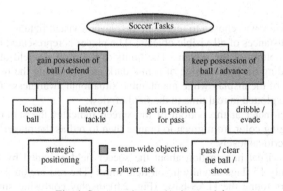

Fig. 1. Soccer task organization model

As shown in the Soccer task organization model, soccer play was broken down into two main team-wide objectives based on the team's state with regards to ball possession. Players work towards attaining these objectives or modes through performing the necessary key tasks. The key tasks named above define the decision-making domain in which we aim to place the ITAS user. So by providing the tools necessary to perform these tasks intuitively and efficiently, the user is left to make decisions as to when, where and how ('how' relating to the parameters of the task rather than it's mechanics) to perform these tasks.

With the exception of communication, which will be implemented in a later version of ITAS, each of the tasks defined in the task organization model requires the application of one or more of three basic skills, awareness of surroundings (objects and environment), maneuvering and ball handling.

These three skills relate almost directly to the three basic commands that RCSS provides; turn, dash and kick. The advanced tasks from the task organization model were designed by combining the use of these commands with interpretations of incoming sensory data to provide the user with higher level commands and a more informative and intuitive display. Conceptual mock-ups of the advanced controls and display features were discussed with our test group of soccer players and revised, before implementation. The tasks identified also helped design the drills that were used to test the different iterations of ITAS throughout development.

As well as following UEL, ITAS's development was split into two stages. An initial version, ITAS-basic, aimed to instantiate all the fundamental interactions between an agent and the RoboCup Soccer Simulator in the simplest form that is comprehensible to the user. This includes displaying real time visual and player state information sent by RCSS and providing the basic commands RCSS accepts such as dash, turn and kick. ITAS-basic is the platform on which more complex versions can be based. The second stage of ITAS is a version that provides the user with more advanced stimuli and control mechanisms. The advanced features in the most recent version of ITAS, as well as features currently in development, enrich the user experience by providing intuitive visual interpretations of the sensory information form the server as well as commands to perform more advanced tasks automatically.

3.2 Display

A birds-eye view was chosen to represent the player's visual field as it best represents the relative distances of all visible objects. The player is represented by a blue or red circle at the bottom of the screen. The limits of the player's field of view are indicated by shading the area outside of it in a dark green, helping the user focus on the relevant area of the display while maintaining situational awareness. Objects seen by the player are plotted in real time in the field of view.

A stamina-meter at the bottom of the screen indicates the stamina level of the player, changing color from green to orange then to red as the player's stamina level decreases to critical levels.

The basic visual information about the soccer field provided by RCSS is rather limited, as only 15 flags actually lie on field lines. The rest are on the outside of the field. Testers found the ITAS-basic (Fig. 2.b) display confusing since the familiar visual queues of a soccer field were missing. The latest ITAS (Fig 2.a), however, calculates and draws missing field lines that aren't indicated as visible objects by inferring them from the visible flags. The inclusion of internal field lines, such as the kick-off circle and center line provides users with even more visual queues that aid in positional awareness.

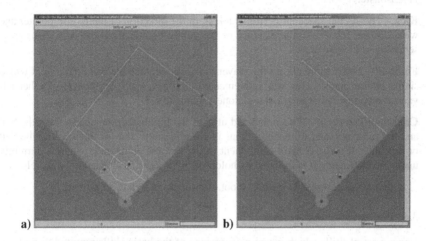

Fig. 2. a) ITAS screenshot b) ITAS-basic screenshot showing only the raw RCSS data

In RCSS, accuracy and consistency of visual information decrease to varying degrees over different ranges. These ranges are visually represented in ITAS using different shades of green in the visible area of the field of view, helping the user make better decisions and have more control over the reliability of what they see.

3.3 Controls

Movement – Instead of the basic dash command, ITAS has a run command which repeatedly sends dash commands as long as the left mouse button is held down. The speed of the running changes dynamically as the mouse pointer is moved closer or farther away from the player representation, allowing for smooth acceleration and deceleration.

Our testers had a hard time turning using the simple RCSS Turn command, as it involves an instant change of orientation in the field. This is analogous of closing one's eyes, turning and then opening them again. One needs to reestablish what it is they are looking at and where they are oriented, as the view of the environment is no longer persistent. This is even more confusing in the alien environment of the simulator. To fix this, we added a panning command. Panning is gradual turning involving many small turns, while the right mouse button is held down. The size of these turns (or the rate of panning) changes dynamically relative to the angle of the mouse pointer to the centre of the field of vision. This allows for a more persistent and less confusing world view.

ITAS also combines running and panning to allow for otherwise unachievable smooth movement. It took our testers a little time to get used to this feature, but was found to be incredibly useful once mastered. This feature allows players to easily move in circles, arcs and figure-8s, something that was very difficult to do in ITAS-basic.

Kick – Hitting the spacebar performs a kick in the direction of the mouse pointer and with a power relative to the distance of the mouse pointer from the on screen representation of the player.

Findball – This feature allows the player to locate and face the ball on the field with a single button press. This was implemented as a necessary component to other advanced features, rather than a stand-alone command.

Chaseball/Intercept – This command utilizes the Findball command to track the ball and the run command to reach it. This is a useful command for intercepting the ball or tackling. It's also a main component of the Dribble command, which is currently under development. It is initiated by holding the 'c' key and aborted upon its release.

Other ITAS commands include shoot, turn-neck, and view adjustment.

3.4 Performance and Testing

Using a set of drills which gauge performance of the basic skills required to perform the key tasks identified in **Fig. 1**, ITAS was continuously evaluated by a test group of soccer players that also have some computer experience. This verification process allows for iterative development of the different features and ensures the involvement of target users throughout development. These drills also serve as a great way to introduce new players to the ITAS interface. The set of drills includes simple drills that each focus on a specific action, such as movement or kicking. In the movement drills, for example, the player was timed while repeatedly traversing one of a set of predefined paths (circle, zigzag, figure-8). Other more complex drills required coordination of several different actions and communication with other players to perform a more complex task. These included passing drills, 'monkey-in-the-middle' and set plays such as cross-ins. Performance on these was gauged by the number of successful attempts at the task and the time taken to complete it. Results from the latest round of tests comparing drill performance between ITAS-basic and ITAS-advanced showed improved performance for part of the movement drills (particularly the figure-8s) and significant improvement for the complex task drills.

Although some games against agent teams have been played, ITAS isn't yet ready for full game testing against stronger RoboCup teams since two important features are still under development, namely Dribbling and Kick-to (which are discussed in the next section).

3.5 What's Next?

Completing ITAS's planned features is the next step. These planned features include Dribbling, Kick-to (which is a command for kicking the ball so as to end its movement within the vicinity of a target location – position of mouse click) and a symbolic team communication tool. Also planned is an online platform for launching RCSS games with cooperating or competing ITAS players and the option of adding agent team players to play with and/or against.

Potential directions for ITAS once initially planned features have all been implemented include developing a hybrid agent (one that can act intelligently as well as accept real-time commands from a user), modifying ITAS to be used as a remote control interface for a RoboCup robot and experimenting with new user input hardware variants to the current mouse and keyboard setup.

4 Related Work

Another project that seems to be taking on the reverse RoboCup challenge is Nishino's OZ-RP (OZ-Real Players) from Japan. While both attempting to solve the same problem, ITAS and OZ-RP differ in their approach to the solution with regards to the display of the player's environment and the player control model.

Instead of limiting the user to the soccer player's perspective, OZ-RP fits incoming data into a global model and provides the user with a full field view. OZ-RP players also send head turn commands periodically so as to have 360° vision. These features have the great advantage of raising the user's positional awareness on the field. ITAS, on the other hand, aims to provide the user with a player oriented view which, while initially takes some getting used to, is a closer representation of a soccer player's experience. While OZ-RP's approach focuses on maximizing the player's performance through providing them with a complete world model, ITAS focuses more on trying to faithfully model the soccer playing experience relative to the player. Each approach has its merits. OZ-RP's approach allows the player to concentrate more on strategic decisions on a team level. ITAS's modeling of the soccer experience makes for a more up hill challenge, but aims to bring authentic human player-level involvement to the simulation league. The lack of a solid world model in ITAS also makes for a better candidate interface for adaptation for telerobotics and remote control in real world dynamic or unfamiliar environments.

The two systems also have different control models for the player's movement. OZ-RP adopts a rather intuitive 'run to click' approach, combining both the dash and turn commands into one atomic action which involves turning to where the mouse was clicked then running forwards until that point is reached. ITAS gives the user control over both running and turning separately. This makes for a more complex movement control model, but it allows for more fluent movement (e.g. arcs and figure-8s).

5 Summary

The primary purpose of the Reverse RoboCup challenge is to take steps towards introducing a human challenger into the RoboCup Simulation League. The Reverse Challenge also has the potential of opening several new areas of research to the RoboCup community. ITAS is an open source project and welcomes the RoboCup community to partake in and benefit from the RoboCup Reverse Challenge in the many creative ways that are possible.

652   T. Hassan and B. Esfandiari

References

1. D. J. Mayhew, The Usability Engineering Lifecycle: A Practitioner's Handbook for User Interface Design, San Diego, CA: Academic Press, 1999.
2. J. Nielsen, Usability Engineering, San Diego, CA: Academic Press, 1999.
3. David Meister, Thomas P. Enderwick, Human Factors in System Design, Development and Testing, Lawrence Erlbaum Assoc; 2001
4. Junji Nishino, http://www.fs.se.uec.ac.jp/~nishino/ozrp/logs/, OZ-RP Website, 2002

SPQR-RDK: A Modular Framework
for Programming Mobile Robots

Alessandro Farinelli, Giorgio Grisetti, and Luca Iocchi

Dipartimento di Informatica e Sistemistica,
Università "La Sapienza", Rome, Italy,
Via Salaria 113 00198 Rome Italy
<lastname>@dis.uniroma1.it

Abstract. This article describes a software development toolkit for programming mobile robots, that has been used on different platforms and for different robotic application. In this paper we address design choices, implementation issues and results in the realization of our robot programming environment, that has been devised and built from many people since 1998. We believe that the proposed framework is extremely useful not only for experienced robotic software developers, but also for students approaching robotic research projects.

1 Introduction

Research on developing autonomous agents, and in particular mobile robots, has been carried out within the field of Artificial Intelligence and Robotics from many different perspectives and for several different kinds of applications, and the development of robotic applications is receiving increasing attention in many laboratories. Moreover, robotic competitions (e.g. AAAI contexts, RoboCup, etc.) have encouraged researchers to develop effective robotic systems with a predefined goal (e.g. playing soccer, searching victims in a disaster scenario, etc.). These robots have been obviously used not only for these competitions, but also for experimenting the research techniques developed within robotic research projects. Moreover, mobile robots are also used for teaching purposes within computer science laboratories and often students are required to work and develop robotic applications on them[1].

This increasing population of robots in the research laboratories and the consequent need for developing robotic applications have started a process of design and implementation of robotic software, that aims in a special way at having a design methodology and a software engineering approach in the development of such applications, which integrates several functionalities and architectural choices that go beyond the scope of conventional robotic applications.

Furthermore, companies producing and selling mobile robots make available to their users development libraries and software tools for building and debug-

---

[1] e.g. CMRoboBits Course at CMU http://www.andrew.cmu.edu/course/15-491/.

D. Nardi et al. (Eds.): RoboCup 2004, LNAI 3276, pp. 653–660, 2005.

ging robotic applications (e.g. Saphira for Pioneer robots [4], OPEN-R SDK for Sony AIBO [5], etc.). These tools are obviously platform dependent and thus they cannot easily be used for building multi-platform robotic systems, and also they usually lack some features that are required from a general purpose robot development toolkit. For instance, the OPEN-R SDK completely lacks facilities for remote monitoring the behavior of the robot. It just support wireless network communication among processes and all the remote information exchange must be explicitly coded. On the contrary, the Saphira environment, although it is specifically implemented for the Pioneer robots, has several facilities for building robotic applications and debugging them also by using a Pioneer simulator and allowing for a graphical display of the robot status.

Finally, a number of open source multi-platform robotic development environments have been realized. For example, OROCOS (Open RObot COntrol Software)[2] is an European project that has recently started with the objective of realizing a framework for developing robot control software under Real Time Linux. This project has many general goals, like independence to architectures used for connecting the components together, to robot platforms, to robotic devices, to computer platforms. The OROCOS project has a long time target and it is currently under development. *Player/Stage* [2] is also a general framework for controlling a robotic system. *Player* supports a wide range of devices, algorithms and viewers, that can be tested through *Stage*, a simulator able to work on complex multi robot scenarios. Each of these devices can be either a server or a client, allowing for a great flexibility in spreading the computation on different machines. However, *Player/Stage* provides only limited support for high level specification of user-defined modules and their interaction. *CARMEN*[3] comprises a set of independent utilities, that communicate with each other through the UNIX inter process communication facilities. This framework has been used for implementing a set of interesting algorithms, but it is mainly suited with the low level activities of the robots (such as navigation and exploration). Also the works in [7,8] are focused on proposing robot middle-ware that are not specific to a given platform or to a particular application domain. In particular, the system presented in [8] is explicitly focused on the realization of soccer applications, while in [7] mostly low level interface issues are addressed.

In this paper we describe a Robot Development Toolkit (RDK) for modular programming of mobile robots. The toolkit we have realized includes a middle-ware that implements all the basic requirements for the development of a typical robotic application, a set of modules implementing the basic functionalities of the robot, and a set of tools that are useful for developing, monitoring and debugging the entire application. In particular the middle-ware implements an infrastructure for: task management, interfacing with the robot hardware, representation of the status of the robot, remote monitoring and debugging.

---

[2] Orocos project, www.orocos.org.
[3] Carmen project, www-2.cs.cmu.edu/~carmen/

Our development toolkit is currently named SPQR-RDK, and is available to be used by robotic programmers[4].

We are currently using our framework for developing different kinds of robotic applications: i) RoboCup soccer [3] ii) RoboCup Rescue [6] iii) RoboCare [1] - a project for developing a multi robot system for assistance of elderly people in a health care house. The development of these applications has given us a real testbed for evaluating the proposed RDK and, by a comparison with the development of similar applications by using a different development environment (in particular, we refer to the robotic soccer application with Sony AIBO robots by using OPEN-R SDK), we have experimented the effectiveness of our toolkit.

2   Design Choices

During the development of our RDK, we have identified a set of fundamental functionalities and a set of software requirements needed for our framework.

As our applications have been developed through the years by different people which were able to work at the application only for a limited period of time, *modularity* and *re-usability* appear to be the main issues to address: the proper division of the code in independent modules exchanging data inside a clear framework ensures to have a coherent software generation, resulting in highly modular and re-usable code. *Efficiency* is also a primary requirement, the middle-ware needed for running the modules must have a minimum overhead with respect to the entire application. Moreover, the hardware computational capabilities must always be considered, posing strict constraints on the implementation choices for our middle-ware; therefore most of the design choices that we have done (e.g. language, operating system, shared memory for information exchange) are motivated by this requirement.

As for functionalities we have identified three main issues to be addressed: i) **Remote Inspection Capability** ii) **Information Sharing** iii) **Common Robot Hardware Interface**.

Remote Inspection is a fundamental functionality for every robotic application. The Remote Inspection mechanism, should allow the developers to use a general mechanism for remote inspecting the internal status of the application, with limited network bandwidth and with minimum computational overhead with respect to the normal execution of the robotic application.

Another important problem that we have faced during our past developments has been the exchange of data among modules. A basic use of shared memory, without any data access policy, is not satisfactory because the management of all the shared data in the program can become very complex. Similarly, the use of message exchanging typically arises the same problems and may also affect modularity of the system, when a module is implemented by including the details of other interacting modules. Therefore, an important functionality for

---

[4] Available from http://www.dis.uniroma1.it/~spqr/.

the RDK is an **Information Sharing** mechanism providing a uniform interface and a policy for sharing data among modules.

When dealing with several different types of mobile bases and sensing devices the independence of the application from the low level details of platforms and devices becomes an important issue. Hence, the development of a **Robot Hardware Interface** has been detected as another important functionality: a uniform interface has to be defined between robot devices and user modules, and hardware configuration is described in a configuration file.

3   Software Architecture and Implementation of the Middle-Ware

The RDK we are presenting in this article is based on a middle-ware that provides the basic functionalities for the development of robotic applications. This middle-ware is composed by a minimum set of modules, common to all the applications that can be developed within our framework. In particular, the middle-ware is made up by the following modules **Robot Hardware Interface**, **Task Manager**, **Robot Perceptual Space** and **Remote Inspection Server** as shown in Figure 1. In the following a description of each of those modules is given.

3.1   Robot Hardware Interface

The Robot Hardware Interface implements a level of abstraction with respect to the specific mobile base in use, providing the user with a common interface for accessing all the robotic platforms and devices. We decided to model this abstraction by exploiting the fact that usually each robotic platform comprises several sensors and actuators (devices), but only one mobile base. For robots and devices we implemented an abstract interface through a class hierarchy; in this way robots and devices of the same kind can be accessed through a common interface, and a user module can thus directly access the information and the services provided by a device, using the more general class needed. Moreover, by enforcing the abstraction on the robot hardware, it is possible to port all the written software on a new mobile base, simply by writing the low level interface.

The Robot Hardware Interface (RHI) module encapsulates the functionalities for accessing the mobile base and the on board devices and provides an abstraction for: i) *mobile robot kinematics*, by implementing the functions for reading

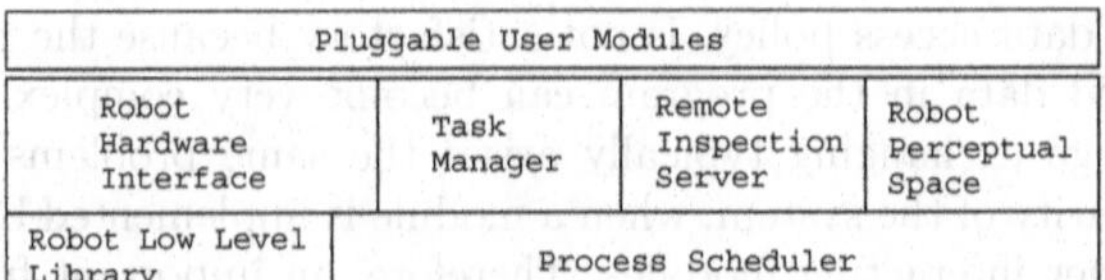

Pluggable User Modules			
Robot Hardware Interface	Task Manager	Remote Inspection Server	Robot Perceptual Space
Robot Low Level Library	Process Scheduler		

Fig. 1. Middle-ware Architecture Layered View

odometry and for controlling the motion that are specific to a mobile platform kinematics model (for example, distinguishing holonomic[5] mobile bases from unicycle-like[6] ones); ii) *mobile base connection*, by providing a standard way to access the mobile base and its specific control functions.

Each mobile base is generally equipped with various kinds of sensors and actuators like sonar rings, laser scanners, cameras, kickers (in the case of our soccer robots) and so on, that are generically defined as *Device*. These devices are connected to the robot and grouped in a set of hierarchical classes.

Both devices and robot drivers can be replaced by simulators or players of real data streams recorded before, allowing for off-line application development and debugging.

3.2   Task Manager

The Task Manager has been designed in order to allow the user to dynamically load his/her modules, to specify their execution features (i.e. execution period, scheduling policy, priority and so on) and to export the information to be shared among them.

A first feature of the Task Manager is to allow the users to easily define the scheduling policy of their modules by wrapping the Linux thread libraries.

Moreover, the Task Manager allows for the exchange of information among modules. When modules need to directly exchange information each other, the simplest solution is to couple them. However, this simple solution has the effect of limiting the software modularity and may results in cyclic references which are difficult to resolve in the linking phase.

Therefore, besides the mechanism of directly coupling two modules, the Task Manager offers another possibility to exchange information, by abstracting on the *type* of information. In fact, if a module needs data provided by some other module, it only needs to know *where* to read such data and *when* the data are available. On the other hand, a module that produces information can easily declare the kind of such information without knowing which user module will use it. This solution grants a complete independence among modules sharing data and it is possible to substitute a module with another, by only ensuring that the two modules produce the same kind of data. This mechanism has been used for sharing information among user modules, as well as between a device and a user module. Notice that the such mechanism requires the use of a shared memory thus limiting the spreading of computation on different machines. However, distributed robotic applications are currently not within the scope of our RDK and solutions explicitly designed for such applications are already provided in other programming frameworks (such as Times tool[7] or Charon[8]).

---

[5] An holonomic robot has three degrees of freedom in its motion.

[6] A unicycle robot has translational and rotational velocity bounded by a given kinematic law.

[7] http://www.timestool.com/

[8] http://www.cis.upenn.edu/mobies/charon/examples.html

3.3   Robot Perceptual Space

The Robot Perceptual Space (RPS) contains all the information known by the robot about the environment, and represents the current knowledge shared by all the modules in the application.

The RPS defines a uniform interface for accessing its data, thus similarly to the Task Manager provides a mean for information exchange. However, the semantics of the information contained in the RPS is different from the information shared through the Task Manager: RPS represents an updated snapshot of the robot perception of the environment, and the information contained in the RPS are specific to the robot application and thus generically useful for all the modules; the information exchanged by user modules through the Task Manager are instead parameters depending on the implementation of such modules and not on the characteristics of the environment.

3.4   Remote Inspection

The lessons learned from the past difficulties in debugging our software yielded to the design of a mechanism for remote control and debug that allows a module to generate and export information that can be received and displayed by a (graphical) remote client application (remote console). Information computed within the user modules are of different kinds and should be represented in different graphical forms: scanner readings, images, sonar readings, detected map features, position hypotheses, etc.

In facing the problem of building a debug interface, a key issue is to consider the high noise level and the latencies imposed by current wireless networks, therefore particular care has to be given in keeping low the bandwidth requirements. According to this consideration it has been designed a sharing mechanism that allows for a flexible run time selection of the information to inspect, in fact avoiding the differentiation of a release version from a debug one. We have thus chosen to implement a publish/subscribe mechanism for debugging information, in order to allow the user for selectively monitor the data of interest.

The Remote Inspection Server (RIS) defined in our middle-ware exports facilities for publishing information that can be monitored by remote clients. The publishing mechanism comprises two steps: The first one is *refresh*, where the RIS copies the information requested by at least one client in a local buffer. The second one is *transmission*, where the RIS performs the transmission of the buffered information to the clients. In this way network latency only affects the communication of the information to the remote host and not the efficiency of the publishing module on the robot. During the normal operation, when it is not needed to monitor the robot behavior in such a deep way, and clients do not request information to the robot, there is no overhead at all, since the Remote Inspection Server detects this situation and avoids useless computation.

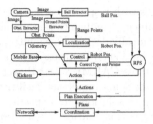   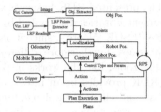

Fig. 2. Real robotic soccer application   **Fig. 3.** Virtual cleaning application

4   Developing a Robotic Application: Pluggable Modules and Supervisor Tools

The development of a robotic application requires the realization of a set of modules implementing specific functionalities that must be appropriately connected together. The middleware realized for our RDK is suitable both for the realization and connection of the application modules.

It is interesting to notice that the composition of an application, in terms of which modules are activated and how they are connected, is simply described in a configuration file. Moreover, once we are satisfied with the robotic application in the virtual environment, our framework allows for an easy interchangeability of modules simulating the behavior of some sensors with actual sensor data interpretation modules, in order to make the application work on a real robot.

As an example in figure 2 and 3 two instances of robotic applications obtained by implementing and connecting modules developed within our framework are shown.

5   Conclusions

In this paper we have presented a framework (SPQR-RDK) for developing modular multi-platform robotic applications, that has been designed for providing modularity, effectiveness and efficiency. This RDK allows a group of programmers to design and implement the modules composing a multi-platform multi-robot application, having both remote control and remote debugging capabilities, with a very small effort, by using a software engineering approach and by focusing on the semantics of the information exchanged among the modules. The main use of our framework is for people (mainly students) that want to develop a solution for a single topic or for a specific application (e.g. localization in an office-like environment, path planning with moving obstacles, multi robot coordination in a soccer domain, etc.), by using available modules for all the other capabilities of the robot. Our RDK provides these programmers with an easy methodological tool for implementing the robotic application and also it allows for easily evaluating the specific application developed under different environment conditions and in comparison with different solutions.

The presented RDK has several advantages with respect to other robotic development libraries distributed by robot producing companies (e.g. Saphira [4], OPEN-R SDK [5], etc.), since it has been specifically designed for multi-platform applications. Furthermore, differently from other general-purpose robotic development tools, like the works in [7,8] or the tools CARMEN and Player-Stage, our RDK provides in an integrated framework some important facilities, such as easy and efficient implementation of modular solutions to a specific robotic problem, remote control and inspection, information sharing, abstraction with respect to the mobile base and the connected devices, and a set of useful tools for developing typical robotic applications.

The SPQR-RDK is continuously increasing in the number of modules that are realized for the different applications that are currently under development within our group, but always maintaining the same middle-ware. This is an important achievement for our group since having several modules that can be combined for building different robotic applications with a small effort, allows for developing different solutions to common robotic problems and to evaluate them in several scenarios and in general to increase over time the quality and the effectiveness of the robotic applications developed.

References

1. S. Bahadori, A. Cesta, G. Grisetti, L. Iocchi, R. Leone, D. Nardi, D. Oddi, F. Pecora, and R. Rasconi. Robocare: an integrated robotic system for the domestic care of the elderly. In *In Proceedings of Workshop on Ambient Intelligence AI*IA-03*, Pisa, Italy, 1995.
2. B. P. Gerkey, R. T. Vaughan, and A. Howard. The player/stage project: Tools for multi-robot and distributed sensor systems. In *In Proc. of the Int. Conf. on Advanced Robotics (ICAR 2003)*, pages pp. 317–323, Coimbra, Portugal, June 30 - July 3 2003.
3. H. Kitano, M. Asada, Y. Kuniyoshi, I. Noda, E. Osawa, and H. Matsubara. Robocup: A challenge problem for ai and robotics. In *Lecture Note in Artificial Intelligence*, volume 1395, pages 1–19, 1998.
4. K. Konolige, K.L. Myers, E.H. Ruspini, and A. Saffiotti. The Saphira architecture: A design for autonomy. *Journal of Experimental and Theoretical Artificial Intelligence*, 9(1):215–235, 1997.
5. Sony. Open-r sdk, http://www.jp.aibo.com/openr/.
6. S. Tadokoro and et al. The robocup rescue project: a multiagent approach to the disaster mitigation problem. *IEEE International Conference on Robotics and Automation (ICRA00), San Francisco*, 2000.
7. H. Utz, S. Sablatng, S. Enderle, and G. K. Kraetzschmar. Miro - middleware for mobile robot applications. *IEEE Transactions on Robotics and Automation, Special Issue on Object-Oriented Distributed Control Architectures*, 18(4):493–497, 2002.
8. Hui Wang, Han Wang, C. Wang, and W. Y. C. Soh. Multi-platform soccer robot development system. In *RoboCup 2001: Robot Soccer World Cup V*, pages 471–476, 2001.

Mobile Autonomous Robots Play Soccer - An Intercultural Comparison of Different Approaches Due to Different Prerequisites

Peter Roßmeyer, Birgit Koch, and Dietmar P.F. Möller

Universität Hamburg, Fachbereich Informatik,
Arbeitsbereich Technische Informatiksysteme,
Vogt-Kölln-Str. 30, 22527 Hamburg
{6rossmey, koch, dietmar.moeller}@informatik.uni-hamburg.de
http://www.informatik.uni-hamburg.de/TIS/

Abstract. In the effort to meet the steadily changing demands of teaching computer science and computer engineering, new methods of learning and teaching are used by which multifarious knowledge and learning techniques can be imparted, practical skills and abilities can be developed and teamwork and creativity are encouraged. A promising attempt is the use of robotic construction kits.

This paper portrays the educational environment that was used at the courses "Hamburg RoboCup: Mobile autonomous robots play soccer" at the University of Hamburg, Germany and "Advanced robotics - Soccer playing mobile autonomous robots" at the California State University of Chico, USA and compares the experiences made during both courses due to intercultural differences.

1   Introduction

For over thirty years the "Epistemology and Learning Group" of the Massachusetts Institute of Technology (MIT) did research about correlations between learning environments and learned skills. One of the results based on the research of Seymour Papert is the idea of using robotic construction kits coupled with user-friendly programming environments [1].

While the utilization of robotic construction kits at schools was analyzed in detail and appreciated [2][3], it was often depreciated as a toy and therefore considered irrelevant in the context of universities [4]. Using "real" robots at universities has the disadvantage of being very expensive so that many students often have to share a single robot. Additionally it can be difficult to motivate the students to work with "real" robots because the orientation time of a complex robot system often requires weeks or even months so the course is nearly over before the students have figured out all the possibilities of the robot. To avoid these obstacles in the courses at the University of Hamburg and the California State University of Chico we decided to use robotic construction kits which are less expensive (and therefore available in sufficient numbers), more flexible and easier to understand.

D. Nardi et al. (Eds.): RoboCup 2004, LNAI 3276, pp. 661–668, 2005.

The opportunity to offer the same course at different universities arose from the USE-ME Project. USE-ME means "US-Europe Multicultural Educational Alliance in Computer Science and Engineering", a cooperation project to promote the development of new student-centered teaching units in computer modelling and simulation with exchanges of students and instructors between European and US universities. For this reason the courses were held by the same instructors under the same circumstances at both universities which makes them easy to compare.

2 Robotic Construction Kits and Programming Environments

To understand the relevance of using robotic construction kits at universities it is important to know the elements contained and the possibilities of programming. In the following this is exemplified through the LEGO Mindstorms robotic construction kit [5].

The LEGO Mindstorms kit contains a programmable RCX-brick (Hitachi H8/3293-microcontroller with 16 KB ROM and 32KB RAM), two touch sensors, a light sensor, two motors and lots of common LEGO bricks. Also included are an infrared sender to transmit data between the RCX-brick and a personal computer, the programming environment Robotics Invention System (RIS) and a construction handbook. The RCX-brick provides three inputs for sensors, three outputs for motors or lamps, five spots for programs, a LCD-display, four control buttons, a speaker and an infrared interface. Figure 1 shows a RCX-brick with two motors, two touch sensors and a light sensor.

The RCX-brick comes with the firmware installed. The firmware is necessary to communicate with a personal computer to load programs from the computer

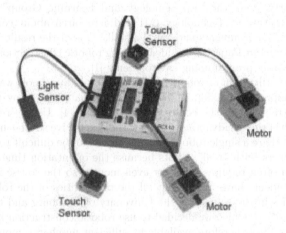

Fig. 1. RCX-brick with sensors and motors

to the robot. Five simple programs are preinstalled so the robot can be tested immediately after just short periods of configuration.

The Robotic Invention System (RIS) is a graphic-based programming environment that works with blocks. Every programming instruction is represented by one block. The blocks are joined by "Drag and Drop" in form of a chain while programming and executed in this order when the program is running. To include sensor-data of the robot, parallel chains of blocks can be used.

Besides the software RIS that comes with the LEGO Mindstorms construction kit and that is directed to children and teenagers without programming skills, several other possibilities to program on an advanced level have been developed by active LEGO online-groups. Most of this software can be downloaded as freeware.

Some examples of programming environments used during the courses in Hamburg and Chico are:

- ROBOLAB: works with a kind of advanced flowcharts, based on LabVIEW, was developed especially for use in schools
- RCX Command Center: with the programming language Not Quite C (NQC), a language similar to C, programs can be written text-based
- LEGO Java Operating System (lejOS): an implementation of a Java Virtual Machine (JVM)

Other programming environments can be found in [6][7][8][9].

3   General Structure of the Course

The overall goal of the courses was to build teams of soccer playing robots according to the rules of the RoboCup Junior league. Two teams, consisting of two robots each, play against each other on a field measuring 122 cm by 183 cm with a wall around it to keep the ball and the robots from falling of. The game lasts for two 10-minute halves with a 5-minute break in between. For more details see [10][11][12].

In Hamburg prerequisites were made by offering the course as a seminar only open to advanced students with intermediate diploma in computer science. Ten students were participating. In Chico a quite similar restriction was made by offering the course in the advanced level. Among the fifteen students participating one lacked the official qualifications but was still admitted and had no problems following the course.

Both courses where held with a quite similar structure. During the first sessions the students got a brief introduction on the general topic of robotics. They also got some first hands on experience with the LEGO Mindstorms kit. The students were then grouped into teams: Each team was given two LEGO Mindstorms kits and they should come up with a soccer-team at the end of the course.

Additional they had to choose one out of a list of relevant special interest topics, work it out and present it to the other students. Afterwards they served as specialists for this topic and answered students questions on their topic for the remaining time of the course.

One example of such a special interest topic is modelling a robot with fuzzy logic. Fuzzy behavior of the robots recognized by the students can be precisely expressed with fuzzy logic. While students work with the robots they find out that the robots do not move on a straight line although both motors run with the same performance. In this situation fuzzy logic can be very helpful to solve the problem. The students learn by means of linguistic variables of the mobile robot to determine the membership function of the linguistic variables and to draw up rules to describe the dynamic behavior of the soccer-playing robots.

Some contests where held during the course to test the performance of ideas and solutions on their real world behavior. Several problems crop up during these contests: the robots could not find the ball, the robots aimed at the wrong goal or had no orientation at all, the robots fell apart when they touched the wall or other robots, etc. Most of the students were stimulated by these problems to extend their work on the project. Finally a tournament was held at the end of the course. Figure 2 shows one of the LEGO Mindstorms robots build by the students.

Fig. 2. Soccer playing LEGO Mindstorms robot

To evaluate students opinions about robotic construction kits being utilized in university education several questionnaires were given to the students during and at the end of the courses. These questionnaires also served to find out about the students understanding and interest for robotics in general and their willingness to learn with interactive elements.

4   Experiences During the Courses in Hamburg (Germany) and Chico (USA)

As shown in Figure 3 neither the students in Hamburg nor those in Chico had any experience with LEGO Mindstorms. Only one student in Hamburg had no experience with LEGO. The experience of the other students resulted in short time for building the robots, very fast adaptation whenever a hardware problem occurred and some rebuilding of the robots to adjust it to specific problems rather than solving them with the software. In Chico a majority of the students had no experience with LEGO at all. This was caused by a great number of students coming from India and China (the same number as students stated to have no experience with LEGO) where LEGO is not as widely spread as a kids toy as it is in Europe and the USA. As a result the robots built in Chico were closer to those shown in the LEGO construction handbook and underwent fewer changes during the course. Adaptations were made in the software rather then changing any hardware.

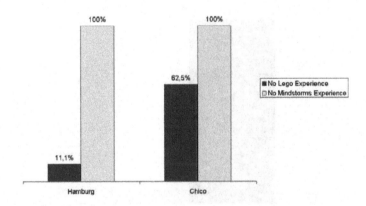

Fig. 3. "Do you have any experience with LEGO or LEGO Mindstorms?"

Interesting results also came from the questions "Was the class as you expected?". Overall students expectations were met total or at least party. But there are interesting differences in the details. The first questionnaire was answered shortly after the presentation of the special topics and before most work was done on the LEGO Mindstorms. The second questionnaire was answered at the very end of the course. While total agreement fell from first to second questionnaire in Hamburg it rose in Chico. This indicates that students in Hamburg rather expect theoretical work than hands on experience in courses while students in Chico seem to have expected less theory and more practical work. This is shown in Figure 4.

A very encouraging result came from the question: "Would you take the class another time?" In Hamburg and Chico all students stated they would do so on

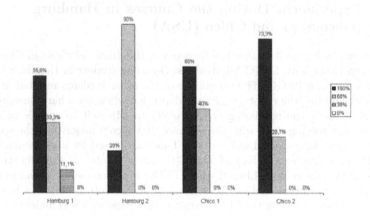

Fig. 4. "Is/Was the course as you expected?"

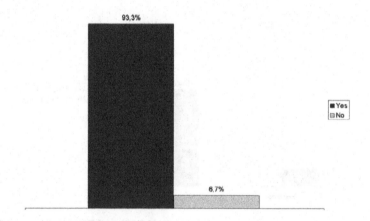

Fig. 5. "Would you like to do more research in robotics?"

both questionnaires. In Chico a large majority would also like to make some more research in robotics (Figure 5), unfortunately this question was not on the Hamburg questionnaire but since a lot of students continued in the field of robotics it seems to would have been quite similar.

Some students in Chico later stated that an introduction to LEGO at the beginning of the course would have been very helpful. This was not given by the instructors because of the assumption it would be irrelevant based on the experience made in Hamburg. If this course should be taught in an "unknown" environment again, the instructors will make this their first question (since experience with LEGO Mindstorms should not be a prerequisite) and adjust the beginning of the course according to the answer.

5   Conclusion

Although the students in Hamburg and Chico had quite different prerequisites they all reached the overall goal of the course to come up with a team of soccer-playing LEGO Mindstorms robots. But according to their different prerequisites they took different approaches and learned different new skills during the course.

As expected those students not familiar with LEGO had built their own robots but still were not as creative with the hardware as those that had played with LEGO since early childhood. But they still started to change the given designs from the handbook to more suitable designs for soccer playing robots.

Those students who tried to solve most problems by changing the hardware learned that this was not always possible and that the software also had to been taken in account. In fact, most problems where solved by a nice piece of software because the hardware of a LEGO Mindstorms robotic construction kit is too simple to solve such complex problems like playing soccer without sufficient use of proper software (meaning more than a few simple "if X then Y else Z" statements).

Regarding the different prerequisites in learning-skills of the students it can be said that most students adopted those skills they lacked before. The ones who had not presented a scientific topic before had to learn gathering information and presenting it in a decent way. Those not used to actively forming a course had to give up their inertness and communicate, on the one hand with other students to build decent robots and programs on the other hand with the instructors to come up with good presentations for the special interest topics.

6   Future Perspectives

The experiences made during the first Hamburg course and the course in Chico already lead to some consequences. Points criticized during the first Hamburg course were tried to avoid in future courses (some self fulfilling like eliminating the waiting time for delivery of hardware). Students taking the course a second time rose the overall performance of the robots because they told the new students which ideas had already proven to be good or bad so the same mistakes were not made again. One problem persisting was the very limited hardware performance of the LEGO Mindstorms robot construction kit. Some students programs were too big for the memory of the RCX-brick or too complex for the Hitachi H8/3293-microcontroller of the RCX-brick to run in decent time. To be able to run even more complex programs there was a need for more powerful hardware. Hence the idea of a continuing course using more complex hardware arose. Since most students focused on building good programs while simply assuming sound working hardware a new approach was made by using SONY AIBO ERS-210A robots. Since the Sony Fourlegged League in Robocup prohibits any changes in hardware this league automatically focuses on the software. So this league meets exactly the demands of students which have participated in the LEGO Mindstorms course and wanted to continue on a higher level of programming robots. Some of the students changing from the LEGO Mindstorms course to the Sony Fourlegged League participate at RoboCup 2004 in Lisbon as "Hamburg Dog Bots".

References

1. Papert, S.: Mindstorms: Children, Computers and Powerful Ideas. New York: Basic Books (1980)
2. Christaller, T.; Indiveri, G.; Poigne, A. (eds.): Proceedings of the Workshop on Edutainment Robots 2000, 27th - 28th September 2000, St. Augustin, Germany, GMD Report 129, GMD-Forschungszentrum Informationstechnik GmbH (2001)
3. Müllerburg, M. (ed.): Abiturientinnen mit Robotern und Informatik ins Studium, AROBIKS Workshop Sankt Augustin, Schloss Birlinghoven, 14. - 15. Dezember 2000, GMD Report 128, GMD - Forschungszentrum Informationstechnik GmbH (2001)
4. Koch, B.: Einsatz von Robotikbaukästen in der universitären Informatikausbildung am Fallbeispiel "Hamburger Robocup: Mobile autonome Roboter spielen Fußball". Diplomarbeit, Fachbereich Informatik, Universität Hamburg (2003)
5. The LEGO Mindstorms website. http://www.legomindstorms.com/
6. Baum, D., Gasperi, M., Hempel, R., Villa, L.: Extreme MINDSTORMS. An Advanced Guide to LEGO Mindstorms. Apress (2000)
7. Baum, D.: Dave Baum's Definitive Guide To LEGO Mindstorms. Apress (2000)
8. Erwin, B.: Creative projects with LEGO Mindstorms. Addison-Wesley (2001)
9. Knudsen, J. B.: The Unofficial Guide to LEGO Mindstorms Robots. O'Reilly (1999)
10. RoboCup Junior Webside: http://www.robocupjunior.org
11. Kroese, B.; van der Boogaard, R.; Hietbrink, N.: Programming robots is fun: Robocup Jr. 2000. In: Proceedings of the Twelfth Belgium-Netherlands AI Conference BNAIC'00 (2000) 29–36.
12. Lund, H.H.; Pagliarini, L.: RoboCup Jr. with LEGO Mindstorms. In: Proceedings of International Conference on Robotics and Automation (ICRA2000), New Jersey: IEEE Press (2000)

From Games to Applications: Component Reuse in Rescue Robots

Holger Kenn and Andreas Birk

School of Engineering and Science,
International University Bremen,
Campus Ring 12, D-28759 Bremen, Germany
h.kenn@iu-bremen.de
http://www.faculty.iu-bremen.de/kenn/

Abstract. Component-based software engineering is useful for embedded applications such as robotics. However, heavyweight component systems such as CORBA overstrain the ressources available in many embedded systems. Here, a lightweight component-based approach is used to implement the system software of the so-called CubeSystem, CubeOS. Since 1998, CubeOS and its component system have been successfully used in various areas from industry projects over RoboCup-related research to edutainment applications. Many of the components used in RoboCup soccer have been carried over in the implementation of the IUB Rescue robots, demonstrating the potential for software reuse.

1   Introduction

Component-based software engineering has emerged in recent years as a widely used approach to simplify software reuse. It relies on a large base of reusable software components and an integrating framework for those components [21]. Such frameworks for application software are CORBA [20] or JavaBeans [22] which alreay have been used in Robocup[16].

The main benefit of software component approaches are the possibility to reuse components within a framework. This leads to a significantly reduced development time both for implementation and testing/debugging if a high number of tested components is readily available.

The same advantages of component-oriented software engineering can also be used in the design of software for embedded systems. With the tendency towards more and more complex embedded systems, handling this complexity in the software design process becomes more important and component-oriented approaches are one solution to this.

Lightweight component architectures try to limit the overhead that is created by the component infrastructure without losing the advantages of the component-oriented software engineering approach and maintaining a maximum of the protection features of a heavyweight component architecture. Examples of lightweight component architectures for implementing embedded operating systems are pebble [3] and eCos [13].

In this paper, a very simple lightweight component-oriented approach is described and it is shown how it has been applied to the design of the system and application

D. Nardi et al. (Eds.): RoboCup 2004, LNAI 3276, pp. 669–676, 2005.

software for robots. The main design goals for this software system have been minimal performance overhead, very modest hardware requirements and the possibility for code reuse by multiple groups and projects.

The CubeOS system described here has been continously developed since 1998 and has been used in several research projects. Its main benefit was the reuse of existing component code for future projects. For example a large part of the low-level control software of the current robots platforms of the IUB Robocup Rescue [7] team is based on existing components of the VUB AI Lab Robocup Smallsize League [6] team.

2   Autonomous Systems

In the recent years, research on autonomous systems, i.e. networked embedded devices has shown the need for reliable energy-efficient low-cost computing platforms. Where it is possible, such as in systems used in the RoboCup [14] Middle-Size Robot League, this computing platform mostly consists of embedded PC hardware, running commercial or free general-purpose operating systems. However, for applications where the physical size and the energy sources of a device are restricted even further, these PC-hardware-based approaches are of limited use. Several other available platforms such as Lego Mindstorms [17] have limited compute ressources that restrict their use to Edutainment applications [2]. The need for a small and energy-efficient extendible platform led to the development of the CubeSystem [10]. Apart from its use as a standalone controller [6] the cube system can also be used in combination with PC hardware to execute realtime control tasks [8, 7].

Together with this new hardware platform, a new approach to system software design based on lightweight components has been pursued. For autonomous systems, the system software has to provide only limited services

- Standard operating system functions such as concurrent thread execution, inter-thread communication and synchronization, time measurement and realtime clock services.
- Interface code for sensor- and actuator devices ranging from simple i/o functions to complex software for computer vision applications
- Ad-hoc network communication service between multiple systems using various communication interfaces, e.g. wired bus systems, radio communication etc.
- Mechanisms that allow the extension of the system by third parties to enable the integration of new hard- and software.

Many of the features mentioned here are available in commercial operating systems for deeply embedded devices. Unfortunately, these come at a significant cost and/or restrictive license conditions. However, in order to benefit from the component-oriented software engineering approach through code reuse and to be able to use CubeOS for various projects with different licensing requirements, an open-source approach was chosen for the implementation of CubeOS. Other Projects such as eCos [13] later followed a similar approach.

3   A Look at the CubeSystem Hardware Platform

The RoboCube hardware platform [10] is using the CPU32 Core [18] as CPU in the
M68332 MCU [19]. It is a 32-bit CISC architecture without MMU or cache. Additionally, the MCU contains functions such as local memory, serial I/O, timer functions and
programmable chip selects that make it suitable for the design of embedded systems. The
RoboCube hardware platform extends the MCU with at least 1 Mbyte of Flash-ROM
and 1 Mbyte of S-RAM. These components form the CPU board.

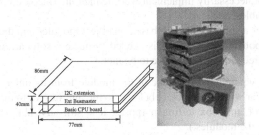

Fig. 1. The physical layout of a RoboCube stack of boards

The CPU board is about 8×8 centimeters in size and about two centimeters in
height. It has two stacking connectors a the edges that carry all bus and power signals
so that multiple boards with similar connectors can be stacked together. These form a
CubeSystem. In a cube system, exactly one CPU board must be present.

Apart from the CPU board, there are various extension boards available such as
memory boards (4Mbytes of SRAM), Bus Controller Boards (2x UART, 2x I^{2}C) and
I/O Boards (24 8-bit A/D inputs, 6 D/A outputs and 16 bidirectional digital I/Os on one
I/O Board).

The users can develop application-specific extension boards.The most basic application-
specific extension boards are so-called base boards that implement application-specific
power supply and I/O. Such application-specific boards have been developed for various
projects [8, 7, 9, 5].

From this description of the hardware, it can be seen that the platform is quite flexible,
not only in the way that users can use different hardware modules of similar functionality but that the hardware platform can be extended with completely new functions. The
system software has to accommodate this by supporting the user in developing software
that works with custom hardware at ease. An unfortunate feature of the CPU is that it
does not contain a memory management unit, i.e. it is unable to offer memory protection services that could be used to separate components, i.e. preventing intentional or
unintentional manipulation of memory used by a different component.

4   The Component Design of CubeOS

According to [23], a software component is a unit of composition with contractually
specified interfaces and explicit context dependencies only. Applications are created by

combining components from possibly different origins. From this it can be concluded that a software component is a unit of independent deployment and of third-party composition. Unlike an object, i.e. an instance of a class in object-oriented languages, a component has no persistent state in itself, so there are no multiple instances of components, either a component is available or not, but it is not available several times.

This illustrates the strong correspondence between software modules and software components. A software module (such as a C object-code module) can implement the code for a software component. But the module itself is not sufficient to form a component since it does not necessarily implement well-defined interfaces, e.g. it does not protect its internal variables.

CubeOS uses a simple approach based on object modules and the standard C linker to form a component system. What is needed to make a software module a software component? Three requirements have to be met:

1. In order to separate the interface of the module from its inner structure, it has to be made clear whether an object belongs to the interface or to the implementation.(However, for grey-box testing purposes, it is still advisable to export the inner structure of the modules.)
2. Since the C linker identifies every object with an unique name, it has to be made sure that no two objects use the same name for any interface or implementation object.
3. From the two former steps, it is clear that a component system can be formed by linking the modules together if each module (and the application program) uses the appropriate interface objects of the other modules. However, this is not sufficient to make sure that the modules can be used independently. To ensure this, modules of an appropriate size have to be defined and their interfaces have to be documented.

CubeOS implements the first and the second requirement by prefixing every object name of a component with the component name, e.g. `KERN_schedule()` is the interface to the kernel scheduler. Internal objects get an additional _ , e.g. the KERN component implements the process table in a private array: `extern struct process _KERN_ptable[];` . This has been considered good practice for coding C-modules but can be used in the same way in a component-based design process.

Unfortunately, neither the C language nor the hardware used support access-protection. Therefore, the method of marking the objects thorugh this naming scheme helps the user to observe the acces rules.

The third requirement cannot be met within the programming language itself but is a requirement for the implementation. The only option for a component system is to encourage the users of making their components reusable by providing an adequate set of tools that make it simpler to do so.

CubeOS uses the software documentation system "Doxygen" for this purpose. Although originally designed to document object-oriented software, Doxygen can be used to document component systems as well. For this, documentation groups with the same name as the component are used. Doxygen includes the documentation in the source code by using specially formated comments.

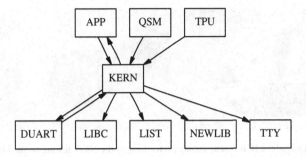

Fig. 2. The graph representing the component interdependencies of the KERN component with arrows pointing from the origin of the call to the called component. The LIBC and NEWLIB interface components represent the calls to the C library, the APP component represents the application program, in this case the CUBEOS test library. The call from KERN to APP is the initial call of main(). Note that calls between other components such as calls from APP to NEWLIB are not shown in this graph

The graph that is shown in Figure 2 illustrates the relation of the KERN component with other components. This graph can automatically be created by the build system from the object code and is helpful for documentation and error analysis.

Often, it is necessary to interface legacy code such as standard or mathematical libraries. Examples of such legacy libraries that have been ported to CubeOS are the XDR libraries and the open-source JPEG compression library.

5   Successful Applications of CubeOS in RoboCup

The CubeOS has been used in various robotics applications (figure 3), ranging from educational activities [2, 5] over basic research [1] to industrial applications [8, 7] . In all projects, the usage of CubeOS proved to be beneficial both in respect to the fast development time of the overall systems as well as in respect to the stability and reliability of the systems. Moreover, the use of the component-oriented approach enables code-sharing between many of the projects mentioned that goes beyond the operating system itself. For example, a number of general-purpose mobile robot control components has been implemented that implement PID motor control, odometric pose tracking and high-level motion commands.

One application is for example within the Small Robots League of RoboCup, the world championship of robot soccer [14, 15]. The CubeOS has been used on robot teams from the Vrije Universiteit Brussel (VUB) and recently from the International University Bremen (IUB). The teams participated in various tournaments, including RoboCup World Championship'98 in Paris, RoboCup World Championship'99 in Stockholm, the RoboCup European Championship 2000 in Amsterdam [12, 11].

When in 2002 a seemingly completely different task within RoboCup was pursued by the IUB team, namely the creation of rescue robots, it turned out that due to the

Fig. 3. Left: One of the IUB rescue robots performing in the NIST testing arena during RoboCup 2002 in Fukuoka, Japan. Center: Two robots designed and programmed by high school students for a robotics competition. Right: The inside core of the RoboGuard base, a commercial semi-autonomous robot for surveillance applications

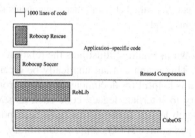

Fig. 4. This diagram illustrates that most software used for the IUB Rescue Robots has been reused from the soccer robots

design of CubeOS, most of the code from the soccer robots could indeed be reused. The concrete numbers are shown in Figure 4. It shows that 97.8 % of the code were be reused.

6   Conclusion

CubeOS demonstrates the use of a lightweight component-oriented design approach for the design of both system and application software on multiple embedded mobile robot platforms used for Robocup research. This represents an advantage over the use of heavyweight component architectures in embedded systems since it reduces ressource usage that is critical for embedded applications such as mobile robotics but still gives the benefits of component-oriented software engineering such as code reuse. The component model of CubeOS is very simple, thereby simplifying code reuse of existing legacy code and the from-scratch implementation of reusable components. CubeOS and its component system have been successfully used since 1998 in various areas from industry projects over RoboCup robotics research to edutainment applications.

References

1. Holger Kenn Andreas Birk and Luc Steels. Programming with behavior processes. *International Journal of Robotics and Autonomous Systems*, 39:115–127, 2002.
2. Minoru Asada, Raffaello D'Andrea, Andreas Birk, Hiroaki Kitano, and Manuela Veloso. Robotics in edutainment. In *Proceedings of the International Conference on Robotics and Automation, ICRA'2000*, 2000.
3. John Bruno, Jose Brustoloni, Eran Gabber, Avi Silberschatz, Christopher Small. Pebble: A Component-Based Operating System for Embedded Applications In *Proceedings of the USENIX Workshop on Embedded Systems* , 1999.
4. D. (Daniele) Bovet and Marco Cesati. *Understanding the Linux kernel*. O'Reilly & Associates, Inc., 2000
5. Andreas Birk, Wolfgang Günther, and Holger Kenn. Development of an advanced robotics-kit for education and entertainment of non-experts. In *1st International Workshop on Edutainment Robotics*, 2000.
6. Andreas Birk and Holger Kenn. Heterogeneity and on-board control in the small robots league. In Manuela Veloso, Enrico Pagello, and Hiroaki Kitano, editors, *RoboCup-99: Robot Soccer World Cup III*, number 1856 in LNAI, pages 196 – 209. Springer, 1999.
7. Andreas Birk and Holger Kenn. A control architecture for a rescue robot ensuring safe semi-autonomous operation. In Gal Kaminka, Pedro U. Lima, and Raul Rojas, editors, *RoboCup-02: Robot Soccer World Cup VI*, LNAI. Springer, 2002.
8. Andreas Birk and Holger Kenn. Roboguard, a teleoperated mobile security robot. *Control Engineering Practice*, in press, 2002.
9. Andreas Birk, Holger Kenn, Martijn Rooker, Agrawal Akhil, Balan Horia Vlad, Burger Nina, Burger-Scheidlin Christoph, Devanathan Vinod, Erhan Dumitru, Hepes Ioan, Jain Aakash, Jain Premvir, Liebald Benjamin, Luksys Gediminas, Marisano James, Pfeil Andreas, Pfingsthorn Max, Sojakova Kristina, Suwanketnikom Jormquan, and Wucherpfennig Julian. The iub 2002 smallsize league team. In Gal Kaminka, Pedro U. Lima, and Raul Rojas, editors, *RoboCup-02: Robot Soccer World Cup VI*, LNAI. Springer, 2002.
10. Andreas Birk, Holger Kenn, and Thomas Walle. On-board control in the robocup small robots league. *Advanced Robotics Journal*, 14(1):27 – 36, 2000.
11. Andreas Birk, Thomas Walle, Tony Belpaeme, Johan Parent, Tom De Vlaminck, and Holger Kenn. The small league robocup team of the vub ai-lab. In *Proc. of The Second International Workshop on RoboCup*. Springer, 1998.
12. Andreas Birk, Thomas Walle, Tony Belpaeme, and Holger Kenn. The vub ai-lab robocup'99 small league team. In *Proc. of the Third RoboCup*. Springer, 1999.
13. The eCos open-source embedded operating system See http://sources.redhat.com/ecos/
14. Hiroaki Kitano, Minoru Asada, Yasuo Kuniyoshi, Itsuki Noda, and Eiichi Osawa. Robocup: The robot world cup initiative. In *Proc. of The First International Conference on Autonomous Agents (Agents-97)*. The ACM Press, 1997.
15. Hiroaki Kitano, Milind Tambe, Peter Stone, Manuela Veloso, Silvia Coradeschi, Eiichi Osawa, Hitoshi Matsubara, Itsuki Noda, and Minoru Asada. The robocup synthetic agent challenge 97. In *Proceedings of IJCAI-97*, 1997.
16. Gerhard K. Kraetzschmar, Hans Utz, Stefan Sablatnög, Stefan Enderle, and Günth er Palm. Miro - Middleware for Cooperative Robotics. In *Andreas Birk, Silvia Coradeschi, and Satoshi Tadokoro, editors, Proceedings of RoboCup-2001 Symposium, volume 2377 of Lecture Notes in Artificial Intelligence, pages 411-416*, Berlin, Heidelberg, Germany, 2002. Springer-Verlag.
17. Henrik Hautop Lund and Luigi Pagliarin Robot Soccer with LEGO Mindstorms LNCS 1604, 1999

676   H. Kenn and A. Birk

18. Motorola, Inc. CPU32 Reference Manual, 1996 see http://www.motorola.com
19. Motorola, Inc. 68332 Users Manual, 1995 see http://www.motorola.com
20. The Object Management Group (OMG) see http://www.omg.org
21. Ian Sommerville Software Engineering, 6th Edition Addison Wesley, 2001
22. JavaSoft, Sun Microsystems, Inc. JavaBeans Components API for Java, 1997
23. Clemens Szyperski. *Component Software, Beyond Object-Oriented Programming*. Addison-Wesley, 1999.

Author Index

Lecture Notes in Artificial Intelligence (LNAI)